T0297799

CAMBRIDGE LIBRARY COLLECTION

Books of enduring scholarly value

Technology

The focus of this series is engineering, broadly construed. It covers technological innovation from a range of periods and cultures, but centres on the technological achievements of the industrial era in the West, particularly in the nineteenth century, as understood by their contemporaries. Infrastructure is one major focus, covering the building of railways and canals, bridges and tunnels, land drainage, the laying of submarine cables, and the construction of docks and lighthouses. Other key topics include developments in industrial and manufacturing fields such as mining technology, the production of iron and steel, the use of steam power, and chemical processes such as photography and textile dyes.

Harbours and Docks

Professor of civil engineering at University College London, Leveson Francis Vernon-Harcourt (1839–1907) drew on considerable practical experience, having worked most notably on London's East and West India docks. The present work was first published in two volumes in 1885. This reissue combines in one volume the text and the plates, including plans and maps of important examples. The topics discussed include natural and artificial harbours; the impact of waves, tides and currents; and general principles of construction. Furthering Vernon-Harcourt's aim to educate readers on both the theory and practice of hydraulic engineering, the work features case studies on specific projects (including their origins and condition at that time), shedding much light on the history and operation of infrastructure that proved essential for the development of modern trade. Of related interest, Thomas Stevenson's *The Design and Construction of Harbours* (second edition, 1874) is also reissued in this series.

Cambridge University Press has long been a pioneer in the reissuing of out-of-print titles from its own backlist, producing digital reprints of books that are still sought after by scholars and students but could not be reprinted economically using traditional technology. The Cambridge Library Collection extends this activity to a wider range of books which are still of importance to researchers and professionals, either for the source material they contain, or as landmarks in the history of their academic discipline.

Drawing from the world-renowned collections in the Cambridge University Library and other partner libraries, and guided by the advice of experts in each subject area, Cambridge University Press is using state-of-the-art scanning machines in its own Printing House to capture the content of each book selected for inclusion. The files are processed to give a consistently clear, crisp image, and the books finished to the high quality standard for which the Press is recognised around the world. The latest print-on-demand technology ensures that the books will remain available indefinitely, and that orders for single or multiple copies can quickly be supplied.

The Cambridge Library Collection brings back to life books of enduring scholarly value (including out-of-copyright works originally issued by other publishers) across a wide range of disciplines in the humanities and social sciences and in science and technology.

Harbours and Docks

Their Physical Features, History, Construction, Equipment and Maintenance with Statistics as to their Commercial Development

LEVESON FRANCIS VERNON-HARCOURT

CAMBRIDGE
UNIVERSITY PRESS

University Printing House, Cambridge, CB2 8BS, United Kingdom

Cambridge University Press is part of the University of Cambridge.
It furthers the University's mission by disseminating knowledge in the pursuit of education, learning and research at the highest international levels of excellence.

www.cambridge.org
Information on this title: www.cambridge.org/9781108072021

This edition first published 1885
This digitally printed version 2014

ISBN 978-1-108-07202-1 Paperback

Clarendon Press Series

HARBOURS AND DOCKS

VERNON-HARCOURT

London

HENRY FROWDE

Oxford University Press Warehouse

Amen Corner, E.C.

Clarendon Press Series

HARBOURS AND DOCKS

THEIR

PHYSICAL FEATURES, HISTORY, CONSTRUCTION EQUIPMENT, AND MAINTENANCE

WITH

STATISTICS AS TO THEIR COMMERCIAL DEVELOPMENT

BY

LEVESON FRANCIS VERNON-HARCOURT, M.A.

MEMBER OF THE INSTITUTION OF CIVIL ENGINEERS
AUTHOR OF 'RIVERS AND CANALS'

VOL. I.—TEXT

Oxford
AT THE CLARENDON PRESS
1885

PREFACE.

Harbours and Docks are treated of in this book in a somewhat similar manner to that adopted for Rivers and Canals in my previous work, with the hope that the two books together may furnish a fairly complete exposition of the principles and practice of hydraulic engineering, as applied to navigation and commerce, both inland and marine.

The two subjects of Harbours and Docks are separately dealt with in two distinct parts; and in each case, general principles, the different methods of construction, and the various accessory works, are first considered; and then concise descriptions are given of several of the most important harbours and docks, both at home and abroad, comprising their early history, progress, development, maintenance, and present condition.

Harbours are classified, both in respect of their form and their method of protection, and also more especially with reference to the types of breakwaters by which they are sheltered; so that comparisons may be more readily instituted between the separate groups, and also between the several examples in each group. The different systems of construction are thus easily contrasted; and the causes of the success, or failure, of special examples of works, resembling each other in construction, can be properly investigated.

Docks do not admit of the same distinct classification as breakwaters; but, nevertheless, a broad distinction can be drawn between tidal and tideless ports. Moreover, the dimensions and details of locks and entrances, the sections and composition of dock walls, the arrangement of quays, and the various appliances for facilitating trade afford ample scope for comparison. Statistics also, concerning the growth of trade and the relative importance of ports, furnish valuable evidence of the general lines of traffic, the conditions favourable to the development of a port, the comparative capabilities of docks, and the prospects of a demand for further extension.

Illustrations are so invaluable for the due comprehension of engineering works, that I devote a special volume to Plates, in addition to several woodcuts dispersed throughout the text. These Plates, moreover, present a novel feature which I trust will materially add to their value; namely, that, in most cases, the various figures in each plate are drawn to the same scale, and several of the plates also are similar in scale; so that not only are the different figures on the same plate comparable at a glance, but also in some instances two or three plates can be likewise contrasted. The scales also are given definite proportions, so that the relation between them is at once perceived. Thus all the Plans on Plates 2, 3, and 5, are drawn to $\frac{1}{30000}$th of the natural scale, and the Plans on Plates 1, 4, 10, 11, and 13, are made half as large again; whilst the Plans on Plate 12 are double the same scale. All the sections of breakwaters, forty-five in number, are drawn to $\frac{1}{600}$th; whilst the sections of dock walls, comprising thirty-two examples, are all made $\frac{1}{200}$th of the natural scale; and the other illustrations are similarly arranged.

Besides embodying in the book the results of personal observation, experience, and practice, extending over a period of twenty years, I have freely sought the co-operation of my professional brethren, which, with rare exceptions, has been most readily given, and for which I beg to tender my most grateful thanks. The assistance thus received, which greatly enhances whatever value the book may possess, is duly acknowledged in the notes; but a few instances of special help demand special recognition. Mr. Druce of Dover, Mr. Cay of Aberdeen, Mr. Broadrick of Leith, Mr. T. J. Long of New York, Col. Mansfield of Galveston, and M. Barret of Marseilles, sent me specially written particulars about their respective ports, which have proved of great service; whilst General Wright the Chief of Engineers of the United States, afforded me similar information about Delaware Harbour. Many details about Madras Harbour were given me by Mr. Parkes; and Mr. J. G. Gamble sent me several reports from Cape Town relating to Table Bay Harbour.

I am indebted to the Library of the Institution of Civil Engineers, and Mr. James Forrest the Secretary, for official records and reports from which descriptions of Portland and Colombo Harbours have been framed. I must also add my best thanks to M. Schwebelé, the Librarian of the Ecole des Ponts et Chaussées at Paris, for the ready aid he has given me in directing my inquiries about French ports.

The limits necessarily imposed on such a book have prevented my entering into very detailed accounts of the various works described, or from alluding to all the principal ports of the world; and my chief aim has been to dwell upon the prominent features, or peculiarities, of the various examples selected, so as to give a comprehensive view of this important and difficult branch of engineering, such as I hope may aid the student in his inquiries and the engineer in his practice, and may offer interesting general information to the trader and the public at large.

L. F. VERNON-HARCOURT.

6 QUEEN ANNE'S GATE, WESTMINSTER.
October, 1884.

CONTENTS.

PART I. HARBOURS.

CHAPTER I.

PRELIMINARY CONSIDERATIONS.

CHAPTER II.

WAVES.

CHAPTER III.

TIDES; CURRENTS; AND CHANGES IN COASTS.

CHAPTER IV.

FORMS OF HARBOURS.

CHAPTER V.

JETTIES AND BREAKWATERS.

CHAPTER VI.

BREAKWATERS FORMED OF A MOUND AND SUPERSTRUCTURE.

CHAPTER VII.

UPRIGHT-WALL BREAKWATERS.

CHAPTER VIII.

JETTY HARBOURS WITH PARALLEL JETTIES.

CHAPTER IX.

HARBOURS WITH CONVERGING JETTIES.

CHAPTER X.

HARBOURS PROTECTED BY RUBBLE MOUND BREAKWATERS.

CHAPTER XI.

HARBOURS PROTECTED BY RUBBLE AND CONCRETE-BLOCK MOUND BREAKWATERS.

CHAPTER XV.

HARBOURS PROTECTED BY A RUBBLE MOUND, AND A SUPERSTRUCTURE FOUNDED BELOW LOW WATER.

CHAPTER XVI.

HARBOURS SHELTERED BY UPRIGHT-WALL BREAKWATERS.

CHAPTER XVII.

HARBOURS ON SANDY COASTS.

CHAPTER XVIII.

LIGHTHOUSES; BEACONS; BUOYS; AND REMOVAL OF SUNKEN ROCKS.

PART II. DOCKS.

CHAPTER XIX.

SITES AND PRELIMINARY WORKS FOR DOCKS.

CHAPTER XX.

DOCK WALLS; PITCHED SLOPES; AND JETTIES.

CHAPTER XXI.

ENTRANCES AND LOCKS; DOCK-GATES; CAISSONS; GRAVING DOCKS; MOVABLE BRIDGES.

CHAPTER XXV.

DESCRIPTIONS OF DOCKS.

CHAPTER XXVI.

GOVERNMENT DOCKYARDS.

CHAPTER XXVII.

FOREIGN DOCKS.

CHAPTER XXVIII.

FOREIGN DOCKS AND RIVER QUAYS.

APPENDIX I.

APPENDIX II.

APPENDIX III.

APPENDIX IV.

APPENDIX V.

APPENDIX VI.

APPENDIX VII.

LIST OF PLATES.

LIST OF WOODCUTS.

ERRATA.

Page 76, line 17 from top, *for* Portland *read* Pentland.
„ 87, „ 13 from top, *for* 800 feet *read* 850 feet.
„ 92, „ 4 from bottom, *for* Ronds *read* Roads.
„ 362, „ 8 from bottom, *for* Rudyard's *read* Rudyerd's.
„ 364, „ 3 from bottom, *for* Lairn *read* Laira.
„ 379, „ 7 from bottom, *for* Hard Deeps *read* Hand Deeps.
„ 395, „ 9 from top, *for* St. Catherine's *read* St. Katherine.

HARBOURS AND DOCKS.

PART I. HARBOURS.

CHAPTER I.

PRELIMINARY CONSIDERATIONS.

Objects of Harbours and Docks. Varieties of Harbours. Instances of Natural Harbours. Necessity for Artificial Harbours. Wind: Velocity Anemometer; Pressure deduced from Velocity; Pressure Anemometer; Maxima Velocities and Pressures; Anemometry; Influence and Prevalence of Winds. Forecast of Weather.

HARBOURS and docks are essential for the safe and expeditious carrying on of the commerce of the world. It is by means of the shelter afforded by harbours that vessels can approach the land; and docks enable them to discharge their cargoes, and to be reladen with merchandise, both safely and rapidly. Harbours also are valuable, from a military point of view, as places from which one vessel, or a whole fleet, can issue at a favourable opportunity for attack, or to which they can run for security or provisions. The term 'harbour of refuge' is applied to that special class of harbours which, from their position and size, serve as a refuge for vessels overtaken by a storm, where they can ride in safety till the gale has subsided. It is essential that harbours of refuge should be accessible at all times; whilst many small commercial harbours can only be entered when the tide is high, and are consequently called tidal harbours. Harbours occasionally form the actual basins in which the vessels discharge and take in their cargoes; but generally they are merely the approach channels, or basins, through which access is afforded from the open sea to interior docks or quays, where better shelter and greater conveniences can be provided. Sometimes also they serve as the entrances by which rivers or ship-canals are approached. The points of importance for a harbour are—good shelter, and a sufficient depth of water; and for large, or refuge harbours, firm anchorage-ground is also requisite.

Natural Harbours. Harbours may be either natural or artificial. The mouths of rivers are instances of natural harbours, such as the Thames, the Mersey, the Humber, the Firth of Forth, the Seine, the Scheldt, the Potomac, and the St. Lawrence: their value, however, is in many cases diminished by the existence of a bar, reducing the available depth of water at their entrance. They form by far the most important and numerous class of natural harbours in affording sheltered access to docks, quays, and wharves, at any suitable points along their course, in some cases several miles inland, as, for instance, at London, Antwerp, and Rouen.

Creeks and bays form more or less good natural harbours in proportion to the depth of water, the enclosure provided, the narrowness of the entrance, and its direction in regard to the worst winds. Milford Haven is an instance of a well-sheltered creek; and the bay of Rio de Janeiro, about thirty-two miles in circumference, with an opening only a mile in width, and surrounded by hills, forms the finest natural harbour in the world. Occasionally the necessary shelter is furnished by an outlying island, as in the case of Portsmouth Harbour and Southampton Water, which are protected by the Isle of Wight. In some localities a moderate amount of shelter is afforded near the coast by outlying sandbanks, which, when deeper water and good anchorage-ground exist inshore, enables vessels to wait till a change of wind, or the rise of the tide, permits them to enter the neighbouring port. The Yarmouth Roads, and the roadstead in front of Dunkirk, are instances of this sort of imperfectly sheltered area; but such roadsteads are only serviceable under certain conditions of wind and weather, and more resemble other somewhat sheltered sites with good anchorage and deep water, such as the Downs, rather than harbours in the strict sense of the term. These outlying sandbanks, however, are also of some value in protecting the entrances to the adjacent harbours from the full force of the sea during an on-shore gale.

Necessity for Artificial Harbours. Frequently, a port having been established in a site possessing some natural advantages, which were sufficient when vessels were small, has to be improved and enlarged in order to provide adequate depth and accommodation for the increasing size of vessels. Also, long lines of exposed coast are sometimes devoid of any shelter; and great losses of life and property may be prevented by the establishment of harbours of refuge. Moreover, no natural harbours may exist in the nearest line of communication with other countries, or be suitably situated for strategical purposes. Under these circumstances, harbours have to be improved or created; and, occasionally, most unfavourable and exposed sites have to be resorted to, in which the skill of the engineer is taxed to the utmost for providing shelter where waves and currents have ruled supreme. As the sea-works which will be described in the following chapters have to be designed to resist, or to evade, the natural forces in action along the coast, it is important to investigate briefly the general nature of these forces before entering upon the consideration of the works themselves.

Wind. The wind, which blows with varying intensity, at different periods, in every quarter of the globe, is the main origin of the difficulties which have to be encountered in the construction and maintenance of harbours. The movements of the air, the causes of its motion, its direction and its force, belong to the science of meteorology, and are fully investigated in treatises relating to that subject[1]. A knowledge of the principal currents of the atmosphere, such as the trade winds, monsoons, &c., their limits, and the periods of their occurrence, and also of the seasons, extent, and violence of cyclones, tornadoes, and other atmospheric disturbances, is chiefly important for navigation. Nevertheless, an acquaintance with these matters is very useful to the marine engineer, as it enables him both to form a conception of the forces with which he may have to contend, and also to judge in what localities, and at what time of year, particular causes of disturbance may be anticipated. Whilst, however, the sailor is mainly interested in the laws of wind and storms at sea, the engineer has to consider them merely as they affect the coast. The violence of a storm depends upon the force exerted by the wind. This force has been estimated by various instruments, in different localities, and with very different results. The instruments most commonly used are Robinson's cup-anemometer[2] and Osler's plate-anemometer[3].

Velocity Anemometer. Robinson's anemometer consists of four metal hemispherical cups, placed at the extremities of two rods which cross each other at right angles and turn freely on a vertical axis. The axis has an endless screw at its lower end, by which its motion is communicated to a series of toothed wheels which register the number of revolutions performed by the cups. A gentle breeze suffices to make the instrument revolve, owing to the greater effect of the wind upon the concave side of the cups than on the convex. Dr. Robinson's earlier investigations led him to the conclusion that the velocity of rotation of the cups was one-

[1] The following books may be consulted with advantage: 'The Physical Geography of the Sea and its Meteorology,' M. F. Maury; 'The Law of Storms,' H. W. Dove; 'Les Mouvements de l'Atmosphere et des Mers,' H. Marié Davy.

[2] 'Transactions of the Royal Irish Academy,' vol. xxii. part i. p. 155.

[3] 'Report of the British Association for 1844,' p. 253. Mr. Osler has also himself written a description of the instrument, which was printed at Birmingham about the year 1839.

third of that of the wind. The results, however, of more recent experiments, by Dr. Robinson and others[1], indicate that the velocity of the wind cannot be exactly expressed by multiplying the velocity of the cups by the simple factor three. The factor has to be varied according to the pattern of the instrument, as larger cups revolve more rapidly; also, a higher factor is required for low velocities. A factor of 2·5 has been found suitable for the cup-anemometer of the large Kew pattern adopted by the Meteorological Office. This type of instrument merely measures the velocity, and the pressure of the wind has to be deduced from it.

Pressure of Wind deduced from its Velocity. Various formulæ have been proposed for expressing the pressure of wind in terms of the velocity. Smeaton deduced a formula $\frac{v^2}{200} = p$, from the results of Rouse's experiments, for winds having a velocity not exceeding fifty miles an hour; where v is the velocity in miles per hour, and p is the pressure in lbs. on the square foot. Other formulæ for expressing the pressure have been proposed at various times, and have been collected together by M. Gaudard[2], in which either the square of the velocity is involved, or the square and single velocity together, or even the cube of the velocity. Some cup-anemometers are furnished with apparatus for keeping a continuous record of the velocities; but, generally, only the mean velocity per hour is recorded, and, in this case, the mean pressure alone can be ascertained by the formula. However, the Committee appointed to consider the question of wind-pressure on railway structures, wishing to deduce the maximum pressure, in the case of high winds, from the data of mean velocity furnished at various places, made some

[1] 'Proceedings of the Royal Irish Academy,' Second Series, vol. ii. p. 427; 'Philosophical Transactions,' 1878, vol. 169. p. 777, and 1880, vol. 171. p. 1055; and 'Proceedings of the Royal Society,' vol. xxxii. p. 170.

[2] 'Minutes of Proceedings, Inst. C. E.' vol. lxix. pp. 121–123.

comparative experiments, at Bidston Observatory near Liverpool, with Robinson's velocity anemometer and Osler's pressure anemometer. From these experiments they arrived at the empirical formula $\frac{V^2}{100} = P$ as the relation between the mean velocity V in miles during an hour and the maximum pressure experienced in that period, the lowest value of V being taken at forty miles[1]. The following table indicates to what extent the values given by the formula correspond with the actual results obtained by experiment:—

Maximum hourly run of the wind in miles.	Maximum pressure in lbs. on the square foot. Experiment.	Formula.
40	14·7	16·0
50	23·7	25·0
60	33·9	36·0
70	48·0	49·0
80	65·5	64·0
90	...	81·0
100	...	100·0

The Committee assumed that the velocity of the cups was one-third that of the wind; so that, for absolute accuracy, a slight modification of their table would be required to make it conform to the results of recent investigations.

Robinson's anemometer possesses the merit of simplicity: it is easily set up; it is kept in order with little trouble; its records are simple to observe, and, in the case of identical instruments, are reliable for comparing the relative force of gales at different places and periods. It is not, however, adapted for measuring the actual wind-pressure with absolute accuracy. Not only is the friction of the working parts of the instrument liable to vary, unless they are kept well oiled and cleaned and sheltered from the weather; but also the conversion of velocity into pressure is subject to error. In the first place, the relation between the velocity of the cups and that of the wind has not been accurately ascertained,

[1] 'Report of the Committee on Wind-Pressure on Railway Structures,' 1881, p. 1.

except for a special type of instrument, and it, moreover, varies with the velocity; and, in the next place, the precise relation between velocity and pressure is somewhat undefined, as no particular formula has been universally accepted. Also, although the pressure and velocity of the wind are both found to increase together in proportion as the wind is relieved from the retarding influences of the earth's surface at higher elevations, it appears probable that at considerable heights, though the velocity may be still further increased, the rarefaction of the air prevents a corresponding increase in the pressure of the wind. Fortunately, the harbour engineer is only concerned about the pressure of wind near the sea-level; otherwise, the above consideration would introduce further complications into a problem already far from simple.

Pressure Anemometer. The inevitable sources of error just referred to, in deducing the pressure of wind from its velocity, may be avoided by employing a pressure anemometer. The instrument usually adopted for this purpose is Osler's anemometer, which consists of a flat iron plate, from 1 foot to 2 feet square, turned by means of a vane so as always to face the wind, and fastened to a strong spiral spring at the back[1]. When the plate is exposed to wind it is forced back, compressing the spring in proportion to the pressure of wind against its face; and as soon as the force of the wind diminishes, the plate is pressed forward again by the spring. Accordingly, the motion of the plate backwards and forwards indicates the varying pressure of the wind. This motion is recorded by a pencil, connected with the plate, which traces a continuous irregular line along a sheet of paper. The sheet is so arranged, and gradually moved along, that the abscissæ of the line indicate the period and duration of the pressures, and the ordinates show the amount

[1] A drawing of this instrument is given in the 'British Association Report for 1844,' pl. 38.

of motion of the plate. The connection between the motion and the actual pressure on the plate is determined beforehand by noting the amount of compression produced on the spring by known weights. This instrument, accordingly, furnishes a continuous record of the actual wind-pressures, and is, in this respect, superior to the cup-anemometer; it is, however, more cumbrous, and the correctness of its indications has been sometimes questioned in the cases of the very high pressures occasionally recorded by it.

Maxima Velocities and Pressures of Wind. Few definite records exist of very high velocities of wind; but the strongest hurricanes have been considered to possess velocities of from ninety to a hundred miles an hour, and this is probably about the extreme limit near the earth's surface, though balloons are stated to have travelled at a greater speed.

The highest observed wind-pressures are those recorded at Bidston Observatory, which stands on a very exposed site 251 feet above mean sea-level. The highest records are as follows: 70 lbs. on the square foot on February 1st, 1868, and September 27th, 1875; 80 lbs. on December 27th, 1868; and 90 lbs. on March 9th, 1871. The Committee on wind-pressures, however, considered that these exceptionally high pressures were very limited in their extent, and that 56 lbs. on the square foot might be safely assumed as the highest limit of pressure over any considerable area.

Anemometry. Meteorologists are by no means satisfied, either with the sufficiency, or with the accuracy of the results hitherto obtained with regard to the velocity and pressure of wind. Indeed, Mr. Laughton, in his Presidential Address to the Meteorological Society in 1882[1], on the history of anemometers, said that no thoroughly satisfactory anemometer had hitherto been designed, and that the recording apparatus was much more perfect than the instruments whose motions

[1] Quarterly Journal of the Meteorological Society,' vol. viii. p. 180.

it recorded. Harbour engineers, however, are not so much interested in the exact determination of wind-pressures, as in the knowledge, which observations in a variety of places might afford them, of the comparative exposure of different sites, a point of great importance in designing harbours. It is towards the furtherance of this object that the best existing methods of anemometry have been briefly touched upon; for if anemometers, of one type, were set up at every harbour in an exposed situation, many valuable statistics with regard to the force of wind in different localities might be readily compiled, which would be most useful to engineers, and might also aid the progress of meteorological science.

Influence and Prevalence of Winds. The effect of wind on structures in the sea is a more complicated problem than for erections on land. Breakwaters, piers, and even lighthouses, have to be built so solidly to resist the sea, that they are in no danger of being overturned by the mere pressure of the wind. The wind is, indeed, the primary cause of the attacks to which these structures are exposed, but its influence is exerted through the medium of the waves which it raises; and therefore the maximum pressure of the wind is not the only element to be taken into consideration in calculating the stability which structures in the sea must possess. The direction of the prevailing winds must be observed, and the seasons at which they generally blow; the distance also which the winds travel with an unimpeded course should be ascertained. The annexed diagrams (Fig. 1), indicating the respective frequency of the various winds at the ports of Havre and Marseilles, illustrate a good method of graphically representing the comparative prevalence of winds at every point of the compass, for any particular locality, frequently adopted on French charts. A knowledge of the most prevalent winds is, moreover, necessary in order that the entrance of a harbour may be made to face in the most convenient direction for vessels, and also in order that the line of the travel of sand, silt, and detritus along the coast may be known and provided for.

CURVES SHOWING FREQUENCY OF WINDS.

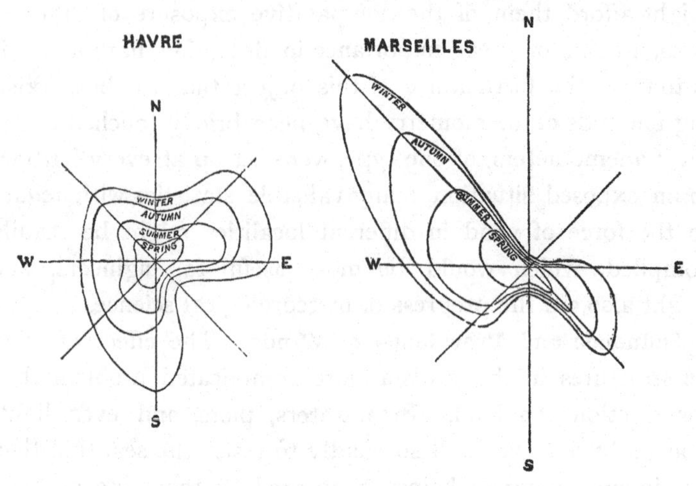

$\frac{1}{10}$ inch for 10 days.

Fig. 1.

Forecast of Weather. In all parts of the world some periods of the year are more favourable to the prosecution of sea-works than others. The working season, as it may be termed, is most clearly defined in places which are visited by regular periodical winds, such as the monsoons; but even in Europe, where the changes of wind and state of the weather appear subject to no definite laws, the force of the wind is, on the average, much less from April to September than during the rest of the year. The period also when violent storms are most liable to occur may be fixed within certain limits. Thus, the hurricane season of the West Indies is between July and October; the typhoons of the China Sea occur between June and November; the cyclones of the Indian Ocean are most frequent in May and November; whilst the hurricanes of the southern seas are experienced between January and April. It might appear to a casual observer that no general rule could be laid down with regard to the occurrence of storms on the shores of Great Britain and France; but, though one year may differ widely from the next in the distribution of storms, it is possible, by combining the observations of a number of years, to deduce some general law as to the prevalence of gales. Thus, the results of observations made during a period of twenty-two years at Alderney, given in Appendices V to VIII of my Paper on Alderney Harbour[1], indicate that, in the English Channel, gales are most frequent, as well as most violent, in December and January; and this is further confirmed by the fact that, not only at Alderney, but also along the British and French coasts, the worst injuries to sea-works generally occur in these months. It is rather commonly supposed that the equinoxes are periods of unusual atmospheric disturbance even in our latitudes; but, though windy weather may be expected about that time, the month of March is on the average hardly more stormy than February, and in September both westerly and easterly gales are much less frequent than in October and November. Assuming that the results of the Alderney observations may be accepted as correct for the British Isles and for France, it is worthy of note, for situations open to the east, that winds from the eastward are more prevalent in March, April, May, and November, than during the other months; whilst easterly gales are, on the average, most frequent in February, March, and November, and very rare in June, July, and August. For places exposed to the west in these localities, it may be noted that westerly winds are most prevalent in July and August; whilst, as regards westerly gales, May, June, and July are by far the calmest months, and December and January the most stormy. Comparing the frequency of easterly and westerly gales together for these parts, it appears that the proportion of days of easterly gales to westerly

[1] 'Minutes of Proceedings, Inst. C. E.' vol. xxxvii. pp. 80–83.

gales is, on the average, as three to seven in the whole year, as one to two in November, as one to three in January, and as one to four in December; whilst May is the only month in which easterly gales are the more prevalent of the two. The barometer is naturally the main guide of the harbour engineer for predicting the arrival and departure of storms; but the aspect of the sea, the appearance of the clouds, and particularly the veering of the wind, furnish valuable indications of the approaching weather. Just as an unusual clearness of the atmosphere denotes the probable approach of rain, so a sudden and unsteady shifting of the wind, and especially the veering of the wind in a direction opposite to the apparent path of the sun or the motion of the hands of a clock, is a fairly certain indication of a coming storm; whilst the gradual veering of the wind in the opposite direction, from south-west towards the north, if accompanied by a rising barometer, marks the termination of a gale on our coasts. Each locality possesses its own special meteorological indications; and it is therefore very important that continuous records should be kept of the movements of the barometer, and of the direction of the wind, at every harbour office, so that statistics may be compiled for forecasting the probable weather at the particular site. The records also and the forecasts published by the Meteorological Office are valuable in indicating the probable direction of the wind, or the prospects of a gale, on any particular day; and the telegraphic intimation from America of a westerly gale sweeping over the Atlantic provides a most timely warning of the possible advent of bad weather within a given period of time, thus enabling preparations to be made against a probable danger, and preventing a storm bursting unforeseen upon the western shores of Europe.

CHAPTER II.

WAVES.

Generation of Waves. Wave Motion: Theory of Vertical Oscillations; Motion of Particles in Orbits; Theory of 'Flot de Fond,' objections to it; Variation of Motion with Depth. Velocity of Waves. Breaking Waves. Size of Waves. Influence of Exposure on the Size. Variation in Length. Ground Swell. Importance of Length. Influence of Depth. Waves of Translation, and of Oscillation. Force of Waves; measured by Dynamometer; shown by damages at lighthouses, beacons, and breakwaters.

Generation of Waves. Waves are produced by wind; by the attraction of the moon and sun, in the form of a tidal wave of great magnitude; and by any sudden disturbance, such as an earthquake.

The tidal wave only affects sea-works by raising and lowering the level of the sea, and by the currents which it produces in contracted channels: it will be referred to in the next chapter when the subject of tides is considered. The commotion sometimes caused in the sea during earthquakes is too uncertain in its occurrence and extent, and too limited in its action, to be investigated with advantage. Waves produced by wind must be carefully studied, owing to their great importance in relation to the form and stability of structures in the sea. The sea, which on a calm day presents a perfectly level surface, is gradually raised by wind into waves which increase in size in proportion to the violence and duration of the gale. When a sudden squall bursts upon a surface of still water, numberless ripples are formed, which soon subside directly the wind abates. When, however, the wind continues blowing, or rises gradually, a series of waves are generated, which appear to roll along in succession as if the wind had imparted its horizontal motion to them. As the waves rise, the wind is enabled to exert a greater pressure against their windward surfaces, and the waves rise higher and roll on with increased rapidity. The progression observed is that of the undulation, not of the particles of the water; just as when a rope is shaken at one end, an undulating motion passes along the rope, whilst each portion of the rope resumes its original position when the undulation has ceased. The force of the wind is transmitted from one particle to the next, and the stored-up force is only manifested when there is no longer a succession of fluid particles to transmit it, as when the wave breaks upon the beach, or encounters a solid obstacle such as a cliff or breakwater. The old familiar example of a row of balls in contact best illustrates this principle of transmission; for when a shock is imparted to the ball at one end of the row, in the direction of the row, its effect is only rendered apparent by the propulsion of the ball at the other end of the row.

Wave Motion. The earlier writers on the subject of wave motion assumed that the particles of water, forming an undulation, merely moved up and down in a vertical line, rising to their highest point under the summit of the crest, and descending to their lowest limit under the bottom of the trough. It was, however, pointed out very clearly by M. Emy, in his book on waves published in 1831[1], that this theory could not possibly represent the real motion of the particles, in a limited depth, and with an incompressible fluid such as water. M. Emy demonstrated that the particles must have some sideways motion, tending towards the crest of the wave when rising, and away from it when descending, and that an orbital motion of the particles would perfectly account for the various forms and motions of waves. This view had been previously indicated by the observations of

[1] 'Du mouvement des ondes, et des travaux hydrauliques maritimes, A. R. Emy.

the brothers Weber on the movements of waves generated in troughs with glass sides[1], and has been proved to be correct by Sir G. B. Airy in his mathematical investigation of wave motion[2]. The orbital motion is the result of the combined action of the horizontal force of the wind, which propels the water forwards on the crest of the wave, and the force of gravity, which draws it down and causes it, in falling, to react on the adjacent particles in front of it, which rise in their turn and are successively subjected to the influence of the wind. Sir G. Airy has shown that the path of the particles, in a vertical plane in the line of the wave motion, may be a circle, or an ellipse with its major axis horizontal; and that the motion of the particles is in the direction of wave under its crest, and in the reverse direction under its hollow. Accordingly, during the passage of an undulation, all the particles subject to its influence are oscillating in circular or elliptical orbits. In water whose depth is considerably greater than the length of the wave, the orbit at the surface is approximately circular; and the orbits of the particles are flatter, and also shorter, in proportion to the depth below the surface.

M. Emy used his demonstration of orbital motion merely as a preliminary step in the explanation of his theory of the existence, under certain conditions, of an under-current, or bottom-wave, which he called *flot de fond*, to which he attributed the injuries produced by waves on structures in the sea. He imagined that wherever a sudden rise, or sort of steep step up, existed at the bottom of the sea, whereby the lowest undulations of a wave were arrested, the upper portions of the undulations, whose lower portions were stopped by the rise, being prevented from continuing their orbital motion, proceeded onwards along the bottom as a sort of

[1] 'Wellenlehre auf Experimente gegründet,' E. H. und W. Weber (Kastner, Archiv Naturl. vii. 1826.)

[2] 'Tides and Waves,' G. B. Airy. (Excerpt 'Encyclopædia Metropolitana.')

under-current, being pushed forwards by the unimpeded undulations above them. The existence, however, of a sudden rise in the level of the bottom is essential, according to this theory, for the production of a *flot de fond*; whereas disturbance is equally caused by waves under the more common condition of a regularly shelving beach. Moreover, no evidence was adduced by M. Emy of the production of such an under-current, in those special conditions, flowing along below a mass of merely oscillating water, except the movement of materials, lying at the bottom, under the influence of waves; and this movement is more fully and satisfactorily explained by the accepted theory of wave motion. The theory gave rise to considerable discussion, and found some advocates; but it does not appear to rest upon any sound basis.

The depth at which wave motion ceases varies with the height and length of a wave. Thus Mr. Scott Russell has stated that waves 9 inches high and 4 or 5 feet long do not sensibly affect the water at a depth of 12 feet; whereas waves 30 or 40 feet long do produce a sensible effect at that depth, though much less than near the surface[1]: also that at a depth of 10 feet below the trough, a wave 10 feet high and 32 feet long only agitates the water to the extent of 6 inches; whilst a wave 100 feet long, and having the same height, produces a motion of 18 inches[2].

Velocity of Waves. The velocity of various kinds of waves, in the same depth of water, is not uniform, but depends upon the length of the wave; and, for a given length of wave, it varies with the depth of water within certain limits. According to Sir G. Airy's investigations, the velocity of very short waves does not depend on the depth, and the velocity of very long waves depends simply on the depth. Tables have been drawn up by Sir G. Airy[3]

[1] 'Report of the Meeting of the British Association in 1837,' vol. vii. p. 447.

[2] 'Minutes of Proceedings, Inst. C. E.' vol. vi. p. 136.

[3] 'Tides and Waves,' G. B. Airy, p. 291. (Excerpt 'Encyclopædia Metropolitana.')

and Mr. Scott Russell[1], giving the velocities of waves having lengths ranging between 1 foot and 1,000,000 feet; but whilst Mr. Russell has verified some of his figures by the results of observations, he has not introduced the element of depth into his tables. Thus, whereas Mr. Russell gives a velocity of 19 feet per second for a wave 100 feet in length, a velocity of 60 feet per second for a wave 1000 feet in length, and a velocity of 189 feet for a wave 10,000 feet in length, these velocities would only be correct, according to Sir G. Airy's formula, for depths of about 19 feet, 190 feet, and 1900 feet respectively, or where the depth is about one-fifth of the length of the wave. From Sir G. Airy's table it appears that the velocity of a wave ceases to increase with the depth beyond a depth equal to the length of the wave; and it is evident, from the form of the second factor of his formula, that, practically, the limit of increase in velocity is reached at considerably smaller depths.

Breaking Waves. Water is broken by the wind when a sudden squall bursts upon a sheet of water, owing to some of the particles being dragged along by the wind before there has been time for the wind to overcome the inertia of the mass of the water and raise it into regular waves. In like manner, the summits of the crests of waves are sometimes broken by a rising wind, which impels forwards the tops of the waves at a more rapid rate than that already attained by the regular undulations. Waves are also occasionally broken by encountering cross currents, or other waves pursuing a different course.

The principal cause, however, of the breaking of waves is their advancing into shallower depths than those in which they were generated. A wave travels along for considerable distances, with little perceptible modification of its form, till it reaches a depth too shallow for the regular undulation to continue unimpeded. As the wave approaches the shore

[1] Report of the Meeting of the British Association in 1844,' vol. xiv. p. 374.

it gradually increases in height and diminishes in length, and its front slope becomes more and more vertical (Fig. 2),

WAVE BREAKING ON BEACH.

Fig. 2.

till at last, the lower undulations being arrested or forced upwards, the upper crest is urged onwards more rapidly by the accumulated momentum, it passes the vertical line, curls over, and dashes against any obstacle in its course, or descends with a heavy thud upon the beach, where it breaks into surf and rushes up the slope till the force of gravity causes it to recoil. Whilst a portion of the force existing in the lower parts of a wave is spent by friction against the beach, another portion is communicated to the crest of the wave when rising and breaking, so that an increased momentum is imparted to the breaking water; and, consequently, a breaking wave deals a heavier and more concentrated blow against a structure than an unbroken wave, which expends its force over a larger area. When a wave breaks in shoal water before reaching the shore, the upper crest only is broken, and a smaller undulation is transmitted towards the beach. Waves invariably begin to break when they enter a depth of water approximating to their height; and Mr. Thomas Stevenson has observed waves, from five to eight feet high, breaking when the depth of water below their trough was twice the height of the wave. The breaking of waves on a steep beach is more violent than on a flat shore, as it takes place more suddenly.

Size of Waves. The size of waves depends, not only upon the strength and duration of a gale, but, more especially, upon the area of water over which the gale acts. Thus the short steep waves raised in the English Channel by a north-east wind bear a marked contrast to the long waves brought in from the Atlantic by south-westerly gales. The heights attained by waves in the open sea have been noted by different observers. Mr. Scott Russell mentions 27 feet as the greatest height of wave he had observed in the British seas; and Sir G. Airy considers 30 to 40 feet to be the extreme height of unbroken waves. Dr. Scoresby, whilst crossing the Atlantic in 1848, observed, at the close of a storm which had lasted thirty-six hours, a series of waves averaging 26 feet in height and 560 feet in length[1], and proceeding at a rate of 32⅗ miles an hour[2]. The highest unbroken waves he observed were estimated to have a height of 43 feet. Mr. Henderson, however, measured waves 50 feet high and from 600 to 1000 feet long[3], which he considered not uncommon dimensions in southern latitudes; whilst Captain Ginn found that the usual velocity of the waves in the Atlantic was 22·3 miles an hour, and 26·8 miles an hour off Cape Horn[4], which shows that the greatest waves occur in the southern seas.

Influence of Exposure on the Size of Waves. The effect of the amount of expanse of water on the maximum size of waves is best illustrated by a comparison of the heights observed in various localities. Thus waves reach a height of from 6½ to 10 feet in large deep reservoirs, and also in the lake of Geneva; in the Mediterranean the waves are seldom higher than 16½ feet; whilst at Cherbourg they attain a height of from 16 to 20 feet, and 23 feet in the Bay of Biscay. In the Atlantic they reach about 40 feet; and off the Cape of Good Hope, where the exposure is complete, waves of from 50 to 60 feet in height have been observed. Endeavours have been made to reduce to a formula the

[1] 'Report of the British Association for 1850,' Second Part, p. 28.

[2] This velocity, amounting to 48 feet per second, corresponds fairly closely to the velocity of 53 feet per second which Sir G. Airy's formula gives for a wave 560 feet long in an ample depth.

[3] 'Minutes of Proceedings, Inst. C. E.' vol. xiii. p. 34.

[4] 'The Physical Geography of the Sea,' M. F. Maury, 14th edition, p. 273.

variation in the size of waves in proportion to the amount of exposure. Thus Mr. Hawksley has determined, by observations, that the height of waves produced in large reservoirs by the heaviest gales in this country may be represented by the formula, $H = \frac{\sqrt{l}}{40}$, where H is the height in yards, and l the length in yards of any one run of wave, or the distance over which the wind acts in one direction[1] Mr. Hawksley considers that this formula may be extended to the case of the open sea, and that the greatest run would be limited to about 100 miles owing to the circular course taken by storms. Mr. Thomas Stevenson has found that the formula, $h = 1{\cdot}5 \sqrt{d}$, represents, with tolerable accuracy, the height of waves during heavy gales, as gathered from a comparison of observations at various places, where h is the height in feet, and d the distance of exposure in miles[2] It must, however, be borne in mind that the line of greatest exposure may not be in the direction from which the strongest gales blow, in which case the exposure is not really so great as it appears to be. It is difficult to ascertain over what distances gales may continue to travel in the same course. It would appear, however, from the length of time that gales have been observed blowing continuously from the same point, and more especially from the exactness with which the arrival of gales on the coasts of Great Britain are now predicted by telegraph from America, that a storm frequently traverses the Atlantic with little change in its direction. Of course, for these long distances, Mr. Hawksley's formula would be inapplicable, and Mr. Stevenson's formula would require some modification of its numerical factor. Though the whirling character of many of the storms reaching these shores from the west, would tend

[1] 'Minutes of Proceedings, Inst. C. E.' vol. xx. p. 361.

[2] 'The Design and Construction of Harbours,' T. Stevenson, 2nd edition, p. 25.

to impede the continuous propagation of the waves throughout the entire path of the storm, nevertheless, the wind would not shift in direction along the line traversed by the outer edge of the storm.

Variation in Length of Waves. The formulæ given above refer only to the height of waves, and there is no definite relation between the height of a wave and its length. It would appear, however, that the length of waves is even more affected by the extent of exposure than the height; for whilst the greatest waves observed by Dr. Scoresby in mid-ocean were from 26 to 43 feet high, their greatest length was from 560 to 790 feet; whereas the waves observed by Sir J. N. Douglass from the Bishop's Rock, off the western extremity of the Scilly Isles, were 1800 feet long, and only 20 feet high[1], indicating that, in travelling from the middle of the Atlantic to the leeward shore, the waves tend to increase in length and to diminish in height. The increase in length is probably produced by the coalition of a succession of smaller waves travelling at somewhat different rates; and the reduction in height must be occasioned by the friction of the particles of water in the transmission of the undulations, and by the resistance offered by the air to their progress, when a gale, owing to a change in direction or diminution in force, ceases to maintain the waves it has raised.

Ground Swell. The waves known under the appellation of *ground swell* are well-known instances of this effect. These waves, which are long and low, generally appear in the English Channel with a light southerly wind, and must be the results of storms which have not reached these latitudes. The ground-swell waves present the same contrast to waves raised by north-easterly gales in the Channel, that the rounded chalk downs of the south coast of England do to the peaked hills of North Wales. The ground swell usually

[1] 'Minutes of Proceedings, Inst. C. E.' vol. xv. p. 8.

appears without any warning of its approach; and sometimes only two or three undulations reach the coast.

Importance of Length of Waves. The maximum height waves are liable to attain, with a given exposure, is required to be known for designing reservoir embankments high enough to be in no danger of being overtopped, which was the object of Mr. Hawksley's formula. The height, moreover, of waves has more effect upon vessels, and produces a greater impression on the imagination, than the length; which may account for the height of waves having been given more prominence than their length. The length of waves, however, is of the most importance to harbour engineers; for both the extent of the horizontal oscillation of the particles, and also the velocity of the transmission of the undulation, depend upon the length. A long low wave is more dangerous to the stability of a breakwater than a short high one, both on account of the greater shock produced by the velocity of the undulation, and also because the agitation of the particles of water is not merely greater horizontally, but, moreover, extends to a greater depth.

Influence of Depth on Waves. It has been previously pointed out that large waves cannot exist in a small depth of water. Accordingly, if the sea is shoal for some distance in front of a breakwater, whatever may be the violence of the winds and the exposure of the site, large waves cannot approach the shore, as they are more or less broken on entering shoal water. On the contrary, when the depth is considerable close inshore, a sudden shoaling produces a tremendous surf upon the beach if the fetch of the sea is great. A shoal at a considerable distance from the shore is equally effective in reducing the waves rolling in from the ocean, even when deep water exists between the shoal and the shore, as the broken waves merely transmit small undulations. Under these conditions, the worst seas reach the shore when the waves are just small enough not to be broken on the shoal, as then they pass over undiminished to the land. Long waves are not so readily influenced by shoals as short waves, for they require a considerable modification in their form before they can break.

Besides examining the general depth of the sea in front of any site for a harbour, it is advisable to investigate the contour lines of the soundings; for when the lines of deep soundings run up in a sort of converging bay towards the coast, the waves from the open sea become concentrated, and larger undulations may approach the shore than if the lines of soundings ran parallel to the coast.

Waves of Translation, and of Oscillation. In discussions concerning the comparative merits of various forms of breakwaters, a distinction has been frequently drawn between waves of translation and oscillation. The advantage, indeed, claimed by the advocates of the upright-wall system, to which reference will be made hereafter, is that the oscillating sea waves which come against it are reflected, and merely exert a hydrostatic pressure due to their height; whereas with other forms of breakwaters they are changed, in breaking, into waves of translation which strike against the wall. The distinction appears to have originated from Mr. Scott Russell's Report on Waves to the British Association in 1844, in which he divides waves into separate classes or orders[1], and places in the first order what he terms *The Wave of Translation,* and in the second order the ordinary sea waves, which he calls *Oscillating Waves.* These names, and the extract from the Report given in the footnote, imply that

[1] 'Some of them are distinguished by always appearing alone as individual waves, and others as companion phenomena, or gregarious, never appearing except in groups. In examining the paths of the water-particles, corresponding differences are observed. In some the water-particles perform a *motion of translation* from one place to another, and effect a permanent and final change of place; while others merely change their place for an instant to resume it again, thus performing *oscillations* round their place of final repose.' Report of the British Association for 1844, p. 316.

the waves of the first class are solitary and have a progressive motion, and that those of the second class appear in groups, and merely move backwards and forwards without any progression; and this is the interpretation that has commonly been attached to them. Mr. Russell defined waves of translation to be solitary waves, which, passing along a fluid, cause every particle of the fluid to be moved slightly forward during the passage of the wave, in the proportion that the size of the wave bears to the cross section of the channel. These particles, having attained their new position, remain at rest after the passage of the wave of translation. He considered, on the contrary, that the ordinary wind waves rolling along in succession are merely oscillatory, and produce no real advance in the position of their particles. Moreover, whilst in a wave of translation the motion is fairly uniform throughout the depth, in an oscillating wave the agitation is greatest at the surface, and diminishes in proportion to the depth below the surface till a point is reached where no agitation is produced.

If Mr. Russell had included in his first order only those waves which are generated by an attraction or impulse acting throughout their depth, such as the tidal wave, and a wave produced in a channel by drawing forward a solid body filling up the sectional area of the channel, there would have been some ground for assuming that waves so generated might possess distinct characteristics from waves produced by the surface action of the wind. This, indeed, is apparently the view taken at the date of the report; as sea waves are distinctly placed under the second order, and even breaking waves are referred to under this heading, though regarded as marking a change into the first order of waves. Subsequently, however, Mr. Russell classed the rollers of the southern seas, the waves from the Atlantic, and ground swells, as waves of translation, though they are not solitary waves, and, except in shallow water, cannot produce at all the same agitation of the particles of water towards the bottom as at the surface[1]. Whilst transferring waves raised by gales from the second order to the first, he endeavoured to maintain a distinction between the great wave of translation, which he claimed to have discovered, and oscillating waves, by retaining the smaller wind waves in the second order. It is evident, however, that if great wind waves are waves of translation, the same must be true of the smaller wind waves, though in a less degree. As soon as it is admitted that wind may produce some translating effect upon water, the distinction between waves of translation and oscillation cannot be maintained. It is clear that wind does impart some motion to water, as exemplified by the well-known currents flowing along the shores of the Mediterranean in the direction of the prevailing winds, and by the heaping up of the water that takes place upon a lee shore during a storm, raising the level of the sea. In reality all wind waves have some motion of translation varying with the size, and more especially with the length of the wave; and it is this motion of translation, or rather transmission of force, which is the cause of their effect upon sea-works. The horizontal motion of

[1] 'Now in deep water there were not only the oscillating surface waves to be encountered, but also those which he had termed waves of translation, forming what were called rollers at the Cape of Good Hope, and when on a smaller scale, known as ground swell. These were a much more troublesome class of waves; it was mainly with them that the engineer had to deal in places open to the Atlantic; and after a storm of some duration at sea they become the deadliest enemies, in the cases of deep water, against which breakwaters for harbours of refuge had to contend. These rollers, or ground swell, did not merely oscillate up and down, and backwards and forwards, and they could not be eluded, or turned back, by giving to the wall a particular curve, suited to the form of a cycloidal oscillation. These great waves of translation constituted a vast mass of solid water, moving in one direction with great velocity, and this action was nearly as powerful at a great depth as at the surface. They resembled the tidal bore of the Hooghly, of the Severn, or of the Dee; they formed a high and deep wall of water, of great weight, moving horizontally with great force, and causing all floating bodies they met with to travel with them with great velocity in the same direction.' Minutes of Proceedings, Inst. C. E., 1860, vol. xix. p. 653.

the wind is gradually imparted to the undulations, till at last a storm which has travelled across the Atlantic may, through the medium of the waves which it has raised, impart a succession of concentrated blows against a breakwater situated in, or near, deep water, in an exposed situation.

In the extract given in the note on the preceding page, Mr. Russell not only considers rollers and ground swells as waves of translation, but he goes a step further, and represents them as masses of water moving onward with great velocity and carrying floating bodies along with them, having therefore the nature of a current rather than of a wave. If this was the case, boats in a ground swell would be carried along; but everyone who has been in a boat during the passage of a ground swell knows that no such motion of translation occurs, and that the boat rises whilst the undulation passes under it, as in the case of ordinary waves. Rollers and ground swells, however, entering shallow water, have a greater power of transmission near the bottom than shorter waves in the same depth, owing to their undulations having been produced by a greater force of win dacting for a long space of time.

A breaking wave has more effect than an unbroken wave, owing to the force of transmission of the whole wave being concentrated in a small mass of moving water: a breaker, however, would have only a slight power if it was not the vehicle of the force of transmission previously existing in the wave, and manifested in the breaker owing to the absence of particles of water to continue the transmission.

Force of Waves. It is very difficult to estimate, with any approach to accuracy, the force that a breaking wave may exert against an obstacle. Mr. Thomas Stevenson, however, designed, many years ago, an instrument, which he calls a marine dynamometer[1], by which he has endeavoured to measure the shock of waves in a way very similar to that by which Osler's plate anemometer measures the pressure

[1] 'Transactions of the Royal Society of Edinburgh,' vol. xvi. p. 23.

of wind. The instrument consists of a flat iron plate, or disc, exposed to the shock of the waves (Fig. 3); the size

MARINE DYNAMOMETER.

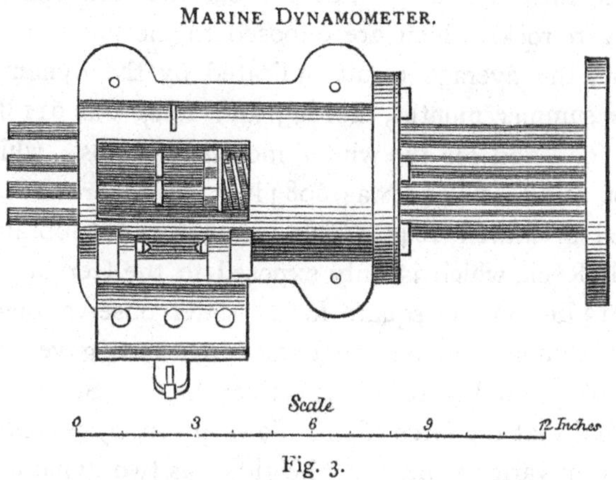

Fig. 3.

of the discs employed averaged six inches in diameter, and the instrument was generally placed at the level of a three-quarter tide. The disc is connected with four rods passing through an iron cylinder which is bolted to a rock or wall. The motion of the rods is controlled by a spiral spring which, winding round the rods, is fastened to the front end of the cylinder at one extremity, and to the rods at the other, so that any inward movement of the rods produces an extension of the spring. An index, consisting of a leathern ring, placed on each rod at the further end of the cylinder, is prevented from following any retrograde motion of the rod by the plate at the end of the cylinder. When the disc is struck by a wave, the iron rods are pushed more or less through the cylinder, and the index remains against the end of the cylinder. As soon as the pressure of the wave on the disc is removed, the spring brings back the rods and disc to their original positions, the indexes travelling forwards with the rods. The rods are graduated, and the value of each division has been previously ascertained by loading the instrument with weights, so that it is only necessary to read off the distance the index has travelled along the rod in order to find out the maximum pressure exerted by the waves against the disc. Mr. Stevenson found that at the Skerryvore rocks, which are exposed to the full force of the Atlantic, the average result indicated by the dynamometer for five summer months, in 1843 and 1844, was 611 lbs. per square foot, and for the winter months 2086 lbs.; whilst the greatest result attained was 6083 lbs. per square foot, which occurred in March 1845. The greatest result obtained at the Bell Rock, which is only exposed to the German Ocean, was 3013 lbs. on the square foot. Later observations, however, at Dunbar and on the Banffshire coast, gave maxima results of 3½ and 3 tons respectively[1]. Mr. Stevenson has also found that the force of the blows given by waves in the same storm varies at different heights; as two dynamometers placed at different levels on the Skerryvore rocks gave different results, the lowest, though put farther out, exhibiting the least pressure. This was due to the waves being somewhat broken by outlying rocks during low tide. It will be found that with most breakwaters there is some particular state of tide when the waves in a storm appear to rise highest, and to break with the greatest force, owing either to shoals outside, or to the form of the foreshore of the breakwater, or to the set of the tidal current.

The force of waves is also manifested in the actual injuries produced by their impact on lighthouses, beacons, and breakwaters. These results, moreover, are not confined to the absolute level of the waves, as the wave in striking against a rock or structure forces up a portion of its water to a height which depends both on the force of the blow and the configuration of the structure. The column, or sheet of water, thus raised, is borne by the wind over the structure, or dashed against its upper portion. The height to which the

[1] 'The Design and Construction of Harbours,' T. Stevenson, 2nd edition, p. 51.

column of water or spray is raised, and the power which this water is able to exert far above the sea-level, are very remarkable. Thus Sir J. N. Douglass has stated, that, at the Longships lighthouse off the Land's End, panes of glass, 2 feet square and ¾ inch thick, had been driven in by the sea during a storm, at an elevation of 86 feet above high water: that, at the Bishop's Rock lighthouse off the Scilly Isles, a fog bell, suspended 100 feet above high water, had been torn from its brackets; and that, at Smeaton's Eddystone lighthouse, the granite gallery, 60 feet above high water, with the lantern and optical apparatus, weighing altogether 61 tons, was lifted by the sea, which indicated that at that height above the sea-level a pressure must have been exerted equal to 1 ton per square foot. The breaking off of the iron tower of the lighthouse on the Calf Rock, in December 1881, at a height of 86 feet above high water, furnishes another instance of the power of the sea. At Alderney, during severe storms, stones were not unfrequently lifted by the waves from the sea slope of the rubble mound of the breakwater and hurled over the top of the superstructure, forty-five feet above the mound; and instances are recorded of stones weighing as much as 9 tons having been thus raised.

The most important and instructive instances, however, of the power exerted by waves are afforded by the actual movement or failure of structures in the sea under the repeated blows of the waves. The injury sustained by the beacon on the Petit-Charpentier rock at the mouth of the Loire, in 1867, forcibly illustrates the power of the blows which may be exerted by waves. The site of the beacon is peculiarly exposed, as not only is it open to the full violence of the Atlantic Ocean from the south-west, but also the funnel-shaped formation of the rocks in front of the beacon tends to lead up the waves and concentrate their force. The beacon, however, was constructed with the utmost care, and built of granite carefully dressed at the beds and joints, and set in Portland cement mortar. It rises 21 feet from its footings embedded in the rock whose summit is 11½ feet above low water of equinoctial spring-tides. The diameter of its base is half its height, namely, 10½ feet; but the batter on its face reduces the diameter of the top to 8½ feet. The summit of the beacon is 13 feet above the highest spring-tides. This beacon, which was built in 1862, was broken right across near its base in January 1867. No other injury occurred to any part of the structure; but the detached portion had been moved an inch to the north-east, whilst maintaining its vertical position. M. Le Ferme, in discussing this singular accident[1], calculates that the force exerted by the impact of the wave which effected the rupture must have considerably exceeded 4800 lbs. per square foot, and probably have approximated to 6140 lbs. per square foot. As the dislocated portion of the beacon was not overturned in the interval that elapsed between the accident and its repair a few months later, which would only have required a pressure of 864 lbs. per square foot, it is evident that no subsequent wave can have exerted even this force during that period. A similar accident, which occurred in 1856 to a masonry beacon, 10 feet in diameter at the base and 33 feet high, situated on a rock in the bay of St. Malo[2], indicates that, as the beacon was not overturned in spite of the perishing of the mortar in the lower joints, the pressure exerted by the waves cannot in this case have exceeded 863 lbs. per square foot, which coincides with the ordinary maximum limit of wave action off St. Nazaire as gathered from the incident of the Charpentier beacon. The difference between the normal maximum force of a storm-wave and the extraordinary power attributed to it in rare instances is

[1] 'Annales des Ponts et Chaussées,' Fourth Series, vol. xvii. p. 387, and pl. 190, figs. 1 to 4.

[2] Ibid., Third Series, vol. xv. p. 188.

so great, that, were it not for the extreme pressures recorded by Mr. Stevenson, and the undoubted effects produced under exceptional circumstances, it would be more natural to attribute the failure, in such a case as the Charpentier beacon, to a gradual disintegration along the line of final fracture, under the repeated blows of the sea, rather than to the result of a single wave-stroke of exceptional vehemence.

The slight deflection of the superstructure of the Alderney breakwater both from line and level, and the repeated injuries it has sustained during long-continued storms, furnish ample evidence of the enormous power exerted by waves on an exposed site combined with a rapid tide-way and deep water.

Numerous instances might be cited of injuries inflicted by waves on sea-works in various parts of the world. For example, during a storm in January 1877, a portion of the promenade wall of Dover Pier was torn off; a length of 150 feet of the sea-wall of the superstructure of Colombo breakwater was deflected by the waves during the south-west monsoon of 1878; the outer arms of the Madras breakwaters were laid in ruins by a cyclone in November 1881; and hardly a year passes without some casualty on the sea-coast being reported.

The most marvellous instance, however, of the power of the sea is afforded by the movement of a mass of masonry set in cement, weighing 1350 tons, which had been built as a termination to the breakwater in Wick Bay. This mass, having been specially designed, in 1871, to protect the outer end of the breakwater which had suffered damage on several occasions, was itself carried away in December 1872. It rested upon two courses of 80 and 100 ton blocks laid upon the rubble base of the breakwater, to form a level foundation, 5 feet below low water; it was about 45 feet wide, 21 feet high, being raised 3 feet above high water, and about 26 feet long. The repeated blows of the waves during the storm gradually turned the huge mass round on its base, and at last tilted it off its foundation on to the inside of the pier. This remarkable result may perhaps be accounted for by the following considerations. The breakwater at Wick is peculiarly exposed, for not only is it open to the Atlantic Ocean, but its outermost portion exactly faces the direction of the worst waves, which appear to be so much increased in size by the rapid tidal current flowing through the Pentland Firth, and the funnel-shape of the bay, that they occasionally attain the extraordinary height of 42 feet, though the depth at low water is only 30 feet. These waves are stated by Mr. David Stevenson to have projected masses of water, from 25 to 30 feet deep, over the top of the parapet of the breakwater, and to have dashed clouds of spray into the air to a height of 150 feet. It is probable that, on the occasion referred to, the surging sea dashed into the interstice between the protecting mass and the extremity of the regular breakwater, thus exerting an enormous pressure against the side of the rubble mass, which produced the primary movement of pivoting. As the mass was gradually turned round, it presented a wider face to the blows of the waves, till at length, whilst the weight remained constant, the surface exposed to the full force of the waves was increased from 500 to 950 square feet; and the whole of this area was probably subjected to the impact of the breaking wave. Moreover, in all such cases, the effective weight of the mass is considerably reduced by being enveloped in water; and possibly the waves, in this instance, may have forced a way under the bottom, as well as at the side of the mass, which would very materially assist the pivoting and subsequent displacement of the mass. Under this combination of circumstances, a result which might appear, at first sight, almost incredible, is brought within the limits of the more extraordinary manifestations of the power of waves.

CHAPTER III.

TIDES; CURRENTS; AND CHANGES IN COASTS.

Tides: Importance of Tidal Rise. Theory of the Tides. Spring and Neap Tides. Effect of Sun and Moon's Declination. Tidal Wave. Variety of Range. Effect of Wind and Atmospheric Pressure. Peculiar Tides. Tidal Observations and Predictions. Progression of Tidal Wave. Establishment of a Port. Age of Tide. Cotidal Lines. *Currents:* Tidal Currents. Bore up Rivers. Direction of Tidal Currents. Transport and Deposit of Sediment. Ocean Currents. Wind Currents. *Changes in Coasts:* Erosion of Points. Silting up of Bays. Travel of Shingle and Sand; Instances. Progression of Deltas. General Remarks.

ONE of the very first points which have to be taken into consideration in the design of harbours or docks, or in the selection of a suitable site for such works, is the range of tide. An approach channel, which at low water, or in a tideless sea, would be quite inadequate in depth, may by the rise of tide be rendered accessible for the largest vessels at high water, as, for instance, the estuary of the Mersey up to Liverpool, and the approach channel to the Cardiff Docks. Again, well-sheltered sites, which at high water might appear suitable for harbours of refuge, are only available for tidal harbours owing to the fall of the tide. Moreover, whilst the uniform water-level of a tideless sea fixes an absolute limit at which work can be executed out of water, the fall of the sea along tidal coasts, during low water of spring tides, enables ordinary work or repairs to be carried out at levels which are quite inaccessible at other times. The tidal flow along rivers is one of the principal causes of the favourable state for navigation of the mouths of many tidal rivers, as compared with those flowing into tideless seas. The currents, also, produced by the ebb and flow of the tide and by wind, cause changes along the coast and in the sandbanks, affect the entrances to harbours, and influence the course of navigation.

TIDES.

Theory of the Tides. The phenomenon of the ebb and flow of the tides was naturally observed in early times, and Herodotus refers to it; but, though Pliny attributed it to the influences of the sun and moon, it was not till the researches of Newton and Laplace had demonstrated the influence of the heavenly bodies, that the real cause of the tides was understood. Indeed, till the principle of attraction pervading the universe was discovered by Newton, it would have been impossible for any true theory of the tides to be established.

It is now universally admitted that the tides are due to the combined attraction of the sun and moon. The relative influence of these bodies at the earth's surface is readily obtained from the two laws of attraction, that all bodies attract one another in proportion to the product of their masses, and that the attractive force varies according to the inverse square of their distances. The sun's mass is 26,550,000 times the mass of the moon, but its distance from the earth is 400 times that of the moon, so that its attractive force is $26{,}550{,}000 \div (400)^2 = 166$ times that of the moon. The effect, however, of the attraction on the waters of the globe depends on the difference in the attractive force at the near and far sides of the earth. Thus the moon, being almost 60 times the earth's radius distant from the centre of the earth, is 59 radii away from the nearest side of the earth, and 61 radii from the farther side, so that the attractive forces on the two sides are in the proportion of $\frac{1}{(59)^2}$ to $\frac{1}{(61)^2}$, giving a difference of one-fifteenth of the total attractive force between the moon's attraction on the near and far sides of the earth. As the sun, however, is about 23,000 terrestrial radii distant from the earth's centre, the difference between its attractive forces on the two sides of the earth is only about $\frac{1}{5750}$ of its whole attraction. Taking the moon's attractive force as unity, the sun's attraction is 166, and then the relative effective attractive forces are obtained as follows:—

$$\text{Moon's effective attraction} : \text{sun's effective attraction} :: \frac{1}{15} : \frac{166}{5750};$$

$$\therefore \text{Moon's effective attraction} = \frac{5750}{15 \times 166} \text{ sun's effective attraction};$$

or, moon's effective attraction = 2·31 sun's effective attraction, or, approximately, as $2\frac{1}{3}$ to 1.

Accordingly the tides are mainly governed by the moon, in spite of the much greater attractive force exerted by the sun at the earth's surface.

The action of the moon's attraction in drawing upthe waters of the earth on the side nearest to it is readily intelligible; but the fact of a precisely similar heaping up of the waters on the opposite side of the globe, at the same period, is not quite so self-evident. The attraction exerted on the earth by the moon is exactly the same as if the whole mass of the earth was concentrated into its centre. Now the moon, being about 4000 miles nearer to the waters lying on the near side of the earth than it is to the centre of the earth, exercises a greater attractive force on these waters than it does on the whole mass of the earth, and consequently produces that heaping up of the waters, or wave, which is manifested as the tidal rise. On the other hand, the moon, being nearer by about 4000 miles to the centre of the earth than it is to the waters on the far side of the globe, exercises a greater attractive force on the mass of the earth than it does on the more distant waters, and accordingly draws away the earth from these waters, thus reproducing exactly the same phenomenon as its direct action produces on the nearer waters, though owing to the greater distance the actual effect is somewhat less.

The sun's attraction of course produces precisely similar results, though smaller in amount in consequence of the sun's distance bearing so much greater a proportion to the earth's radius than the distance of the moon does. In fact, though the sun exerts a much greater pull on the whole earth and the waters upon it than the moon can, yet the differences between the force of the pull at the surfaces and at the centre are much smaller than in the case of the moon. Consequently, although the attractive power of the sun on the earth is 166 times that of the moon, the considerable difference between the attractive forces of the moon at the near side, the centre, and far side of the earth, causes the moon's attraction to be twice and a third more efficient than the sun's attraction in raising the waters of the globe. The tides, therefore, follow the apparent passage of the moon round the earth, due to the earth's rotation; and two lunar tides are experienced in the course of the day.

Spring and Neap Tides. Though the moon revolves round the earth in 27 days and 7 hours, the progression of the earth in its orbit increases the interval between the new or full moons to 29 days 12 hours 14 minutes, which is the interval known as the lunar month. During each lunar month the sun and moon are once in conjunction, that is, approximately in a line with the earth on the same side of it, and once in opposition, or almost exactly on opposite sides of the earth. In both these cases the effective lunar and solar attractions are combined, and produce an increased rise known as a spring tide, and a corresponding fall owing to the concentration of the waters in the tidal wave and consequent drawing away of the waters from other parts. When, however, the sun is in a position at right angles to the moon as regards the earth, the lunar tide is reduced by the opposing action of the sun, by the extent of the solar tide, and is called a neap tide. Now, regarding the tides as waves traversing the oceans in the wake of the moon and sun, during a spring tide the lunar and solar waves have their crests superposed, and their troughs therefore coincident, producing a large rise and a low fall; whilst in a neap tide, the trough of the solar wave is under the crest of the lunar wave causing a diminished rise, and as the crest of the solar wave rises over the trough of the lunar wave there is a corresponding reduced fall.

Variations in Distance of the Sun and Moon. As the moon and the earth move in elliptical orbits, the distances of the moon and sun from the earth vary, and produce corresponding alterations in the attractive forces. The axes, however, of the moon's orbit are constantly shifting, performing a complete revolution in 8·85 years; accordingly, the attractive force of the moon does not change regularly like that of the sun according to the season of the year, but goes through a cycle of changes in the above-named period. The greatest and least distances of the sun from the earth are approximately in the proportion of 93 to 90, or a difference of $\frac{1}{30}$th of the total distance, so that the relative attractive forces at these distances are as $(30)^2$ to $(31)^2$, or as 900 to 961, which is approximately as 15 to 16. The eccentricity of the moon's orbit is greater than that of the earth, its greatest and least distances from the earth being approximately as 253 to 222, or a difference of about $\frac{1}{8}$th of the total distance, and its attraction at these distances is as 3 to 4. Accordingly, the variation in the attractive force of the moon is much more than that of the sun; and, moreover, the effective attraction undergoes a considerably greater alteration from a similar change of distance in the case of the moon than in that of the sun, owing to the much smaller proportion which the moon's distance bears to the earth's radius.

The moon's attraction passes through all the phases of its variations in the course of each month's revolution; but the effect of the variation in attraction, due to distance, on the tidal range is constantly changing, both owing to the change in the relative position of the axes of the moon's orbit to the sun produced by the revolution of the earth in its orbit, and also in consequence of the cycle of changes which the axes undergo in relation to the earth, already referred to. Accordingly, whilst the variation in attraction due to the sun's change of distance is regular, and the maximum occurs in our winter, when the earth is in the part of its orbit nearest to the sun; the greater variation due to the moon's change of distance is irregular, the maximum occurring successively at spring and neap tides, and at a variable position in relation to the earth corresponding to the position of the major axis of the moon's orbit.

Changes of the Sun and Moon's Declination and their Effects. The changes which the apparent path of the sun across the earth undergoes in relation to the equator, according to the seasons, exercise a most important influence on the tides. When the sun shines vertically down on the equator, and its apparent path, accordingly, coincides with the equator, it exerts the greatest effect on the waters of the globe. At those periods, which occur at the equinoxes in March and September, the sun's declination, or the angle which its apparent path makes with the equator, is zero. The sun's declination, however, is constantly varying, reaching its maxima, of 23° 27½′, on the 21st of June and 21st of December, when the sun shines vertically on the tropics of Cancer and Capricorn respectively. The sun, accordingly, exerts its greatest influence on the tides at the period of the equinoxes in March and September.

The moon's orbit is only inclined at an angle of about 5° 9′ to the ecliptic, or plane of the earth's orbit, so that at full and new moon the moon's declination is almost the same as that of the sun. Accordingly, the moon also, when new and full, has most influence on the tides at the equinoxes. As the maxima of the attractive forces, due to the position of the sun and moon over the equator, are thus combined at the equinoctial spring tides, these tides are the highest and lowest in the year. The case, however, is exactly reversed at neap tides; for the moon at her quarters reaches her greatest declination during the equinoxes, and is over the equator at the summer and winter solstices when the sun's declination attains its maximum. Though the inclination of the moon's orbit has a maximum variation of only 10′, its nodes, or the points where the orbit intersects the ecliptic, have a retrograde motion and perform a complete revolution in 18·6 years, thus furnishing another element of variation in the relative positions of the sun and moon to the earth, which has to be taken into account in calculations of the tides.

The Tides; their Normal Rise, their Origin, and Extent. Although the variations in the attractive forces mentioned above complicate the calculation of the attraction actually exerted at any particular period, yet there would be no difficulty in ascertaining the rise of tide at any particular time and place on the assumption that the earth's surface was entirely covered by water; and it is readily seen that the tides would rise higher near the equator, than in the temperate latitudes, where the sun and moon never shine down vertically, and where the nearer and farther sides from these bodies are less distant than at the equator. Newton calculated that the solar tide at the equator would rise 2 feet if unimpeded; and taking the lunar tide as $2\frac{1}{3}$ times this rise, the corresponding spring tides would rise $6\frac{2}{3}$ feet, and neaps $2\frac{2}{3}$ feet. Sir G. Airy, however, taking into account the diminution of the attractive effect due to the earth's rotation, calculated that the tidal wave, following the moon over the ocean-covered surface of the earth, would have a height of 0·61 feet (sun's attraction) + 1·34 feet (moon's attraction) = 1·95 feet at spring tides, and 1·34 − 0·61 = 0·73 feet at neaps, or about 2 feet, and 9 inches respectively[1].

The passage, however, of the tidal wave is impeded by the continents which separate the oceans; and the irregular configuration of the shores, and the variations in depth, produce numberless anomalies in the progress and rise of the tides. In the first place it appears that the tidal wave is only fully developed in large oceans, such as the Pacific and Atlantic Oceans, where the passage of the moon over them occupies a

[1] 'Tides and Waves,' G. B. Airy. (Excerpt 'Encyclopædia Metropolitana'), p. 355.

sufficient time to permit the attractive force to generate the wave effectually; whereas on smaller seas detached from the large oceans, such as the Mediterranean, Caspian, Black, Baltic, and other seas, the moon's transit over them is too brief to enable her full influence to be imparted to the waters. Thus the tidal wave is found on the shores bordering the Pacific and Atlantic Oceans, and in the seas connected with them; whilst in the Mediterranean, the largest of all inland seas, the tidal rise is scarcely perceptible except where augmented by the configuration of the coast-line.

Dr. Whewell considered the Pacific Ocean to be the cradle of the tides, where, owing to its wide expanse, the attractive forces were alone able to exert their influence, and that even the Atlantic Ocean derived its tides from that source. It appears, however, to be more consistent with observed facts to admit that the tide is produced both in the Atlantic and Pacific, by the generation of the tidal wave in the middle of their vast areas, which is then propagated towards their shores. Moreover, though the Mediterranean is always classed as a tideless sea, it is not entirely free from tidal changes, as there is a rise of about a foot at Leghorn, between one and three feet in the Gulf of Venice, and somewhat more in the Gulf of Tunis. Even in Lake Michigan, a tidal rise of three inches has been discovered; so that the tidal influences extend also to inland seas and lakes, though in these cases they are so small as to be liable to be masked or neutralised by other operating causes.

Irregularities in the Tidal Range, and their Causes. The wave of the ocean tides on approaching smaller and shallower seas, or in advancing up bays, gulfs, estuaries, or creeks, has its velocity reduced and consequently its height increased, so that a very small initial tidal rise in mid-ocean may, under certain conditions, be largely augmented on approaching the coast. Moreover, the tidal wave, on reaching a projecting headland, is frequently split up into two distinct waves, which, in the case of an island, may follow separate channels and finally meet again, producing in some places an increased undulation or higher tidal rise, and in other places nearly neutralising each other. Thus, whereas the extreme range in mid-ocean is estimated at only about 2 feet, it increases on approaching the British coast, attaining 16 feet at spring tides off the Scilly Isles. The tidal wave, moreover, from the Atlantic is divided off the south-west coast of Ireland and the Land's End into three portions; the main wave continues its course along the west coast of Ireland, whilst two branches diverge into the Irish Sea and the English Channel respectively (Fig. 5, page 50). Thus there are three distinct propagations of the tide along the shores of the United Kingdom; and the meeting of these waves, after having performed their different partial circuits of Great Britain and Ireland, produce great varieties in the rise of tide, which are still further enhanced by the numerous bays and estuaries by which our islands are indented. In fact observation shows a range of tide varying from almost nothing at Courtown on the eastern coast of Ireland up to about 30 feet on the Lancashire coast. The obliteration of the tide at Courtown is due to the interference of the ocean tidal wave, travelling southwards after having passed round the northern coast of Ireland, which meets the branch tide coming up St. George's Channel at that place; and the increased range on the opposite English coast, north of Holyhead, is produced by the combination of the two waves projected against the eastern boundary of the Irish Sea. A similar sort of neutral focus is met with in the English Channel at Portland, where the range is reduced to 6¾ feet; whilst the rise of spring tides in the Bay of St. Malo, on the opposite French coast, attains the unusual height of 35 feet, which is caused by the concentration of the tidal wave in the bottom of the bay, where its progress is suddenly checked by the abrupt contraction of the channel opposite the projecting point of Cape La Hogue.

The tides of the North Sea along the eastern shores of Great Britain are produced by the ocean wave which has passed to the west of Ireland and Scotland, and rounded the Shetland Isles. The range of tide gradually increases as it progresses southwards, till it reaches a maximum of 23 feet at the Wash, which forms a sort of bay enclosed by the Lincolnshire and Norfolk coasts. The projection of the Norfolk coast appears to divert the wave, and the tidal range is rapidly diminished till it attains a minimum at Yarmouth of 6 feet at springs. From this point it begins to increase again, owing probably to the gradual contraction of the North Sea southwards, and the influence of the branch tide coming through the English Channel and passing along the French, Belgian, and Dutch coasts. The tide from the north, which has taken 20 hours in travelling from the south of Ireland, meets a channel tide of 12 hours later origin near the straits of Dover, where the spring range attains 18¾ feet. This latter tidal wave, passing through the straits and expanding along the southern extremity of the North Sea, gradually decreases in height; so that whereas the rise of spring tides at Calais is 20½ feet, it is 18 feet at Dunkirk, 15½ feet at Ostend, and 6½ feet at the mouth of the Maas[1].

The above facts demonstrate what great varieties in tidal range may be found within a narrow compass; and similar differences exist in various parts of the globe, though the small number of observations and the wide expanse of seas prevent definite conclusions being formed as to the exact causes of all these diverse results.

The Gulf of Mexico, though open to the Atlantic, has an average rise of tide of only 14 inches; whilst on the Pacific side of the Isthmus of Panama the tide rises from 8 to 21¼ feet. At Tahiti, in the middle of the Pacific Ocean, the tidal range is not quite 1 foot; and on the southern portion of the east coast of South America, the estuary of the river La Plata is devoid of tidal oscillations, owing to conflicting

[1] The rise of tide at the various harbours and docks described in subsequent chapters are given in Appendix No. 1.

tides, though somewhat further south, in the gulfs of St. George and Santa Cruz, tides of over 50 feet have been observed.

The tides along the coasts of Asia and Australia, and in the intervening archipelago, present similar varieties. The range is very small along the south-west coast of Australia, and does not exceed 9 feet on the southern coast, but reaches to over 30 feet towards the north. At Sumatra the range has decreased again to 3 feet, and there is no rise at the northern end of the straits of Malacca. Whilst at Madras the extreme range is 6 feet, and only 3 feet at Cochin, it reaches 17 feet at Bombay, and at the entrance to the Persian Gulf it is 9½ feet.

The effect of bays and gulfs in concentrating the tidal wave, and consequently increasing its range, is well illustrated in the Mediterranean Sea, where an imperceptible tide, out at sea, is rendered apparent at the top of the gulfs of Venice and Tunis, and even in the wider bay of the Ligurian Sea. The largest known tidal rise occurs in the innermost recess of the Bay of Fundy, between Nova Scotia and New Brunswick, where the tide has a range of over 60 feet, although it rises only 9 feet at the entrance of the bay.

The tidal wave generally experiences an increase in height when advancing up estuaries of rivers having a funnel-shaped form. Thus, at the entrance to the Bristol Channel the rise at spring tides is only about 18 feet, whilst at Swansea it is 27 feet; it attains 40 feet at the mouth of the Usk, and about 50 feet in the Wye at Chepstow. The Thames has a rise of 16 feet 2 inches at Sheerness, which increases to 20 feet 9 inches at London Bridge; the Scheldt has a rise of 13 feet 7 inches at Flushing, and 15 feet 2 inches at Antwerp; and the St. Lawrence has a rise of about 5 feet at its mouth, and 14 feet at Quebec. The gradual contraction in width, and diminution in depth, produce this augmentation of the tidal range; and the actual rise of the tide on rivers at any particular time is affected by the amount of the fresh-water discharge.

Effects of Wind and Atmospheric Pressure on the Tides. The wind exerts considerable influence on the tidal rise by raising or depressing the sea-level along coasts. Thus a strong on-shore wind will increase the normal rise of the tide, and prevent its falling to its full extent, whilst an off-shore wind produces a precisely opposite result; and a wind blowing along, or against, the path of the tidal wave has similar effects. The action of wind is still more marked when blowing along an estuary or tidal river; and the occasional inundations of the Thames at London are due to the concurrence of a spring tide with a strong easterly wind which impels the tidal waters up the river, and thus an abnormal rise is produced from which the low-lying districts near the river are not efficiently protected. Where the rise of tide is very small, a strong wind has frequently more influence in altering the level of the water than the tide, and completely masks the tidal oscillations.

Changes in atmospheric pressure, by altering the local pressure on the sea, are found to affect the tidal rise. The extent to which the tide is raised or lowered by a change of pressure may be ascertained by observing the difference between the actual level of the mercurial barometer and its mean level, and this amount, below or above, multiplied by 13, the density of mercury as compared with salt water, is assumed to give generally, with tolerable accuracy, the excess or deficiency of rise due to the variation from the mean atmospheric pressure. It is, however, liable to be affected by local causes; and Sir J. Lubbock found that a rise of 1 inch of the barometer produced 7 inches depression of the tide at London, and 11 inches at Liverpool; and 13½ inches has been shown to be the equivalent at Bristol.

Peculiar Tides. It has been already stated that two tides occur in the course of a day, and this is the case in most places. Theoretically there should be a difference at certain times in the rise of the two tides, owing to the declination of the sun and moon causing a difference in the attractive force exercised on the opposite sides of the globe, and it is known under the name of diurnal inequality. Frequently no difference is perceived, whilst in some places it is so marked as evidently to be due to other causes. Along the coasts of Europe the diurnal inequality is very small, but it is found in most cases on the shores bordering the Pacific Ocean. On the west coast of North America it is considerable, amounting to about 3 feet at both high and low water; it does not exist at Panama, but reappears again further south. There is a large inequality along the southern coast of Australia; and both at King George's Sound at the south-western extremity, and in the Gulf of Carpentaria at the north, the inequality is so great that sometimes only one tide is discernible in the day. Along the coasts of New Zealand, where the spring tides rise from 7 to 14 feet, the diurnal inequality only amounts to half a foot. At Singapore the difference is sometimes 5 feet, the average rise at springs being 7 feet, so that one of the tides is scarcely perceptible. In the Gulf of Mexico the small rise of 14 inches occurs only once a day. Various inequalities are also observed in the intervals of time between two successive tides.

These anomalous results are produced by the interference of tidal waves having different heights, and generated in different parts of the ocean, which, moreover, are modified by the intricate configurations of continents and islands, and the variety in depth of the waters that they traverse.

Tidal Observations and Predictions. The tidal phenomena are so complicated in their origin that it would be impossible to calculate the tidal rise at any station from the statistics of another one. All additional knowledge about the tides must be obtained from direct observation, and till these observations have been taken at a great number of places, the course of the tides and the causes of their variations must remain open to conjecture. Tidal observations are frequently confined to noting the levels of high and low water in the daytime on a vertical graduated plank, fixed in the water to a known datum at some sheltered spot, and recording the times at which they occur. A far better system of observation, which should be adopted wherever it is practicable, consists of a float moving vertically in a well communicating with the water outside by only a small aperture some distance below low water, so as to prevent any wave-oscillations in the well. A wire passing over a pulley has one end attached to the float, and the other to a counterpoise weight keeping it stretched. The pulley is turned by the motion of the float, and its axis communicates this motion to an apparatus by which a movement proportionate to the rise and fall of the float is recorded continuously on a sheet of paper wrapped round a revolving cylinder. The tidal motion is thereby exhibited throughout its whole course as a curved line, of which the ordinates represent the heights, and the abscissæ

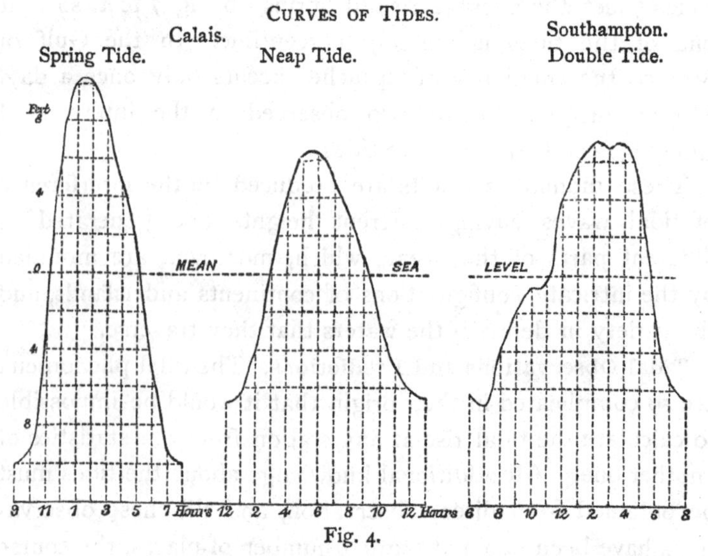

Fig. 4.

the times. The curves in Fig. 4 indicate the results obtained with a self-registering tide-gauge. The first curve represents

an equinoctial spring tide at Calais, and the second a neap tide. The third curve shows the well-known double tide of Southampton, due to the tidal wave, passing to the south of the Isle of Wight, coming round the island and going up Southampton Water after the direct undulation has begun to fall. It also indicates very clearly the difference in rate of the rise and fall. A description of the most perfect tide-gauge instrument of this self-registering type has been given by Sir W. Thomson in his Paper on 'The Tide-Gauge, Tidal Harmonic Analyser, and Tide Predictor[1].' Though it is impossible to determine the range of tide in a place where no observations have been made, it is quite practicable to compute from observations of any known date what should be the rise of tide at any future period. These tide-tables were formerly calculated by a formula; but now the various tides, corresponding to the changing positions of the sun and moon, can be graphically drawn, by an elaborate instrument called the Tide Predictor, provided the data are furnished of previous observations. Of course all these tidal predictions are subject to the unknown variations that may be produced at the time by wind or atmospheric disturbance.

Progression of the Tidal Wave. The velocity of long waves, as previously mentioned (p. 16), depends on the depth of water. It may therefore be anticipated that the very long tidal undulation, whose crest represents high water and its trough low water, will progress more rapidly in mid-ocean than in shallow seas, and in contracted channels, where friction along the shores aids in impeding its motion. The rate of progress has been ascertained by observing the interval of time which elapses between the occurrence of high water at two places whose distance apart is known. The tidal wave appears to travel at the rate of about 530 miles an hour in ocean depths of 5000 fathoms; at about 400 miles an hour in the North Atlantic Ocean, whose depth does not generally

[1] 'Minutes of Proceedings, Inst. C. E.' vol. lxv. p. 2.

exceed 3000 fathoms; at about 60 miles an hour in depths of 50 fathoms; at about 45 miles an hour along the English Channel, and only 15 miles an hour in depths of 5 fathoms.

Establishment of a Port. The time of high water does not, in most places, coincide with the passage of the moon across their meridians. Newton accounted for this by the supposition that the delay was due to the inertia of the mass of the sea when attracted by the moon. It is, however, clear that the interval which elapses between the passage of the moon and the occurrence of high water, is mainly due to the time occupied in the propagation of the undulation from the place in mid-ocean where it originates to the place of observation on the coast. This depends on the path traversed by the wave, and the depth of water through which it travels; and it is evident that even round the British coast high water must occur in different places, on the same day, at every period of the twelve hours, as the northern tidal wave occupies twenty hours in making the partial circuit of the British Isles. As it is important that the time of high water should be known for every port, some common standard of comparison has to be chosen. Now, as the time varies with the position of the moon, it has been arranged that the time of high water shall be expressed in terms of the period which elapses between the moon's passage across the meridian, when full or new, and the occurrence of the next high water at the particular port. This period of time is called the *establishment* of the port. The interval varies somewhat with the changes in relative position of the moon and sun. Dr. Whewell accordingly proposed to adopt the exact mean of all the intervals, which he termed the *corrected establishment*, and which forms a better basis for scientific investigations. The ordinary establishment is, however, more easily intelligible, and is the one employed for navigation.

Age of Tide. The establishment must not be confounded with the age of a tide. The rise of a tide on the shores of

Great Britain is not due to the relative positions of the sun and moon on the day on which it occurs, for the highest tides take place more than two days after the conjunction or opposition of the sun and moon. It appears therefore that the tidal wave is about two days old on reaching our coasts, and half a day older along the east coast of England. The age of the tide varies in different parts of the world according to the rate of propagation of the tidal wave, and the distance it has to traverse. It can only be ascertained by observing the interval between the time of full and new moon and the occurrence of the highest tides; but it is important for works on the sea-coast, as well as for navigation, to know exactly when the highest and lowest tides may be expected.

Cotidal Lines. It occurred to Dr. Whewell that by grouping together those places where high water occurs at the same time, and connecting them by a line on a map, it would be possible to represent graphically, by a series of lines, the actual progress of the tidal wave. These lines he termed cotidal lines. Each line would, indeed, represent the crest of the tidal wave at the particular time when high water occurs at the different places connected by the line. To carry out this idea with absolute accuracy, it would be necessary to have numerous observations in the sea, as well as along the coast. Observations, however, of the height of tide cannot be obtained in the middle of the sea, and therefore the tracing of each cotidal line must depend on a few isolated observations of the time of high water. The map given in Fig. 5, taken from Sir G. Airy's modification of Dr. Whewell's chart of the cotidal lines for each hour round the British coasts, furnishes a very clear illustration of the general manner in which the three branches of the ocean tidal wave, referred to on p. 41, travel along the shores of these islands. This is the chief value of such a chart, as it can only be regarded as a rough approximation to the delineation of the actual crests of the tidal waves, for which no data exist except along the coast; and it cannot represent the various smaller branches which are produced by the numerous headlands and bays, or illustrate the intricate commingling and interference of the

COTIDAL LINES OF THE BRITISH SEAS.

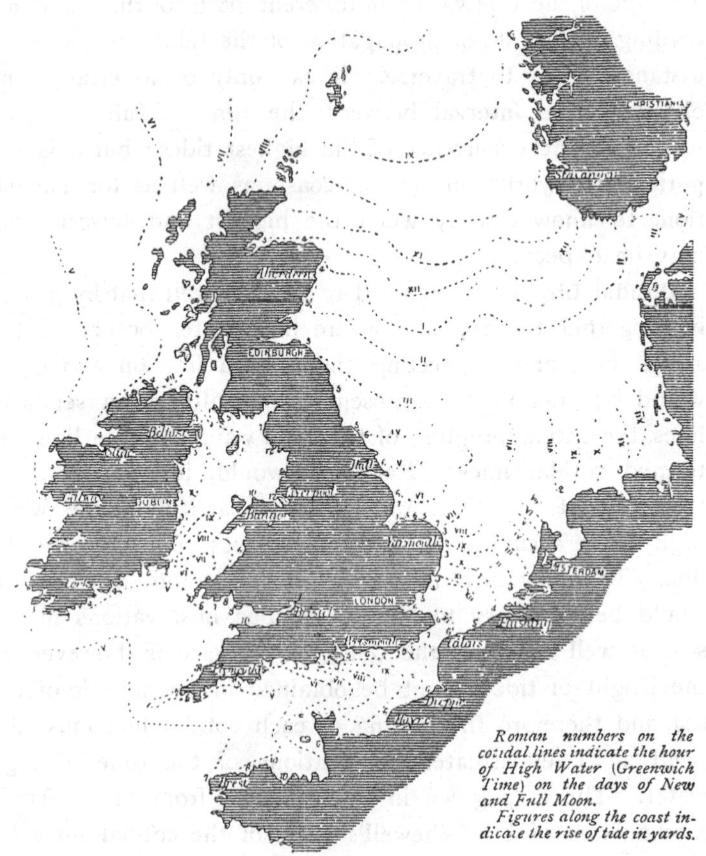

Roman numbers on the cotidal lines indicate the hour of High Water (Greenwich Time) on the days of New and Full Moon.
Figures along the coast indicate the rise of tide in yards.

Fig. 5.

main branch waves meeting near the Straits and in St. George's Channel. The cotidal lines are drawn in a more forward position in mid-channel, as the tidal wave progresses more rapidly in deeper water and away from the friction of the coast.

Dr. Whewell endeavoured to extend this system of graphic representation to the chart of the globe[1]; but the observations are too few and far between to render these cotidal lines even approximately reliable. The crowding together, however, of the cotidal lines occurs where the greatest oscillations are observed, as for instance on the southern portion of the east coast of South America, and the north-eastern coast of Australia. This is in accordance with theory, for the closing together of the hourly cotidal lines indicates a diminished rate of progress in the tidal wave, occasioned by shoaling water, which would be accompanied by an increase in height.

CURRENTS.

The study of the currents surrounding a coast is almost as important as that of the tides, for the selection of the direction for the entrances to harbours must be guided by a knowledge of the prevalent currents. The existence of a strong current in front of a harbour necessitates caution in the design of its approaches, lest the current should prejudicially affect the entrance or exit of vessels, though this has become a less vital question since the decline of sailing vessels and the substitution of steamers. The effect, moreover, of the littoral currents on the maintenance of the depth at the entrance is of the highest importance, especially upon sandy shores. Thus, the direction of the tidal currents in Dublin Bay have been largely instrumental in assisting the improvements in the channel over the bar, effected by means of converging embankments; whilst the currents in front of the Dunkirk jetties are considered to be prejudicial to the deepening of the entrance-channel to the port by means of sluicing.

[1] 'Philosophical Transactions,' 1833. p. 147.

Currents generally result from disturbances of level, and the consequent rush of water under the action of gravity to restore the equilibrium. They are also engendered by a variation in the temperature, and consequently in the density, of water. They may, moreover, be occasioned by long-continued winds.

Tidal Currents. Tidal currents are produced by the disturbance of equilibrium occasioned by the attraction of the sun and moon. In mid-ocean, where the rise is very small, the action is chiefly confined to a transmission of the undulation, and little current is perceptible; but when the channel along which the tidal wave is propagated becomes contracted, and especially in bays, gulfs, and estuaries, where the rise is considerable, the rush of water in filling and emptying these receptacles of tidal water is great. Thus, the tidal undulation is converted into progressive motion; and where the approach channels are impeded and shallow, as amongst the Channel Islands at the entrance to the bay of St. Malo, the currents attain a velocity during spring tides of from 7 to nearly 10 miles an hour. These rapid tidal currents are called races: one of the most noted of these is the Race of Alderney, between that island and the French coast; and still more rapid races are met with amongst the islands off the west coast of Scotland, the race through Coryvreckan running at a rate of 11 miles an hour, and the race in the Pentland Firth off the northern coast having a velocity of over 10 miles an hour.

Bore up Rivers. The tidal flow is also very rapid up estuaries like the Severn, where the rise is considerable. In some of these rivers, where flat sandbanks encumber the entrance, the rise of tide at the height of spring tides is accompanied by a phenomenon known as a *bore*. The tide, instead of flowing up gradually, has to rise considerably at the entrance before it can overcome the resistance to its flow offered by the wide expanse of sands, and appears suddenly in the form of a wave travelling up the river with considerable rapidity; and after its passage the water is found to have risen some feet, and the current which till its arrival was flowing downwards has suddenly changed its direction. This phenomenon occurs on the Severn, the Seine, the Hooghly, the Amazon, and other rivers. The bore attains a height of about six feet on the Seine at Caudebec, where the contraction of the river-channel favours the increase in height of the travelling wave; and on the Amazon it is said to appear as two or three waves in succession from 12 to 15 feet high. On the Severn the bore rises about 6 feet near Newnham; and on the Hooghly it is over 7 feet in height.

Direction of Tidal Currents. The terms *flow* and *ebb*, commonly applied to the rise and fall of the tide, appear to imply the existence of a current in one direction during the rise, and in the contrary direction during the fall of the tide. This, indeed, is generally true of the tidal current on rivers, where the currents are most easily and frequently noticed; but it has only a limited application to the tidal phenomenon in the sea. In fact, even on rivers, the reversal of current, or the period of slack water, does not coincide with the times of high and low water: the tide continues to flow up after the highest level has been reached, and the river may be observed to be still flowing downwards after a very perceptible rise has taken place in the water-level. In the sea, as much variety may be found in the direction, and period of reversal, of the currents as in the rise of tide itself. A knowledge of the flow and velocity of the currents can only be obtained from careful observation; but, fortunately, as regards these, the Admiralty Charts furnish valuable information for those parts of the seas which are frequented by vessels. No general rules can be laid down as a guide in these matters, as each place has its special peculiarities. Sometimes the direction of the current may vary through every point of the compass, as in some parts adjoining the Channel Islands. In some places, as at Courtown, there are currents with hardly any variation in level; whilst at others there is a considerable rise and fall without any current, as for instance at the meeting of the two tides near the Straits of Dover, and this point of meeting varies with the period of the tide between Beachy Head and Dover, and is influenced by the wind. At Havre, after the main tidal current has begun to ebb, there is no fall of tide for about an hour, owing to the arrival of a later undulation from a different direction; whilst at Southampton, as was noticed on p. 47, and represented in Fig. 4, the first ebb is arrested, and a second flow appears.

The periods of the rise and fall of the tide often exhibit some difference, and this causes a difference in the velocity of the flow and ebb, and, moreover, the velocity varies at different states of the tide.

Transport and Deposit of Sediment. The transporting power of a current depends upon its velocity; whilst deposit results from a diminution of velocity, and is greatest during slack water. The currents carry along detritus from the cliffs, sand from the shores and banks, and silt from the rivers. Any obstruction causes them to deposit a portion of the material with which they are charged. It is important therefore, in constructing works, to interfere as little as possible with sediment-bearing currents, and to prevent their entrance into a harbour if practicable, as the whole of their sediment would be deposited in the sheltered area.

It is for the same reason of vital interest to admit the tidal current as far up a river as possible, so that the period of slack water may be reduced to a minimum, and to render its flow regular by regulating the channel. Barriers across rivers, by arresting the upward flow, produce slack water at an early period of the tide, and cause the tidal waters to deposit their sediment in the channel.

Ocean Currents. Large currents are produced by the difference of the temperature at the equator and near the poles. The heated water, having its density reduced, flows away towards the pole, being replaced by an undercurrent of cold water travelling towards the tropics. The Gulf Stream is the best known of these oceanic currents. It originates in the Gulf of Mexico, and in travelling northwards is impelled by the prevailing winds towards Europe, and passes by the west coast of Ireland. It very materially affects the climate of western Europe; it brings fogs and rain, and is the path by which tempests cross the Atlantic and beat against our western coasts.

Wind Currents. The action of wind in producing waves, in raising the water-level, and in affecting the rise of tide, has been already noticed. The power of wind to produce currents is liable to be masked by tidal influences, and by frequent changes in direction; but in places where the tidal range is slight, and where the wind blows for some time continuously in one direction, currents are found which may be traced to this origin. Thus the trade-winds produce, or accelerate, currents in the ocean; the monsoons produce currents which are readily observed along the coast; and the well-known currents along the shores of the Mediterranean are due to the action of the prevailing winds.

Evaporation by reducing the water-level, and the influx of rivers by raising it, influence the currents at the entrances of inland seas.

CHANGES IN COASTS.

Any considerable changes in the shore-line have a most important influence on the maintenance of a port. The Cinque Ports furnish a most striking illustration of decay due to this cause. Out of the five original ports, Dover alone maintains its importance; Hastings is only available for vessels of 100 tons; whilst Romney, Hythe, and Sandwich have ceased to be ports. Marshes exist between Romney and the sea. Hythe was a port of considerable importance in the eleventh century, but in the time of Queen Elizabeth it had become choked up with sand, and the sea is now three miles distant from the old Roman port. Sandwich, reported to have been the most famous harbour of England in the time of Canute, the most ancient of the Cinque Ports, and continuing to be a great port up to the beginning of the sixteenth century, became closed in the seventeenth century, and is now some distance from the sea which continues to recede.

Gradual changes are taking place in many parts of the world; in some cases large areas are slowly rising, whilst others are being depressed; and occasionally volcanic eruptions, and earthquakes, produce sudden alterations, such as happened in the catastrophe of August 1883 in Java and the adjacent straits. The sea is gradually encroaching on the shores of Yorkshire, Norfolk, and Suffolk, and is receding along parts of the western coast of England and Belgium. Considerable modifications have been produced in the coast-line by the sudden inroads of the sea, such as have frequently occurred along the coast of Holland by the bursting of the dykes during storms, and of which a permanent record exists in the Goodwin Sands, which formed in early times the eastern sea-boundary of Kent, but, owing to neglect in maintaining the banks, were submerged by the sea in the year 1100.

Erosion of Points, and Silting up of Bays. Most of the changes referred to above are either so gradual, or so limited to certain districts, as to possess merely a geological or historical interest; but there are changes from two sources, which must not be overlooked in considering the possibilities of maintaining or improving harbours. These two sources of change are, the erosion of points, and the silting up of bays.

Water is popularly said to try to find its own level, but it not only does this, it tends also to level every obstacle that comes in its course. Every track, and every hollow or mound, on a sandy shore, are obliterated when covered by the rising tide. Currents, winds, and differences in the hardness of the strata forming the shores, modify the result; but the tendency manifests itself everywhere. Coasts are indented by numerous bays and creeks, the results of ancient geological formations, or the consequences of the more rapid degradation of the softer strata. The action of the sea tends to reduce the shore to a regular line; it dashes against the projecting cliffs, and carries the débris into the sheltered bays. If the materials falling from the cliffs are too large to be transported by the currents, they are gradually ground into shingle or sand by the continuous motion of the surf, and are finally removed. Where the rocks are very hard, the action of the waves is very slow; and where the coast is much exposed, or the currents rapid, little silting-up occurs, the bays and creeks undergo little alteration, and deep water often exists near the shore. When, on the contrary, the cliffs are composed of clay, chalk, or other easily disintegrated material, the erosion proceeds rapidly; and large beaches of shingle, sand, or silt, are found in the sheltered parts of the shore.

The sands which exist in Solway Firth, in Morecambe Bay, in front of the Mersey estuary, in Sandown Bay, in the Wash, in Tees Bay, in Dublin Bay, and along most indented coasts, and the silting up of the estuaries of the Dee, the Ribble, the Seine, and the Maas, and the harbours of Brading, Romney, Pevensey, Winchelsea, and Sandwich, furnish examples of the action of the sea in filling up sheltered places along the coast with detritus worn away from projecting cliffs.

The sea is aided in its work of destruction by springs, rain, and frost. The water flows into cracks in the cliffs, it gradually increases the apertures, and exerts a hydrostatic pressure; the cliffs are disintegrated by the admission of moisture, and the action of frost, till at last a landslip occurs, and the fallen material is readily removed by the sea. The formation of the undercliff on the south coast of the Isle of Wight is due to this cause.

The amount of material which annually falls into the sea from the erosion of cliffs is very large, as may be gathered from the following indications: The cliffs on the Yorkshire coast between Bridlington and Spurn Point, a distance of 36 miles, recede, on the average, 2¼ yards annually. Near Cromer, on the Norfolk coast, the cliffs are wasting away at a rate which averages nearly 5 yards in a year; and the site of Old Cromer is covered by the sea. Along the south coast between Dover and the Isle of Wight, the yearly waste of the cliffs has been estimated at about 5,900,000 cubic yards. Cape Grisnez loses a quarter of a yard annually. The wear of the cliffs of the department of the Lower Seine amounts to a width of one foot a year, which for a length of 142 miles, and an average height of 190 feet, gives 5,300,000 cubic yards as annually deposited in the sea. The Calvados cliffs lose 8 inches in width, on the average, in a year, for a distance of 68 miles, amounting to a loss of 1,700,000 cubic yards.

The above instances, taken within a very restricted area, show sufficiently what large masses of material are in course of removal from cliffs in various parts of the world, and go to advance the shores, to form or enlarge sandbanks, or to fill up deep places in the sea. It is during violent storms that the greatest encroachments occur. Thus, whilst the average annual erosion of Cape La Hève is about 2 yards, during a tempest in 1862 as much as 16 yards were removed. The average rate of erosion, moreover, not only depends upon the nature of the rocks, but also upon the direction and force with which the waves beat upon the coast, which may be modified from time to time by the actual changes in the adjacent coast-line, and in the depths in front produced by the action of the sea.

Travel of Shingle and Sand. A casual inspection of a shingly or sandy beach might give the impression that the only effect produced by the waves was a periodical raising or lowering of the beach, according to the state of the weather. A careful examination, however, shows that, on most shores, the shingle and sand are constantly progressing along the coast. Sometimes the travel is in one direction, and at other times in another; but, in general, there is a preponderating tendency in one direction, depending upon the line of the strongest tidal current, and the point from which the strongest and prevalent winds blow. On a straight coast-line there is little evidence of this motion; but directly an obstacle is presented to the along-shore progression, the sand or shingle heaps up against the side of the obstruction facing the direction from which it comes.

This action is frequently employed for protecting shores against erosion by the projection of groynes of timber, stone, or concrete, from the shore, which, arresting the travel of the shingle, form an artificial foreshore in front of the coast. Unless, however, the groynes are continued along the whole length of the coast exposed to erosion, the protection of one portion may lead to the greater injury of the part beyond; for the shingle, being stopped at the place where the groynes are erected, no longer goes on to supply the place of the shingle which is being carried away further along the coast, and this portion is accordingly denuded of its supply of shingle till the accumulation of shingle, having reached the extremities of the groynes, begins again to travel along the coast. Groynes for this purpose have been erected at Dover, Hastings, Brighton, and other seaside towns where the enhanced value of the land, and the proximity of houses, render it necessary to protect the shore from erosion. Sometimes also groynes are formed where the encroachment of the sea might be injurious to navigation, of which Spurn Point at the entrance to the Humber is an instance.

The travel of shingle is from north to south along the east coast of England, as manifested by the sand and shingle having actually blocked up the ancient outlets of the rivers Yare and Alde, which originally flowed into the sea at Yarmouth and Aldborough, and forced them to find an exit 2¾ miles, and 10 miles, respectively, from their old mouths. At Lowestoft Ness, on the Suffolk coast, the projecting point forms a natural groyne which has gradually grown out by accession of sand and shingle washed from cliffs further north. The direction of the movement of the sand and shingle is due to the flood-tide coming from the north being stronger than the ebb, and to the north-east wind having the greatest influence along a shore which is protected from the west.

On the south coast of England the travel of detritus is from west to east, the south-west winds from the Atlantic being both the most prevalent and strongest winds, and the flood-tide from the ocean being stronger than the returning ebb. Accordingly, along this coast the accumulation of sand and shingle takes place on the western side of artificial groynes, or of those natural groynes of which Portland Bill and Dungeness are such notable instances. To the west of Portland Island is that marvellous accumulation of shingle known as the Chesil Bank; and on the west of Dungeness there is a vast accumulation of shingle which is constantly renewed by fresh accretions.

The actual method of formation of the Chesil Bank has formed the subject of considerable discussion, owing to the larger pebbles being found towards the eastern extremity of the bank; whereas, the travel of shingle being from the west, it might have been anticipated that the largest stones would have been arrested first. It is impossible to accept the assumption that the larger stones have travelled further along the beach than the smaller pebbles; and the result can only be accounted for by the larger stones having come from a different point, or, as seems more probable, that the larger stones in travelling from the west have been rolled southwards into deeper water, and that it is only the force of the waves, beating upon the most exposed south-easterly extremity of the bay, that can dislodge them and carry them up on the shore.

At Dungeness, the travel of the shingle is not only checked by the projection of the point, but it is also checked in rounding the point, as the spit progresses gradually seawards as well as eastwards to the extent of about 6 yards annually. This result is probably due to a conflict of the tides and waves.

The amount of shingle carried along by the littoral drift cannot be estimated merely from the increase in the bays, as a considerable quantity is worn into sand by the constant attrition, and is carried beyond the limits of the shingle beaches. That the quantity is very large is amply shown by the amount of shingle thrown up on the Chesil Bank, during a gale in 1852, in a single day, having been estimated by Sir J. Coode at 3,500,000 tons; and 18,000 cubic yards of shingle are reckoned to arrive yearly at Havre, and 40,000 cubic yards at Dieppe.

The limit of the shingly beach is reached when the littoral current becomes too feeble to carry it along, or when the distance from the source of supply causes it to wear away in its transit. Thus shingle ceases at Shingle End on the east coast of Kent, and is arrested at Cape Hourdel on the French coast. The sand, however, washed from the coast, and also produced by the destruction of the shingle, is conveyed much further, and, owing to its lightness, is readily affected by the slightest influences of wind, wave, or tidal current. In the English Channel, it is carried by the more powerful north-easterly current through the Dover Straits into the North Sea, where the current, in expanding under the shelter of the French and Belgian coast, deposits its burden of sand against any obstacle in its course, or in sheltered creeks and estuaries. The Dogger Bank is supposed to be due to a state of comparative quiescence at that part of the North Sea, caused by the clashing of the North Sea and Channel tides, which favours the deposit of sand; and the numerous sandbanks lying off the southern shores of the North Sea result probably from a checking of the Channel current. The jetty harbours of the North Sea, when projected from the shore, act like groynes and arrest the travelling sand, which causes a progression of the foreshore. The sand-bearing current finds also sheltered places in the estuaries of the Scheldt and the Maas where it leaves its sediment.

The littoral currents, which cause the travel of shingle and sand along coasts, are generally produced by the combined action of the waves and tidal current; and the action is naturally most powerful when the waves and tide act in the same direction, as during a flood-tide in the English Channel with a south-westerly gale. It would be difficult to determine the relative efficiency of these two causes. The tidal action is only due to the difference in force between the flow and the ebb, but it acts twice a day. The wind veers about, but the most violent storms on exposed coasts generally come from a particular quarter, and it is these storms, occurring perhaps at distant intervals, which raise the greatest waves and produce marked changes on the sandbanks and coasts. Thus, whilst the wave-action along a coast is the most readily perceived, the constantly recurring tidal-action must not be overlooked, especially when the currents are rapid, and the difference in power of the flow and ebb clearly marked.

Along shores exposed to periodical winds, the action of the waves on the beach alternates; and observations extended over at least a year are necessary to determine in which direction the greatest movement occurs. For instance at Madras, the waves raised by the south-west monsoon drive the sand along the coast in a northerly direction, whilst the north-east monsoon causes it to travel in the opposite direction; so that an accumulation of sand raised by one monsoon is removed by the following one, and the resultant effect can only be determined by noticing what change has been produced at the close of the second monsoon, as the effect of either separately would lead to most erroneous conclusions.

Progression of Deltas. The influence of the sea upon changes in coasts may be said to be universal; but there is another action constantly at work, which though comparatively very restricted in its effect on coast-lines, yet, within certain limits, exerts a most important influence. Every river carries with it sediment which it has collected in its course, and tends to deposit when its current is checked on flowing into the sea. In tidal rivers, however, this tendency is somewhat neutralised by the ebb and flow of the tide, which disperses the silt and prevents its accumulating in large quantities at the mouth. When, however, a river flows into a tideless, or almost tideless sea, the greater part of its sediment is deposited at its mouth which splits up into several diverging shallow channels, called a delta, favouring the more thorough deposition of the silt; and this delta is gradually pushed seawards by successive accumulations of sediment, at a rate varying with the amount of material brought down by the river and the depth of the sea in front. The largest rivers, bringing down the greatest mass of detritus, form the most extensive deltas; and the Nile, the Danube, the Po, the Rhone, the Ganges, and the Mississippi, are prominent examples of delta-forming rivers. All these rivers, with the exception of the Ganges, flow into tideless seas; and in the case of the Ganges, when the river is in flood, its waters overpower the tidal current, and thus during the period that it brings down the mass of its sediment, it resembles a river flowing into a tideless sea. The combined waters of the Ganges and Brahmapootra have been estimated to discharge on the average, 235,000,000 cubic yards of solid matter annually; but this great volume is small as compared with the discharge of the Mississippi, which is computed to be more than 3000 times greater.

The small delta of the Tiber has entirely obliterated the ancient port of Rome at Ostia, which is now 2 miles from the coast; the alluvium from the Po has converted the old seaports of Adria and Ravenna into inland towns; and the delta of the Rhone threatens to block up the Bay of Foz into which both the Arles and St. Louis canals open.

The progression of a delta would efface any closed harbour lying within its range; and the fine alluvium of which a delta is composed, projecting out beyond the regular coast-line, is liable to be conveyed a considerable distance along the shore by littoral currents. It is, accordingly, very important not to construct a closed harbour within the influence of deltaic alluvium; and this objection was made to the site of Port Said harbour, but fortunately, though a turbid current does impinge upon the western breakwater, the increase of deposit is so slow as not to cause anxiety for the maintenance of the harbour.

General Remarks. The rise of tide is an important consideration in determining a site for a harbour, as a good rise of tide provides an increased depth over shallow banks and bars during high water, and furnishes a means for scouring the entrance-channel. The directions of the currents are also of importance, as they may materially affect the maintenance of a harbour, and aid or hinder the entrance and exit of vessels. The most vital question, however, in the design of a harbour relates to the changes which are taking place along the adjacent coast, how they may affect the maintenance of the harbour, and whether they can be controlled or evaded. The tides and currents can be easily observed; but the changes in the adjacent coast are much less discernible, and necessitate a careful examination of previous surveys, or observations, so as to deduce the future probable changes from the experience of the past.

The effect of deltaic influences has been already noticed; and when the fate of Sandwich and other ports, the progression of the foreshore on the north side of Lowestoft Harbour (Plate 4, Fig. 9), and the west side of the jetty channel of Dunkirk (Plate 1, Fig. 7), the rapid silting up of the Seine estuary by the material from the cliffs of Calvados, the growth of alluvial deposit in the estuaries of the Scheldt and the Maas, the diversion of the outlets of rivers like the Yare and Alde, and the accretion of land at one place and its washing away at others, are considered, it is evident that an investigation into the changes in the coast can be no more neglected, in considering the proper form for a harbour and its prospects of maintenance, than the amount of exposure, and the force and direction of the waves, in designing the cross section of a breakwater.

CHAPTER IV.

FORMS OF HARBOURS.

Classification of Harbours—(1) *Estuary Harbours:* Works referred to. (2) *Harbours with Backwater:* Parallel Jetties; Sluicing Basins; Converging Jetties; Lagoon Harbours. (3) *Harbours partly sheltered by Nature:* Bays sheltered by Detached Breakwaters; by Breakwaters from the Shore; Remarks. (4) *Harbours protected solely by Breakwaters:* Converging Breakwaters; Two Entrances; Causes of Construction; Form. (5) *Peculiar Forms of Harbours:* Detached Breakwater protecting Entrance; parallel to the Coast. *Entrances:* Position; Width; Instances.

THE classification of harbours may be conducted on two principles; either according to their form, or according to the construction of the breakwaters that protect them. By the first system, harbours would be grouped together according to the local peculiarities of their sites, which determine the form, or plan, of the harbour, without any reference to the method in which the sheltering breakwaters are constructed. (Plates 1 to 5.) By the second system, the arrangement of groups would depend solely on the type of breakwater adopted, and not on the general plan of the harbour. (Plates 6 and 7.) Both systems have their advantages; but as, in designing the plan, the discretion of the engineer is in a great measure limited by the natural conditions of the site, the type of breakwater to be chosen is generally by far the most important consideration in the design of a harbour, as upon this, its security and cost mainly depend. Accordingly, whilst in the present chapter, harbours will be classified with regard to their form; in the descriptions of harbours (Chapters X to XVI), they are grouped together in relation to the type of breakwater which shelters them, with the exception of Jetty Harbours, described in Chapters VIII and IX, in which special class, form is of more importance than the construction of the jetties. (Plates 1 and 2.)

Classification of Harbours. Harbours, when considered merely with reference to their form, may be conveniently divided into five separate classes:—1. Estuary Harbours; 2. Harbours with backwater; 3. Harbours partly sheltered by nature; 4. Harbours protected solely by breakwaters; 5. Peculiar types of Harbours with detached breakwaters. Besides these, there are the purely natural embayed harbours, of which Rio de Janeiro has been already quoted as an instance; but as they require no improvement at the hands of engineers, they present no interest from the constructive point of view.

1. ESTUARY HARBOURS.

The mouths of rivers constitute a large class of more or less natural harbours, presenting every variety of condition and accessibility. Some, like the Potomac and St. Lawrence, form excellent harbours and approach channels to the interior; whilst others are so impeded by bars and shoals as to require the utmost skill of the engineer for their improvement, of which the Danube, the Mississippi, the Seine, and the Maas, are notable instances. As, however, this subject has been fully dealt with, in the chapters on the improvement of rivers, in my book on 'Rivers and Canals[1],' it is unnecessary to enter upon it here. The amelioration of their condition is indeed of the utmost importance, not merely to render them more serviceable as harbours of refuge in stormy weather, but also to enable vessels of larger draught to get up to the ports along their banks. In some cases the works of improvement are essentially river-works, such as the jetties of the Mississippi and the training walls of the Seine; in others they form a sort of connecting link between river and harbour works, as, for instance, the breakwaters at the mouth

[1] 'A Treatise on Rivers and Canals,' L. F. Vernon-Harcourt, pp. 227–322.

of the Tees; whilst Tynemouth Harbour (Plate 4, Fig. 10) is in reality a regular refuge harbour at the mouth of the river Tyne.

2. HARBOURS WITH BACKWATER.

This class comprises the numerous harbours situated at the mouths of small rivers, lagoons, or creeks, to which they owe their origin, and by which their depth is more or less maintained. The various jetty harbours on the North Sea coast furnish early examples of this class; whilst, more recently, large harbours of this type have been constructed with converging jetties or piers.

In the early times of commercial enterprise, sites for harbours were very naturally selected which were readily accessible to vessels of the small draught then employed, and which provided some natural shelter from the sea. Small rivers and creeks accordingly furnished suitable positions for the establishment of ports in those days, where the entrance-channel was maintained by the outflow of the land waters, or by the ebb and flow of the tide. In the first instance, these places were merely natural harbours maintained by natural means. The progress, however, of trade led to the adoption of vessels of larger draught; and the shallow entrances of early times were found inadequate to meet the growing demands for increased depth. Moreover, not only did the original depth become insufficient, but even this depth itself was in some cases reduced, owing to the enhanced value of land in the neighbourhood of a flourishing port having induced the inhabitants to resort to reclamations of the tide-covered areas, which diminished the tidal scour at the outlet, as particularly exemplified at Calais and Ostend. (Compare Fig. 12, p. 149, and Fig. 14, p. 156, with Plate 1, Figs. 5 and 10.)

Parallel Jetties. In order to maintain and improve the entrance-channel, longitudinal jetties were constructed along each side of the channel, which served to direct and concentrate the fresh water and tidal currents along the channel, increasing its depth, and tending to scour away the bar which formed at the outlet. These jetties, however, when projecting from a shore along which there existed a natural drift of sand or shingle, acted like groynes, and, arresting the travel of material along the coast, produced an advance of the foreshore, which eventually compromised the depth at the extremity of the jetty channel. This difficulty was met by a periodical extension of the jetties, till at last, when the drift was considerable, the jetties extended a long distance in front of the port, as may be seen at the present day at Dunkirk. (Plate 1, Fig. 7.) It was, indeed, attempted to reduce the accumulation of deposit by making the upper portion of the jetties open timber-work, so that, whilst marking the entrance-channel, it might interfere as little as possible with the drift along the coast. As, however, even the open jetties presented some obstruction to the currents, and the solid portions had to be raised somewhat above the level of the beach to prevent the filling in of the channel, the progression of the foreshore was only delayed, and not arrested. Accordingly, whilst the system of parallel jetties is still employed for preserving the entrance-channel of several ports of this class, it has been recognised that the periodical extension of these jetties produces a merely temporary advantage, and that, whilst the length of the channel and the cost of maintenance are thereby increased, no permanent increase in depth is secured.

Sluicing Basins. When experience proved that parallel jetties were inadequate to provide the whole additional depth at the entrance necessitated by the growing requirements of trade, the aid of tidal scour was sought, which had proved in some cases of considerable value in early times for maintaining the depth at the entrance. As the low-lying marshes and creeks, which had formerly provided natural receptacles of tidal water, had been either reclaimed, or so much raised by long-continued deposit of silt as to be no longer serviceable, artificial basins had to be formed. These sluicing basins have been constructed near the entrance-channel by enclosing a portion of the adjacent low strand with embankments, and building sluice-gates across the opening leading

REVOLVING SLUICE-GATE.

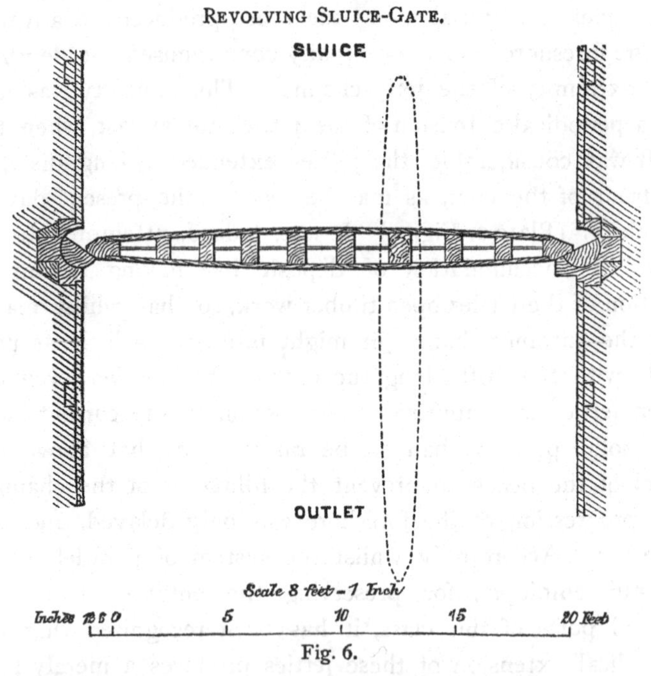

Fig. 6.

into the jetty channel, so that the tidal water can be admitted and let out as desired. Dunkirk and Ostend were provided several years ago with sluicing basins (Plate 1, Fig. 10): a sluicing basin was opened at Honfleur in 1881, having an area of 143 acres, formed in the estuary of the Seine (Plate 12, Fig. 1); and a still larger sluicing basin is in course of construction on the low foreshore at Calais, enclosing an area of 225 acres. (Plate 1, Fig. 5.) These basins are smaller than those originally furnished by nature at Calais and Ostend; but their efficiency is increased by being situated near the outlet of the jetty channel, and by the whole mass of water, retained in them at high water, being under perfect control, so that it can be all let out at low water, whereby its scouring power is much augmented. They can, moreover, be protected, in a great measure, from silting up by only admitting the tide, during springs, previous to sluicing operations, instead of allowing each tide to bring in its burden of silt which is deposited during slack water.

In many instances the sluice-gates are constructed so as to turn on a nearly central vertical pivot. (Fig. 6.) The gate, when closed, is kept in place by abutting at one end against the projecting straight side of a vertical semi-cylindrical post fixed close against a recess in the side wall of the sluice-way, but capable of being turned on its axis. By giving this post a quarter of a revolution the gate is free to revolve; and the pressure on the larger half overcoming the pressure on the other half, the gate turns to a position in a line with the channel, as shown in dotted lines in Fig. 6, and releases the water in the sluicing basin.

Converging Jetties or Piers. Parallel jetties form the distinguishing characteristic of the harbours mentioned above, to which sluicing basins have been added in some cases. There is, however, another description of harbour, which must be included in the class of harbours under consideration, where the jetties converge towards the outlet instead of being parallel, and thereby a sort of sluicing basin is enclosed by the jetties themselves. Aberdeen, Dublin, and Charleston are good instances of this form of harbour. (Plate 1, Fig. 12, and Plate 2, Figs. 1 and 9.) The object of the convergence of the jetties is to concentrate the scour of the fresh and tidal waters at the entrance, where a bar naturally tends to form. The results realised by this system have been very satisfactory, as will be seen in the sequel (Chapter IX). The jetties, indeed, in these cases have been made solid; but the obtuse angle which they make with the shore-line prevents their being very conducive to the advance of the foreshore, and a state of equilibrium is sooner reached.

Lagoon Harbours. There are some harbours which, in their present condition, resemble the state of Ostend and Calais in former times. In these cases the marshes and large creeks are still in existence, and the entrances to the harbours are maintained by the tidal scour produced in the emptying and filling these extensive areas. Of these lagoon harbours, as they may be termed, the most notable instance is Venice: the harbour of Kurrachee is similar in type (Plate 2, Fig. 5); and a true lagoon harbour exists on the Irish coast at Wexford. These harbours are liable to constant deterioration from silt or sand brought in from the sea and deposited in the still water of the lagoon, as occurs at Wexford; and also, in some cases, from the detritus brought down by the rivers flowing into the lagoons as well, to which source of injury the Venetian lagoons are exposed.

The Malamocco approach-channel to Venice has been deepened by projecting stone embankments into the sea, which resemble in principle the North Sea jetties, but are solid and on a larger scale. (Plate 2, Figs. 7 and 8.)

The entrance to Kurrachee Harbour has been contracted, protected, and deepened, by a breakwater and groyne. (Plate 2, Fig. 6.)

At Wexford, however, no works of importance have been carried out for improving the entrance; whereas extensive reclamations inside the lagoon have diminished the depth.

Though the rise of tide at Venice is only about 2 feet, it is compensated for by the wide expanse of lagoon over which the tide spreads; and a good scour is produced at the entrances, the ebb being reinforced by the fresh-water discharge.

Remarks. The harbours comprised in this division present marked features of resemblance, both in their present or former condition, and in the methods adopted for their improvement. Forming thus a very distinct class by themselves, they may be grouped together, and will be described in Chapters VIII and IX under the general heading of 'Jetty Harbours.'

3. HARBOURS PARTLY SHELTERED BY NATURE.

In many places, projecting points or bays furnish some natural protection; and it is only necessary to supplement this natural shelter by a breakwater, facing the exposed direction, to convert the site into an excellent harbour. These sites are naturally selected for the formation of a harbour, if they exist within a reasonable distance of the locality to be accommodated. Moreover, as places of importance on the sea-coast rarely spring up where no natural shelter exists, it is generally in such situations that the demand for improved protection arises.

Bays sheltered by Detached Breakwaters. Next to an estuary where vessels can run a long distance inland for shelter, the best protection provided by nature is a bay protected on each side by headlands, and only requiring shelter at its entrance to form a perfect harbour. This natural advantage exists at Plymouth and at Cherbourg; and in each case the necessary additional shelter has been provided by a detached breakwater across the entrance to the bay, not connected at either end to the shore, but leaving entrances between each extremity and the coast for the entrance and exit of vessels. (Plate 3, Figs. 5 and 8.) Delaware Harbour has a similar method of protection; but in this instance the bay is so extensive that much larger works would be needed to render the shelter equally complete. (Plate 3, Fig. 10.)

A detached breakwater furnishes the main artificial protection to Portland Harbour in Weymouth Bay. In this case, however, one of the entrances is between the detached breakwater and the end of another shorter breakwater connected with Portland Island; whilst the other entrance is the full width of the harbour, as it faces the sheltered northern part of Weymouth Bay. The peculiar configuration of the coast at this place gives Portland Harbour a character quite distinct from other embayed harbours.

The form of the harbour at St. Jean-de-Luz is very similar to Plymouth and Cherbourg upon a smaller scale. (Plate 3, Fig. 9.) The shelter at the entrance to the bay is, however, secured by three breakwaters, two projecting from the coast on each side, and a central detached breakwater. By this arrangement the entrances are in deeper water than could be found near the shore.

The bay in which the harbour of Alexandria is situated, about 6 miles across and nearly 1½ miles deep, is not much enclosed at the ends, and the entrance therefore is too wide to stretch a breakwater across it. Accordingly, about a third only of the bay has been enclosed by a detached breakwater, bending in towards the shore at the south-western end, to protect the harbour in that direction, and almost connected with the land at the other extremity. (Plate 3, Fig. 1.)

Bays sheltered by Breakwaters from the Shore. Frequently a bay is too wide, or the depth inshore too shallow, for it to be suitably sheltered by a detached breakwater with entrances at each side. It then becomes advisable to protect only a portion of the bay, and to project one or more breakwaters from the shore, which is more economical in construction, and also affords more convenient access after completion.

The harbours of Genoa and Barcelona are instances of well-sheltered bays protected by breakwaters carried out from the shore at each extremity, instead of by a single detached breakwater, and with the entrance accordingly facing the centre of the bay. (Plate 5, Figs. 3 and 10.) The harbour of Algiers is similarly sheltered, though the bay in this case is less deeply indented (Plate 5, Fig. 1); and the harbour at Oran belongs to the same type. (Plate 5, Fig. 15.) At Ramsgate, a small inlet is sheltered by two breakwaters extending to low water (Plate 4, Fig. 4); and Whitehaven Harbour is formed in a similar manner. (Plate 4, Fig. 11.)

Holyhead Bay in Anglesea, and Braye Bay in Alderney, are both protected by single breakwaters projecting from the shore. (Plate 4, Figs. 1 and 5.) Neither of these, however, must be accepted as models of the kind of protection that would have been designed if harbours of the present size had been contemplated from the commencement. Both harbours were at first proposed to be of a comparatively small size, and protected by two breakwaters from the extremities of the inner bays; in both cases the design was enlarged during construction, and the second breakwater abandoned. Owing to the alterations of the plans, the direction of the breakwaters had to be altered in each case, so that in being extended they might not approach too close to the rocks on the opposite side of the bay; and consequently there is an unsightly angle in each breakwater, appearing to protrude into the harbour and diminish its area, which is most marked at Holyhead, but which formed no part of the original designs. The configuration of Holyhead Bay has enabled the single breakwater to afford excellent shelter; but at Alderney the absence of an eastern breakwater has been severely felt in easterly gales.

The bay of St. Nicholas at Bastia has been converted into a harbour by a single breakwater from the shore. (Plate 5, Fig. 6.)

At Dover and Cape Town, short breakwaters starting from projecting points afford a limited shelter within the large areas of Dover and Table Bays: but the Admiralty Pier at Dover (Plate 4, Fig. 2) forms only a small part of an extensive scheme; and the extension of Table Bay breakwater is in progress (Plate 4, Fig. 8), and proposals have been made for forming an enclosed harbour by a second breakwater to the south-east of the existing one.

It is possible to obtain adequate tranquillity in a small portion of Mormugao Bay, on the west coast of India, by the construction of a straight breakwater at right angles to the coast, owing to the naturally well-sheltered condition of the estuary at that site. (Plate 5, Fig. 13.)

Fraserburgh Bay is too extensive to be converted into a harbour, except for a harbour of refuge; so that a creek at one extremity of the bay has been efficiently sheltered, for the purposes of a fishing harbour, by a single breakwater (Plate 4, Fig. 14); and the port of Malaga is similarly protected. (Plate 5, Fig. 9.)

The deep recess of Wick Bay appears peculiarly suited for the site of a harbour (Plate 4, Fig. 13); but the waves, lashed into fury by the great exposure and rapid currents of the Portland Firth, seem to acquire increased intensity in rolling up the funnel-shaped bay, so that a breakwater, which was in course of construction across the bay, as shown on the plan, has been converted into a mass of ruins.

The entrance to Newhaven Harbour is being protected from the west by a single curved breakwater, in course of construction (Plate 4, Fig. 7); but in this case the bay is hardly sufficiently recessed for full shelter to be provided by one breakwater alone, and the entrance will be exposed to south-easterly gales.

It was originally proposed to form Colombo Harbour with only a western breakwater, curving towards the shore at its extremity; but in order to have the entrance situated in a good depth of water, and at the same time to provide the additional shelter found necessary towards the north, a detached breakwater has been added to the design. (Plate 4, Fig. 3.)

Remarks. Some natural shelter exists in all the harbours referred to above; but it will be noticed that the amount of the shelter varies considerably. Thus, whilst at Cherbourg, Plymouth, Wick, Genoa, and Barcelona, the entrance alone of a complete bay requires protection; at Holyhead, Table Bay, and Alexandria, only a portion of the extensive bays in which the harbours are situated can be utilised, though the existence of the bay diminishes considerably the exposure; and, lastly, at Dover, Newhaven, and Colombo, projecting points of the coast, rather than regular bays, are taken advantage of for the site of a harbour.

The natural protection made use of in these instances diminishes the extent of breakwater needed to enclose a given area; but it does not follow that because a harbour is partly sheltered by nature, the breakwater for completing its protection may not be as much exposed as in purely artificial harbours. The force of the waves, depending upon the exposure and the depth, and the angle at which the largest waves strike the breakwater, may be as great as on an unprotected site, provided the strongest winds do not blow from the protected quarter. The disasters that have occurred at Wick breakwater furnish a striking exemplification of this fact.

A detached breakwater, though more costly to construct, possesses one important advantage in sheltering enclosed bays; namely, that, by leaving an opening at each extremity of the bay, a current is generally produced through the harbour, which prevents the tendency towards silting up.

4. HARBOURS PROTECTED SOLELY BY BREAKWATERS.

Occasionally harbours have to be formed where no natural protection exists; and this generally occurs on flat sandy shores, as deep rocky coasts are usually indented. The harbour is, under such circumstances, not merely exposed to injury from the sea, but it is also liable to act as a groyne and arrest the sand or shingle travelling along the shore in the direction of the prevalent winds and currents.

Harbours formed by Converging Breakwaters. These harbours are usually formed and sheltered by two breakwaters, projecting from the shore at some distance apart, and converging at their extremities so as only to leave an opening just adequate for the safe entrance of vessels. Kingstown, Howth, Madras, Tynemouth, Oswego on Lake Ontario, Tarragona, Ymuiden, and Port Said, are instances of this type of harbour. (Plates 4 and 5.) Oswego Harbour, indeed, is situated at the mouth of the Oswego River, but it is quite distinct from the jetty works on that river; and though Tynemouth Harbour presents some resemblance to the jetty harbours already referred to (Plate 2), yet it has been purposely constructed on a much larger scale than would have been needed for the improvement of the Tyne, in order to provide a refuge harbour for that exposed coast.

Harbours with two Entrances. In some instances a detached breakwater is constructed between the extremities of the two breakwaters starting from the shore, with the object of providing two entrances, which are very serviceable when they face in different directions, as then one or other of them can be used according to the direction of the wind. The new harbour now being constructed at Boulogne has been designed in this way (Plate 5, Fig 8); and the harbour extension works at Buffalo on Lake Erie are being carried out in a similar manner. (Plate 5, Fig. 12.) Also in the scheme proposed for Dover Harbour in 1847, two entrances were provided by the same method. (Plate 4, Fig. 2.)

Causes of Construction of purely Artificial Harbours. Various reasons have led to the construction of this class of harbours, where no natural protection indicated any suitability of site. At Tynemouth, the harbour furnishes the double advantage, of improving the access to the Tyne, and protecting the coasting trade. Boulogne, Oswego, Buffalo, and Lowestoft harbours, are substitutions for the jetty harbours in those places, which had proved inadequate in depth and size.

Howth and Kingstown harbours were bold attempts to obtain, by means of an enclosed harbour on a sandy coast, a suitable depth for the trade to Dublin, which, at the period of their construction, it was feared could not be realised at the mouth of the Liffey: the first proved a failure, the second a success. Only imperative necessity could have led to the construction of harbours at such apparently unfavourable sites as those on which Port Said, Madras, and Ymuiden harbours are situated. The necessity of providing a deep water outlet for the Suez Canal was the cause of the formation of a harbour, on a shallow sandy shore, within the influence of the turbid current coming from the delta of the Nile, on a spot where the progression of the foreshore on the western side, and the silting up of the harbour, were confidently predicted. At Madras, the well-known littoral drift of sand from the south during the prevalence of the south-west monsoon prevented, till recently, any measures being taken for affording the important town of Madras, and its surrounding districts, any kind of protection for their commerce on that surf-beaten strand. The flat sandy shore of the North Sea along the Dutch coast appeared so unfavourable to the construction and maintenance of a harbour at Ymuiden, that in 1825 the circuitous route of the North Holland Canal was chosen for the trade of Amsterdam, in preference to the direct line through Lake Y, of which at last the growing requirements of Amsterdam, and the competition with Rotterdam, necessitated the adoption [1].

Forms of purely Artificial Harbours. The object of this class of harbours is to enclose as large a space as practicable with a given length of breakwater. For this purpose the square form is the best; and this is approximately attained at Madras, Boulogne, Kingstown, Howth, and Michigan. Other considerations, however, come in, which tend sometimes to modify the form, such as the depth of water in which the entrance should be placed, and the best lines for the break-

[1] 'A Treatise on Rivers and Canals,' L. F. Vernon-Harcourt, vol. i. pp. 177–185.

waters so that they may be most economically constructed, be least liable to the direct shock of the waves, and interfere as little as practicable with the travel of sand. Accordingly, the oblong form is adopted in some instances, as at Buffalo, Oswego, Chicago (Plate 5, Fig. 5), and the design for Dover; whilst in others, the converging form is preferred, of which Port Said, Ymuiden, and Tynemouth, are examples. The square form encloses the largest area with a minimum length of breakwaters; the oblong form protects a longer line of coast, and diminishes the depth and exposure of the outer breakwaters; whilst the converging form enables a greater depth to be reached with shorter breakwaters, and interferes somewhat less with the littoral currents.

5. Peculiar Types of Harbours with Detached Breakwaters.

There is a class of harbours which, as regards their natural condition, might be included in one or other of the last two classes, but whose methods of shelter differ essentially from those hitherto considered. This class comprises two distinct types, in both of which a detached breakwater forms an important feature. In the one type, the detached breakwater is situated in front of the entrance to a harbour sheltered by breakwaters in the ordinary way, so that two approach-channels are provided, and the entrance to the harbour is protected. Civita Vecchia, Leghorn, Cette, Odessa, and Chicago harbours, are instances of this arrangement. (Plate 3, Figs. 2 and 3, and Plate 5, Figs. 4, 5, and 16.) In the second type the detached breakwater forms and shelters the harbour; but, instead of stretching partly across a natural bay, it runs parallel to a straight line of coast, and protects a series of moles or jetties projecting at right angles from the land, which serve as quays. Marseilles Harbour is the finest example of this type; and Brest and Trieste harbours are similar in design. (Plate 3, Figs. 4, 6, and 7.)

Detached Breakwater protecting Entrance. The detached breakwater, in the instances quoted above, furnishes merely an additional protection to the shelter already provided, and changes the direction of the approach to the harbour as well as providing a double one.

At Civita Vecchia, the width of the entrance, as compared with the size of the harbour, must have caused a considerable motion in the harbour, with an on-shore wind, before the construction of the detached breakwater.

The detached breakwaters at Chicago, Leghorn, Cette, and Odessa, not only protect the entrances to the inner harbours, but also form a sort of outer harbour with moderate shelter.

Detached Breakwater parallel to the Coast. The object of this type of breakwater is to form a port in which vessels can lie to discharge and take in their cargoes. It is therefore placed parallel to the coast, at a convenient distance; and the space between the breakwater and the shore is divided into a series of basins by solid jetties, or wharves, projecting at right angles to the shore, and leaving an interval between their extremities and the breakwater to allow of the entrance and exit of vessels. These jetties, the intervening quay walls along the shore, and the inner side of the breakwater, form the quays of these basins. (Plate 3, Figs. 4, 6 and 7, and Plate 7, Figs. 7 and 15.)

This arrangement is specially suitable for Mediterranean ports, as the absence of tide dispenses with the necessity of inner docks with gates; and the comparatively moderate exposure enables the breakwater to be used as a quay. The same principle has been adopted at Brest, which lies in a well-sheltered site; but the rise of tide there renders it somewhat unsuitable, and the construction of a regular dock along the foreshore has been proposed. (Plate 3, Fig. 6.)

ENTRANCES TO HARBOURS.

The most important point next to the site of the harbour, and the design of the breakwaters, is the entrance to the harbour. Facility of access is, indeed, as essential as the shelter provided by the harbour.

Position of Entrance. The entrance has to be placed in an ample depth of water, as, owing to its exposure to waves, its normal depth is reduced by about half the height of the waves during a storm. Also, in tidal seas, it is important that the entrance should be accessible, if possible, at every state of tide, or at least for as long a period as circumstances admit during every tide. Accordingly, the entrance is generally situated at the outer extremity of the harbour; and the distance to which the jetties or breakwaters are extended from the shore is generally determined by the depth to be attained at the entrance.

The entrance is usually placed either facing directly out to sea, or pointing in the direction of the worst wind; for though the run of waves into the harbour is thus greater than if the entrance was placed in a more sheltered position, sailing vessels find it easier to make the entrance by running straight in, than if they had to weather the outer part of the breakwater and then tack into a sheltered side entrance. Two entrances have the advantage of enabling vessels to select either of them, according to the state of the weather; but they tend to diminish the tranquillity of the harbour, and they are consequently undesirable unless the harbour is large, or unless they are provided, as previously explained, by an outer detached breakwater, as at Leghorn and Cette. (Plate 3, Fig. 3, and Plate 5, Fig. 16.)

Width of Entrance. Occasionally, owing to the naturally well-sheltered position of the roadstead, it is possible to afford an ample width of entrance without affecting the tranquillity of the harbour. Thus, at Portland (Plate 3, Fig. 11), the main entrance to the harbour faces the coast, and is as wide as the harbour itself. This arrangement is possible owing to the peculiar position of Portland Bill in relation to the general coast-line. These wide entrances are so easy of access that vessels can readily enter such a harbour, even though the entrance is necessarily turned away from the worst winds; in fact, the manœuvring required for this purpose is more like rounding a headland than entering a harbour. Holyhead Harbour furnishes another instance of a wide, but sheltered entrance (Plate 4, Fig. 5); and similar examples are found at Alexandria, Oswego, and Buffalo. (Plate 3, Fig. 1, and Plate 5, Figs. 11 and 12.) When a breakwater is carried nearly straight out from the shore, and furnishes the only artificial shelter to a harbour, as at Newhaven, Table Bay, Wick, Malaga, and Mormugao (Plate 4, Figs. 7, 8, and 13, and Plate 5, Figs. 9 and 13), it forms really a sort of projecting point protecting the bay within; and there is no regular entrance to these harbours.

A detached breakwater sheltering a bay naturally furnishes two entrances between its extremities and the shore on each side; and these entrances cannot be made narrow without placing them too near the land and in shallow water, unless two breakwaters are projected from the shore on each side of the bay, as at St. Jean-de-Luz (Plate 3, Fig. 9), entailing three structures instead of one. At Cherbourg and Plymouth, however, the large extent of the bays compensates for the great width of the entrances. (Plate 3, Figs. 5 and 8.) At St. Jean-de-Luz, it would have been preferable to dispense with the detached breakwater, and form a single entrance, if the Artha Rock did not stand in the centre of the opening of the bay. The entrances to Marseilles, Trieste, and Brest, are naturally situated round the ends of the detached breakwater which a vessel has only to turn to come within the shelter of the outer harbour; and more perfect shelter is provided in the inner basins by the projecting jetties. (Plate 3, Figs. 4, 6 and 7.)

Considerable facility of access exists in the class of harbours where a detached breakwater shelters an outer roadstead and the entrance to an inner harbour, of which Civita Vecchia, Leghorn, Odessa, Chicago, and Cette have been cited as instances. (Plate 3, Figs. 2 and 3, and Plate 5, Figs. 4, 5 and 16.)

The determination of the proper width to be given to the entrance becomes a matter for very careful consideration in the case of harbours enclosed by two or more breakwaters.

In harbours with parallel jetties, the width has to be regulated by the size of the jetty channel. It is inexpedient to increase the width at the entrance by a funnel-shaped mouth, as at Ostend (Plate 1, Fig. 10), for the swell entering the harbour is thereby increased. The widths, however, of the ends of the jetty channels at Dieppe and Calais are being made somewhat larger by putting back the eastern jetty in each case, so as to be more in a line with the inner channel, making these entrances 250 feet and 400 feet respectively. The entrance to the long jetty channel of Dunkirk is only 200 feet; but this is acknowledged to be too narrow, and suggestions have been made for constructing a totally new channel both wider and more direct. The Ostend entrance is 360 feet wide; whilst the widths at Havre and Honfleur are 200 feet and 130 feet respectively.

The widths of the entrances to these ports must not, however, be regarded as any guide to what might be desirable, as they have been dictated by the existing conditions of the old harbours; and, moreover, the entrances are more or less sheltered by outlying sandbanks. It is also advantageous to keep the channel as narrow as possible, for promoting the efficiency of the sluicing current.

At Malamocco, where the extensive Venetian lagoons

more than compensate for the small tidal range, the entrance-channel has a width of 1520 feet.

In converging jetty harbours, a greater width of entrance is generally given than in the case of the parallel jetty harbours. (See Plates 1 and 2.) These harbours are of more recent origin, and consequently designed for a larger class of vessels; they also commonly extend into deeper water, and are frequently more exposed. In these cases the width has to be a compromise between conflicting conditions. The object of these jetties is to secure a deep entrance: if the entrance is made very wide, a bar is liable to form from the inefficiency of the scouring current; if the entrance is very narrow it induces a rapid current, which indeed maintains the depth, but is prejudicial to the entrance of vessels, and a narrow passage is naturally objected to by seamen. The entrance to Galveston Harbour (Plate 2, Fig. 12), if the jetties are carried out as proposed, will afford an example of an extreme width of 8500 feet. The width of 2200 feet at Kurrachee is due to local circumstances. Charleston, the finest jetty harbour in the world, is designed to have an entrance 2000 feet wide. The north and south walls at Dublin Harbour converge to a width of 980 feet: the entrance in this case is within the shelter of Dublin Bay, and a narrower width would have engendered a too rapid current for vessels; whilst the scour induced by the convergence has considerably lowered the bar in front. The entrance to Aberdeen Harbour is 1100 feet wide; and the south outlet at Sunderland is 220 feet. At Newburyport, the width between the converging jetties is 1000 feet, which appears excessive for the size of the harbour; and probably a better depth might be gained by a narrower entrance.

It is evident that the choice of width in these cases must be guided by the rise of tide, and the tidal area enclosed by the jetties, together with the discharge of fresh water where these jetties are situated at the mouth of a river. The instances given above show how various are the widths given to the entrances of jetty harbours.

The width of entrance for enclosed harbours can be determined by more definite rules. It should depend upon the area of the harbour, and the height of the greatest waves reaching the entrance, varying directly with the former and inversely with the latter. Mr. T. Stevenson has drawn up the following formula for ascertaining the reduction produced in waves, at a definite distance within the harbour, after passing through an entrance of a given size:—

$$x = H\frac{\sqrt{b}}{\sqrt{B}} - \frac{\left(H + H\frac{\sqrt{b}}{\sqrt{B}}\right)\sqrt[4]{D}}{50};$$

where x is the reduced height of wave within the harbour at any point A; H, height of wave at entrance; b, breadth of entrance; B, breadth across harbour measured along arc with radius D and passing through A, where D is the distance from the entrance to A[1]. Mr. Stevenson has applied his formula to a few harbours, and has found that the results obtained afford a fair approximation to recorded observations; but the only large harbour he selected was Kingstown, which is tolerably sheltered; and it appears that the reduction of waves in Madras Harbour is much less than the formula would give. More extended observations are undoubtedly needed before this important question can be based upon calculation.

Turning now to the more definite results of actual practice, we find considerable variety in the widths of entrance adopted, depending doubtless, in a great measure, upon the size of the harbour, the exposure of the entrance, and the requirements of trade. Here again, a compromise has to be made between the conflicting views of engineers and sailors. Nautical men mainly look for great facility of access, and always

[1] 'Edinburgh Philosophical Journal,' vol. liv. 1853–54, p. 378.

urge a great width of entrance at the outset; whereas engineers are bound to consider also the tranquillity inside, which is incompatible with an unduly wide opening. Not unfrequently, when the former views prevail, regret is felt too late at the motion in the harbour.

The small harbour at Ramsgate, having an area of 30 acres, has an entrance 200 feet wide. Whitehaven, with an area of 24 acres, has a width of entrance of 510 feet; and at Lowestoft, the opening is 145 feet for an area of only 11½ acres. The following are the widths of entrance, and areas, of some of the more important harbours: Kingstown, 750 feet, 250 acres; Madras, 550 feet, 220 acres; Colombo, 800 feet, 500 acres; Ymuiden, 800 feet, 250 acres; Barcelona, 870 feet, 380 acres; Chicago, 500 and 800 feet, 455 acres. A width of 450 feet was originally proposed for Madras, and would improve the tranquillity of the harbour; and a width of 600 feet was designed at first for Colombo, and it remains to be seen, when the breakwaters are completed, whether the smaller width might not be more advantageous.

The smaller side entrances at Portland, Oswego, and Buffalo are 400, 350, and 300 feet, with corresponding areas of 2700, 220, and 550 acres. The new harbour at Boulogne is to have two entrances, 800 and 500 feet in width, its area as designed being 650 acres; whilst the large harbour originally proposed for Dover in 1847, with an area of 600 acres, was to have two entrances, 700 and 200 feet wide respectively. (Plates 3, 4, and 5.)

The harbour at Tynemouth is not yet completed, and the precise width of its entrance has been left indeterminate (though designed for 1000 feet); for it is proposed to stop the breakwaters provisionally at a distance apart of 1800 feet, and subsequently reduce the width as the results obtained may render advisable. As the harbour has an area of 240 acres, the trial width of entrance is large; but this is not merely the case of an enclosed harbour, it is also the outlet of the river Tyne, and besides providing an ample entrance for the very extensive trade of the river, it is also important not to check at all the free admission of the tidal waters, or the discharge of the river floods.

The width of the harbour has even a still greater influence on the reduction of waves than the area; for when once a wave, after passing through the entrance, has expanded to the full width of the harbour, very little further reduction occurs. Thus, the form of Barcelona Harbour (Plate 5, Fig. 10) is less favourable for wave reduction than that of Kingstown and Madras; whilst the expanding form of Galveston, Dublin, Tynemouth, Port Said, and Ymuiden, continues the reduction of the wave to the beach, and should be more efficient than differently shaped harbours of equal area. If, for instance, it was desired to construct a harbour having a greater tranquillity than Madras, without altering the size or position of the entrance, it would be necessary to widen it laterally, as any extension of the harbour seawards would have but little effect in reducing the entering waves. The forms of Chicago, Oswego, and Buffalo harbours are specially favourable for stilling the waves entering through the seaward entrances. (Plate 5, Figs. 5, 11 and 12.)

The width of entrance should also vary according to the depth in which it is placed, as well as with the size and exposure of the harbour, for deep water enables larger waves to approach the coast. Thus, the width at Madras, in 8 fathoms, should naturally be more restricted than at Kingstown in a depth of 4 fathoms, and even than at Colombo in a depth of 6½ fathoms, with less exposure and a larger area. Wide entrances are only admissible in shallow water and in sheltered situations; elsewhere, even moderate entrances do not ensure stillness in the harbour, as experienced at Kingstown, Madras, Barcelona, and other harbours. Works are in progress at Barcelona for forming an inner harbour to secure additional tranquillity.

The new entrance at Genoa, 2300 feet in width, is large for the size of the harbour; but this is an instance of a peculiar form of harbour, protected by overlapping breakwaters, which is favourable for checking the ingress of waves, but presents a rather winding course for the entrance of vessels. (Plate 5, Fig. 3.)

CHAPTER V.

JETTIES AND BREAKWATERS.

Definitions. *Jetties:* Mound and Open Superstructure; Timber and Stone; Cribwork; Fascinework; Wave-Breaker; Wave-Screen; Floating Breakwaters. *Breakwaters:* Various Types. Mound Breakwater; Rubble Mound; Rubble and Concrete Blocks; Concrete Blocks; Fascines. *Construction of Mound Breakwaters:* Deposit of Rubble; of Concrete Blocks. Remarks on Mound Breakwaters.

THERE would be some difficulty in drawing an exact line of demarcation between jetties and breakwaters, with reference to their construction. The jetties, indeed, at the entrances to what I have termed the Jetty Harbours of the North Sea are quite distinct from breakwaters as commonly understood; but the section of the Malamocco jetties, or moles, is exactly like the sections of some breakwaters. It is easier, however, to define jetties by the object for which they are designed, rather than according to the nature of their construction.

Definition of Jetties. Jetties, in sea-works, may be defined to be constructions, whether partly open or solid, which are built for the purpose of guiding or concentrating the current of a river, or of the tide, when emerging into the sea. They may provide a certain amount of shelter as well; but their main object is to keep open and deepen the entrance-channel of a harbour, or the mouth of a river.

Definition of a Breakwater. A breakwater is essentially, as its name implies, a structure for providing shelter from the waves. It differs, accordingly, from a jetty in sheltering a space of water, instead of maintaining a channel, and is generally situated in more exposed positions.

According to these definitions, the works at the mouths of

the Maas, the Danube, and the Mississippi, are jetties; for their main purpose is to guide these rivers and deepen the channel across their bars, though they have also to resist the waves along the coast. Sometimes, however, the works at the mouths of rivers are more extensive than needed for the mere improvement of the river, being designed to form a harbour as well. This is partially the case at the mouth of the Tees, and still more so at Tynemouth, so that the term *breakwater* may fairly be applied to the piers at the mouth of the Tyne; and the works in progress to protect Tees Bay are not inappropriately called the Gare breakwaters.

The word *pier* is sometimes applied indiscriminately to jetties and breakwaters, for the jetties of the North Sea harbours are called occasionally piers; the same term is commonly given to the works at the Sulina mouth of the Danube; the breakwaters at Tynemouth and Ymuiden are frequently described as the Tynemouth piers, and the North Sea piers; and the Dover breakwater is known as the Admiralty Pier. It would seem preferable to reserve the term *pier* for structures specially erected for landing-places and promenades.

JETTIES.

Mound and Open Superstructure. The form of jetty regularly adopted at the North Sea jetty harbours, as well as at other ports, consists essentially of a series of timber, or, sometimes, iron frames embedded in a rubble mound, or pitched slope, raised slightly above the level of the beach. (Plate 1, Figs. 2, 4, 6, 8, and 9.) They are expressly designed to allow the littoral currents to pass along unimpeded. Whilst chiefly marking the line of the entrance-channel, and aiding in maintaining it, they also partially divide the comparatively small waves which reach the shallow shores on which these jetty harbours are situated, and afford some shelter to vessels passing along the entrance-channel.

Timber and Stone Jetties. In some instances the jetties of these North Sea harbours have been made more solid; thus the lower half of the inner portion of the west jetty of Calais harbour has been built of solid masonry, with an open timber superstructure on the top. The jetties at Dunkirk were formerly partially constructed of rubble stone enclosed in a timber casing, which, however, was found troublesome to keep in repair; and a similar construction was more recently adopted for the Port of Blyth jetties[1]. These jetties, as well as the more open kinds, fulfil satisfactorily the purposes for which they are intended, namely, marking, preserving, and protecting the channel, directing the scouring current during sluicing, and serving as piers of access to the lights at their extremities; but they are not calculated to afford protection from heavy seas.

Cribwork Jetties. Timber cribs filled with rubble stone are commonly used for training the outlets of the harbours situated at the mouths of rivers flowing into the large inland lakes of North America. The harbours of Sheboygan on Lake Michigan, and Great and Little Sodus on Lake Ontario have been improved by regulating their outlets with cribwork jetties. Where timber is cheap, and small stone easily procurable, this type of jetty is very convenient; and, as the cribs are strongly constructed, this form of jetty has every prospect of being more durable than those at Dunkirk, and in the event of decay it can be cheaply rebuilt.

Fascinework Jetties. The use of fascines for works in the sea has been gradually extended. A fascinework dyke was formed, in 1825, at the harbour of Niewediep, for concentrating the current from the Zuider Zee and forming a deep channel from the outlet of the North Holland Canal to the Texel Ronds. Fascinework was subsequently employed for the much more exposed jetties at the mouth of the Maas, and has been used for the great jetties at the mouth of the Mississippi. Fascinework jetties are now being constructed

[1] 'Minutes of Proceedings, Inst. C. E.' vol. xviii. p. 73.

for forming the jetty harbours of Charleston and Galveston. (Plate 2, Figs. 1 and 12.) Sections of these jetties (Plate 2, Figs. 2 and 13) show the nature of these constructions. The mattresses of fascines are cheap, easily constructed, and readily floated into place. Moreover, owing to their flexible and light character, they readily adapt themselves to an irregular or soft sandy bottom, and can be heightened as they settle into place. They are consolidated and weighted with stones; and appear to be more durable than might have been anticipated, owing to the fascines becoming soon imbedded in sand and silt. The width given to the base of these jetties renders them very stable; and, by keeping them as low down as compatible with proper shelter, they are little affected by the waves.

Wave-Breaker. When the waves are blown by the wind directly on-shore, they run up the jetty channel, which, if converging as at Ostend (Plate 1, Fig. 10), or curved as at Dieppe (Plate 1, Fig. 1), is liable to intensify the undulation. To provide against the introduction of this swell into the harbour itself, the jetty is made more open for a certain distance; and a basin with a shelving bottom is constructed behind it. (Plate 1, Fig. 1, and Plate 12, Fig. 2.) The wave, in travelling up the channel, finds an exit through this aperture, and expends itself by expanding into the basin and running up the shelving beach. The iron frames placed for this purpose along a short length of the Dieppe jetties are shown on Plate 1, Fig. 3.

The French apply the term '*briselame*' to this portion of the jetty; but, in reality, the open jetty does not act as a wave-breaker, it merely provides for the withdrawal of the wave from the channel; and the wave subsides actually in the basin, and breaks upon the sloping beach. The basin is in fact a stilling basin, and fulfils the same function precisely as the recess of the Potato Garth near the mouth of the river Wear[1].

[1] 'Rivers and Canals,' L. F. Vernon-Harcourt, vol. i. p. 240.

Wave-Screen. Proposals have been made for stilling waves by constructing a sort of sloping gridiron, which the waves, in passing over, drop through in a divided form, as through a screen[1]. Iron screens of this type have indeed been constructed; but apparently the results obtained have not warranted the extension of the system. Theoretically, the wave-screen may possess some merit in cutting up the wave and stilling it by the aid of gravity; but, practically, the structure is not strong enough to resist the impetus of heavy waves, as the wave motion must be stopped by the screen if it drops through the interstices.

Floating Breakwaters. Various schemes have been suggested, from time to time, for arresting waves by means of floating breakwaters moored in position. It has been imagined that the undulation, being on the surface, might be stopped or reduced considerably by an obstacle at or near the surface; and thus the cost of building up a breakwater from the bottom could be saved. Though, in the case of large waves, the undulatory motion is not simply superficial, yet, undoubtedly, the power of the waves would be greatly diminished if the upper portions could be arrested in their progress; and the gain in dispensing with a solid structure founded on the bottom of the sea would be very great.

A floating breakwater was moored in front of the entrance to the port of Ciotat with the object of protecting it from the swell. The breakwater consisted of a double row of long buoys chained together, and so arranged that the spaces between the buoys of the one row were covered by the buoys of the second row. The buoys, however, broke from their moorings during storms, and moreover did not arrest the waves, and they were consequently removed.

Other forms have been tried or proposed, such as rafts of timber; and also a series of beams floating vertically, by being moored at one extremity, in imitation of weeds and reeds

[1] 'Minutes of Proceedings, Inst. C. E.' vol. xviii. p. 85 and vol. xix. p. 649.

growing in the water. These attempts, however, have not been successful in adequately reducing the waves; for these kinds of constructions are liable, either to be shattered, or to afford little shelter, or, if strong enough to offer real resistance to the waves, to drag or break their moorings. Thus the Great Eastern, which may fairly be regarded as a huge floating breakwater, broke from her moorings when exposed to a heavy sea off Holyhead. The whole strain of stopping the waves is in fact thrown, in this case, upon the moorings; and they would have to be made enormously strong to be able to bear it.

The force of waves is so great, as indicated by its effects in moving huge masses, that no fragile floating moored construction could possibly oppose an adequate resistance. The accumulated power of the wind, acting through the medium of the waves, cannot be evaded, but must be met; and this can only be effectually accomplished by a solid breakwater, of which the various forms, and their advantages, have now to be considered.

BREAKWATERS.

The size and force of waves have been shown to vary greatly with the exposure and depth of water; and as the strength of a breakwater must be made proportionate to the forces it has to resist, the exposure of the site, the depth of water in front, and the direction of the worst waves, have to be carefully ascertained previous to designing a breakwater. The direction of the waves does not always coincide with that of the wind; as the configuration of the coast, the shelving of the beach, and the deflection caused by obstacles are liable to modify the path of the waves. A neglect of due attention to these considerations has led to serious disasters. A breakwater, which has proved successful in one place, may prove quite inadequate in another of greater depth or exposure; and the different parts of the breakwaters of the same harbour need strengthening according to their position.

The form, or type, of breakwater to be adopted depends upon the nature and supply of materials available for its construction: an abundant supply of stone admits of the employment of a rubble mound; whilst a deficiency of stone renders it necessary to resort to a concrete-block mound, or an upright wall.

Various Types of Breakwaters. There are three very distinct types of breakwaters: (1) Mound of rubble or concrete blocks; (2) Mound with Superstructure; (3) Upright Wall.

Each of these types include certain minor varieties. Thus, the first class comprises a rubble stone mound, a rubble and concrete-block mound, and a concrete-block mound, to which may be added a mound of fascine mattresses: the second class contains a rubble mound with superstructure commencing at low water, and a rubble mound with superstructure founded below low water; and the third class, an upright masonry wall, and a concrete wall.

MOUND BREAKWATERS.

Rubble Mound. The simplest kind of breakwater is formed by simply tipping a mound of rubble stone into the sea till it reaches the surface. The waves approaching the mound are stopped by it, and break upon the mound as upon a shingly beach. The effect of the sea is to lower the top of the mound, and flatten its sea slope. This action extends from the top to from 12 to 20 feet below low water. The mound has to be constantly fed with stone till the sea has formed a natural slope, ranging between 5 to 1 and 10 to 1 from the top of the sea slope down to where the waves appear to lose their influence, and forming from thence to the bottom a steeper slope of about 1 to 1: the harbour slope assumes the form of about 1 to 1. When the mound has been formed to what may be termed its natural slope under the action of the waves, varying in flatness according to the exposure, it undergoes little alteration from the bottom up to near low water. The portion, however, of the mound above that level is very liable to injury, as the waves in storms tend to lower it, either drawing the stones down the slope by their recoil, if the mound is raised some height, or, more commonly, washing the stones over the mound into the harbour.

This system of construction was adopted at several ancient harbours. The Cherbourg breakwater was also formed, in the first instance, in this manner. It was, however, found that the sea lowered the rubble mound down to about low-water level, which prevented the breakwater from adequately sheltering the harbour; and the mound was not rendered stable and thoroughly efficient till the present superstructure had been erected upon it.

At Plymouth, the rubble mound was introduced with similar results, and underwent constant injuries till the portion near and above low water was protected by pitching. (Plate 6, Fig. 1.) The upper part of the mound forming the detached breakwater at Portland is protected by large stones, and the rubble breakwater at Delaware has been similarly secured. (Plate 6, Figs. 2 and 6.) It will however be seen, in the descriptions of these breakwaters, that the protection of pitching and large stones does not maintain them intact; but that occasional injuries are inflicted by the waves during violent storms, necessitating regular maintenance varying in extent from year to year.

A mound of rubble stone is a suitable method of construction when there is an abundant supply of stone in the neighbourhood, when the space occupied by the wide base of the breakwater is a matter of indifference, when there is no necessity to afford shelter along the top or close to the inner side of the breakwater, and provided the sea slope between high and low water is adequately protected. There is always some wear and tear in a rubble mound where the waves produce any motion; and there is liable to be a travel of the stone along the breakwater in the direction of the strongest run of waves and currents, similar to what occurs on a shingly beach. The attrition of the stones rounds them, and reduces their weight, rendering their displacement easier; whilst the removal of stones by the waves from one part of the mound to another, and even round the end, weakens the places from which the stones are taken. This loss, which occurs more or less in every rubble mound, requires replacing by occasional deposits of fresh stone.

Pitching, set in cement, as adopted for protecting the weak part of the rubble mound at Plymouth, furnishes a smooth surface on which the dashing up and recoiling waves can exert little force. Unfortunately, when once a stone of the pitched surface gets displaced, either by settlement, by the washing out of cement, or by the impact of a breaking wave, the sea rapidly widens the breach thus formed, which, unless promptly repaired, would soon extend along the breakwater and leave it bare of protection.

Large blocks of rubble stone on the sea slope, above low water, form a simpler protection than pitching; and the injury from the displacement of some of the blocks is not liable to spread rapidly.

Rubble and Concrete-Block Mound. As waves have less effect in moving bodies in proportion to the depth, and as the inner parts of a breakwater are protected by the outer portions, it would be theoretically correct to place the smallest stones in the centre of the mound, to graduate their size as the mound is raised, and to put the largest materials on the top and sea slope of the mound. Though this entails the sorting of the materials, it has been strictly carried out in the construction of the mounds of several Mediterranean breakwaters. (Plate 7, Figs. 4, 7, 12, 15, and 16.) Most of these breakwaters have the additional protection of a superstructure; but Algiers breakwater is an instance of a mound built up on this system. Frequently rubble blocks are not obtainable of an adequate size, or in sufficient quantities, to protect the upper portion of the sea slope; and then, as at Algiers, concrete blocks serve as a very convenient substitute. Concrete blocks can be made of any size, and therefore adapted to the forces they have to resist; and, by care in their manufacture with the best Portland cement, they can be made very durable. They are less liable to injury when forming the sole covering of the exposed portion of the mound, as at Algiers and Marseilles (Plate 6, Fig. 9, and Plate 7, Fig. 16), than when they are intermingled with rubble stone, as at Cherbourg and Tynemouth (Plate 7, Figs. 1 and 9), for in the latter case the blocks are battered by the easily moved smaller rubble.

The breakwater at Alexandria is a fine example of this type, differing somewhat in its methods of construction from the other Mediterranean harbours referred to. (Plate 6, Fig. 14.) Instead of forming first a core of small rubble and covering it with successive layers of larger material, the mound of large concrete blocks was first laid on the sea side of the breakwater, the interstices between the blocks were then filled with small rubble, and finally large rubble stones were placed on the harbour slope. This system involves the use of a considerably larger number of blocks; but, by forming the protecting portion first, the rest is deposited under shelter, and is in no danger of being washed away during construction. The smaller stones consolidate the blocks, and the larger rubble adds to the mass of the breakwater.

Concrete-Block Mound. Occasionally no rubble stone is employed, and the mound is entirely composed of concrete blocks. The breakwater at Biarritz, the principal portion of the Port Said breakwater, and the mound at Leghorn, consist exclusively of concrete blocks. (Plate 6, Figs. 11 and 13, and Plate 7, Fig. 6.) This arrangement is adopted where rubble stone is difficult to procure, and where consequently concrete blocks form the cheapest mound. A mound thus formed is not so compact as the mixed type of mound, and consequently the waves, rushing into the interstices, are more liable to shift the blocks; and sand in suspension may pass through the breakwater and settle in the harbour. The system, however, possesses the advantage of forming a mound with the least possible quantity of material; it is sufficiently stable if the blocks are made of adequate size, and by degrees the interstices are filled up with deposit.

Fascine Mound. The employment of fascines for the construction of jetties has been already alluded to (p. 92); but some of these fascine jetties are sufficiently extensive and exposed to be classed under the head of breakwaters. Thus the jetties at Galveston (Plate 2, Figs. 12 and 13), Charleston (Plate 2, Figs. 1 and 2), and the mouth of the Maas, are breakwaters composed of fascine mounds loaded with stones. These fascine mounds are naturally resorted to where brushwood is plentiful, and stones scarce; they become compact by the accumulation of silt; they are pliable, and intertwined, and therefore not readily injured by waves, and their broad base renders them very stable.

Construction of Mound Breakwaters.

Formation of Rubble Mounds. A rubble mound breakwater is constructed either by tipping from wagons, as in forming a railway embankment, or by depositing the rubble stone from hopper barges. The latter plan has the merit of rapidity in execution, as any number of barges can be used for the work, depositing the stone along an extended line; whereas the rate of progress with wagons is limited to the width available for roads on the breakwater above high water.

Staging is also frequently used for enabling the wagons to tip over a wider base, and at a higher level above the range of the tides and waves. The staging used for tipping the base of the Holyhead breakwater is shown on Plate 8, Figs. 9 and 10. It has the disadvantage of being liable to injury from the waves, and it is not generally available for detached breakwaters, though at Portland the staging was carried across the eastern opening for forming the outer rubble mound breakwater. (Plate 3, Fig. 11.) The inner portion of the mound of the south-western breakwater at Boulogne has been formed by merely tipping it from wagons running out on the mound from the shore; but this method proved too slow as greater depths were reached, so that an inner harbour has been formed within the protection of the root of the south-western breakwater, in order to serve as a shelter for barges which are to be used for the future in depositing the base. The rubble base of the Alderney breakwater was carried out from the shore by wagons for the first two years, but after that barges were employed, a special harbour being formed to receive them. All the materials for the rubble mound at Plymouth had to be conveyed by boats to the site. At Cherbourg, the first modern instance of a rubble mound breakwater on a large scale, an attempt was made to form the breakwater by floating out hollow timber truncated cones, and filling them with small stone, upon the same principle as the cribwork now largely employed in America; but, owing to the damage these cones sustained from storms, the system was abandoned, and the mound was subsequently completed by deposit from barges.

Deposit of Concrete-Block Mounds. The formation of a mound of concrete blocks requires greater care, and more powerful appliances, than in the case of rubble stone, as the blocks are larger and consequently heavier than the rubble, and therefore more liable to be fractured in depositing. The best method of placing the blocks is to carry them out to the site between two pontoons, or to convey them in a barge and sling them out by means of a crane or shears erected on the barge. The former method has been adopted at St. Jean-de-Luz, Genoa, and Leghorn, whilst the latter plan was followed at Alexandria (Fig. 15, p. 223). When the blocks are to be situated at or near high-water level, they can be formed within timber casing in the place they are to occupy, and in such a case they can be constructed of any size; but a favourable season of the year must be selected for their construction, and they must be protected from the wash of the waves till they are completed and till the concrete has had time to set.

Remarks on Mound Breakwaters. The most perfect type of mound breakwater consists, undoubtedly, in a due combination of rubble stone and concrete blocks. By placing rubble stone at the base, and raising it on the sea slope to the limit below the sea level at which it will stand at 1 to 1, and then using larger stones or concrete blocks, increasing in size towards the top, as a covering to the sea slope, whilst retaining the smaller rubble for the interior and the harbour slope, it is possible to construct a mound having slopes throughout of 1 to 1, and therefore containing the least possible amount of material. Such a mound, moreover, is compact, and proportioned at all parts of the sea slope to the varying power of the waves; and by using large concrete blocks from low water upwards on the sea side, the structure can be rendered perfectly stable and permanent. It has been proved that there is no mechanical difficulty in depositing blocks weighing 20 tons on a mound, as this was successfully and rapidly accomplished at Alexandria; and the possibility of moving and accurately placing much larger masses, under favourable conditions, has been demonstrated by Mr. Stoney with his huge monoliths of 350 tons for forming the quays in Dublin Harbour. (Plate 8, Figs. 11 and 12.) Much must depend on the conditions of the site, both as to the facility and rapidity with which suitably sized blocks can be deposited, and whether the extent of the works justifies a large expenditure on plant; but the success which has attended this mixed system at Algiers, Alexandria, and Port Said, exhibits its practical value; whilst the failure of the block mound at Biarritz, and the constant maintenance necessary in rubble mounds, indicate the importance of protecting the exposed portion of the sea slope with blocks of adequate size.

CHAPTER VI.

BREAKWATERS FORMED OF A MOUND AND SUPERSTRUCTURE.

Advantages and Disadvantages of a Superstructure. Two Classes. (1) *Superstructure founded at low water:* Protection; Construction; Form. (2) *Superstructure founded below low water:* Objects of Design; Depth to which Waves affect Base. Construction of Superstructure; Erection from Staging; with Cranes; Comparison of systems of erection. Sloping-block system; advantages, and improvements. Concrete-block Wave-breaker. Remarks.

IT has been seen in the preceding chapter that the weak point of a rubble or concrete mound breakwater is the portion from near low water upwards; and that an objection to this type of breakwater is the large amount of material which it requires, increasing greatly with the depth of water, more especially in the case of a rubble mound with a very flat slope, such as Cherbourg, Plymouth, and Table Bay. With the object of remedying these defects, a superstructure was erected on the mound at Cherbourg; and a similar method of construction has been adopted in numerous instances. (Plate 7.)

Advantages of a Superstructure. A superstructure protects the top of the mound by its weight; it prevents the waves from washing over the mound and carrying the materials of the sea slope into the harbour; and it enables a larger amount of protection to be afforded with less material. It also admits of the formation of a sheltered roadway along the breakwater, giving access to any part, and to a light at the extremity. It enables also the inner side of the breakwater to serve as a quay in places where the exposure is not great, or during fine weather.

Both as regards construction and maintenance, as well as shelter and utility, the mound surmounted by a superstructure appears to be greatly superior to a mound protected merely by pitching or large blocks, and would undoubtedly be universally adopted on soft foundations, and in deep water where a rubble base is essential, if this system also did not present some deficiencies in actual practice.

Disadvantages of a Superstructure. The superstructure has to be built upon the yielding foundation of the base, which is only gradually consolidated under the action of the sea. In deep water, where the base is high, the settlement is considerable and long-continued; and, moreover, unequal settlement occurs, as each additional length of superstructure is laid upon it, under the weight of the superincumbent mass. This settlement tends to produce cracks and fissures in the superstructure, which become the sources of serious damage, and even breaches, under the action of the waves and the compression of the air inside them.

Moreover, the actual foundations of the superstructure are liable to be attacked by the waves and undermined, if laid within their influence at or near low-water level. This can indeed be remedied by placing the foundations of the superstructure at a greater depth; but blocks below low water cannot be cemented together, and therefore in their unconnected state, unless most carefully laid and free from unequal settlement, are specially liable to displacement.

Again, whereas a rubble mound gradually breaks the force of the wave, and allows a portion to pass over into the harbour, the superstructure opposes this partial movement of translation, and has, under certain conditions, to resist the whole force of the wave breaking upon the mound. Indeed if the superstructure is to be a thoroughly efficient protection to the weak part of the mound, it must bear the chief brunt of the waves, not by letting the breaking waves penetrate and expend their force, as they do in the interstices between the large blocks of a mound, but by opposing a solid face to the shock.

These are the difficulties with which designers of this type of breakwater have to contend; and it is to the endeavour to overcome them in various ways that the different constructions shown in Plate 7 are due.

Division of the Mixed Type of Breakwater into two Classes. The breakwaters composed of a rubble mound and superstructure may be divided into two classes: (1) Superstructure founded at low-water level; (2) Superstructure founded below low water.

The object of the first system is to commence the protecting superstructure at as low a level on the rubble mound as is consistent with the structure being built in a solid mass cemented together. As, however, in this case, the mound has to be raised up to low-water level, it is necessarily somewhat exposed to the action of the waves.

In the second system, an endeavour has been made to place the foundations of the superstructure at a low enough level for the sea to have no influence on the rubble base, which necessitates laying the lower courses of the superstructure without cement.

In the first case, the stability of the mound and the foundations of the superstructure are somewhat compromised for the sake of the solidity of the superstructure; whilst in the second, the solidity of the lower courses of the superstructure is sacrificed for the maintenance of the mound, and for the security of the superstructure from undermining. In both cases the weak point lies near low-water mark; but in one case the sea is most liable to attack the mound, and in the other the superstructure. The amount of material required for the base is considerably larger in the first case; but this may be set off against the increased cost of a superstructure founded below low water.

1. SUPERSTRUCTURE FOUNDED AT LOW-WATER LEVEL.

In the earlier designs of the mixed system of breakwaters, the superstructure was commenced at low-water level, owing to the difficulty which was experienced in those days in laying blocks below the water[1]. Cherbourg, Portland, Holyhead, St. Catherine's, and Marseilles, as well as the first designs at Alderney and Tynemouth, are instances of this type of construction. (Plate 7.)

Protection of the Base and Foundations of the Superstructure. The endeavour in the first instance, both at Cherbourg and Tynemouth, was to place the superstructure on the top of the rubble mound, raised to low water, without any additional protection. It, however, soon became evident that small rubble would not remain on the sea slope at low-water level; and the lowering of the mound threatened the undermining of the superstructure. The foreshore was accordingly raised a little against the face of the superstructure at Cherbourg; and the top of the sea slope has been protected, both at Cherbourg and Tynemouth, by large concrete blocks. At Portland and Holyhead, large rubble blocks have been deposited on the sea slope of the breakwater, raising the mound to high-water level, and thus practically converting the breakwater into a rubble mound with merely a backbone formed by the solid superstructure. The exposure at these places is sufficiently moderate to enable large rubble blocks to remain fairly intact on the sea slope. The slope, however, at Cherbourg, where not protected by concrete blocks, has to be maintained by a yearly addition of rubble; and the outer portion of the slope at Holyhead has been weighted with old chains, for a short distance from the superstructure, to keep

[1] The introduction of the diving-bell, used for the first time, in harbour works, by Smeaton at Ramsgate Harbour towards the close of last century, did little to facilitate the laying of foundations below low water, for the diving bell is a cumbrous machine, it admits of only a slow rate of progress, and is not suited for exposed situations.

the rubble stone in place and prevent its travelling around the head.

In several Mediterranean harbours, the sea foreshore is protected by a mound or layer of concrete blocks; and the breakwaters resemble very closely in construction those at Alexandria and Biarritz, with the single exception of the mound being surmounted by a slight superstructure for protecting the quay. The superstructure, indeed, at Marseilles, Trieste, and Ile Rousse (Plate 7, Figs. 5, 7, 15, and 16) forms rather a parapet to the quay than a regular superstructure; for the mound possesses sufficient stability without it, and forms the natural breakwater to the harbour. This peculiar form, however, of the mixed type is rendered necessary by the breakwater having to serve as a quay to the port, and not merely as a shelter. It is consequently essential for the waves to be thoroughly broken and arrested before they reach the superstructure; otherwise they would render the quay unapproachable in any rough weather, owing to the amount of water that would dash over the superstructure. The main object of a superstructure in these instances is to protect the quay; the mass of the breakwater is not reduced, nor is any important addition of stability imparted to the mound. In proportion as a superstructure is valuable as a protection to the mound, or as a shelter to the harbour, in the same degree it is less suitable for a quay. It is impossible for a superstructure to arrest the waves, in an exposed situation, without being enveloped in a cloud of spray or sheet of water. The superstructures at Cherbourg and St. Jean-de-Luz (Plate 7, Figs. 1 and 21) are too efficient as portions of the breakwaters to be suitable for regular quays; whilst at Marseilles and Trieste, and in a lesser degree at Portland and Holyhead, the outer protection is so ample that the superstructure suffices to shelter the quay. Gravel remains intact on the quay level at Holyhead; whereas the quay level at Cherbourg has to be paved with pitching, and the cannons standing on it are sometimes overturned by the sea; and at Alderney, the quay pitching set in cement is occasionally rooted up by the waves which dash over the high promenade wall.

The mound at Leghorn (Plate 7, Fig. 6) consists entirely of concrete blocks, possessing the merit of greater stability with less material; but the wide interstices between the blocks render it less suitable for the foundation of a superstructure, and, moreover, its solidity would appear to render such a structure unnecessary. At Port Said, a superstructure formed part of the original design, but it was abandoned as being superfluous.

At other breakwaters, the mound has been more or less protected by concrete blocks: they entirely cover the mound at St. Jean-de-Luz (Plate 7, Fig. 21); and are merely placed on the most exposed part of the sea slope at Bastia and Oran. (Plate 7, Figs. 4 and 12.) The concrete blocks are useful in preserving the top of the mound before the erection of the superstructure, and they afterwards serve to protect the superstructure from being undermined.

Construction of the Superstructure. The erection of the superstructure in these instances presents no special difficulties, as by taking advantage of low water at spring tides it can be entirely built out of water. When the rubble mound is high, it is important to allow as much time as practicable for the mound to consolidate before erecting the superstructure upon it. Settlement must necessarily occur more or less; and any injuries resulting from it should be promptly repaired.

Where the mound has been deposited from staging, the same staging serves for the erection of the superstructure. (Plate 8, Figs. 9 and 10.) If barges only have been used for forming the mound, then special staging has to be provided; or cranes may be employed for setting the larger face blocks, and the backing can be built up with rubble. When concrete-in-mass is adopted, it is only necessary to provide suitable panels for the frames.

Advantage has to be taken of the lowest tides for laying the foundations; and a large proportion of cement must be used in the mortar for the lower courses, so that it may set before the rise of the tide. Medina and Roman cement used to be employed for this purpose, on account of their rapidly setting qualities; but it has been found that these cements are liable to deteriorate after a lapse of time, so that the advantages they offer in the first instance are dearly purchased. The manufacture of Portland cement has, within recent years, been carried to such a state of perfection that it combines rapidity of setting with great durability; and it is only necessary to regulate the admixture of sand to obtain any desired quickness in setting. When Portland cement is mixed with one or two parts of sand, it forms a very strong mortar for facework, and the joints may be pointed with neat cement if exposed to the rising tide. The wash of the waves may be kept off freshly built work by covering it with sacking weighted with stones.

The foundations of the superstructure have to be laid during fine weather, in the summer months, upon a length of base which should have been subjected to the action of the sea for at least two winters; and the work should be rapidly raised to quay level. Each length of superstructure has to be specially protected at the end, to secure it from disturbance by the waves till the succeeding length can be joined to it; and the termination of each season's work has to be strongly secured to prevent its being damaged by the winter storms. The upper portions of the superstructure can be completed during favourable weather in the winter time. This period is also utilised in making blocks, quarrying and dressing stone, repairing machinery, and other works of preparation for the next season's work.

Form of Superstructure. A great variety of forms have been adopted for the superstructure. (Plate 7.) Occasionally, where the sea slope is raised high and is well protected by large blocks, the superstructure consists merely of a high thin wall, as at Marseilles and Ile Rousse. Sometimes the superstructure is broader and less high, stronger but affording less shelter, as at Bastia and Oran, but still somewhat rudimentary. A more developed superstructure has been built at Leghorn, with a broad concrete base and high masonry wall.

A superstructure has often roadways at two different levels, and the wall which carries the upper or promenade roadway on the sea side acts as a parapet to the quay level. The upper roadway enables material to be tipped on to the sea slope when required for maintenance, and the lower roadway serves as a quay. At Cherbourg, the upper roadway is narrow, and the whole superstructure has been built solid. (Plate 7, Fig. 1.) Usually the superstructure is composed of two walls with filling between; the sea wall carries the upper roadway, and the harbour wall retains the filling on the harbour side and serves as the quay wall. Holyhead, Portland, and St. Catherine's (Plate 7, Figs. 2, 10, and 11) are instances of this type of construction, which enables a wide quay to be formed at a comparatively small cost. The superstructure at Genoa is more massive; it has been carried right across the top of the rubble mound, the inner portion has been built in a series of arches, and a sheltered pathway and storehouses have been provided along the breakwater under the arches. (Plate 7, Fig. 3.) The tendency in the more recent breakwaters is to return to the narrower and solid form of superstructure adopted at Cherbourg, leaving the superstructure to protect itself except at the base, and dispensing with a sheltered quay, of which Boulogne and St. Jean-de-Luz are instances. (Plate 7, Figs. 8 and 22.)

2. Superstructure founded below Low Water.

The two main objects aimed at in placing the foundations of the superstructure below low water are, to reduce the amount of material required for the breakwater, and to render the base unassailable by the sea. The first object is readily attained by reducing the height of the base, the vertical superstructure replacing the large mass in the slopes of the base, as amply exemplified by comparing the sections of Marseilles and Odessa (Plate 7, Figs. 16 and 24), and still more those of Holyhead and Madras (Plate 7, Figs. 2 and 17), situated in similar depths of water. The accomplishment of the second object depends upon the depth to which the foundations of the superstructure are lowered, varying according to the exposure and the size of the materials composing the top of the base.

Depth to which Waves affect the Rubble Base of a Breakwater. It was formerly supposed that the waves had no influence upon large rubble at a depth of 12 feet below low water of spring tides; and this is the depth to which the Alderney superstructure was carried when foundations at low water proved unstable on that exposed site, though subsequent settlement on the mound increased this depth to between 15 and 18 feet. (Plate 7, Figs. 13 and 14.) Even then, however, it was found necessary to protect the lower courses of the superstructure by raising the mound, on the sea side, to low-water level. At this higher level the mound was invariably lowered during the winter months, though covered at the top with large blocks of rubble; so that neither is the ideal type realised in this instance, nor is the necessity of constant maintenance avoided. At Tynemouth, a similar alteration has been made in the design, the superstructure having been founded at low-water level near the shore; whereas it is now being carried down to a depth of about 20 feet below this level. In this case also the rubble mound has been raised above the foundations of the superstructure; but, in order to avoid constant maintenance, the sea slope near the face of the superstructure has been protected by a double row of concrete blocks. (Plate 7, Fig. 9.) This forms an important improvement on the Alderney design; but here, the bottom being only a few feet below the foundations of the superstructure, the mound rather serves as a foundation and protection to the base of the superstructure, instead of forming a distinct portion of the breakwater as it does at Alderney, Manora, Madras, Colombo, and other breakwaters of this type. To render the mound secure from disturbance in such an exposed situation as Alderney, it would have been necessary to keep it down to a depth of at least 20 feet below low water.

The ideal type appears to have been attained at Odessa, on the tempestuous shore of the Black Sea, where the superstructure is founded at a depth of 22 feet (Plate 7, Fig. 24), and the base is raised about 4 feet higher. At Kurrachee, the base is raised to 14 feet below low water; and it appears that there the mound remains uninfluenced at that depth, owing partly to the moderate exposure of the site, and partly to the lowness of the superstructure, which allows a portion of the waves to pass over it, reducing their recoil from the superstructure, and consequently their influence on the rubble base. (Plate 7, Fig. 19.) The Colombo superstructure is placed on a rubble mound 20 feet below low water; and the mound, which is raised a little higher against the sea face, appears to be stable. (Plate 7, Fig. 23.) The base of the Madras breakwaters was terminated at a depth of 22 feet below low water, being much more exposed than the Manora breakwater, and having a slightly higher superstructure. (Plate 7, Fig. 17.) Though more exposed than Colombo, the depth at which the Madras base was stopped seems to have been ample, except at the rounded corners of the breakwaters. The undermining, however, which occurred at these parts during the severe cyclone of 1881, may be traced to the peculiar sidelong rush of the waves occasioned by the form of the breakwaters, and not to the ordinary action of the waves. Moreover, the materials of the rubble mound appear to have been smaller than the average, and some portion not hard enough.

The experience gained at Odessa, Kurrachee, Colombo, and Madras, shows that it is quite feasible to design this mixed type of breakwater so that no maintenance of the rubble base is necessary. The variation in the stability of the base according to the site is amply indicated by the fact that, while at the land-locked St. Catherine's breakwater, on the east coast of Jersey, the rubble mound has required no additions for many years, though rising above low-water level (Plate 7, Fig. 10), a depth of 22 feet below low-water level has proved barely sufficient for the Madras base.

Construction of Superstructure. The building of the superstructure requires great care in either of the two classes of this mixed type of breakwater, but more especially where the lower courses of the superstructure, being laid below low water, cannot be cemented together. At Alderney, and other breakwaters, the work below low water was accomplished by laying accurately squared blocks by means of helmet-divers, placing generally granite ashlar at the face, and concrete blocks at the back, and fitting and bonding them together as carefully as possible. As, however, it was very important to raise a certain length to the level of low water each fortnight during the working season, so that the masons might commence on the cemented courses as low as possible at each spring tide, and also in order that the blocks under water might be secured by the weight of the masonry on them, it has been customary to found the superstructure in suitable lengths, leaving a break at the close of each fortnight's work; and each season's work was terminated by a carefully protected scar. Accordingly, the superstructure was composed of a series of lengths, formed at different periods, between which unequal settlement of the base necessarily occurred; and a still more marked division was formed at the close of each season's work. The junctions formed so many weak points in the superstructure, where serious disruptions tended more or less to occur. Above low water these cracks could be filled up, unless extending far into the interior of the superstructure; but below low water no means could be used for filling up the crevices, and the waves rushing in caused a compression of the air inside them. This resulted in the forcing out of face-stones about low-water mark, which were also liable to drop a little owing to the lowering of the base. When once some face-stones have been displaced, the enlargement of the breach in the fissured breakwater is merely a question of time; and during long-continued storms the damage may extend rapidly, without a chance of repairing it at an early stage. The progress of the injury is specially rapid when the superstructure consists of a sea and harbour wall with rubble filling between them, as then the sea has only to penetrate the comparatively thin sea wall, or to remove the quay paving, to reach the hearting, which is readily displaced. Accordingly, this form, which was adopted in the earliest designs of Alderney and Colombo, has been abandoned for superstructures founded below low water; for the dry rubble hearting is a source of weakness, and is only useful as a cheap method of forming a wide quay, whilst a quay is of little service in rough weather where these superstructures are much exposed. A relic, however, of the former sea and harbour walls may still be noticed in the later superstructures of Alderney and Tynemouth (Plate 7, Figs. 9, 13, and 14), where the foundations are deeper near the sea and harbour faces. This system, probably due to motives of economy, promotes unequal transverse settlement and the consequent formation of fissures; it is harmless on the shallow base at Tynemouth, but at Alderney the results were very damaging. The foundations of the more recent superstructures have been carried level right across, and this practice should be invariably followed.

Erection of Superstructure from Staging. The superstructures of the earlier breakwaters were erected from staging resting on piles secured in the rubble base. The staging carried gantries with crabs for moving and placing the blocks across the whole width of the superstructure. This staging was gradually carried forward as the work progressed, extending from the finished portion of the breakwater to the base prepared for the diving courses. The inner piles were built into the masonry, but the rest of the woodwork was removed from over the finished work, and used again at the further extremity.

The fixing of the piles had to be carefully done with the aid of divers, and required calm weather. At the close of each season's work the staging remained exposed to the winter storms, and not unfrequently portions were washed away, involving not merely the loss of the plant, but also delay in the commencement of the following season's work. This system, accordingly, though simple, was not wholly satisfactory.

Laying Blocks with Cranes. A sort of balance crane was used at Alderney for lowering the piles for the staging during the progress of the works, and it was subsequently employed for lowering stones for the repairs of breaches. This machine, called a "Samson," was formed of balks of timber, 76 feet long, trussed above, and turning horizontally upon wheels. It was placed upon a frame which ran on rails, laid to a 15 feet gauge, along the breakwater. One end carried a balance weight, and the other end served as a jib, and supported the load which was raised or lowered by a chain running over a travelling sheave, and worked by a crab. The effective overhang of the jib from the outer support was 30 feet, and the machine could lift a load of 4 tons. This machine appears to have been the first instance of the application of an overhanging crane for sea works, which has since been so greatly extended and used directly for the erection of superstructures.

An overhanging travelling crane was employed for laying the concrete blocks of the breakwaters at Ymuiden (Plate 5, Fig. 2, and Plate 6, Fig. 19), after staging had been tried unsuccessfully, owing to the washing out of the screw piles by the scouring away of the sand round them. The blocks employed weighed from 6 to 12 tons.

About the same period, Mr. Parkes designed a special overhanging balance crane, of larger dimensions, for depositing the 27-ton concrete blocks of the Manora breakwater on the new sloping system which he inaugurated there in 1870[1]. (Plate 8, Figs. 1 and 2.) This crane, called a "Titan," could suspend a load of 27 tons out to a distance of 26½ feet from the outer point of support, and move it transversely a distance of 12 feet. The crane was balanced at the tail end by the engine and a tank, and it was further secured by chains fastened to lewises in the finished work. A traveller and crab, worked by the engine, moved the suspended block at the other end of the crane. The crane ran on eight wheels, and could be moved backwards and forwards along the breakwater by the same engine. The staging of the crane was so constructed that the engine could run between its supports, and bring the block underneath the crab.

Precisely the same system is being employed for laying blocks of 33 tons in the superstructure of the Colombo breakwater; and a larger Titan of the same type is in use at the Mormugao breakwater for depositing blocks of 37 tons. These cranes, however, are not designed to turn round; but a smaller revolving steam crane, capable of lifting 15-ton blocks (Plate 8, Fig. 7), has been designed by Sir J. Coode for the Port Alfred Harbour Works, which, like the previous block-setting cranes, has been made by Messrs. Stothert and Pitt of Bath.

A somewhat similar crane has been constructed for laying the wave-breaker 20-ton blocks at Mormugao, as the Titan cannot stretch sideways to deposit them.

A balance steam crane, which combines the power and overhang of the Titan together with the revolving motion

[1] 'Minutes of Proceedings, Inst. C. E.' vol. xliii. p. 1.

of the smaller cranes, has been designed by Mr. Messent for the Tynemouth northern breakwater. (Plate 8, Fig. 8.) This "Goliath," as it has been designated, has an overhang of 75 feet from the outside bearing wheels to the centre of the load, and is capable of lifting 40 tons. The "Goliath" resembles in principle its comparatively small predecessor, the Alderney "Samson," but with every part greatly magnified, strengthened, and improved, together with the introduction of steam power. This machine has been only recently completed, but it appears likely to satisfy every requirement of a block-setting crane. At the present time (1884) the old and the new systems may be seen side by side at Tynemouth; for whilst the Goliath is being used for the extension of the north breakwater, the outer end of the south breakwater is surmounted by staging. It is, however, probable that when the success of the new crane is assured, another one will be provided for the continuation of the southern breakwater. The Goliath spans the width of the breakwater, and whilst one set of wheels runs on rails placed on the coping of the parapet of the promenade, the other set runs on rails near the harbour edge of the quay level, the difference in level being adjusted by a framing placed on the wheels on the harbour side, and there is space underneath the body of the crane for wagons to run freely along the quay level.

Comparison between Staging and Cranes. The great advantage possessed by block-setting cranes over staging for building a breakwater is, that whilst staging must be left exposed to storms, the cranes can be run back into shelter on the approach of bad weather. Also, cranes, though costly in the first instance, dispense with the great labour, cost, and delay of erecting staging in the sea. It is true that cranes have not wholly escaped injury from the sea, as some of the Ymuiden breakwater cranes were washed away, till the expedient was resorted to of running them to the shore every night, and the Madras Titans perished with the superstructure of the outer arms. In the one case, however, the loss was due to neglect of precautionary measures, and in the latter instance to too great a confidence in the security of the superstructure. Moreover, block-setting with cranes has proved an expeditious, as well as a cheap method, as will be seen in Chapter XV, in the accounts of the Manora and Madras breakwaters. The cranes also enable a short time of fine weather to be utilised at periods when the probable advent of stormy weather would preclude the extension of staging. Lastly, cranes have been made powerful enough to lift weights with ease which have never been entrusted to staging, and the size of blocks in sea works is an important element in successful construction.

Sloping-Block System of Construction. Two defects exist in the ordinary system of building and bonding a superstructure: namely, the long length of stepping back required at the termination of the work of each length, necessitating temporary protection, and the dislocation which occurs from unequal settlement on the mound. To obviate these defects, Mr. Parkes devised the system of dividing the breakwater into a series of sections, each free to settle independently of the others without any disruption, and each massive enough to withstand the blows of the waves. This system was adopted in the construction of the Manora superstructure. (Plate 8, Figs. 1 and 2.) The superstructure is divided into a series of transverse sections, each 4 feet 6 inches thick, composed of six blocks in two symmetrical rows of three blocks, each placed one on the other, the blocks being each 8 feet high and 12 feet wide. (Plate 7, Fig. 19.) The blocks are laid with an inclination of 76° to the horizontal, so that, whilst leaving each section free to settle on the rubble mound, there might be no tendency to tip forward during construction, and that each block might be more easily slid into position. A similar principle was adopted about the same time at the extension of the Kustendjie breakwater, and is now being employed at Colombo and Mormugao. The incline, however, of the sections has been made 47¾° at Kustendjie, and 70° at Mormugao. (Fig. 7.) The Madras superstructure resembles very

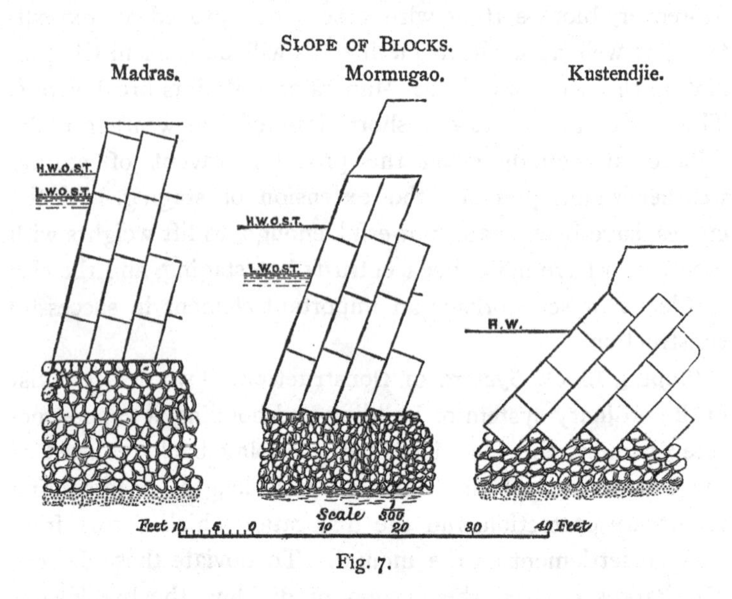

Fig. 7.

closely that of Manora (Plate 7, Fig. 17), and it has been constructed in precisely the same way. The plan of inclining the blocks is a development of a method adopted previously by Mr. Walker and others, of sometimes laying the dry masonry of breakwaters at an inclination of about 60°, for the special purpose of enabling the stones to find their own bearing without dislocating the structure.

The blocks in the Manora breakwater were laid at first without any connection whatever between them, so that the superstructure practically was composed of two independent walls, placed side by side, consisting of three tiers of blocks each, the lower tiers being kept in place by the weight of the blocks above them, but the top tier being unsupported except by their own weight. It seems to have been considered that the weight of these blocks (27 tons), opposing a side face to the sea of only 36 square feet, would ensure their stability. Before the breakwater had proceeded far, it was found that the central longitudinal joint tended to open from the effects of unequal settlement on the mound, some of the top blocks on the harbour side were displaced by the waves as the sea rose over the breakwater and, descending on the central joint and the inner top blocks, gradually forced some of them over into the harbour, and the corresponding outer top blocks were slightly driven inwards by the waves. The insecurity of the top blocks was remedied by inserting a stone joggle between the top block and the one below it.

At Madras, the blocks in each vertical row were connected together by a tenon and mortise, formed at the top and bottom respectively of each block, taking the place of the stone joggle of the Manora superstructure. No attempt, however, was made to connect together the sea and harbour rows of blocks in each section, so that the two walls remained separate as before. (Plate 7, Fig. 17.) The same action took place on the top row of blocks at Madras as had previously occurred at Kurrachee; and the greater force of the waves at Madras overcame the resistance offered by the tenon, and displaced the harbour blocks. The design proposed for reconstructing the outer arms of the Madras breakwaters will remove the disconnection between the sea and harbour walls by cramping the top blocks of each section together, and by a solid capping of concrete all along the top. (Plate 7, Fig. 18.)

The superstructure of the Mormugao breakwater has been specially designed to secure the advantages, and avoid the defects of the Madras superstructure. (Plate 7, Fig. 20.) The sloping system, the division into sections, 4½ feet thick, the tenons and mortises, and the laying of the blocks with a Titan are retained; but the slope is made 70°, instead of 76°, to ensure more perfect stability. (Fig. 7.) The blocks are somewhat larger than those at Madras, varying in size from 37 tons for the larger blocks (17 ft. × 8 ft. × 4½ ft.) to 36½ tons in the truncated bottom blocks, and 28½ tons for the shorter blocks (13 ft. × 8 ft. × 4½ ft.); their method of suspension is shown in Fig. 8. The bottom course consists of two similar blocks, each 17 feet long; but a short and a long block are placed together on each side alternately in the other courses, so as to break bond and to do away with a vertical longitudinal joint. The connection, thus provided between all the blocks of the same section, is further ensured by cramping together each pair of top blocks with iron bars, 2 inches square, whose ends are turned down into the blocks and fixed by oak wedges. Moreover, in addition to the tenon and mortise, the top block is secured to the block below it by two dowels, each composed of a bundle of four double-headed rails, 7 feet long, penetrating 3½ feet into each block; and a concrete joggle of the same length is inserted in the central joint between the four upper blocks. (Fig. 9.) Both dowels and joggles are secured, when in place, by grouting; and then the orifices, by which they were inserted, are filled up with liquid concrete. The upper blocks are eventually to be further connected and secured by a capping of concrete-in-mass, laid all over the top of the superstructure after the settlement on the rubble mound has ceased. This mass of concrete is to be raised higher on the sea side, both forming a parapet, and affording additional weight and strength to the exposed side. These various precautionary measures seem likely to bring the sloping-block system to a state of perfection, bonding together every block of each separate section, whilst leaving it free to settle independently on the mound till it has reached its final position, when all the

SUSPENDED BLOCK.

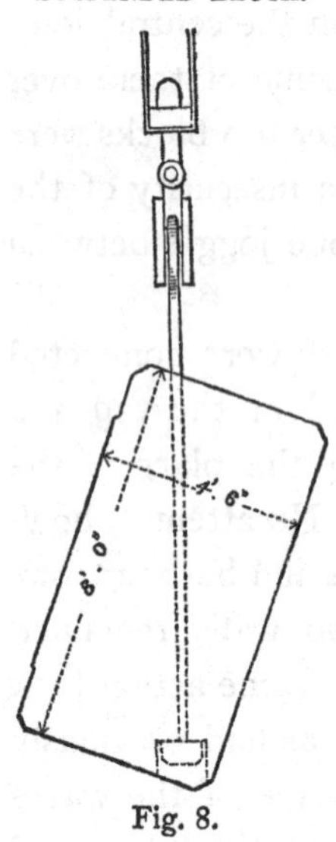

Fig. 8.

sections will be connected into a compact superstructure by the concrete capping.

CONNECTION OF TOP BLOCKS AT MORMUGAO.
Elevation and Plan.

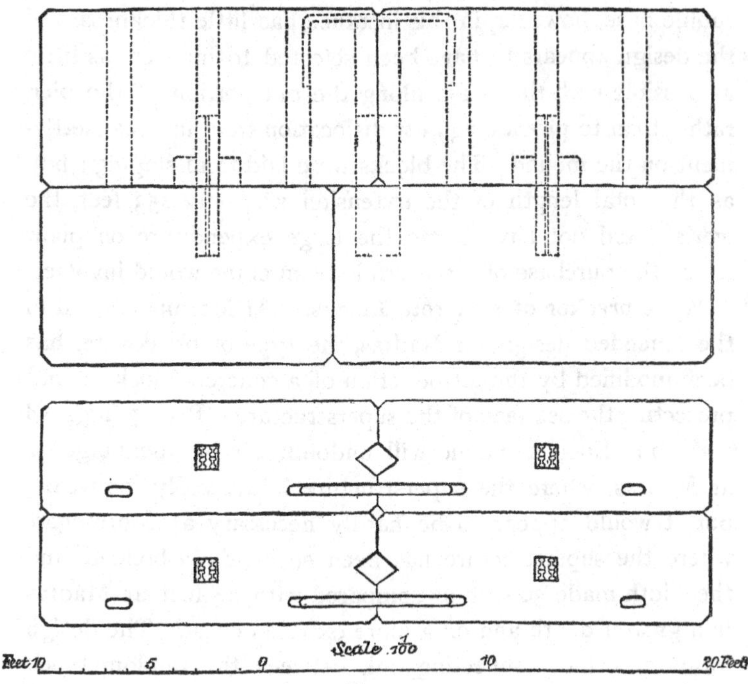

Fig. 9.

The sloping-block system appears to have been adopted by Mr. Liddell at Kustendjie simultaneously with its inauguration at Kurrachee, but quite independently. In fact, the design at Kustendjie differs from the Manora superstructure in the much flatter slope given to the blocks, in their being carried right across the superstructure, and in having a thick capping of concrete-in-mass above the water-line. (Fig. 7.) The blocks, weighing about 30 tons, were made 6 feet high and 5 feet thick, but diminishing in length from 18 feet at the base to about 13 feet at the top, thus forming a batter on each face of the superstructure[1]. Each tier consists of four blocks resting upon a rubble base at a depth of 16 feet. The flatter angle at which the blocks are laid gives greater stability to the individual blocks, but it would be unfavourable as regards the independent settlement of each tier. The rubble base, however, in this instance has little thickness, and the design appears to have been selected to interfere as little as possible with the traffic along the old portion of the pier, rather than to provide against dislocation from unequal settlement on the mound. The blocks were laid from staging; but, as the total length of the extension was only 253 feet, the work would not have borne the large expenditure on plant which the purchase of a powerful steam crane would involve.

Wave-breaker of Concrete Blocks. At Mormugao, and in the amended design for Madras, the type of breakwater has been modified by the introduction of a concrete-block mound protecting the sea face of the superstructure. (Plate 7, Figs. 18 and 20.) Such a mound will undoubtedly be advantageous at Madras, where the superstructure is avowedly too weak; but it would appear to be hardly necessary at Mormugao, where the superstructure has been so carefully bonded, and the width made 30 feet, as compared with 24 feet at Madras in a greater depth and on a more exposed coast. The design forms a sort of connecting link between the random block mound and the mixed type under consideration, and sets aside the theory of the advantage of a vertical wall in preventing waves of oscillation being converted into waves of translation. The object, indeed, of the concrete-block mound is to break the wave before it reaches the superstructure, and thus diminish the shock; but it should be possible to build a superstructure able to stand unsupported, as the experience of Manora, Colombo, and other breakwaters indicate.

Remarks on Composite Breakwaters. The mixed type described in this chapter forms the largest and most varied

[1] 'Minutes of Proceedings, Inst. C. E.' vol. xxxix. p. 142.

class of breakwaters. It is adaptable to almost any depth and any nature of foundations, and can be varied considerably according to the supply of materials available. A glance at Plate 7 shows the numerous modifications of which it is susceptible. The base varies from the simple rubble mound of Portland, Holyhead, and Alderney, to the concrete-block mound of Leghorn and St. Jean-de-Luz; whilst the superstructure varies from the rudimentary type at Trieste, Oran, and Marseilles, up to the fully developed form at Boulogne, Manora, and Colombo. Where the superstructure is founded at low-water level, the protection on the sea slope varies from almost nothing at St. Catherine's, to a thick layer of blocks at Ile Rousse and Marseilles; and where the superstructure is below low water, its defence varies from nothing at Colombo and Odessa, to a large mound of concrete blocks at Mormugao. This class, indeed, includes instances of every possible modification, from the Oran type, which approaches most nearly to the simple rubble and concrete mound breakwater arranged in the most perfect manner, to the closest approximation to the upright-wall system at Colombo and Odessa.

Though the rubble and concrete mound, and the upright wall, may occasionally prove the most suitable forms under special conditions, it is probable that, in the future as in the past, the mixed system will have the most extended application; and, of all the varieties of the type, the best appears to be a superstructure founded some 20 feet below low water upon a simple rubble base, and formed with large concrete blocks laid by overhanging cranes upon the sloping-block principle, securely connected vertically and transversely, and capped with concrete-in-mass after settlement has ceased.

CHAPTER VII.

UPRIGHT-WALL BREAKWATERS.

Use and Advantages of the Upright-Wall type of Breakwater. Limits of its Application. Forms: Two Walls with intermediate Filling; Timber Casing and Rubble; Solid Wall. Methods of Construction: Blocks deposited from Staging; Blocks deposited by Cranes; Concrete in Bags; Concrete-in-Mass within Frames. Comparison of Methods. General Remarks on Breakwater Construction. Concrete in Breakwaters.

Use and Advantages of Upright-Wall Breakwaters. Where materials are scarce, the depth moderate, and the bottom hard, and where it is desired to use the breakwater as a pier, it has occasionally been found expedient to erect the breakwater in a solid mass from the bottom, dispensing altogether with a mound, and giving the breakwater the form of a thick and almost vertical wall. (Plate 6, Figs. 15–21.) This form has been naturally adopted at several small tidal harbours where the foundations are above low-tide level, as at Scarborough, or so little below it that there is not much difficulty in laying the bottom courses, as at Ramsgate and Whitehaven. (Plate 6, Fig. 20.) It is important in such cases that the base should not take up much more space than the top which is used as a quay, and that vessels should be able to get close alongside the breakwater.

When the bottom is several feet below low water, it entails both the cost of building under water, and also the weakness of uncemented blocks. These difficulties, however, have been overcome in the case of superstructures founded upon a rubble mound as much as 20 feet below low water, as described in the previous chapter; and a firm bottom is a better foundation than a yielding base. The weakest point of an upright wall founded below water is close to the level of low water, where the uncemented blocks are liable to be forced out by the waves compressing the air through the joints; but it shares this defect with the superstructure in the second class of mixed breakwaters just considered, whilst it is free from a base tending to settle and sometimes requiring maintenance. Moreover, though it cannot be admitted that waves coming against an upright wall are merely oscillating waves, and inflict no blow upon the structure, as urged by the two ardent partisans of the upright-wall system, yet undoubtedly a shelving rubble base tends to break the wave as it approaches the superstructure which, when situated in a certain depth of water, is liable to receive the more concentrated shock of the breaking wave. This constitutes the main advantage of the upright wall over the best forms of the mixed type. Another merit of the system is the comparatively small amount of material required; but this is balanced by the greater cost of erecting a solid structure under water.

Limits of Application of the Upright-Wall System. The enhanced expense would preclude the erection of an upright wall in deep water. The greatest depth in which an upright wall has been founded is in 40 feet of water at Dover; and the great cost in this instance does not furnish an inducement for imitation elsewhere. (Plate 4, Fig. 2, and Plate 6, Fig. 21.)

A chalk bottom, as at Dover and Newhaven, is specially favourable for the foundations of a wall; but shifting sand forms an unsuitable site. Accordingly a concrete mound was adopted at Port Said, though materials were scarce; but the difficulty has been overcome at Ymuiden by a thin layer of rubble which, sinking a little into the sand, forms a broad and unassailable foundation upon which an upright wall has been erected. (Plate 6, Fig. 19.) The objection to a hard foundation is the necessity of levelling it before placing blocks upon it, which would prove a costly and slow operation in very uneven and hard rock.

Forms of Upright-Wall Breakwaters. Where this type of breakwater is situated in shallow water, and is intended to serve a quay, the plan of a sea and harbour wall with filling between is very naturally adopted, as exemplified by the Whitehaven section. (Plate 6, Fig. 20.) This method, though suitable enough for small breakwaters situated little beyond low-water mark, and therefore not much exposed, would be quite inadequate in deeper water.

A more frail design was, however, proposed many years ago, consisting of rubble stone enclosed within a timber casing. The system had been carried out at the Port of Blyth for some jetties extending to about low-water mark; and the suggestion was made that it might be advantageously adopted for breakwaters in deep water. Similar structures were subsequently tried in more exposed situations, and failed, as might have been anticipated. The Americans have indeed constructed large breakwaters of cribwork (Plate 5, Figs. 5, 11 and 12); but the timber framing is made very strong, the breakwaters are not situated in deep water, and only on the comparatively little exposed shores of their large inland lakes. Moreover, where timber is so cheap, and shelter urgently needed for a rapidly-increasing trade, it may be expedient to construct breakwaters of limited durability which can be quickly completed.

We have already seen, in the previous chapter, that practical experience points to the expediency of omitting loose hearting in the construction of superstructures which are much exposed; and the same applies to the upright-wall breakwater which, except in sheltered situations, should be made solid throughout. This principle, it will be observed, has been followed in all the sections of this type given in Plate 6, with the exception of the special case of

Whitehaven. They are very similar in form, with a slight straight batter on both faces, and with a parapet wall on the sea side except in two instances. (Plate 6, Figs. 15 to 19, and 21.) The breakwaters, however, at Ymuiden are protected by a wave-breaker of concrete blocks like the composite breakwater at Mormugao.

METHODS OF CONSTRUCTION.

Although the forms of the upright-wall breakwaters are fairly uniform, their methods of construction present some important differences.

Blocks deposited from Staging. The Dover Pier is the oldest and most important instance of the upright-wall system. It was commenced at the same period that the mixed breakwaters were started at Holyhead and Alderney, and the rubble mound breakwater at Portland. Staging was at that time the method by which structures in the sea were built, and it was naturally adopted at Dover. The bottom, consisting of chalk, was dressed to a level foundation by the aid of diving-bells; and then the blocks, consisting of granite ashlar on the face and concrete blocks inside, were lowered from the staging. The employment of a diving-bell is a somewhat tedious process; and as a considerable portion of the structure had to be built below low water, reaching out to a depth of about 45 feet, and the staging was often damaged at the outset by storms, and occasionally afterwards, the progress was slow, and the work was also costly. The difficulties, delays, and cost of the work were, however, due to the depth of water, and not to any defects in the system of construction. The breakwater has been well and compactly built by this method; and, though a portion of the parapet was injured by a storm in 1877, the work generally has proved perfectly stable. (Plate 6, Fig. 21.)

The concrete blocks of the Aberdeen South Breakwater below water were deposited from staging by 25-ton steam cranes, and put in place with the help of divers, in the same manner as the superstructures of some of the mixed type of breakwaters were constructed, but with the advantage of a more solid foundation. (Plate 8, Figs. 3 and 4.) The staging and cranes were also used for placing the concrete in mass from low-water level to the top of the breakwater, and for depositing the concrete in bags to protect the sea toe of the breakwater. (Plate 6, Fig. 15.) This work, being only in a maximum depth of about 20 feet, was much more rapidly executed than the Dover breakwater; and the staging being strongly constructed was able to withstand the action of the sea in that moderate depth without injury.

Blocks deposited by Overhanging Cranes. There is no instance on record of an upright breakwater having hitherto been constructed on the sloping block principle, though that system would be as applicable to the simple vertical wall as to superstructures of the mixed type of breakwaters. Overhanging cranes have however been employed, as previously mentioned (p. 116), in the construction of the Ymuiden breakwaters at the entrance to the Amsterdam Canal, owing to the instability of the staging on the easily scoured sandy bottom. The concrete blocks, from 6 to 12 tons in weight, were laid with the ordinary bond; and the work was expeditiously and successfully completed. Two Titans having been swept off the northern breakwater by the sea, a steam travelling crane was substituted, which could be brought back to the shore every night and its safety thus ensured.

Concrete in Bags dropped from Skips, and from Hopper Barges. A considerable portion of the delay experienced in the construction of the Dover breakwater was due to the great care and time necessary to form a perfectly level foundation for the lowest course of blocks to rest upon.

As the levelling of the chalk bottom at Dover proved

a tedious operation, it would have been a considerably slower and more difficult work to level the granite rock upon which portions of the Aberdeen breakwaters rest. Accordingly Mr. W. Dyce Cay fitted the breakwater to suit the rock, instead of shaping the rock to the section of the breakwater. (Plate 6, Fig. 15.) This he accomplished by depositing bags of liquid concrete from skips upon the irregular bottom, which was previously cleared by divers of loose stones and sand. The bags leave the concrete free to adapt itself to the uneven bottom; and the canvas covering protects the concrete from the wash of the sea. The skips used at the South breakwater, holding from 5 to 16 tons of concrete, protected and controlled the bags in their passage through the water, and released them, when close over their proper position, by means of a hinged movable bottom opened by a trigger worked from above[1]. By making the bag rather larger than the skip, sufficient freedom is given to the concrete inside to adjust itself to the irregularities of the bottom. The depositing of the bags was guided by divers, who also levelled the foundation layer by inserting small bags in any holes that they met with, and rammed down or cut off the freshly laid concrete where the bags rose too high.

The foundations for the concrete blocks forming the lower part of the South breakwater at Aberdeen were laid so expeditiously along the shore portion on the granite rock, that the same system of bags was adopted for the remainder of the breakwater resting on a softer base composed of boulders and gravel, and clay mixed with gravel. The base of this portion of the structure was protected from undermining by a concrete apron formed of 100-ton bags, tipped over the side, from a large wooden box, in order to form a toe to the sea face of the breakwater. (Plate 6, Fig. 15, and Plate 8, Figs. 3 and 4.) The necessity for this protection has been rendered evident from the undermining, and subsequent

[1] 'Minutes of Proceedings, Inst. C. E.' vol. xxxix. p. 127.

removal by the sea, of some of the concrete blocks from under the concrete-in-mass along a short portion of the breakwater, just beyond the rock foundation, where the apron had not been laid.

Concrete blocks, weighing from 7½ to 25 tons, were laid upon the concrete-bag foundations from staging, and bonded in the ordinary manner in horizontal courses 4 feet thick. The blocks were carried up to 1 foot above low-water neap tides; and the remaining 18 feet in height of breakwater consists of concrete-in-mass deposited within frames in long lengths. The unconnected blocks are kept in place by the weight of the structure above them. If, however, settlement occurred, some of the blocks might be relieved of the superincumbent weight, and be liable to be displaced by the waves near the level of low water, so that though this system has succeeded at Dover and Aberdeen where the bottom is firm, similar structures on a yielding foundation might sustain damage from the sea. This consideration led Mr. Cay to an extension of his system of bags in the construction of the northern breakwater at Aberdeen.

The North Pier of Aberdeen Harbour, up to between 2 and 3 feet above low water of ordinary spring tides, is composed entirely of 50-ton bags dropped into place from a hopper barge. (Plate 6, Fig. 16.) The bottom along the site of the work consisted of a layer of shifting sand, from 3 to 5 feet thick, overlying boulder clay. The sand was first removed by a bucket dredger, and the bottom course of bags was immediately deposited on the dredged area, and the next course of bags was laid as soon as possible on the others, thus raising the foundations of the breakwater 7 feet, and placing them above the limit of the deposit of sand.

The bags were deposited from an iron hopper barge (Fig. 10), 55 feet long, 20 feet broad, and 8 feet deep, containing a hopper well having an average length of 24 feet, a breadth of 6 feet, and a depth of 5 feet 7 inches. This well held 50 tons of concrete (16 cubic feet weighing 1 ton), and it enlarged a little towards the bottom to facilitate the

HOPPER BARGE FOR DEPOSITING CONCRETE BAGS.
Aberdeen Harbour.

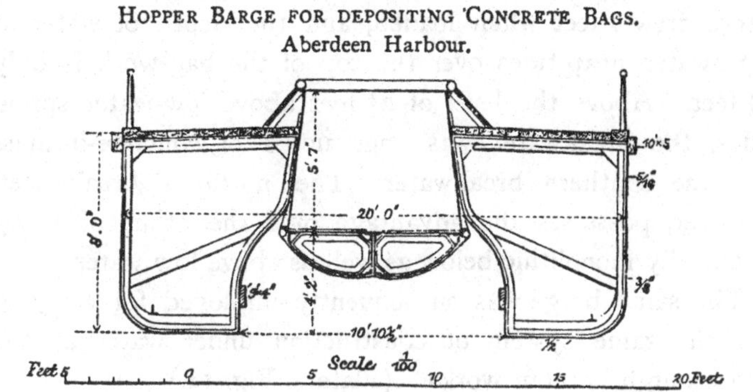

Fig. 10.

discharge of the bag. The bottom of the well was closed by two tubular flaps, which, when opened, formed a continuation of the sides of the well and did not project below the bottom of the barge. These flaps were kept shut by supports at their ends, which could be simultaneously released by the withdrawal of a pin for discharging the bag[1].

The jute cloth forming the bag was fitted into the well of the hopper barge, and as soon as the bag had been filled with freshly-made concrete the barge was towed into position, and during the journey the bag was closed by a flap of cloth sewn over its mouth, and directly the barge was over the required spot, the bag was discharged by opening the flap doors at the bottom of the hopper.

The two foundation courses of the northern breakwater at Aberdeen were made broader than the general section, forming an apron beyond each face of the breakwater in order to prevent undermining from the waves; and the bags in these courses were laid longitudinally, the centre of the upper bags being laid over the joints between the lower

[1] A description of the barge, and of the depositing of the concrete bags, given me by Mr. W. Dyce Cay, will be found in Appendix II.

bags. (Plate 6, Fig. 16.) The remaining courses of bags were laid transversely to a narrower width; and some difficulty was experienced in depositing the top course, as the barge drew 7 feet when loaded, and the depth of water at high-water neap tides over the top of the bag-work is only $6\frac{3}{4}$ feet. Above the level of $2\frac{1}{2}$ feet above low-water spring tides, the breakwater was constructed of concrete-in-mass like the southern breakwater. The northern breakwater, however, possesses the advantage over the other of being practically monolithic below as well as above low water.

The same barge was subsequently employed for carrying out the same system of construction under water at the Fraserburgh harbour works. (Plate 4, Fig. 14.)

The breakwater at Newhaven is being constructed on precisely the same principle (Plate 6, Fig. 17), but the hopper barge has been made large enough to carry a bag of concrete weighing 104 tons. (Fig. 11.) This increase in size has

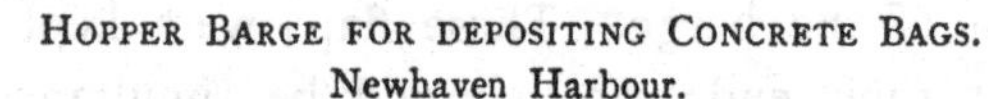
HOPPER BARGE FOR DEPOSITING CONCRETE BAGS.
Newhaven Harbour.

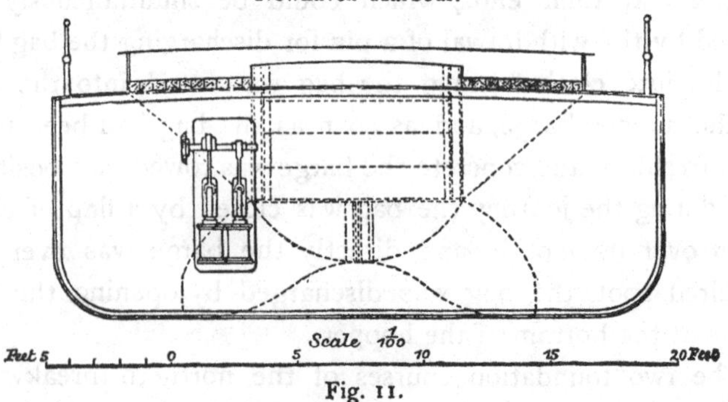

Fig. 11.

been adopted in order that all the bags may be laid transversely, and reach across the full width of the breakwater. The bottom in this case is chalk, thus resembling Dover; but the breakwater will only extend into 18 feet of water instead of 45 feet.

Concrete-in-Mass within Frames. Occasionally concrete-in-mass has been deposited within frames below low water. This plan was successfully adopted many years ago for forming quays at Marseilles. About the same period, in 1849, a breakwater designed on the mixed system was commenced at Fiume in Austria, whose superstructure was built with concrete-in-mass founded on a rubble mound at a depth of 22¾ feet below the water-level. The harbour is situated in a somewhat sheltered bay on the Adriatic Sea, but it is exposed to the Sirocco from the south-east; and the breakwater extends into a depth of 71½ feet. Large bottomless frames, 31 feet long, 23½ feet wide, and 24 feet high, were deposited on the top of the mound, and filled with concrete which was sufficiently consolidated in from 15 to 20 days for the frames to be removed and used for another length of superstructure. The mass thus formed contained 640 cubic yards, weighing about 1150 tons. The series of masses thus constructed had intervals between each of about 20 feet, which were subsequently closed by boards at both ends, and filled up with concrete. The whole superstructure thus forms a monolith of concrete which has remained intact under the action of the waves.

The concrete, when deposited under water, is generally passed through a canvas tube, or discharged from a skip with a movable bottom. The canvas shoot admits of a more rapid rate of deposit; but if the concrete consists of particles of various dimensions, as is commonly the case, the bigger stones fall first to the bottom and lie together in a heap, which deranges the homogeneity of the concrete. The lowering of the concrete in skips is a more tedious process, but produces more satisfactory results.

This system was carried out in the construction of a portion of the Rosslare Harbour works near Wexford, commenced in 1872. The landwards part of the structure forming the harbour consists of piers carrying iron girders for a line of rails, serving as an approach road to the breakwater. The bottom consists of rock, with sand, clay, and boulders, overlying it in places. The piers have been built of concrete-in-mass deposited by skips with movable bottoms. Wooden casing was erected round the site of each pier, and was lined inside with canvas, and the concrete was deposited within the casing, and brought up to the level of low water. Two pyramidal frames were then erected upon this concrete base and filled with concrete, forming the pillars for supporting the longitudinal girders of the viaduct. Each pier, accordingly, forms a concrete monolith, whose base is about 22 feet by 11 feet, resting upon the hard bottom. The viaduct extends into a depth of 16 feet at low water, and the deepest of the piers weighs about 220 tons. Two of the earlier piers were cracked during a storm by unequal settlement, owing to the diver having neglected to remove entirely the loose and soft material at the base; but these piers were rebuilt, and the remainder proved quite satisfactory.

The breakwater and quays erected at Buckie Harbour in 1877–78 were constructed in a similar manner upon a rock foundation. (Plate 6, Fig. 18.) Timber casing was placed on, and fitted to the rock, and lined with jute cloth. The concrete in this instance was tipped straight into the casing from ballast wagons, the greatest depth of water in which it was deposited being from 7 to 10 feet.

Large Blocks deposited by Floating Shears. A method by which huge blocks, weighing 350 tons, are built on shore, and are then lifted, conveyed, and deposited in place by floating shears, has been adopted by Mr. Stoney in raising the foundations of the Dublin north quay and basin walls to the level of low water. (Plate 8, Figs. 11 and 12.) Having carried out this system with very satisfactory results in this comparatively sheltered site, he has proposed its extension to the construction of breakwaters, for enabling upright walls to be built expeditiously with blocks of such a large size as not to be liable to be disturbed by the sea. As, however, this method has not hitherto been applied to structures in the sea, with the exception of a beacon which Mr. Stoney erected in Dublin Bay, its further description must be reserved for Chapter XX, dealing with quay walls.

Comparison of the foregoing Methods of Construction. The employment of staging for the construction of breakwaters possesses the advantage of enabling several portions of a long length to be carried on simultaneously. Also when the divers are at some distance below the water, their operations are not much affected by the waves; and as the cross-pieces of the staging have to be placed out of the reach of the waves, the men can work on it at times when they could not stay on the breakwater. The first of these advantages has been urged in favour of staging as compared with overhanging cranes, which can only work at the end of the breakwater; and the second, as compared with barges which are subject to any motion of the sea. It has been found, however, that although only one overhanging steam crane can be employed for laying blocks, instead of an indefinite length of staging, the method of building with these cranes and sloping blocks is more rapid than the other. Moreover, overhanging cranes possess the great advantage of being capable of being brought back to shelter on the approach of a storm.

Barges cannot deposit concrete bags with sufficient accuracy in rough weather, yet their rate of progress has proved satisfactory at Aberdeen, the extension of the North Pier, 1000 feet long, having been constructed in two years in a depth of 17 feet at low water. The great merits of the system of concrete bags are, that levelling an uneven bottom is dispensed with, and that a monolithic structure can be formed. The bag system of foundations was combined with the ordinary block-setting from staging at the Aberdeen south breakwater, but it was subsequently superseded by the regular concrete-bag system in the later work at this harbour. The bag system does not make as smooth or as regular a wall, below low water, as the block system; it might require a larger proportion of cement, and it does not furnish an occupation for the staff of workmen during the winter as block-making does. Rough weather also prevents the barges working, and is liable to disturb slightly the concrete on the surface of newly deposited bags near low-water level. These, however, are the only objections to this system which possesses such great counterbalancing advantages; and it will probably be extensively adopted in the future, where the bottom is firm, and the depth of water moderate. It has not hitherto been tried on a yielding foundation, and does not appear suited for irregular settlement.

A combination of concrete-bag foundations with the sloping-block system seems well adapted for sites with a rocky uneven bottom, for whilst dispensing with the delay of preparing a bed for the blocks, a greater rapidity of execution would be ensured, especially in exposed situations.

Sloping blocks of large size are specially suitable for yielding foundations. It was at first supposed that block-setting with a Titan at the extremity of a breakwater would be comparatively slow; and on this account Mr. Parkes adopted the highest safe level of 12 feet below low water for the foundations of the Manora superstructure. The experience, however, at Kurrachee demonstrated the rapidity of the system, and, accordingly, Mr. Parkes placed the bottom of the Madras superstructure in a depth of 22 feet in order to accelerate the rate of construction, and to save material in the rubble base. When the upper courses of the sloping blocks are efficiently connected together in each section, as at Mormugao (Fig. 9, p. 123), and capped with solid concrete at the top, this system ought to compare very favourably with other methods for the simple upright wall, as well as for the composite breakwaters.

The method of depositing concrete in frames below water requires great care, and necessitates the use of a large proportion of Portland cement. For large works, it may be considered to have been superseded by concrete in bags; but for small works, especially where not much exposed, it possesses the merit of requiring little plant, and of forming a smooth monolithic mass. It was specially suitable for the Rosslare piers and the Buckie breakwater, and might be advantageously applied to many similar works where stone is scarce and the foundations are firm.

The plan of depositing huge blocks of rubble masonry and concrete by means of large floating shears, with great accuracy and upon a levelled foundation, has yet to be tested in exposed situations. To lay these large masses with nicety would require much calmer weather than is needed for depositing concrete bags from hopper barges, and therefore only a limited number of working days could be reckoned on during the year. Moreover the necessary special plant is expensive, and a place of shelter would have to be provided for the large barges carrying the shears. The system would, therefore, be still more inapplicable to small works than hopper barges for depositing concrete bags; and the rate of progress would have little prospect of being equal to that of the sloping-block system, or even the concrete-bag foundations. A breakwater, however, constructed of these blocks side by side would possess the elements of great stability without the necessity of frequent works of maintenance.

General Remarks on Breakwater Construction. Each of the three types of breakwaters, described in this and the two preceding chapters, possesses its own special advantages, and none can be regarded as universally applicable to the exclusion of the other two. Many years ago, when less experience had been gained in the construction of breakwaters, and the necessity for better harbour accommodation was beginning to be keenly felt, both the rubble mound and the upright wall had their respective partisans; each held the system they advocated to be the only advisable one, whilst both were opposed to the composite system which at that time found few adherents. At the present day, it must be acknowledged that each of the three systems has its respective merits and defects; and it is only by weighing carefully the local conditions of each site that a proper selection can be made. The rubble mound and the upright wall have only a limited application; whilst the mixed system of mound and superstructure is more easily adapted to a variety of conditions and sites. (Plates 6 and 7.)

The best type of mound breakwater is that in which the size of the materials is suited to the difference of exposure of the several parts. Alexandria, Algiers, and the last design of Port Said (Plate 6, Figs. 9, 12, and 14), are the best examples of suitable arrangement with least expenditure of material; though on the lower portion of the sea slope at Algiers, and on the harbour slope at Port Said, concrete blocks might have been dispensed with if small rubble had been readily procurable.

There are so many varieties of composite breakwaters, each suited to somewhat different conditions, that it would be difficult to select the most perfect types. Of well-arranged mounds with superstructures fitted for quays, Marseilles and Trieste are prominent examples. (Plate 7, Figs. 7 and 15.) When a breakwater is intended to serve as a quay, or as a pier of access, it is necessary to protect it by raising the sea slope to high-water level, as well as by a superstructure, like at Holyhead and Portland (Plate 7, Figs 2 and 11), or by a wave-breaker of concrete blocks so as to break the wave thoroughly before it reaches the superstructure, as at Mormugao. (Plate 7, Fig. 20.) If, however, a sheltered quay is not requisite, then the type of the Manora and Colombo breakwaters (Plate 7, Figs. 19 and 23) is preferable, as a stable breakwater can be thus constructed with less material and without entailing maintenance. The concrete-block wave-breaker, adopted at Ymuiden (Plate 6, Fig. 19) and Mormugao, strengthens the structure; but it is probable that a carefully bonded and rather wider superstructure could be made equally strong with less material, on a less extended base, and with less prospect of expenditure in maintenance.

Amongst upright-wall breakwaters, Dover stands out prominently from its size, depth, and importance (Plate 6, Fig. 21); but if an extension of the existing breakwater was decided upon, it would be wise to follow to some extent the method of bag foundations so successfully executed at Aberdeen and Newhaven. (Plate 6, Figs. 15, 16, and 17.)

Alexandria, Madras, and Aberdeen are instances, in their different types, of great rapidity of execution, by slinging blocks from barges, by depositing sloping blocks with Titans, and by the use of concrete bags. Each of these methods is worthy of imitation, and each appears the best that has been hitherto devised for the three distinct types of breakwaters.

Concrete in Breakwaters. One of the most noticeable features in modern practice, relating to the construction of breakwaters, is the large and increasing employment of concrete. It is believed that the Romans used concrete for this purpose; but in the earliest breakwaters of recent times, namely, Cherbourg and Plymouth, rubble stone alone was employed. M. Poirel, however, introduced concrete at Algiers in 1834, and since that time its use has been gradually extended, till, at the present time, few breakwaters are constructed entirely without it. This is mainly due to the great improvements in the manufacture of Portland cement, enabling the hardness and permanence of the blocks to be ensured. It is also due to the facility with which large and square blocks are manufactured, and the ease with which holes are left for the lifting bolts (Fig. 9, p. 123), in moulding the blocks, by inserting blocks and laths of wood within the frames before the concrete is put in. The bolts, which are T-shaped, pass through the block (Fig. 8, p. 122); and the block of wood forms a recess within the bottom of the block in which the head of the bolt can turn, and a quarter of a turn enables it to grip the block, whilst another quarter turn releases it. There are two lifting bolts for each block. (Plate 8, Fig. 2.)

Blocks are generally left to harden for a period of from one to three months before they are placed in the work. When, however, concrete is used in bags, or in mass within frames, it has to be deposited whilst it is fresh, and left to set when in place. This difference might appear to give the advantage to blocks; but fortunately Portland cement can set quite well in water, and provided the concrete is protected from the wash of the waves, it seems not to be impaired by having to consolidate under water. The proportions of the materials used for making concrete at various breakwaters is given in Appendix III, together with the cost.

Soft stone is not suitable for breakwaters, and hard stone entails considerable labour in dressing; lewis holes have to be driven for the lifting bolts; and the stone varies in size. In all these respects concrete blocks have an advantage over stone, and concrete can now be manufactured sufficiently hard to be suitable for face-work. In many places also stone can either not be procured at all, or not of adequate size.

These considerations amply account for the very general adoption of concrete in breakwaters in one form or other, whether as a mere protection of the top of the sea slope, as at Cherbourg, Cette, and Boulogne; or forming an integral part of the breakwater, as at Algiers, Alexandria, and Port Said; or composing the entire mound of the mixed breakwaters at Leghorn and St. Jean-de-Luz; or the backing of the Dover Pier; or the superstructures at Madras, Mormugao, and Colombo; or, finally, making up the whole breakwater at Ymuiden, Aberdeen, and Newhaven.

CHAPTER VIII.

JETTY HARBOURS WITH PARALLEL JETTIES.

DIEPPE. BOULOGNE. CALAIS. GRAVELINES. DUNKIRK. NIEUPORT. OSTEND. MALAMOCCO. KURRACHEE. LOWESTOFT.

Dieppe Harbour: Original Condition; Present State; Jetties; Movement of Shingle; Dredging. *Boulogne Harbour:* Description; Jetty Channel. *Calais Harbour:* Early History; Jetties; Sluicing Basin; Entrance Channel. *Gravelines Harbour:* Description. *Dunkirk Harbour:* Site; Development; Jetties and Sluicing Basin; Entrance Channel. *Nieuport Harbour:* Description; Jetty Channel. *Ostend Harbour:* Early History; Jetties and Sluicing Basins; Peculiarities of Site. *Malamocco Harbour:* Site; Jetties; Improvement. *Kurrachee Harbour:* Situation; Improvement Works; Results. *Lowestoft Harbour:* Description; Works. Parallel Jetties at River Mouths. General Remarks.

THERE are two forms of Jetty Harbours, namely, those whose entrance channels are guided and protected by parallel jetties; and those in which the jetties, starting wide apart at the shore, converge towards the entrance. The first type will now be described, and the second will be reserved for the succeeding chapter. Their peculiar features have already been referred to in Chapter IV; and the past and present condition of some of these harbours will now be described, together with the works which have been carried out for improving and maintaining the depth at their entrances. The harbours along the northern coast of France will be first noticed, as they furnish the most characteristic instances of entrance channels with parallel jetties.

DIEPPE HARBOUR.

Origin and Condition. The original port of Dieppe was formed by the outlet channel of the rivers Arques and Bèthune, whose depth was maintained by the tidal waters, which, flowing over a large extent of land up to Arques during the flood, produced a considerable scour at the outlet during the ebb. The shingle, however, washed from the neighbouring cliffs, and borne along eastwards both by the prevalent westerly winds and the stronger run of the flood tide coming up from the ocean, tended continually to block up the port, and drove its channel towards the east. About the thirteenth century the inhabitants made a new eastern channel, and the western channel was blocked up with shingle covering a large area upon which a portion of the town now stands.

Construction of Jetties at Dieppe. After a time an endeavour was made to fix the channel by jetties; but the shingle gradually advanced, and eventually rounded the jetties. The jetties had then to be prolonged, but they were periodically injured by the sea; and at last, in 1616, a portion of the east cliff fell, and choked up the entrance. The channel was thus driven further eastwards to the base of the new line of cliff, in which position it forms the existing entrance to the port. This channel also was protected by jetties formed of rubble stone and timber-work; but, owing to the frequent breaches made by the sea, dressed stone was adopted in 1725; and some years later, when the eastern timber jetty was swept away during a storm, a low stone jetty was constructed in its place to direct the sluicing current.

The jetties have been extended, at various periods, as much as 1640 feet; but the beach has regularly kept pace with these extensions, and has ended by outflanking the western pierhead.

Lanblardie designed a scheme, in 1776, for quite a new channel in the centre of the valley, pointing in the most favourable direction, with a sluicing basin opening in the same line for maintaining it. The sluiceway was commenced in the following year; and in 1787 a protecting mole was constructed, consisting of a timber framing filled with shingle, 40 feet high and 1050 feet in perimeter. This mole, however, was destroyed by the sea six years afterwards; and, on the advice of Perronet, the scheme was abandoned and the former channel maintained.

A low groyne, projecting in front of the eastern jetty, was lengthened several times to direct the sluicing current, but did not avail to arrest the inroad of shingle; so that at last the east jetty had to be extended.

Conditions of the Site of Dieppe Harbour. The sandy beach, seawards of the line of shingle, has a gentle slope; and a depth of 20 feet at low water[1] is only reached at a distance of 2000 feet beyond the extremity of the west jetty. There are no sandbanks near the entrance; and, though there is no shelter in front, the anchorage is good. The only impediment to shipping is the shifting bar of shingle across the entrance. The prevalent winds are the south-west and west, which are also the most stormy. The rise of spring tides is 28 feet.

State of the Port of Dieppe. The existing state of the port, with its jetties and inner works, is shown on the plan[2]. (Plate 1, Fig. 1.) Dieppe Harbour, like most of the other jetty harbours along the coast, comprises the following distinct parts: the entrance channel with its parallel jetties, and the sluicing basin with its sluice-gates for maintaining the depth of the channel; the outer, or tidal harbour, expanding out at the inner extremity of the entrance channel for giving access to the various basins; and the basins or docks which are entered through locks, or entrances, provided with gates for maintaining a high-water level. There are also some minor works, such as groynes for regulating the coast-line, and stilling basins at the sides of the channel

[1] Whenever low water alone is mentioned, it is to be understood as referring to low water of ordinary spring tides.

[2] This plan, and the sections of the jetties and dock wall, were given me by M. Alexandre, the engineer of the port.

for reducing the swell. This port accordingly, like many others, combines both harbour and dock works. The exposed jetties and tidal harbour, together with the accessory sluicing basin, constitute the harbour; whilst the sheltered inner basins form the docks. The harbour portion of the port is the most important and interesting; for the prosperity of the port, and the size of the vessels frequenting it, depend greatly upon the depth at the entrance, which can only be improved by judiciously designed artificial aids, whereas the extension of the docks to meet the growing requirements of trade is comparatively simple.

Jetties at Dieppe. The inner portion of the jetty channel is lined by solid masonry quays, except at one place on the east side where an open jetty of ironwork has been formed, communicating with a stilling basin for reducing the swell following the concave side of the channel. (Plate 1, Fig. 1, at A; and Fig. 3.) The jetties along the outer portion of the channel near the pierheads are open, with stilling basins behind for receiving the entering waves, the older western jetty being made of timber. (Plate 1, Fig. 4.) The outer jetty channel was formerly somewhat irregular in width, having a contraction of 100 feet opposite the centre of the stilling basins; but the eastern jetty has been put back so as to afford a uniform width of 246 feet.

As the travel of shingle is towards the east, the western jetty has alway been extended further out than the other. Till recently it was as much as 187 feet in advance, and it is now being prolonged 160 feet more. (Plate 1, Fig. 2.)

Movement of Shingle at Dieppe. The shingle travelling along the coast passes round the west jetty, and settling along its inner face used to obstruct the channel to a certain extent, except when temporarily removed by the sluicing current. The travel of shingle has been partially arrested, before reaching the west jetty, by a series of groynes constructed along the shore to the west. The supply of shingle in motion has been estimated to amount to 31,000 cubic yards annually, and occasionally it reaches 40,000 cubic yards. This constitutes a formidable obstacle to the maintenance of the entrance; and the new pierhead of the west jetty will only extend a short distance below low-water mark. Fortunately, however, the removal of large quantities of shingle for ballast has greatly relieved the port from the accumulation which used to block it up.

Dredging at the Entrance to Dieppe Harbour. In addition to the rectification of the jetty channel, it is to be deepened by dredging to a depth of 10¼ feet below low water of ordinary spring tides. The new works also in the port, by enlarging the area of the basins and providing a more direct line for the sluicing current, will assist the maintenance. At present the deep channel can only be made 82 feet wide between the jetties; and the scouring current loses its power beyond, so that a bar exists which is 1¼ feet above low water. As the dredging will have to extend about 450 feet beyond the extremity of the west jetty, it will be necessary to maintain the improved channel by periodical dredging, as the sluicing current, even when strengthened, would have little effect on a channel beyond the jetties and in the proposed depth of water.

BOULOGNE HARBOUR.

Description. The port of Boulogne is situated at the mouth of the small river Liane, whose estuary has been absorbed into the basins, and which in early times formed, and now serves to maintain the harbour. The port was established during the Roman occupation, and it has maintained its importance to the present time owing to its being in the direct line between London and Paris. Jetties were formed during the last century along the entrance channel, but the outlet was encumbered by the sand brought along the beach, and by the sediment from the river. Accordingly, a sluicing basin and tidal harbour were formed, at the beginning of the present century, in the bed of the Liane. The jetties were reconstructed in 1835; the sluicing basin was subsequently made more efficient by placing a dam with gates across its outlet; and, lastly, the formation of a dock raised the port to the position it occupied previous to the commencement of the new harbour works. (Plate 5, Fig. 8.)

Jetties, and Entrance Channel at Boulogne. The south-west jetty extends out beyond the other to facilitate the entrance of vessels; it has been made solid in masonry up to high-water spring tides, with a superstructure of open timberwork. The north-east jetty is solid along the inner portion, and consists of open timberwork for the outer 1000 feet. The current from the sluicing basin of 163 acres has been able to maintain the channel between the jetties at a depth of 5½ feet below low-water spring tides; but the bar beyond the pierheads is sometimes a foot above low water. The available depth at high-water neap tides has not exceeded 16½ feet, and, consequently, only a tidal service can be maintained with England. Dredging, however, by means of sand-pumps, has recently been resorted to for improving the depth; and when the large refuge harbour, in course of construction, is finished, full advantage will be gained of its favourable situation for a regular packet service across the Channel.

Remarks on Boulogne Harbour. Though Boulogne is free from outlying sandbanks, and has a better rise of tide than the ports to the east, it has not achieved much success as a jetty harbour, and presents more interest as an instance where a jetty harbour is being superseded by a large closed harbour on a sandy coast. The consideration of the new harbour works, as distinct from the jetty harbour, will be postponed till Chapter XIV.

CALAIS HARBOUR.

Origin and History. The next harbour along the coast is Calais. A channel across the beach provided an outlet for the land waters, and its depth was maintained also by the influx and efflux of the tide from a large lagoon running parallel to the coast. It was first used as a harbour in the tenth century, and in the thirteenth century it became the port of communication with England, as it affords the shortest passage across the Channel. The state of the harbour in 1610 is shown on Fig. 12. The long jetty constructed

CALAIS HARBOUR IN 1610.
Scale $\frac{1}{40000}$.

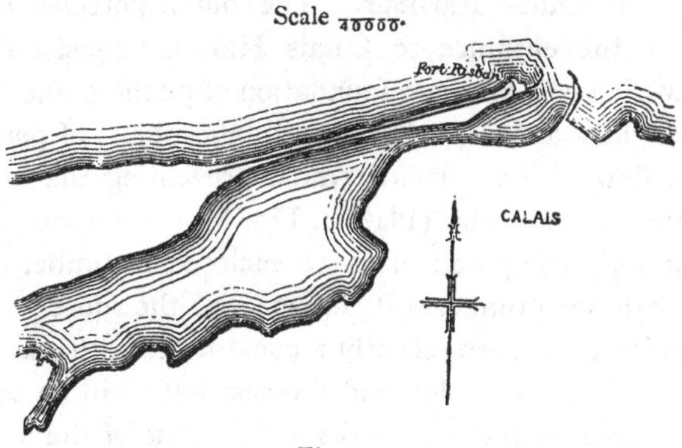

Fig. 12.

in the fifteenth century on the top of a sandbank, and terminating at Fort Risban, served to protect the harbour; but it also promoted the extension of the sandbank by interference with the littoral currents; and encroachments on the lagoon reduced the tidal scour. The low-water mark, accordingly, receded gradually, and at the beginning of the eighteenth century extended about 500 feet in front of the entrance. The jetties, of which the commencements are shown in Fig. 12, were then prolonged to obtain an increase in depth, but this occasioned an advance of low-water mark to the extent of 820 feet. This progression of the shore, and the gradual reclamation of the lagoon, produced a deterioration of the entrance channel. In 1821 another attempt was made to improve the entrance by carrying the jetties across a sandbank in front. Between 1838 and 1842 a sluicing basin of 140 acres was formed on the site of the old harbour, and the jetties were again extended, reaching to the point where they now terminate. (Plate 1, Fig. 5.) This last extension was followed, as before, by an advance of the foreshore; so that a proposition in 1845 for a further prolongation of the jetties was rejected, as it had become evident that this method of improvement afforded only a temporary relief, and led to the progression of the shore, as may be observed by a comparison of Fig. 12, and Plate 1, Fig. 5.

Jetties at Calais Harbour. The outer portions of both jetties at the entrance to Calais Harbour consist of open timberwork, resting upon a foundation of pitching and fascines raised slightly above the beach, with the object of permitting the free flow of the currents, whilst preventing the inroad of sand into the channel. (Plate 1, Fig. 6.) The inner portions were formerly composed of stone enclosed in timber framing below, with open timberwork above; but the shore portion of the west jetty has been recently reconstructed with masonry up to high-water spring tides, and the east jetty will be considerably modified by the new works on that side of the harbour.

Sluicing Basin at Calais. Advantage is being taken of the low flat beach to the east of the entrance channel for forming a large sluicing basin, by enclosing 225 acres with embankments[1]. (Plate 1, Fig. 5.) This new basin is much nearer the entrance than the former sluicing reservoirs, so that, in addition to a large increase of tidal water available for scouring the channel, the efficiency of this new volume of water will be much greater; and by means of a straight cut and a suitable modification of the east jetty, the sluicing current from this basin will be discharged fairly straight into the entrance channel.

Entrance Channel of Calais Harbour. The width of the outer jetty channel was made 330 feet, but the modification of the east jetty will increase it to 400 feet. The external

[1] A recent plan of Calais Harbour was given me by M. Guillain, engineer-in-chief of the port.

channel could be generally maintained, by the former sluicing arrangements, at a minimum width of 130 feet, and a minimum depth of 5 feet below low-water spring tides. This channel has only sufficed to provide a regular service of small mail steamers between this port and Dover. The new sluicing basin will effect a considerable improvement; and dredging by sand-pumps is being employed for deepening the channel beyond the jetties, so that the port should be soon available for a larger class of steamers.

GRAVELINES HARBOUR.

Description. The harbour of Gravelines is situated at the mouth of the river Aa, about midway between Calais and Dunkirk. It is less favourably situated than Calais, as the rise of tide is smaller, the sea in front is more impeded by sandbanks, the shore has a flatter slope, and the greater depths recede further from the land. The entrance channel was formerly maintained by the tidal ebb and flow in the estuary, as well as by the floods of the river; but reclamations have diminished the tidal receptacle. The outlet of the river is guided by parallel jetties of rubble, which have been gradually raised to high-water level to prevent the sand from the beach filling up the channel. Though the jetties are longer than those at Calais, they do not even reach out to low-water mark; and the sluicing basins, formed out of the old trenches round the town, are too far from the entrance to be adequately efficient.

DUNKIRK HARBOUR.

Conditions of the Site. The range of tide decreases to the east of the Straits of Dover along the southern and eastern coasts of the North Sea, which reduces the influence of the tidal ebb and flow, and diminishes the volume available for scouring. Thus, whereas the rise of spring tides is 20½ feet at Calais, it is only 17¾ feet at Dunkirk. Moreover, on emerging from the Straits into the wide expanse of the North

Sea, there is a notable increase in the sandbanks towards the east, till they attain a maximum nearly opposite Dunkirk. The site is, accordingly, very unfavourable to the formation of a deep harbour; though the outlying sandbanks protect the entrance, and form a roadstead.

Origin and Development of Dunkirk Harbour. The port appears to have been established in the tenth century. Its origin was due, like that of other jetty harbours, to the existence of a channel forming an outlet for the fresh waters of the surrounding district, and maintained by the natural tidal reservoirs of marsh lands in the neighbourhood. Jetties were constructed at an early period along the banks of the channel for guiding and concentrating the current; and the state of the port in 1640 is shown in Fig. 13. At that period a sandbank, dry at low water, but detached from the shore, stretched across the entrance. Eventually the jetties were extended across the sandbank, which soon after was connected with the mainland by the silting up of the intermediate channel, producing a considerable advance of the shore.

DUNKIRK HARBOUR IN 1640.
Scale $\frac{1}{40000}$.

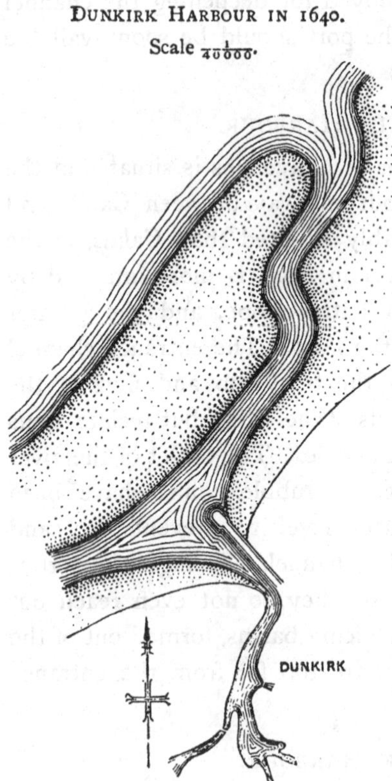

Fig. 13.

Jetties and Sluicing Basins at Dunkirk. The jetties are similar in construction to those already described, and a section of them is given in Plate 1, Fig. 8. They have been extended at various times, having now a length of about 2500 feet. These extensions, however, have not succeeded in attaining deep water at the entrance, for the foreshore has advanced on both sides after each prolongation. This advance has been greatest on the western side, from which quarter the greatest littoral drift of sand comes, borne along by the prevalent and strongest winds and by the more powerful flood-tide. The sandy foreshore, indeed, emerges at low-water spring tides right in front of the western pierhead. The uselessness of further extensions of the jetties has been recognised since 1840; and the lengths of the jetties have therefore been left unaltered in spite of the menacing condition of the western shore. Sluicing reservoirs were originally provided by the trenches of the old fortifications, and more recently a large sluicing basin was constructed on the western side of the entrance channel in front of the town. The site of this sluicing basin has now been taken for the large dock extensions in progress; and the plan of the harbour[1] (Plate 1, Fig. 7) shows the state of the port when these works shall have been completed.

Entrance Channel to Dunkirk Harbour. Up to 1877 the entrance channel was maintained solely by sluicing; but though the jetty channel could be thus maintained at a depth of from 5 to 8¼ feet below the lowest spring tides, the depth decreased to 1½ feet at the shallowest part of the channel beyond the jetties. In 1877 the system of dredging a deeper channel beyond the pierheads to deep water was commenced; and since 1881 three sand-pump dredgers have been employed at this work, a fourth being added in 1882, when over 52,000 cubic yards of material were removed in a single month. By this means the external channel has been deepened several feet, and also widened considerably[2].

[1] This plan, and several particulars relating to the recent condition of the harbour were given me by M. Eyriaud-des-Vergnes, the engineer-in-chief of the port.

[2] The quantity dredged in 1883 amounted to 280,600 cubic yards in the actual

The improvement of the entrance channel is of the greatest importance to the prosperity of the port, for the tendency of the present day is to increase the tonnage and draught of vessels, and unless a port can keep pace with the growing requirements of trade, in all probability vessels will abandon it for ports having better accommodation.

It has been suggested that a new, direct, and wider channel should be cut across the beach at right angles to the line of coast; for the present channel, having been extended in the line of the original outlet, inclines considerably towards the west, and is therefore much longer than necessary. It is also too narrow for the enlarged trade, having a width of only 200 feet between the jetties, which does not admit of two of the largest class of vessels passing easily when there is a strong current.

NIEUPORT HARBOUR.

Description. The sea opposite the Belgian coast is not so much encumbered by sandbanks as in front of Dunkirk, but the rise of tide is less, and the inclination of the strand is flat, making the greater depths recede considerably from the shore. Accordingly, neither of the two Belgian jetty harbours at Ostend and Nieuport are at all favourably situated for the maintenance of deep-water channels. The tide formerly flowed up a creek to a considerable distance inland of the site of Nieuport; but the creek has been gradually silted up and reclaimed, so that, like the old seaports of Hythe and Sandwich, Nieuport is now two miles from the sea-coast. The only vestige of the creek is the tortuous channel of the river Yser, which forms the approach to Nieuport. This channel is maintained by aid of sluices, which retain the fresh waters

channel, chiefly for widening it towards the west, and 225,300 cubic yards in a trench formed to the west of the jetties, above low-water mark, to arrest the sand which used to drift into the jetty channel. By this means the depth of the entrance channel has been maintained at from $9\frac{3}{4}$ to 13 feet below low-water spring tides; whereas the lowest dock sill is only $6\frac{1}{2}$ feet below low water.

draining from the adjacent districts and discharge them at low water.

Jetties and Entrance Channel at Nieuport Harbour. The outlet of the Yser is regulated by parallel jetties, consisting of open timberwork built upon mounds pitched with brickwork and rising two or three feet above the sandy beach. The west jetty, however, does not prevent the inroad of sand blown along by westerly winds, which tends to block up the channel; and the sluices are so far away from the outlet that, except in flood-time, there is a depth of only a few inches at low water at the end of the jetty channel. The construction of the jetties has produced an advance of the foreshore, as in other places; and low water is in a line with the ends of the jetties.

OSTEND HARBOUR.

Early History. The first navigable channel was formed in 1445, to the west of the town, and was maintained by discharging the fresh and tidal waters from interior ditches through sluices in the dykes protecting the low marsh lands. This channel, having deteriorated, was abandoned about 1571, when the existing channel was constructed through a breach which the sea had made in the protecting dyke. This breach admitted the tide into a tract of low-lying land, 7400 acres in extent, which thus provided a large sluicing reservoir for the maintenance of the channel. Nearly half this area was reclaimed in 1628, which soon led to the deterioration of the channel, as the other half had already had its capacity considerably reduced by being gradually raised by the deposit of silt at every tide. The position of the channel, and the extent of the flooded land previous to this reclamation, are shown in Fig. 14. The silting up of the channel was arrested, in 1662, by the readmission of the tide over a portion of the reclaimed lands; and the current produced thereby was so efficient that the harbour was restored to an excellent condition. Fears, however, being entertained that the current would undermine buildings and jetties alongside the channel, and the land-

OSTEND AND ENVIRONS IN 1612.
Scale $\frac{1}{100000}$.

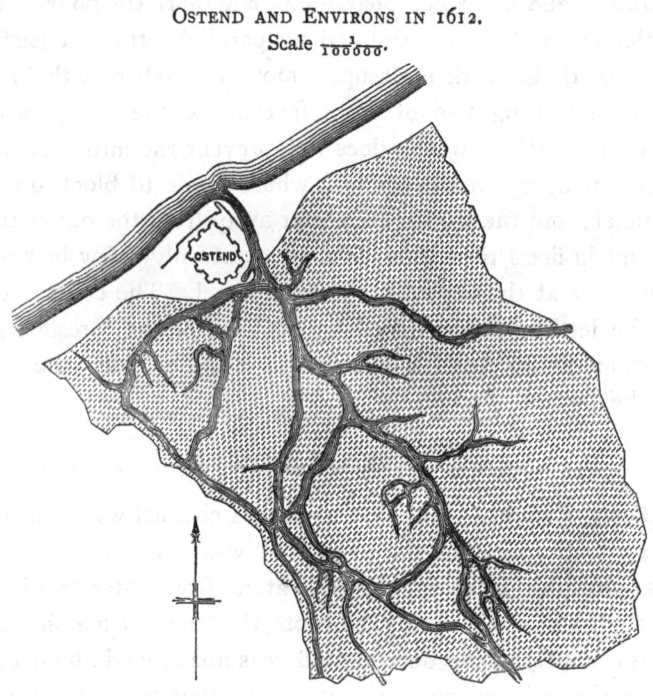

Fig. 14.

owners urging the restoration of their land, the tide was again excluded from this portion in 1700. The channel naturally began to silt up again; and another attempt to deepen it in 1730, by readmitting the tide into the lowest polder, was less successful, as the tidal capacity had been greatly reduced by the constant deposit of silt. The submerged lands gradually became useless for scouring by being raised by deposit, and they were finally reclaimed.

Jetties and Sluicing Basins at Ostend. When the port lost its natural reservoir of tidal scour, other means had to be resorted to for maintaining its entrance channel. Accordingly, early in the present century, the parallel jetties for regulating the outlet channel were prolonged; the first sluicing basin was opened in 1810, and another added in 1821. The channel was narrowed, to improve the scour, by reconstructing the west jetty in 1833–37, and prolonging the east jetty. These improvements secured a depth of 5¾ feet at low water over the bar at the entrance, and 11½ feet inside the harbour. Further improvements were commenced in 1851, which consisted in connecting the two sluicing basins, in forming a low stone mound along the west jetty, and in constructing a new sluicing basin. This last sluicing basin has its opening directed exactly in the line of the entrance channel, and being nearer the end of the channel has been able to maintain a depth of 7 feet 10 inches over the bar at low water. These works, together with some inner docks, have placed the harbour in its present condition (Plate 1, Fig. 10); and dredging is being employed for lowering the bar beyond the jetties. The jetties are very similar in construction to the east jetty at Calais. (Plate 1, Fig. 9.)

Peculiarities of the Site of Ostend. As the coast-line projects somewhat at Ostend, the sea tends to encroach rather than to recede, so that the foreshore has not advanced like at other jetty harbours; and a series of groynes have been formed along the beach to the west of the harbour to protect the shore. It would, however, be impossible to attempt to reach deep water by prolonging the jetties, as the Stroombank extends in front of the entrance at a short distance from the shore, and the navigable channel is between this bank and the end of the west pierhead.

Though a regular mail service has existed for many years between this port and Dover, the bar has sometimes hindered the entrance and exit of the steamers; but dredging will provide a remedy for this defect.

MALAMOCCO HARBOUR.

Conditions of the Site. The port of Venice is approached by channels traversing extensive lagoons. Though the rise of tide, which is almost imperceptible in the Mediterranean, never exceeds 4½ feet in the recess of the Gulf of Venice, yet, acting over the wide expanse of the Venetian lagoons, it produces a considerable scour, and it is aided by the discharge of the fresh waters during the ebb. The lagoons, however, are gradually being reduced in capacity by the sand brought in by the flood tide during storms, which is deposited during slack water, and also by the detritus brought down by the rivers, by the erosion of the banks, and by the refuse from the towns bordering them. This loss of scour was mainly felt at the outlet of the passes leading into the sea, for there the scouring current was diminished in velocity, and was moreover deflected by the littoral current flowing southwards. A bar was formed right across the entrance to the passes, owing to the checking of the outflowing silty current from the lagoon and the littoral sand-bearing current, which at Malamocco reduced the depth to 10½ feet at low water.

Formation of Jetties at Malamocco. During the period of the Venetian Republic, groynes were projected a short distance from the shore with the object of checking the effect of the littoral current, but they merely sufficed to protect the coast. The construction of two parallel embankments was proposed by MM. Prony and Sganzin during the occupation of Venice by Napoleon I; but only the root of the northern embankment was accomplished at that time. The northern embankment was recommenced in 1840; but the southern embankment was not begun till 1853, as the construction of both simultaneously would have impeded the navigation. The works were completed by the Italian government. The northern jetty, or embankment, has a total length of 7200 feet; and the southern jetty, in prolongation of one of the old groynes, is 3230 feet long. The least navigable width between the jetties is 1545 feet. (Plate 2, Fig. 8.)

Section of Jetties at Malamocco. The amount of sand brought along the coast by the littoral current is small; so the main purpose of the jetties is to concentrate and prolong the current from the lagoon, and prevent its conflicting with the littoral current before reaching deep water. The jetties, accordingly, have been made solid, and consist of a rubble base raised to ordinary high-water level with a masonry superstructure on the top. (Plate 2, Fig. 7.) The base was deposited from vessels by the aid of shears. The stone was brought from Istria, and consisted of pieces weighing between 1½ and 4 tons. The superstructure was built on the top of the mound, after an interval of four or five years had been allowed to elapse for settlement; it is 13 feet wide and 6½ feet high, and its top is 4½ feet above the highest tides. In spite of the small size of the rubble base, and the height to which it has been raised as regards the sea-level, it is little affected by the waves, owing doubtless to the small power they exert in that sheltered site and on a gently shelving beach. The maintenance of the north jetty is effected by the deposit of about 1100 cubic yards of rubble annually.

Improvements of the Malamocco Channel. The scour between the jetties has removed the bar, and produced a minimum depth of 30 feet below the lowest tides. The checking of the littoral current has reduced the depths on the northern side; but this diminution does not extend beyond depths of between 16 and 20 feet, so it has not hitherto affected the pass. The internal channels have been deepened by dredging; and their depth is to be maintained at 18¾ feet below the lowest tides, which is 21⅓ feet below mean high water.

KURRACHEE HARBOUR.

Situation. The entrance to Kurrachee Harbour resembles in several respects the pass of Malamocco. It is situated at the outlet of a large backwater, or lagoon, 18 square miles in extent, which flows into the Arabian Sea, where the ordinary rise of spring tides is only 8¾ feet. The waves during the south-west monsoon used to carry the sand across the entrance, forming a bar which half blocked up the channel and drove it towards the eastern shore, rendering it both circuitous and shallow. (Plate 2, Fig. 5.)[1] The harbour does not possess the advantage of a regular fresh-water discharge, as no river except the Layari flows into it, and this river is dry nearly all the year. The sea-bottom, on the other hand, shelves rapidly, so that deep water approaches the coast.

Improvement Works at Kurrachee. In order to direct and strengthen the tidal current along the entrance channel, and to induce scour across the bar, a groyne has been formed on the east side of the harbour running parallel to the channel, and another longitudinal groyne on the west side where the land is low. Manora Point, at the western extremity of the channel, rises high above the sea, and serves as a natural continuation of the western groyne. To prevent the inroad of sand across the entrance, and to direct the tidal current over the bar into deep water, a breakwater has been carried out from Manora Point for a distance of 1500 feet, and into a depth of 30 feet at low water. (Plate 2, Fig. 6; and Plate 7, Fig. 19.) This breakwater also protects the entrance to the harbour during the south-west monsoon. The removal of the bar has been assisted by dredging. A new channel, 2½ miles long, has been formed in the harbour, which, besides being useful for small craft, increases the tidal capacity of the harbour; and a creek leading out of the lagoon has been closed by an embankment, thus compelling the whole scouring current to pass through the harbour entrance.

Results of the Works at Kurrachee Harbour. The effect of the works on the bar and channel is indicated by the

[1] 'Minutes of Proceedings, Inst. C. E.,' vol. xliii, plate I.

comparative charts of 1859 and 1879. (Plate 2, Figs. 5 and 6.) The entrance channel has been considerably straightened, and its depth increased 6 feet; so that the outer harbour is accessible at all times for the largest class of vessels, and the inner harbour is much more commodious for the local craft.

LOWESTOFT HARBOUR.

Description. The fishing harbour of Lowestoft, situated on the Suffolk coast about 10 miles south of Yarmouth, furnished formerly the nearest parallel, on the English coast, to the French and Belgian type of jetty harbours. It was originally formed by cutting a channel from a low-lying marsh, called Lake Lothing, to the sea. This lake served as a tidal reservoir for maintaining the channel; and parallel wooden jetties were carried out across the beach, on each side of the channel, to a depth of 12 feet at low-water spring tides. The scour, however, produced in the channel by the ebbing of the tide from Lake Lothing, though guided by the jetties, was not powerful enough to contend effectually against the drift of sand and shingle from the north across the entrance which formed a bar.

Improvement Works at Lowestoft. A larger harbour, accordingly, was subsequently constructed, enclosed by piers projecting at right angles to the shore on each side of the old entrance channel, and about 650 feet apart. (Plate 4, Fig. 9.) These piers were built solid, of timber framework filled with rubble; the northern pier converges by two kants towards the southern pier, leaving an entrance between the pierheads, 155 feet wide, which faces slightly to the south of east and therefore somewhat away from the littoral drift. Sand nevertheless deposits to a large extent in the harbour, and has to be constantly removed by dredging. The projection of the piers has also produced an advance of the foreshore; for though the pierheads were built in a depth of 20 feet at low-water spring tides, there is now only a depth of about 11 feet at the entrance. The drift also from the north threatened to overlap the end of the northern pier, so a jetty has been built out from it in a north-easterly direction, with the object of arresting the travel of sand; but the low-water mark has reached the extremity of this jetty, and the entrance can be only kept clear by dredging with sand-pumps.

PARALLEL JETTIES AT RIVER MOUTHS.

The improvement of the outlets of rivers by means of parallel jetties resembles very closely the class of works, described above, for forming and deepening the entrance channels to ports. This system has been applied to numerous rivers with more or less success. Thus the mouth of the Yare has been fixed by jetties; the entrances to Newhaven Harbour on the Ouse, and Shoreham on the Adur, have been similarly regulated; and partly open jetties have deepened and protect the mouth of the Adour near Bayonne. Moreover the jetties at the mouths of the Maas, the Danube, and the Mississippi, rank amongst the most important and successful works on the sea-coast for removing bars and deepening the outlets of rivers. Whilst, however, the objects of the works are the same, and whereas the progression of the foreshore has to be avoided, and tidal scour is in some instances available for deepening the channel, yet the flow of the river itself has the most important bearing on the works. Accordingly, such works are more properly classed in the category of river improvements, which have been already dealt with in 'Rivers and Canals[1].'

General Remarks on the Parallel Jetty System. The similarity of the methods of improvement of the entrances to the ports on the southern shores of the English Channel and the North Sea is evident from the preceding descriptions

[1] 'Rivers and Canals,' L. F. Vernon-Harcourt, pp. 241–244, 274–277, and 303–322.

of these harbours, and from an inspection of Plate 1. The consecutive modifications also carried out in these ports are remarkably similar. Thus, in each case, some natural advantage of site has led to the establishment of the port in the first instance; and the entrance has then been guided and regulated by jetties. The value of land increased with the growth of the port, and reclamation of the tide-covered lands resulted, to the detriment of the channel which thereby lost a considerable portion of its natural source of tidal scour. In one instance alone, that of Ostend, was an attempt made, after the silting up of the entrance had been experienced, to restore the land to the tide. In the other cases, and eventually at Ostend also, the remaining means of tidal scour such as trenches and ditches, and also the fresh-water discharge, were utilised as much as possible by retaining them with sluice-gates till low tide, and then releasing the pent-up waters when their scouring efficiency is the greatest. The jetties also, in most cases, were extended in the hope of reaching deep water, which proved fruitless owing to the progression of the foreshore with each extension of the jetties. Next artificial sluicing basins were formed to provide a larger mass of water for sluicing, with the additional advantage that the issuing current was nearer and better directed for scouring the entrance. Lastly, dredging with sand-pumps is being largely employed for deepening the channel beyond the jetties.

Though the parallel jetty system has not proved successful in providing a deep entrance without constant works of maintenance, yet Boulogne Harbour is the single instance where it is being superseded by another design. Moreover it has sufficed for the establishment of a regular mail service at Calais and Ostend, and for the development of a large and increasing trade at Dunkirk.

Experience has shewn how jealously encroachments on the tide-covered lands should have been prevented, and the uselessness of prolongations of the jetties.

M 2

Sluicing and dredging are the two means by which the entrances to these ports may be maintained and improved. They are both needed, as they possess distinct functions. Sluicing is very efficient within the jetty channel, where the sluicing basins are large and well situated, as at Calais, Ostend, and Honfleur (Plate 1, Figs. 5 and 10, and Plate 12, Fig. 1); and the channel would be too narrow to admit a dredger without impeding navigation. The efficiency of the sluicing current, however, diminishes, and soon ceases, after passing the pierheads; and, consequently, sluicing alone is unable to maintain the depth any distance beyond the jetty channel.

It is outside the jetties, from the pierheads to deep water, that dredging is required. Till the adoption of the sand-pump dredger with a flexible tube, it was impossible to continue dredging operations when there was any swell; but now work can be continuously carried on in exposed situations when the waves do not exceed 3 feet. This dredging, moreover, can be accomplished at a very low rate, as the actual cost price of the work alone has been estimated at about 3*d.* per cubic yard at Dunkirk and Calais, the price paid to the contractor at Calais being 6½*d.* per cubic yard conveyed a mile out to sea.

The harbour entrances at Malamocco and Kurrachee are somewhat differently circumstanced to the jetty channels of the North Sea. The backwaters by which these entrances are maintained have a far greater capacity for tidal water, in spite of the small rise of tide on their coasts, owing to the great expanse covered. Accordingly, the tidal scour is much more efficient, and can be best directed by solid piers. Moreover deep water is reached at the ends of the piers, so that no channel has to be formed beyond the pierheads. Also the littoral drift, in these instances, is small, and coming into fairly deep water the progression of the shore would be in any case slow. Dredging was resorted to at Kurrachee for facilitating the scour in the first instance, but is not required for maintenance; and it has not been used at all for the pass between the piers at Malamocco.

These entrances are liable to deterioration from the silting up of their backwaters by material introduced during storms from the sea, and at Malamocco by deposit of detritus from the land as well. In artificial sluicing basins, this source of injury is somewhat guarded against by excluding the tide except when the basin has to be filled for sluicing; but this precaution cannot be adopted for natural lagoons. Nevertheless, judicious dredging inside can restore the tidal capacity, and at the same time improve the channels for navigation; and this course has been adopted both at Kurrachee and Malamocco.

Parallel jetty harbours are one of the most difficult class of harbours to design and maintain successfully; and though the success of the works at Malamocco and Kurrachee cannot be disputed, the modification of the jetty harbours of the North Sea has been advocated. They appear, however, to render important services to commerce at a moderate cost; and the extension of sluicing basins, and dredging, tend considerably to improve their condition. It may be doubted whether parallel jetties form the best design that could have been adopted if the work was begun afresh; but, except where a refuge harbour is needed, the expense of a thorough transformation such as is being effected at Boulogne could not be entertained, and at some sites, such as Dunkirk and Ostend, with their outlying sandbanks, it is probable that such an attempt would not meet with success.

CHAPTER IX.

HARBOURS WITH CONVERGING JETTIES.

DUBLIN. ABERDEEN. CHARLESTON. GALVESTON. NEWBURYPORT. SUNDERLAND. YMUIDEN.

Introductory Remarks. *Dublin Harbour:* Situation; Construction of Jetties; Improvement of Entrance; Peculiarities. *Aberdeen Harbour:* Early Improvements; Converging Breakwaters; Bar. *Charleston Harbour:* Description; Jetties over the Bar; Results. *Galveston Harbour:* Description; Jetties; Results. *Newburyport Harbour:* Description; Jetties. *Sunderland Harbour:* River Entrance; South Entrance. *Ymuiden Harbour:* Converging Breakwaters; General Remarks.

THE system of parallel jetties at harbour entrances, though very serviceable in regulating the entrance channel and guiding the sluicing current, exhibits some deficiencies. The channel does not provide a reservoir of tidal water, which has to be furnished by a separate sluicing basin at the side; it is generally too narrow to admit of dredgers working in it; it merely serves as a passage, and not as a harbour for vessels; and, moreover, it affords no provision for the reduction of the entering waves, except such as can be accomplished by special stilling basins at the side. This last defect is of less importance where the entrance is protected by outlying sandbanks, as at Dunkirk and Ostend, or where the worst waves come sideways to the entrance channel, as at Malamocco and Kurrachee; but where the waves come in nearly straight on shore, this defect would seriously compromise the tranquillity of the port. Also the jetties projected at right angles to the shore, in situations where the littoral current is densely charged with sand or silt, present a serious obstacle, if solid, to the progress of the current, and consequently favour a rapid accumulation on the beach, or, if open, offer a ready place of deposit in the channel itself.

These deficiencies may be avoided by forming solid converging jetties, which provide a reservoir of tidal water close to the entrance, afford ample room for the reception of vessels or the working of dredgers, reduce the entering waves, also in some cases by being kept low allow the littoral current to pass over unchecked, and in every case, owing to their sloping course, present less obstacle to the regular flow of the current round the entrance.

DUBLIN HARBOUR.

The works undertaken at the entrance to Dublin Harbour for improving the channel across the bar have been already described in 'Rivers and Canals' with reference to the river Liffey[1], and therefore only those points will be noticed here which specially relate to converging jetties.

Situation. Large quantities of sand, brought in from the sea, have gradually accumulated on the sheltered western shore of Dublin Bay. The river Liffey, which flows into this bay, is small, and has little influence on the state of the harbour entrance, either by the deposit of detritus, or by producing scour in flood time. The river, indeed, forms the channel of the harbour, dividing the flat sandy beach into two strands, called the North and South Bulls; but the tendency of the sandy deposit is to form a continuous strand all across the western shore of the bay, which is only partially prevented by the tidal flow in the river, and accordingly a bar exists at its outlet. At one period the improvement of the entrance to the harbour was considered so hopeless that Howth and Kingstown harbours were successively constructed to provide a port for the city of Dublin in another position.

[1] 'Rivers and Canals,' L. F. Vernon-Harcourt, p. 237.

Construction of Jetties towards Dublin Bar. The outlet channel of the river was regulated by the construction of a rubble embankment along its southern side, called the Great South Wall, completed in 1796. (Plate 2, Figs. 9 and 11.) This solid jetty served also to protect the channel in south-westerly gales, and prevented the inroad of sand from the South Bull. Owing, however, to the absence of scouring power in the river, even when regulated, no improvement was effected by this work upon the bar. In order to deepen the channel across the bar, another rubble embankment was formed in 1820–25, starting from the northern side of the estuary at a distance of nearly 1½ miles from the southern embankment and converging towards it so as to leave an entrance of only 1000 feet between the pierheads. (Plate 2, Figs. 9 and 10.) A large tidal area has been thereby enclosed, additional shelter and tranquillity afforded, and the scour concentrated and directed over the bar. The northern jetty has been only raised to half-tide level along its outer portion so as to allow the earlier half of the ebbing current to flow over it, as this portion of the current would possess little scouring effect, and would be prejudicial to navigation through the narrowed entrance; and the sand, which it might have carried away, would have been partially brought back by the flood tide owing to the peculiar set of the currents in Dublin Bay.

Improvement of the entrance to Dublin Harbour. The construction of these solid converging jetties, 3⅓ and 1¾ miles in length respectively, has gradually lowered the bar from 6 feet below low water in 1819 to 16 feet in 1880. This remarkable result has been aided by dredging in the inner channel, and by the removal of ballast from the north strand within the northern jetty, thus increasing the tidal capacity of the area enclosed by the jetties, which has at the same time been carefully guarded from reclamation. It is probable that if dredging with sand-pumps was resorted to on the bar, the ultimate limit of improvement that could be maintained by tidal scour would be sooner reached.

Peculiarity of the Site of Dublin Harbour. The entrance to Dublin Harbour, like the North Sea jetty harbours, is situated on a flat sandy beach, and, like them, has been improved by tidal scour. Whereas, however, these jetty harbours are situated on a coast where little alteration occurs in the foreshore, except when artificial impediments to the littoral currents are formed; Dublin Harbour stands in a land-locked bay where continuous silting up takes place without being hindered, or appreciably accelerated, by projections from the western shore. The sand, coming from the sea into the comparatively tranquil bay, would deposit, whether the jetties existed or not. It is possible that they may somewhat concentrate the area of deposit on the strands which they partially shelter; but their absence would not arrest the progression of the beach, and their presence does not materially hasten it, otherwise the greatest advance would be close to their sides. Accordingly, the projection of jetties in such a situation is not open to the objections that may be urged against the extension of the North Sea jetties; and it is probable that, at some future time, the progression of the low-water mark of the North and South Bulls will necessitate an extension of the Dublin jetties.

Aberdeen Harbour.

The harbour of Aberdeen consisted originally of the natural channel of the river Dee at its outlet into the North Sea. The tidal waters flowing up the river about two miles, in conjunction with the fresh waters, maintained with difficulty a passage over the sandy beach during north-easterly gales; and a bar existed at the mouth of the river.

Early Improvements. About 1773, by Smeaton's advice, the river was trained, so that the river floods might have more effect on the bar, by the construction of a pier extending out from the shore on the northern side of the estuary to a distance of about 1200 feet from high-water mark. It was expected that the drift of sand along the shore from the north would be either entirely stopped, or carried out to such a distance that the main tidal current would convey it away, and that thus it would be prevented from blocking the entrance to the harbour. The scour produced by this work, and by a southern pier which formed a bulwark parallel to the river on the opposite side, maintained a depth of about 2½ feet over the bar at low-water spring tides. By the advice of Telford, an extension of the North Pier for 865 feet was completed in 1816; and a southern breakwater was built out from the south shore towards the end of the North Pier, and completed in 1815, in order to concentrate the scour across thc bar and increase the depth in the channel. These works are shown on Plate 1, Fig. 11, giving a plan of the harbour in 1867 shortly before some new works were commenced[1]. Dredging was also resorted to in 1841 for lowering the bar which, below a superficial deposit of sand, consisted of boulder clay; and the depth at the entrance, which had been 2½ feet at low-water spring tides before any works were commenced, was increased to between 7 and 9 feet by 1851.

South Breakwater at Aberdeen, and Extension of North Pier. In order to provide a larger harbour with better shelter, as well as a wider entrance in deeper water, new works were commenced in 1870, consisting of a new South Breakwater starting from the southern extremity of the estuary and stretching due north towards the North Pier, and the extension of the North Pier into deeper water in a straight line from the existing pier. (Plate 1, Fig. 12.)

The South Breakwater was completed in 1873; and the extension of the North Pier was commenced in 1875, and

[1] Charts from which the plans of Aberdeen Harbour, in 1867 and 1879, have been made were furnished by Mr. W. D. Cay, the harbour engineer.

finished in 1877. They have both been built of concrete on the upright-wall system, and are somewhat similar in section (Plate 6, Figs. 15 and 16); but different methods were adopted for their construction, which have been already described in Chapter VII, and will be further considered in Chapter XVI.

The works have so far accomplished the objects for which they were designed, in having provided a larger harbour sheltered from south-easterly gales, and in giving an entrance 830 feet wide transversly to the channel, and 1100 feet between the pierheads, in a depth of about 15 feet at low-water ordinary spring tides, in place of a shallower entrance only 450 feet wide. The bar also, which existed within the entrance in 1867, has been removed by dredging, so that a channel has been formed in the outer harbour exceeding 12 feet in depth. The present entrance is very large in proportion to the size of the harbour, as may be seen by comparing it with Charleston, Dublin, and even Tynemouth (Plate 2, Figs. 1 and 9, and Plate 4, Fig. 10); and therefore it cannot have much influence in reducing the entering waves, or promoting tidal scour. The extension of the North Pier, besides assisting in sheltering a larger area, is specially serviceable in preventing the drift of sand from the north entering the harbour, as the 1½-fathom line overlapped the pierhead before the extension.

Bar at Aberdeen. The existence of a bar at the mouth of the Dee has formed one of the main obstacles to the improvement of the entrance to the harbour. The earlier works were directed towards increasing the scour across the bar; but as it partially consisted of clay and boulders, dredging had to be employed to assist in lowering it. The position of the bar, in 1867, inside the entrance seems to indicate that it tends to be raised by sand brought in by the sea, settling where the agitation of the waves in storms is reduced; for if it was mainly formed by the river deposit it would be more liable to form in a wider part of the estuary, or outside the entrance, where the velocity of the outgoing current is checked. This view is also borne out by the fact that about 40,000 tons of sand are deposited on the bar every winter, when the influence of the waves is greatest.

The bar is lowered by dredging in the summer, so that sometimes in October it is 12 feet below low-water spring tides; but it gradually rises during the winter to 7 or 8 feet below low water. After attaining that limit, further deposit only increases the width, and not the height of the sandbank. Mr. Cay says that floods very rarely lower the bar; and it is evident, from his estimate of 2 million cubic feet of discharge per minute during exceptional river floods, that they generally bear too small a proportion to the sectional area of the estuary at the bar[1] to have much scouring effect. He attributes the limit observed in the height of the bar to the influence of the waves, passing over the bar at that depth, and to the action of the tidal current.

The river brings down a considerable quantity of silt; and the deposit in the estuary and on the bar amounted to 152,000 tons in 1879. This has to be removed by dredging.

The recent pier works, being extended considerably beyond the site of the bar, are not calculated to assist in removing it further than by sheltering the dredgers working on it. They also, by placing the entrance in deeper water, reduce the amount of sand entering the harbour in a given width; but the increased width of entrance provides, at the same time, a freer admission for the sand-bearing waves.

Works in the Port of Aberdeen. From a comparison of the charts of 1867 and 1879, it will be observed that since 1867 the river Dee has been diverted, and a basin for the herring boats has been formed in the old bed of the river. (Plate 1, Figs. 11 and 12.) The tidal harbour and dock have

[1] Mr. W. D. Cay states that the least sectional area of the channel at low-water ordinary spring tides, with the bar only 8 feet below the water, amounts to 3500 square feet.

also been deepened. These works furnish a more regular and direct channel for the river, and increase the tidal capacity of the harbour, and therefore, besides improving the port, assist the maintenance of the entrance channel.

CHARLESTON HARBOUR.

Description. The entrance to Charleston Harbour is situated on the coast of South Carolina, at the mouth of a large natural tidal-basin 15 square miles in extent. Though the average range of tide is only 6 feet at springs and 4¼ feet at neaps, yet, owing to the great tidal capacity of the basin and the inflowing rivers, the mean discharge during the ebb amounts to over 3,655,000,000 cubic feet, of which little more than a fifth is land water. The maintenance of the entrance is, consequently, chiefly due to the tidal water flowing into and out of the vast natural basin. A bar, however, has been formed by wave-action and littoral drift during north-easterly storms, extending right across the entrance about 1¾ miles out; it stretches southwards for ten miles joining the shore at each end, and is about 1¾ miles wide. From four to six shifting channels have always existed over the bar, the deepest of the channels having a depth of from 12 to 13½ feet at mean low water.

Improvement of Charleston Bar Channel by Jetties. A scheme was proposed by General Gillmore, in 1878, for improving the depth of one of the channels across the bar, by means of converging jetties, which are in course of construction[1]. (Plate 2, Figs. 1 and 2.) The jetties, formed of log and fascine mattresses loaded with stone, start from the shore on each side of the entrance, at a distance of nearly 2¼ miles apart, and converge to a width of about 1900 feet over the bar. In order to interfere as little as possible with the littoral

[1] 'Report of the Chief of Engineers (United States) for 1878,' p. 554; 1879, p. 731; 1880, p. 921; and 1881, p. 1043.

drift, to maintain the bar in its present position, to avoid the progression of the shore-line, and to admit the free influx of the tide, it is intended to keep the inner half of the jetties at a low level, especially in the central deep water; whilst the outer portions, lying in the direction of the flood current, may be raised without presenting any obstruction to the flow. By this arrangement the filling of the basin at each tide will be secured, and the surface currents will not be modified; whilst the ebbing tide will be gradually more and more concentrated as the tide falls, and forced through the trained channel. The original estimate was £370,000, but the jetties are being made longer than at first proposed.

Progress and Result of the Works at Charleston. The northern jetty was commenced at the end of 1878; it reached a length of 14,360 feet in 1881, and no further work was done up to the beginning of 1884. The south jetty attained a length of 14,130 feet towards the close of 1883. (Plate 2, Fig. 1.) General Gillmore informed me that though the jetties had undergone some settlement, they were in a satisfactory condition, that sand is accumulating against the mattresses protecting them from the teredo, and that scour had become so manifest by the beginning of 1883 that steps were being taken by means of spur jetties to keep the current away from the foundations of the south jetty. Scour was stated to have commenced in the report of 1881; but the jetties will require extending and raising at the outer ends before the full effects can be produced. Dredging in the channel will also be necessary to obtain the required depth of 21 feet, where only 11½ feet previously existed; but tidal scour should afterwards suffice for maintaining it.

Galveston Harbour.

Description. The situation of the entrance to Galveston Harbour is somewhat similar to that of Charleston Harbour.

The entrance forms the mouth of an extensive inland creek, or bay, having an area of 455 square miles, called Galveston Bay.

The rise of tide in the Gulf of Mexico is very small, not exceeding 3 feet; and the actual rise at any period is much influenced by the wind. Owing to the effects produced by the wind, the highest and lowest recorded levels have a difference of 6·1 feet, which may be regarded as a fair extreme annual range. Though the rise and fall of the water is so slight at Galveston, yet, owing to the wide extent over which it takes place, and the narrow opening through which the flow occurs, a deep channel is naturally maintained at the entrance to the bay having a depth of 40 feet. This Bolivar Channel branches into several channels in the bay, one of which forms the approach to Galveston Harbour which is obstructed by an inner bar. There is an outer bar obstructing the communication between Bolivar Channel and the deep water of the Gulf, stretching in a semicircle from Bolivar Peninsula on the east to Galveston Bay on the west.

The bars consist of fine sand, which also forms the upper strata of the shore and adjacent islands and coast. This sand is of the nature of quicksand, when wet it is readily shifted by currents, and is blown along by the wind when dry. Great changes, accordingly, are continually occurring along the coast, owing to the littoral currents produced by the wind.

The inner bar has been formed since 1841 by the gradual protrusion of Fort Point, the depth over it having decreased from 30 feet in 1841 to 9 feet in 1867; but training works increased its depth to 12 feet in 1872. The outer bar is formed by drift sand brought along by the littoral currents and deposited where they come in conflict with the current issuing from Bolivar Channel. This bar underwent no permanent changes in depth between 1841 and 1873, having remained about 12 feet below the water, but it has moved slightly outwards.

Improvement Works at Galveston Harbour. Owing to the shifting character of the sands on both the inner and outer bars, it was evident that no deepening by dredging would be permanent without the aid of training works to induce scour and to prevent the inroad of sand.

Major Howell proposed a scheme, in 1873, for improving the depth over both the bars by contracting the channel across the inner bar, by a continuation of the city dyke to the Bolivar Channel, and by training the outer channel by jetties from the land, over the bar, into 18 feet of water[1]. (Plate 2, Fig. 12.)

The magnitude of the work of building about seven miles of jetties into the Gulf, in from 5 to 18 feet of water, and exposed to breaking waves during high winds from various quarters, necessitated a cheap method of construction in a situation where stone had to be procured from a distance, and where the trade of the port did not warrant a large expenditure. Moreover, the unstable nature of the bottom precluded the adoption of a heavy structure.

It was intended that the jetties should be only raised a few feet above the bottom, so that being submerged they might not interfere with the filling of the bay at each rise of tide, upon which the scour depends, and yet be high enough to direct the lower ebb current.

Construction of the Galveston Jetties. The work was commenced in 1874 by the prolongation of the city dyke inside the bay; and in 1877 the Bolivar or northern jetty was begun. The shore portion of the northern jetty was formed with a double row of piles for a length of 513 feet; but its continuation, and the inner dyke, were made in the first instance of a double row of gabions, 6 feet in diameter and 6 feet high. The gabions consisted of a wickerwork frame, coated with cement and filled inside with sand. The inner jetty had to be strengthened by placing the gabions in two rows connected by cross walls and supported on the channel side by spur dykes,

[1] 'Report of the Chief of Engineers (United States) for 1874,' p. 732.

at short intervals, which collected the drift sand and prevented undermining. This jetty was gradually covered with sand, and thus sustained no damage from a storm in 1875 which damaged the city dyke. After the restoration of this dyke, and the completion of about 2200 feet of jetty, the inner bar was scoured away to a depth of 20 feet at mean low water.

The gabion jetty was extended 9700 feet on the northern side; but though the gabions were carried in a single length of 12 feet right across the jetty, and were subsequently placed upon reed mats extending 6 feet beyond and ballasted with concrete blocks, the gabions were undermined, some were displaced by storms, and others sank down into the sand[1]. The result was that the northern jetty ceased to have much effect on the channel, and that the system of gabions had to be abandoned.

It was evident that a light broad base was needed for floating the jetties on the soft sand foundation, and to prevent undermining. Accordingly Colonel Mansfield had recourse to the system which M. Caland had so successfully carried out at the mouth of the Maas with mattresses of fascines weighted with stone[2]. He adopted a simpler and cheaper method of construction, without pitching and stakes, and with small stones placed in pockets in the mattresses, so that they might remain undisturbed by the sea. A section of the mattresses is given in Plate 2, Fig. 13; and the jetty as consolidated is shown in Plate 2, Fig. 14. This form, besides being cheap, is said by Colonel Mansfield to resist the action of the sea without injury, and with little settlement[3]. The southern jetty was commenced in this manner in July 1880, and in two years 20,777 feet were constructed. The width of the

[1] 'Report of the Chief of Engineers (United States) for 1880,' p. 1225.

[2] 'Rivers and Canals,' L. F. Vernon-Harcourt, pp. 62 and 274, and plate 2, fig. 2.

[3] A plan and sections of the jetty were sent me by Col. Mansfield, the engineer in charge of the works.

lower mattresses has been made from 60 to 120 feet according to circumstances; and out of 741 mattresses floated out and laid in the jetty, only three were carried away. A short trial length of 90 feet of mattress jetty was added to the northern jetty during this period, and has remained undisturbed. During the year ending 30th June, 1882, the material deposited in the jetty consisted of 84,086 cubic yards of mattresses and 27,307 cubic yards of ballast, giving a total of 111,393 cubic yards, which cost 12*s.* per cubic yard.

Result of the Works at Galveston. The depth over the inner bar has been maintained at 20 feet, a gain of from 8 to 9 feet in depth. The improvement of the channel over the outer bar was only about 2 feet by the beginning of 1883; but as a great portion of the south jetty had been recently built, it was anticipated that, on the arrival of the northerly winds which blow the waters out of Galveston Bay towards the south, a considerable deepening of the channel would result. As the greatest ebb current occurs with a northerly wind and tends towards the south, and as consequently no channels have ever been formed across the bar to the north of the Bolivar Channel, Colonel Mansfield regards the south jetty as by far the most important one for training the channel over the bar, and that the extension of the northern jetty may be deferred till the southern jetty is completed and its effects have had time to manifest themselves. The southern jetty, besides directing and concentrating the main southerly ebb, protects the channel from the great northerly drift of sand during flood tide; and since its construction the sand has been rapidly accumulating on the southern side between the jetty and the beach, extending out a distance of two miles, and likely soon to emerge from the water. Colonel Mansfield proposes to keep this sand from falling into the channel by raising the jetty to mean tide level, or perhaps to high tide. He does not anticipate further advance of the foreshore; but this rapid progression in so short a time might well occasion some anxiety as to the maintenance of the channel, and a fear of its gradually creeping round the jetty, which would then need extension. It is possible that the foreshore may attain a definite limit of advance; but the rapidity of the change renders it unlikely that this point has yet been nearly reached.

The peculiar southerly tendency of the strongest ebb, and the northerly drift of sand, render the southern jetty the most important; but probably the northern jetty would be very serviceable in concentrating the current, especially if made to converge more towards the other, it would aid in fixing the channel, and secure the channel from any drift of sand from the north. If the scour induced by the southern jetty does not produce the expected depth of 20 feet over the bar, in place of the original 12 feet, then the northern jetty must be added to aid the improvement; and this jetty would, moreover, render the approach to the harbour more secure during the strong northerly winds, just as the shelter of the south jetty has already proved advantageous during southerly winds for vessels crossing the bar.

To promote the deepening of the channel, Colonel Mansfield intends to scrape and harrow the bar. This system has been tried at the mouth of the Mississippi, and other places, with little success; but it is possible that as the sediment brought down by the rivers is deposited in the extensive enclosed bay before reaching the outlet channel, the method may be more efficient in this position where the scouring current is not already charged with silt. Dredging with sand-pumps would, however, be a more rapid and probably a more effective means of lowering the bar. The chief merit of stirring up the sand on the bar is that the scour would be sure to maintain any increase in depth produced by its removing the sand; whereas deepening by dredging might exceed the limit that could be maintained by scour.

NEWBURYPORT HARBOUR.

Description. The entrance to Newburyport Harbour is somewhat similarly situated to the entrances of Charleston and Galveston just described. It lies in the contracted outlet of a wide expanse of flats, at the mouth of the Merrimac river. A depth of from 20 to 30 feet is maintained at the narrow outlet, and for a short distance beyond, by the scour through the opening; but further out a bar exists, over which the channel had a minimum depth of only 7½ feet at mean low water in 1880. This channel has been shifting and variable in depth; and changes have also occurred in the line of the adjacent coast[1]. (Plate 2, Fig. 4.)

The mean rise of tide is only 7½ feet; but its scouring power is enforced by the narrow opening through which it passes, and by the extent of flats and creeks inside.

Converging Jetties at Newburyport Harbour. In order to fix the channel over the bar, and to obtain a channel not less than 13½ feet in depth at low water, it was decided, in 1880, to construct two converging jetties of rubble stone raised 4 feet above mean high water, so that the scouring current might be directed and concentrated over the bar. (Plate 2, Figs. 3 and 4.) The northern jetty is to be 2910 feet long, and the southern jetty 1500 feet, leaving an opening 1000 feet wide between their extremities. The cost of the work has been estimated at £24,000, and it was commenced in 1881. It is anticipated that when the works in progress are fully completed there will be a depth of 17 feet at the entrance, at low water, formed and maintained by the scour through the contracted opening.

SUNDERLAND HARBOUR.

River Entrance. The old entrance to the port of Sunderland was by the river Wear, the outlet of which was many

[1] 'Report of the Chief of Engineers (United States) for 1881,' p. 501.

years ago guided by parallel jetties across the sandy beach[1]. These works improved the mouth of the river, but a depth of only 4 feet at low water was attained; and the jetties produced an advance of the foreshore, like the similar works at the entrance channels of the foreign North Sea ports already described in Chapter VIII. The jetties, accordingly, at the mouth of the Wear were not prolonged; and when docks were constructed at Sunderland, a new southern outlet was provided affording a better depth of water.

Converging Piers at the South Entrance, Sunderland. The piers forming the outer harbour and southern entrance to the Sunderland Docks converge at a sharp angle to the line of coast, so that, though wide apart at their commencement, they leave a width of only 225 feet between the pierheads. (Plate 12, Fig. 6.) As the chief littoral drift comes from the north, the entrance has been made to face south-east. The sloping line of the piers interferes as little as possible with the littoral currents, whilst placing the entrance in deeper water, and forming a projecting point which induces scour. Moreover the plan of the outer harbour is specially favourable for stilling the entering waves; it gives easy access to the two dock entrances, and affords ample area for dredging.

Though no natural backwater exists, as in the cases previously described, for a tidal reservoir, the docks and outer harbour form artificial sluicing basins which concentrate their scouring current through the narrow entrance.

YMUIDEN HARBOUR.

Converging Breakwaters. The problem of forming a harbour on the flat sandy shore of the North Sea near Amsterdam, which for a long time prevented the shortest line of water communication being chosen for that port[2], has been solved

[1] 'Rivers and Canals,' L. F. Vernon-Harcourt, pp. 240 and 241.

[2] Ibid., pp. 177–185.

by the construction of solid converging breakwaters. (Plate 5, Fig. 2; and Plate 6, Fig. 19.) In this case no available scouring current exists, except the tidal water entering the harbour and the adjacent reach of canal, as the level of the Amsterdam Ship-Canal within the North Sea locks is kept low for the drainage of the surrounding low-lying district, and the land waters are discharged by pumps at the opposite end of the canal. The solid concrete breakwaters start at a distance apart of 3750 feet, and converge so as to leave an opening of only 850 feet at the entrance, with a depth of about 28 feet at low water.

The projection of solid piers on a sandy coast has naturally led to an advance of the shore-line near the breakwaters; but this line appears to have reached a new position of equilibrium before getting out far enough to compromise the depth at the entrance. The depth inside the harbour, which tends to diminish by the deposit of the fine sand brought in by the waves and blown along the beach by the wind, is maintained by dredging.

General Remarks on Converging Jetty Harbours. The foregoing descriptions indicate that harbour entrances formed by converging jetties have, in general, attained a better depth of water than those where parallel jetties have been adopted. The scouring current is concentrated against the shallowest point of the channel; and whatever shoals are formed in the wider sheltered area inside are easily removed by dredging. Artificial outlets have been formed at Sunderland and Ymuiden, but in the other instances the works have been directed to improving existing channels.

The general principles followed at Dublin and at Charleston (Plate 2, Figs. 1 and 9), at Newburyport and at Sunderland (Plate 2, Fig. 4, and Plate 12, Fig. 6), are very similar. At Dublin and Sunderland, the tidal scour at the entrances has been created by the works; whereas at Charleston, Galveston, and Newburyport, large natural reservoirs of tidal water exist, which enable a small rise of tide to exercise a very important influence in deepening and maintaining the entrance. At Ymuiden, the small amount of tidal scour is compensated for by the depth into which the entrance has been carried. The Aberdeen breakwaters have been carried beyond the bar, and afford shelter for dredging operations; but they do not at present approach close enough to produce scour at the entrance, which, however, like at Ymuiden, has been carried into deep water. At Charleston, the channel is being carefully trained by jetties right across the bar; whereas at Galveston one jetty has hitherto been mainly relied upon for improving the channel.

The system of producing scour by a single jetty can only be resorted to in very exceptional circumstances; and it may be doubted whether the completion of the low northern jetty will not eventually be found expedient at Galveston. The method is, however, being tried by Col. Mansfield at Aransas in Texas, where he is training the channel across the bar by a single concave jetty; but the southern direction of the outlet channel running almost parallel with the coast, and the prevalence of northerly winds driving the issuing current against the jetty, render this a peculiar case.

The selection of the materials for forming these jetties has been guided by circumstances. Thus whilst fascine mattresses have been chosen at Charleston and Galveston, as combining cheapness for extensive works with lightness, and a broad base on an unstable foundation, the simple rubble mound has been adopted at Dublin and Newburyport, and upright concrete walls have been built at Aberdeen and Ymuiden.

All these converging jetties have been made solid, but they have not been raised to a uniform height as regards the sea-level. At Aberdeen, Ymuiden, and Sunderland, the piers have been raised high to afford better shelter, and to prevent the inroad of sand; and it seems probable that the Galveston south jetty will have to be raised above high water for a similar purpose. At Dublin, on the contrary, and at Charleston, the jetties have been kept low in places to ensure the filling of the estuary at every tide.

The main superiority of the converging jetty system consists in its small interference with the littoral currents, in tending to induce scour round the projecting point formed by the piers, in extending the tidal reservoir area, and in concentrating the current at the most critical point of the channel.

CHAPTER X.

HARBOURS PROTECTED BY RUBBLE MOUND BREAKWATERS.

PLYMOUTH. PORTLAND. DELAWARE. HOWTH. KINGSTOWN. TEES. CETTE. TABLE BAY. BREST.

Plymouth Harbour: Site; Breakwater, construction, maintenance. *Portland Harbour:* Situation; Breakwaters, construction, disturbance, cost, compared with Plymouth. *Delaware Harbour:* Situation; Breakwaters, progress, maintenance; State; Improvements proposed; Compared with Plymouth and Portland. *Howth Harbour:* Site; Breakwaters; State. *Kingstown Harbour:* Situation; Breakwaters; State; Compared with Howth. *Tees Breakwaters:* Construction; Remarks. *Cette Harbour:* Description; Breakwater. *Table Bay Harbour:* Situation; Design; Breakwater, extension, progress and cost; Inner Works; State. *Brest Harbour:* Description; Commercial Harbour, breakwater and jetties, peculiarity, cost. Review of Rubble-Mound Breakwaters.

THE rubble mound is the simplest form of breakwater; it offers no scope for ingenuity in construction, as it opposes a mere shapeless mass of stone to the sea, though, under the action of the waves, it gradually assumes more definite slopes, depending upon the exposure of the site and the size of the material on the face of the slopes. It can only be adopted where there is an abundant supply of stone, and where the area occupied by the breakwater is of no consequence; and its weak point is between high and low water on the sea slope.

PLYMOUTH HARBOUR.

Site. Plymouth Sound is an extensive bay into which the rivers Tamar, Laira, and Plym flow. It possesses excellent natural shelter on the north, east, and west, but it was formerly open to the south. The centre of the entrance to the Sound is somewhat obstructed by various rocks; so that the best approaches were always on the east and west sides of the bay. The anchorage inside is good, the bottom being composed of blue clay. (Plate 3, Fig. 8.)

As the Sound was exposed to winds between E.S.E. and W.S.W., it did not afford shelter for ships during high winds between those quarters; and the inner harbours of Hamoaze and Cattewater, in the estuaries of the Tamar and the Plym, had to be resorted to.

Plymouth Breakwater. The idea of protecting the Sound by a breakwater was started in 1806; and after numerous proposals had been made and investigated, the design of Mr. Rennie, consisting of a detached breakwater of rubble stone, was adopted in 1811. (Plate 3, Fig. 8; and Plate 6, Fig. 1.) The advantages of this position are that this portion of the entrance to the Sound was obstructed by rocks, and that the eastern and western channels are left intact, thus affording the original facilities for the entrance of vessels, and interfering as little as possible with the currents in the bay.

The original intention was to raise the mound merely to half-tide level, as it was considered that this height would be sufficient for shelter, and that the deposit of silt inside would be prevented by allowing the waves to break over it.

The breakwater consists of a central portion facing nearly due south, 3000 feet long, and two side arms, on each side, slightly inclined towards the north and having a length of 1050 feet on the top, making the total length of the breakwater 5100 feet exclusive of the slopes at the ends. The object of the angular line was to prevent an accumulation of waves, and to offer less interference to the currents. The extremities are circular, with a lighthouse on the western end and a beacon on the other. The opening left on the western side is 4300 feet, and 2200 feet on the eastern side, with depths of from 5 to 7 fathoms.

Construction of Plymouth Breakwater. The work was commenced in 1812. The stone was procured from quarries close to the estuary of the Plym; it was run in wagons on to specially designed vessels, capable of holding four rows of wagons, from which it was discharged, on reaching the breakwater, by means of a tilting platform at the stern of the vessel[1]. There were ten of these special vessels of about 80 tons burden, and forty-five smaller ordinary vessels employed on the work. The deposit of rubble was commenced in the centre, and carried forward gradually towards each end. A portion was raised above low water in 1813; and it was decided, in 1814, to raise the whole mound to 2 feet above high-water spring tides, so as to ensure a more perfect protection for the smaller class of vessels. The annual deposit of stone averaged about 220,000 tons up to 1821, when the greatest deposit of 373,000 tons was accomplished. After this the deposit was continued at a slower rate for some years, averaging 73,000 tons annually between 1822 and 1829. Up to 1824 the mound was formed on the sea side to a slope of 3 to 1, and was not much affected by the waves, except during a very severe gale in 1817 when the rubble stone was displaced between high and low water mark along a length of 600 feet, and a portion containing stones of from 2 to 5 tons was thrown over the breakwater on to the harbour slope, reducing the inclination of the sea slope to 5 to 1. A violent storm, however, occurred in 1824, and threw a large quantity of stone on to the harbour slope along a distance of 2400 feet of the completed western portion of the breakwater, flattening the sea slope again between high and low water.

As experience had proved that a 3 to 1 slope on the sea side was not stable during violent storms, it was decided, in 1825, to form a 5 to 1 slope on the sea side, and to pave it with granite set in cement and firmly connected from low water to the top. (Plate 6, Fig. 1.) This alteration entailed a considerable addition to the mass of the mound; and between 1830 and 1833 as much as 634,000 tons of stone

[1] 'An Account of the Breakwater in Plymouth Sound,' Sir John Rennie, p. 20.

were deposited. Subsequently to this it was only necessary to deposit stone to make up deficiencies in the mound, due to displacement or consolidation; and up to 1847 the average annual deposit for this purpose was 25,500 tons, the total amount placed in the mound from the commencement to that date being 3,620,000 tons.

The total cost of the work was £1,500,000, including the lighthouse at the western end with its foundations carried up solid from low water.

Maintenance of Plymouth Breakwater. As the breakwater at Plymouth is exposed to the Atlantic on the south-west, it is from that quarter that the worst waves roll in, and it consequently is on that side that the greatest damage was done to the breakwater during construction, and that the principal need for maintenance occurs now. The waves tend to displace the rubble foreshore down to 10 feet below low-water spring tides, and up to about half-tide level on the western portion of the breakwater, and especially when recoiling against the western face of the vertical foundation of the lighthouse. The recoil drags the foreshore down seawards; but the greater portion of the displaced rubble is thrown over the breakwater from the sea slope, or driven round the western extremity from the face of the lighthouse foundation. The breakwater, accordingly, requires to be maintained by regular deposits of stone. The foreshore of the central and eastern portions of the breakwater are comparatively little disturbed by the waves, as the worst waves from the south-west do not strike them so directly as the western arm, and pass over the top of the breakwater.

Portland Harbour.

Situation. The site of Portland Harbour, like that of Plymouth, is peculiarly well adapted by nature for a harbour. The position of Portland Island, forming a bay facing the general coast-line, gives shelter to a spacious roadstead from the west and south; and the bay only needed artificial shelter from the south-east to form one of the finest harbours in the world. (Plate 3, Fig. 11.) Portland was naturally one of the places selected when the construction of national harbours along the coast, for refuge and defence, was decided upon in 1847.

Portland Breakwaters. Shelter has been provided for Portland Harbour from the south-east and east by two breakwaters; an inner breakwater connected with Portland Island, 1700 feet long, and an outer, or detached breakwater, 6400 feet in length, lying to the north-east of the inner one. (Plate 3, Fig. 11.) There is an opening 400 feet wide left between the two breakwaters affording an entrance from the south-east; and the main entrance to the north extends across the whole width of the harbour, and is reached by rounding the northern extremity of the detached breakwater; it is amply sheltered by the mainland on the north. The total area sheltered is 2130 acres.

As an abundance of stone could be procured close at hand from the Portland quarries, the system of a rubble mound was naturally resorted to at Portland (Plate 6, Fig. 2), in spite of the difficulties that had been previously experienced at Plymouth, and the damages which occurred in the still earlier similar construction at Cherbourg. The stone, however, used at first at Cherbourg was of small size, and therefore was readily disturbed; and the situation of the Portland breakwater differs in one important respect from that of Plymouth. It has been pointed out that the chief damage at Plymouth has been due to south-westerly storms, in which direction it is open to the Atlantic Ocean; whereas Portland, being further up the English Channel, is more sheltered, and the breakwater, being protected by the land from the west, is merely exposed to the east, where the waves have only the fetch of the Channel. Owing to this difference, it has not been necessary to pave the sea slope of the Portland mound;

and in consequence of the smaller rise of tide and therefore the smaller extent of the flat slope, it has been possible to give the mound at Portland a less width than at Plymouth, though situated in a greater depth of water. (Compare Plate 6, Fig. 2 with Fig. 1.)

The inner breakwater has been constructed on the mixed system, having a rubble base, and a superstructure founded at low water to serve as a quay; and it will therefore be referred to again in Chapter XIII. (Plate 7, Fig. 11.)

Construction of the Outer Breakwater at Portland. The outer breakwater at Portland resembles the breakwaters of Cherbourg and Plymouth in being detached. As, however, it is only separated from the inner breakwater by an interval of 400 feet, the deposit of the mound from staging with wagons run out from the shore, by temporarily bridging over the space left for the entrance, was more easily accomplished than by barges. The wagons, after being filled in the quarries, were let down inclines, and run out to sea on wooden staging carrying five lines of way and supported on screw piles. The piles were formed of two creosoted beams fastened together by bolts and bands; and the iron screw fastened to their lower end was made to penetrate the clay bottom to a depth of 6 or 8 feet. The piles were placed 30 feet apart. The wagons were fitted with a movable end so adjusted that, on the arrival at the place of deposit, by the release of two triggers the wagon tipped its contents into the sea. Ridges of stone were thus formed along the top of the mound under each road of the staging; and the waves gradually levelled these ridges, and consolidated the mound to stable slopes. This breakwater, like the one at Plymouth, is level at the top, and has a flat slope on the sea side down to about 12 feet below low water; and the harbour slope, and the lower portion of the sea slope, are much steeper, having an inclination of about 1¼ to 1.

Convict labour was employed for quarrying the stone and

loading it into the wagons; and the contractors conveyed it to the breakwater, and deposited it. The staging was raised 18 feet above high water, which placed the roadway out of reach of the waves, and rendered the work independent of the weather. Arrangements were provided by the contractors for depositing the stone at the rate of 3000 tons per day; but the supply furnished from the quarries by the convicts was irregular. The mixture of large and small stone was deposited in the mound just as it came from the quarries, a due proportion of large stones of from 3 to 7 tons being maintained. The mound was thus made compact; and the large stones protected it from disturbance when consolidated.

Progress of Works at Portland. Preliminary works were commenced in 1847; and the first deposit of stone in the inner breakwater took place at the end of 1849. The staging had been carried across the opening, and the outer breakwater was begun in 1852. The annual deposit of stone varied considerably, owing to the variation in the number of convicts employed in the quarries; it reached a maximum of 553,400 tons in 1860, and was only 39,700 tons in 1863, the yearly average being 494,000 tons between 1853 and 1860, after which it diminished to an average of 137,000 tons annually till 1866, when, the breakwater being nearly completed, the deposit of stone was confined to making up deficiencies in the mound, averaging each year 26,000 tons up to 1871.

The total amount of stone deposited in the mound, from the commencement in 1849 up to 1871, was 5,731,000 tons[1]. The only portions of the outer breakwater not composed of rubble stone were the two pierheads at each extremity, which were built of masonry and founded 24 feet below low water, their foundations being protected by large stones.

[1] One ton of stone deposited in the mound is reckoned to measure 20 cubic feet; and as 15 cubic feet of Portland stone, on the average, weighs one ton, the interstices in the mound amount to about one fourth of the whole mass.

The damage during construction was confined to the washing away of some of the staging during south-easterly gales, and the subsidence of a portion of the masonry of the northern pierhead owing to an inequality in the strata at the bottom.

The rate of progress varied, both owing to variation in the supply of stone, and also to changes in depth; but the average annual advance of the breakwater was about 450 feet, and on two occasions as much as 1000 feet of staging were erected in a year. When the removal of the staging was commenced in 1867, arrangements were made for completing the deposit of stone on the outer breakwater by means of barges.

Disturbance of Mound at Portland. Though the breakwater has had ample time to consolidate, and though the exposure is comparatively moderate, the slopes are still liable to injury from the sea between high and low water levels. In easterly and southerly gales the stones of the outer slope are displaced by the recoil of the waves; whilst during north-easterly gales the waves break over the mound and disturb the inner slope. The stone thus displaced amounts sometimes to between 500 and 3000 tons.

Cost of the Works at Portland Harbour. The total cost of the harbour works, from their commencement up to 1871, amounted to £1,034,000, which is equivalent to £127 per lineal foot of breakwater. This cost does not appear to include the convict labour. Such labour, however, has the drawback of being uncertain, and of requiring special arrangements. Thus the contractors had to provide and maintain plant sufficient to discharge 3000 tons of stone daily; but frequently a portion of this plant had to lie idle owing to deficiencies in the supply of stone.

Comparison of Cost of Portland and Plymouth Breakwaters. The cost of the Portland breakwater naturally contrasts favourably with that of Plymouth, which averaged £294 per lineal foot, for it was constructed at a time when much greater experience had been gained in large works; it was formed by stone run out from the shore, instead of being conveyed in barges which are hindered by bad weather; it has required no paving, owing to its less exposed situation; and convict labour was provided at a merely nominal cost. The depth of water, indeed, was more favourable at Plymouth, where there is a maximum of 7 fathoms as compared with 10 fathoms at Portland; but this advantage is to a great extent neutralised by the greater rise of tide at Plymouth, 15½ feet as compared with 6¾ feet, necessitating a much greater length of flat slope.

DELAWARE HARBOUR.

Situation. The harbour of Delaware is situated at the south-eastern extremity of Delaware Bay, which is an extensive estuary, or creek, into which the Delaware river flows. The harbour is formed in a secondary bay, protected to the south and west by the mainland of Delaware, to the north by the opposite coast of New Jersey, and to the south-east by Cape Henlopen forming the south-eastern extremity of the bay. The bay, however, was open to the north-east; and owing to the great length of the estuary in a north-westerly direction, and the drifting ice from the Delaware river, it needed protection also from that quarter.

Delaware Breakwaters. Artificial shelter has been provided for the harbour by two breakwaters facing north-east and north-west, called the 'Breakwater,' and 'Ice-breaker' respectively, and separated by a gap of about a quarter of a mile[1]. (Plate 3, Fig. 10.) The breakwater is 2556 feet long on the top, and the ice-breaker 1359 feet; they both consist of rubble mounds formed of stones weighing from ¼ ton to 7 tons, brought by boats from the nearest

[1] A plan and sections of the breakwaters, as well as some information about the works, were supplied me by the Chief of Engineers, U. S. Army.

quarries on the mainland. (Plate 6, Fig. 6.) The smaller stones were placed in the base and interior of the mound, and the large stones on the top and slopes. The mound has been raised 14 feet above low water, the rise of tide being 11 feet.

Progress of Breakwaters at Delaware. The breakwaters were commenced in 1829; and 835,000 tons of stone were deposited in the following ten years, giving an average annual deposit of 83,500 tons, at a total cost of £391,000. After this, the works were carried on irregularly till 1869, when they were completed as they exist at present, an average of about 9000 tons having been deposited annually during the last three years of the works. The total expenditure up to 1869 amounted to £442,300.

The average width of both breakwater and ice-breaker is 22 feet on the top, and 160 feet at the base.

Maintenance of Delaware Breakwater. It was estimated, in 1869, that a sum of £5100 would be required for repairs in a period of fifteen years; but it appears that up to 1883 no appropriation had been made for that purpose. As recently, however, as 1879, the mound was reported to be still showing signs of movement[1]. In 1878, the foundations of the lighthouse at the north-western extremity of the breakwater had to be repaired and extended; and several other parts of the breakwater have settled. It appears from the section that some of the rubble has been carried over the breakwater into the harbour. (Plate 6, Fig. 6.)

State of Delaware Harbour. As Delaware is the principal harbour of refuge along that exposed coast, it is largely resorted to by coasting vessels; and the number that seek its shelter has increased so far beyond any anticipation, that it cannot at times accommodate half the vessels that desire to use it. Its value and capabilities are considerably reduced by the swell which enters through the wide gap

[1] 'Report of Chief of Engineers (United States) for 1879.' p. 453.

between the breakwaters. Moreover, whilst its shelter is yearly becoming of more importance, its depth is gradually decreasing by the deposit of silt. This is mainly due to the introduction of silt through the gap by the waves, and also to the checking of the sweep of the tides across the harbour by the current through the gap. The deterioration has also been promoted by neglect in removing wrecks in the harbour, which have each furnished a nucleus towards the formation of a shoal. The closing of the gap was, accordingly, to be commenced in 1883; and it is estimated that the work will be completed in five years. It is believed that the closure of the gap will prevent further deposit in the harbour, and remove the silt from the neighbouring shoals; it will also nearly quadruple the area protected by the breakwaters.

Design of Breakwater across the Gap at Delaware. In order to reduce the amount of material required for closing the gap, and to dispense with any maintenance, it has been decided to form the connecting breakwater (shown by dotted lines on the plan and section) with a concrete superstructure resting upon a rubble mound of granite. (Plate 3, Fig. 10; and Plate 6, Fig. 6.) The estimated cost is £114,600.

A timber bridge is first to be constructed across the gap, the outer piles of which will serve to support the framing for the concrete superstructure. The rubble base will next be deposited from the bridge, right across the gap, and raised to 12 feet below low water, with stones of from 1 to 3 tons, 48 feet wide at the top and with slopes of 2 to 1. Upon this base the concrete superstructure will be built, 24 feet high, 24 feet wide at the base, and 12 feet at the top. The concrete is to be deposited in framing, in 16-feet lengths corresponding to the bays of the bridge, connected by tie bolts placed in pipes through the concrete, so that both framing and bolts can be used again for successive lengths.

The simple rubble-mound system will, accordingly, be replaced by the mixed system; and the breakwater will present the peculiar combination of a mixed breakwater connecting two rubble mounds.

Comparison of Delaware Breakwater with Plymouth and Portland. The rubble mound at Delaware has a much steeper sea slope, from the top down to a few feet below low water, than at either Plymouth or Portland, indicating that either the exposure must be less, or the stone protecting the sea slope must be larger. (Compare Plate 6, Fig. 6, with Figs. 1 and 2.) As the harbour is almost on the Atlantic, it is probable that the breakwater is somewhat sheltered by the northern coast of the bay, and that north-easterly storms are not severe in those parts. The general depth of the site of the breakwater is about 5 fathoms, somewhat less than Plymouth.

The cost of the breakwater was only £113 per lineal foot; less than Portland, and little more than a third of Plymouth breakwater. This result, where labour is dear and the work was spread over a number of years, and with breakwaters at some distance from the land, must be due to the much smaller mass of the mound, owing mainly to the steep slopes, and also to the less depth.

In spite of the comparatively small mass of rubble per lineal foot, it has been deemed expedient to abandon the system previously carried out there, as well as at Plymouth and Portland, and to resort to the mixed system of rubble mound with superstructure. This new breakwater will form an interesting example of a superstructure formed with concrete-in-mass deposited in framing under water on a rubble mound; the only previous instance being that of Fiume breakwater on the Adriatic. (See p. 135.)

HOWTH HARBOUR.

Site. The harbour of Howth in Ireland is situated to the north of the hill of Howth which forms the north-eastern extremity of Dublin Bay. It was hoped that, being placed beyond Dublin Bay, it would be exempt from the silting-up influences taking place in that bay, and would, consequently, afford a better and more permanent depth of water than could be obtained in the bay itself; whilst the proximity of the site to Dublin, and its favourable position for communication with England, gave a good prospect of its becoming an important Irish port. The area enclosed is 52 acres.

Breakwaters at Howth. The harbour is formed by two breakwaters projecting from the shore in a north-easterly direction, and slightly converging. The eastern breakwater, at a distance of 1300 feet from the shore, turns towards the north-west for a length of 1150 feet, and almost overlapping the straight western breakwater 2020 feet long, leaves an opening between their extremities of 300 feet pointing to the north-west.

The breakwaters consist of a rubble mound with a facing of granite on the harbour side to serve as a quay, and are very similar in section to the Kingstown breakwaters subsequently erected. (Plate 6, Fig. 3.) Though the works are not large, and are fairly sheltered, they present an interest as having been begun before the Plymouth breakwater, and therefore the earliest instance of the application of the rubble-mound system to a harbour of any importance on the British coast. They were commenced in 1807, and only completed in 1825; their cost was £346,600, which is equivalent to £77 per lineal foot. This comparatively low cost was due to the small depth, not exceeding 11 feet at low water, the connection of the breakwaters with the shore, and the proximity of the quarries.

State of Howth Harbour. The entrance is sheltered on the north by the island called Ireland's Eye, it points away from the Irish Channel towards the coast, and is out of the run of the tide. This position, whilst favourable to the tranquillity of the harbour, has been prejudicial to its maintenance; for sand readily accumulates in this sheltered area, and there is no scouring tidal current across the entrance. The projection, moreover, of the western breakwater has arrested the travel of sand in the bay. Accordingly the low-water mark has progressed on the western side of the harbour, so that it now extends out two-thirds of the distance along the western breakwater; and there is a depth of only 7 feet at the entrance. The sand also, travelling along from the west, readily enters the harbour, where it has deposited to such an extent that about two-thirds of its area is dry at low water.

KINGSTOWN HARBOUR.

Situation. Kingstown Harbour is situated on the southern shore of Dublin Bay, and, like Howth Harbour, was established for the purpose of providing a port for Dublin with a better depth than the bar at the mouth of the Liffey afforded. Though placed in Dublin Bay, it is sufficiently to the east of the South Bull strand to be, at present, beyond the influence of the gradual advance of the western foreshore of the bay. The tidal currents, moreover, sweep across its entrance, which is situated in a depth of 22 feet at low water, so that no deposit occurs in front of the entrance; and the waves in that depth do not introduce much sand into the harbour.

Breakwaters at Kingstown. The harbour has an area of 250 acres, and is nearly symmetrical in form, being enclosed by two breakwaters, 4280 and 5160 feet long respectively, projecting at right angles from the shore, about 4000 feet apart, and then converging by a series of angular bends, so that an opening of only 750 feet is left between their extremities. (Plate 4, Fig. 6.) The breakwaters consist of a rubble mound of granite, in pieces weighing from $\frac{1}{4}$ ton to 10 tons and upwards, brought from quarries three miles distant, and run out on a tramway from the shore. The mound was consolidated by the sea till the sea slope assumed an inclination of 5¼ to 1 in its exposed part. The sea slope was then pitched to protect it from the action of the waves; the mound was raised to shelter the roadway on the harbour side; and a sloping quay was formed on the harbour slope. (Plate 6, Fig. 3.) The work was commenced in 1817, and completed in 1836 at a cost of nearly £700,000, corresponding to about £74 per lineal foot of breakwater. The eastern pierhead was founded on the natural bottom of rock and firm sand by means of a diving bell.

The sea slope of the eastern breakwater, being open to the Irish Sea, is liable to be disturbed in severe north-easterly storms, and has occasionally needed repairs.

State of Kingstown Harbour. As the entrance points somewhat towards the exposed quarter of north-east, the interior of the harbour is not as tranquil as might be desirable during north-easterly gales. This was attributed by Sir John Rennie to the width of entrance having been fixed at 750 feet, instead of 450 feet as designed, and to the reduction in area of the sloping beach in the harbour by a railway embankment. Sand brought in during north-east storms has reduced the depth in the south-western corner of the harbour; a slight shoal has formed just inside the entrance, owing to an eddy produced round the pierheads; and a small advance of the foreshore has occurred outside the eastern breakwater; but with these exceptions the depths inside and adjacent to the harbour have not been affected. Kingstown Harbour, accordingly, has proved a successful instance of a closed harbour constructed on a sandy coast, owing to the scour of the currents across the entrance, the depth in which the entrance is situated, and the absence of sandbanks in front from which material for deposit might be collected by the entering waves.

Comparison of Howth and Kingstown Harbours. The harbours of Howth and Kingstown were both designed by

Mr. Rennie; they have both been constructed as closed harbours, with similar breakwaters; and though Kingstown has the advantage in area, Howth is beyond the silting-up influences of Dublin Bay, whilst both are situated on sandy shores. Kingstown, however, has maintained its depth; whereas Howth has silted up. The situation of Howth Harbour is not in the course of the tidal currents, and its western breakwater has acted as a groyne producing an advance of the sandy foreshore; whilst the sandbanks in front of its shallow entrance have furnished a supply of material which the waves readily deposit in the tranquil waters of the harbour. At Kingstown, on the contrary, the tidal currents scour round the breakwaters and prevent deposit; and the introduction of sand is obviated by means of a deep entrance pointing seawards, at the cost of some tranquillity inside.

TEES BREAKWATERS.

The breakwaters which are being constructed for protecting the estuary of the Tees, and promoting the deepening of the entrance, form part of a scheme of river improvement, and as such have been treated of in 'Rivers and Canals'[1]: they will, when completed, furnish an example of the converging jetty system of which instances have been given in the last chapter; and they also belong to the type of rubble-mound breakwaters. It is under this last aspect that they may be suitably referred to here.

Construction of the Tees Breakwaters. The iron furnaces of Middlesbrough and the adjacent Cleveland district, to which the Tees mainly owes its trade, furnish a large supply of slag which is an encumbrance to the iron-masters. This refuse, coming in blocks of about 3 tons from the furnaces, has been utilised for the formation of the breakwaters. The blocks of slag are turned out at the works on to iron trollies,

[1] 'Rivers and Canals,' L. F. Vernon-Harcourt, p. 293, and plate 19.

which convey them along a railway to the end of the break-water, where they are tipped into the sea. The breakwaters, accordingly, consist of mounds of slag, for the removal of which the iron-masters pay a sum about sufficient to defray the cost of carriage; so that the material is actually deposited in the mound at a nominal cost.

The South Gare breakwater, commenced in 1863, was approaching completion in 1883. It has a length of 12,810 feet, and its sea slope is protected by a concrete wall from a little below low water up to high-water level. (Plate 6, Fig. 8.) This solid protection is specially needed, as the blocks of slag are broken up when exposed to the waves. The movement of the slag towards the estuary at the end of each temporary termination is very marked on the inner side of the breakwater, where a series of ridges extend out at intervals from the line of the inner slope, indicating the effects of the storms of each winter on the end of the slag tip. The permanent extremity is protected by a head of concrete blocks. Nearly 4 million tons of slag have been deposited in the South Gare breakwater, which extends for a considerable distance over shallow sandbanks and terminates in a depth of about 4 fathoms.

The North Gare breakwater is in progress, and it is estimated that it will be completed in 1891, nine years from its commencement; its length is to be about 6000 feet. An opening of 2400 feet is to be left between the extremities of the breakwaters, which are 3¼ miles apart at their commencement.

Remarks on the Tees Breakwaters. The chief feature of these breakwaters is their formation with hard heavy refuse material which practically costs nothing. The slag forms an excellent backing, but, unfortunately, owing to the fracture of the large blocks when disturbed by the waves, it is not well suited to resist the action of the sea, and has to be protected by more durable material.

The rate of progress is about 650 feet annually, depending upon the supply of slag that can be obtained from the furnaces.

The effect of the converging breakwaters on the bar will not be ascertained till the northern breakwater approaches completion; but the large tidal capacity of the estuary within the breakwaters should create a considerable scour; and the breakwaters will afford shelter to dredging operations in the inner channel, will prevent the waves washing the sand from the adjacent banks into the channel, and will render the access to the river much safer. Sand appears to have accumulated a little in the angle between the southern breakwater and the shore, but not hitherto to an extent to cause anxiety about the maintenance of the harbour; and the direction of the breakwaters, and their projection from the shore-line, are not likely to affect prejudicially the sweep of the littoral currents.

CETTE HARBOUR.

Description. Cette is a port of some importance on the Mediterranean, and forms the outlet of the Languedoc Canal. Its harbour is purely artificial, being formed by two converging breakwaters from the shore, which shelter the inner harbour and the entrance to the docks and canal; whilst a curved detached breakwater protects the entrance to the inner harbour, shelters an outer roadstead, and provides two entrances to the port. (Plate 5, Fig. 16.)

Detached Breakwater at Cette. The outer breakwater is similar in construction to the breakwaters at Kingstown, being formed of a rubble mound with a roadway on the top protected by raising the mound; but the harbour slope, not being required for a quay, is left at its natural inclination, and the sea slope is protected by large concrete blocks near the water-level. (Plate 6, Fig. 7.)

A shoal appears to have formed under the shelter of the concave side of the breakwater.

TABLE BAY HARBOUR.

Situation. Table Bay, situated on the south-west coast of Africa to the north of the Cape of Good Hope, has a wide opening towards the west of about 21 miles. Cape Town stands in the southern recess of the bay which is exposed to the north-west, but is well sheltered from the west to south-east by the land, and is more or less protected round to the north by the opposite shores of the bay. Though the bay is not exposed to the huge waves which are encountered south of the Cape, where the winds blow uncontrolled over the vast expanse of the southern Pacific Ocean, yet it is open to the Atlantic Ocean, and the wrecks used to be frequent in the winter months.

Design of Table Bay Harbour. Various schemes were proposed for the protection of shipping coming into the bay; and at last Sir John Corde's plan of a straight breakwater running out from the shore, at a point a little to the north of Cape Town, for a distance of 3250 feet, to provide shelter from the north-west, was commenced in 1860. (Plate 4, Fig. 8.) The recess even of the bay is too extensive and wide for more than a limited shelter to be provided by such a work; but it was calculated that the breakwater would protect an area of 867 acres from north-west winds, which are much the most prevalent and severe. The area thus sheltered is somewhat exposed from north round to south-east; but to the north, the land is not more than 21 miles distant, and only 5 miles away in the line of the breakwater; and winds from those quarters are rare, and never very high. Moreover, an outer and inner basin, where vessels can lie in perfect shelter, formed part of the scheme; and it was proposed that the western side of the breakwater should be used during easterly winds, but it appears that vessels never lie on that side.

The original design has been modified by giving a southerly

bend to the outer portion of the breakwater now in progress, whereby the deficiency of shelter from the north-east will be remedied. (Plate 4, Fig. 8.) More efficient means of shelter have been under consideration, consisting in the formation of a southern pier parallel to the breakwater, and about 1500 feet from it, with return arms from the pier and the breakwater, thus enclosing an area of about 62 acres, the entrance to which, 240 feet in width, would be protected by the extended bend of the breakwater. This proposal has, however, been postponed till the breakwater extension is completed, and its extent of shelter ascertained.

Table Bay Breakwater. The breakwater consists of a rubble mound deposited from staging, like the Portland breakwater[1]. (Plate 6, Fig. 4.) The stone was procured from the excavation of the inner basin; but as the supply of sufficiently large stone from this source was less than had been anticipated, the breakwater was temporarily terminated at a distance of 1900 feet from the shore in 1868. Gales in 1862 and 1865 damaged the staging, and washed down a portion of the mound. In the latter gale seven bays of staging were washed away, thirty more were injured, 270 feet of the head were washed down, and, altogether, 60,000 tons of stone were displaced. The repairs of the damage done by this one storm occupied six months, and cost £3000.

The first portion of the breakwater only extends into a depth of 5 fathoms; but the extension of 1800 feet, now in progress, will reach a depth of 8 fathoms. Even in the smaller depth, the base of the mound is as wide as that of the Portland breakwater in double the depth of water; and the sea slope very nearly corresponds with the Plymouth section, though the rise of tide at Table Bay is 5½ feet, only about a third the rise at Plymouth. (Compare Plate 6, Fig. 4, with Figs. 1 and 2.) The exposure at Table Bay is probably very similar to that of Plymouth; but the absence of pitch-

[1] A section of the breakwater was sent me by Mr. Jenour, the late resident engineer.

ing at Table Bay necessitates a flatter slope. The temporary end was protected by 80-ton concrete blocks formed in place.

The mound appears to have gradually assumed a stable slope under the action of the sea, having an inclination, on the sea side, of about 9 to 1 from low water to 13 feet below. Stone has occasionally been deposited on the outer slope, and round the end, from the excavations for a graving dock; but this has been done rather to strengthen the mound than merely to maintain it.

Extension of Table Bay Breakwater. Though the first length of breakwater has afforded more shelter than might be imagined from its shortness and the extent of the bay, owing mainly to the infrequency of northerly winds, yet during severe storms serious casualties continued to occur to the shipping. A Commission was, accordingly, appointed in 1878 to report upon the subject, and urged the extension of the breakwater. The sanctioned extension is shown by dotted lines on the plan (Plate 4, Fig. 8); it was commenced in 1881. In design and mode of execution it is precisely similar to the earlier portion of breakwater; but as it will extend into deeper water, and be more exposed, it is probable that the sea slope will eventually assume a still flatter inclination. The stone is being procured from a site suitable for a future basin, the smaller and lighter portion of the excavation being used for reclaiming some land suitable for wharves along the protected shore, or for ballasting vessels, and the harder and larger portions only are being deposited in the mound.

Progress and Cost of Table Bay Breakwater. The first 1900 feet of the breakwater were executed in about eight years, giving an average rate of progress of 235 feet annually; and it is expected that the extension of 1800 feet will occupy about nine years, which would be a yearly advance of 200 feet, or rather a better rate of progress than formerly, allowing for the additional depth in which it has to be constructed.

The report of 1859 stated a period of ten years, and a cost of £396,475, for the whole length of 2350 feet then contemplated and the inner works, on the assumption that convict labour would be available. Free labour, however, has had to be employed; and the estimated cost has been considerably exceeded. The estimate for the extension of 1800 feet is £286,200, or an average cost of £153 per lineal foot. The total cost of the construction of the breakwater, docks, basins, jetties, stores, etc., amounted to £804,800 up to the end of 1881, exclusive of the cost of a graving dock and the breakwater extension[1]. Probably the increased cost was due to some extent to the much flatter slope assumed by the mound at Table Bay than its more sheltered prototype at Portland, which has necessitated a much larger supply of stone than a section based on the experience of Portland would indicate.

Inner Works at Table Bay. The accessory works to the breakwater for providing facilities for the trade of the port are shown on the plan. (Plate 4, Fig. 8.) They consist of a coaling jetty, extending out from the inner face of the breakwater; an outer basin of 6 acres sheltered by piers; an inner basin, or dock, of 10 acres; a slip-way, and a graving dock 530 feet long. Additional dock accommodation will be provided by the excavations for material for the breakwater extension; and the large outer harbour contemplated would provide increased shelter for vessels.

State of Table Bay Harbour. The harbour is at present deficient in shelter from the north, but this will be corrected when the bend is carried out; and if this should not prove sufficient for the increasing trade, a prolongation of the bend will enlarge the sheltered area.

There has been no indication of the accumulation of silt in the site of the harbour, either before or since the con-

[1] The total expenditure on the construction and maintenance of the harbour works amounted to £1,437,200 up to the end of April, 1884.

struction of the breakwater. The only currents in the bay are produced by the wind; and the Salt river, which flows into the south-eastern end of the bay, does not affect the coast in the vicinity of Cape Town. Accordingly, the projection of the breakwater produces no interference with prevailing currents; and there is no prospect of any reduction in the depth of the harbour.

BREST HARBOUR.

Description. At the extreme western projection of the French coast, stretching out as far towards the Atlantic Ocean as the Lizard Point, there is a large inlet into which three rivers flow, and which is separated from the sea by a channel a little more than a mile wide at its narrowest part. This channel has, for the most part, a depth of at least 11 fathoms, being scoured by the ebb and flow of the tide in the extensive roadstead of Brest. The naval harbour of Brest is situated in the estuary of the Penfeld, on the north side of the bay; and till recently merchant vessels were allowed the use of a certain length of quay on the same river, but their movements were much impeded by the vessels of war which had the precedence.

Commercial Harbour of Brest. Various designs were proposed, from 1769 downwards, for the formation of a commercial harbour, quite distinct from the naval station, a little to the east of the mouth of the Penfeld. At length, in 1869, just one hundred years after the first scheme, the plan now being executed was approved. (Plate 3, Fig. 6.) The design includes a tidal harbour, of 101 acres, which has been made, and a dock which is proposed. The tidal harbour is enclosed by jetties, serving as quays projecting from the shore which has been embanked, and by a detached breakwater on the southern or outer side, leaving entrances in front of the main jetties on each side, 460 and 390 feet wide respectively.

Detached Breakwater and Jetties at Brest. The protecting breakwater is 3117 feet long; it consists of a rubble mound, with a slope of about 7 to 1 on the sea side, paved with stones of ⅔ to 1 ton from about half-tide to high-water level; there is a roadway on the top protected by a slight parapet, and a pitched 1 to 1 slope on the harbour side set in mortar towards the extremities of the breakwater[1]. (Plate 6, Fig. 5.) The mound was deposited from iron hopper barges.

The jetties connected with the land are very similar to the breakwater, with the exception that, on the harbour side, they are faced with a quay wall founded, about 20 feet below low water, on a rubble base.

The rubble mound affords a broad foundation for the breakwater and jetties on the silty bottom. With the exception of a rise of tide of 21 feet, and a certain amount of exposure through the approach channel to the bay, the breakwater is peculiarly favourably situated, and should need no maintenance beyond repairing the settlement which occurs on the silty foundation. The sea slope, indeed, appears flatter than necessary, except to reduce settlement and secure the base from scour.

Peculiarity of Brest Harbour. The tidal harbour of the commercial port of Brest resembles in plan the harbours of Marseilles and Trieste, on the Mediterranean, which are not exposed to the fluctuations of the tide, and where, consequently, open basins, sheltered by an outlying breakwater, are very suitable. At Brest, on the contrary, the vessels are subjected to a considerable tidal range, which must be inconvenient for trade; though adequate depth has been secured, at all times of tide, by dredging down to 24½ feet below low water. The proposed dock will render it possible for vessels to be placed out of reach of the tide, as at other tidal ports.

Cost of Brest Tidal Harbour. The total cost of the tidal basin has been estimated at £930,000. The detached break-

[1] A section of the breakwater was furnished me by M. Fénoux, engineer-in-chief of the harbour.

water cost the moderate sum of £24⅓ per lineal foot, and the deeper quays £21⅓ per lineal foot.

Review of Rubble-Mound Breakwaters. The section of a rubble-mound breakwater affords a very correct indication of the exposure of the site, provided similar sized rubble is employed. Thus a glance at Plate 6 suffices to show that the exposure at Plymouth and Table Bay is much greater than at Portland and Delaware. The two first are open to the Atlantic on the west, from which quarter the most prevalent and worst winds blow, whilst Portland and Delaware breakwaters face towards the east, which is a less stormy quarter, and Portland breakwater lies in a bay in the English Channel. If the Portland mound had been situated to the west, instead of to the east of Portland Bill, it would have presented a very different appearance; it would have required a flat slope on the sea side like the Plymouth breakwater, and would have proved a more reliable model for the slopes of the Table Bay mound. The slope of the east breakwater at Kingstown is somewhat flatter than that of Portland, owing to its more exposed situation and the greater rise of tide. The width at the base of the Kingstown breakwater is increased by the roadway on the top, and the height of the mound protecting it. The very flat slope of the Brest breakwater is attributable to the small size of the material used, the rise of tide, and the necessity for a broad base on the silty foundation. The section of Delaware breakwater presents a striking contrast to that of Brest, and indeed to all the rest, in the steepness of its sea slope. This indicates that gales from the north-east cannot be severe in that locality, and that the waves are reduced by outlying shoals. Moreover, the breakwater has needed repairs, though protected on the surface by large stones.

The range of tide forms an important element in considering the slope of a rubble mound; for, since the weak portion of the mound is near the water-level, where it is exposed to the breaking and recoiling wave, a great variation in the tide increases the extent of slope exposed to disturbance. In this respect, Brest is the least favourably situated of the examples given, but it is well sheltered and in shallow water. Plymouth combines the disadvantage of a rise of tide of 15½ feet with great exposure and a depth of 7 fathoms. Table Bay, with a similar exposure and rather less depth of water hitherto, has a rise of tide of only 5½ feet, and is therefore somewhat more favourably situated. The greater depth at Portland is amply compensated for by the small rise of tide, and more especially by the very moderate exposure.

The three most important points which have to be considered in designing a rubble-mound breakwater are: exposure, rise of tide, and depth.

Three methods have been adopted for protecting the rubble mound: (1) Large stones deposited on the surface of the sea slope, as carried out at Delaware and Table Bay; (2) Pitching the sea slope, which has been adopted at Plymouth and Kingstown; (3) Depositing large concrete blocks on the exposed part of the mound, of which Cette is an instance.

A rubble mound is seldom perfectly stable, and is liable to gradual disintegration or displacement under the action of waves. Even the large stones at Portland and Delaware, and the pitching at Plymouth and Kingstown, do not preserve these breakwaters from the necessity of maintenance; and though huge concrete blocks might prove more effectual, these also are exposed to damage from the hammering of the rubble against them.

CHAPTER XI.

HARBOURS PROTECTED BY RUBBLE AND CONCRETE-BLOCK MOUND BREAKWATERS.

ALGIERS. PORT SAID. BIARRITZ. ALEXANDRIA.

Algiers Harbour: Early History; Breakwaters, protection with Concrete Blocks, extension, construction, comparison with Plymouth and Portland sections; State. *Port Said Harbour:* Description; Breakwaters, construction, method of depositing Blocks; Condition. *Biarritz Harbour:* Breakwater, composition, failure. *Alexandria Harbour:* Description; Breakwater, construction, mode of depositing Blocks. Comparison between Algiers, Port Said, and Alexandria Harbours. Advantages of the Concrete-Block Mound.

THE formation of rubble-mound breakwaters dates back to the time of the Romans; and the construction of such breakwaters as Portland and Table Bay consists merely in adapting the mechanical appliances of the present day towards the more rapid and economical execution of works on the ancient system, but on a more extended scale. It has been seen, in the instance of Cette breakwater, that concrete blocks were used to remedy the defective stability of the rubble mound. This system, however, has been applied, not merely to the protection of the mound, but also to enable it to be constructed with steeper slopes, and consequently with less material. The merit of this improvement on the simple rubble mound is due to M. Poirel, a French engineer, who introduced it in 1834 for repairing the old rubble mounds of Algiers harbour[1].

ALGIERS HARBOUR.

Early History. A little island, situated a short distance from the shore in front of the town of Algiers, was united

[1] 'Mémoires sur les Travaux à la Mer, M. Poirel, 1841.

with the mainland in 1530 by a rubble embankment, 574 feet long and 118 feet wide at the top. As this embankment faced north, it protected the port in that direction. A breakwater of rubble stone was also carried out from the island in a southerly direction, for a distance of 410 feet, to provide shelter from the east. (Plate 5, Fig. 1.) These two mounds, together with the island, formed a sheltered basin of about 11 acres. Both the embankment and the breakwater needed constant maintenance, as the rubble was disturbed by the waves. The breakwater, standing more in the sea and facing the worst winds, was most exposed to injury; and the Turks used to employ a large number of slaves, and expended large sums annually, in making good the ravages of the sea.

After Algiers was taken by the French in 1830, the foundations of the buildings on the embankment, and of the pierhead on the breakwater, were repaired and strengthened; but the sea still attacked the mounds on which they stood.

Protection of Mounds at Algiers with Concrete Blocks. When M. Poirel took charge of the works in 1833, he collected a number of blocks of stone, averaging between 2½ and 5¼ cubic yards, the largest the quarries could produce, which he deposited to the extent of 8000 cubic yards on the slopes of the mounds. During the following winter these blocks were disturbed by the sea, the slopes were flattened to an inclination of 6 to 1, and some of even the largest blocks were thrown into the harbour. It was manifest that still larger blocks were necessary for withstanding the waves; and as it was not practicable, either to quarry larger blocks, or to transport them from the quarries, concrete blocks were resorted to. The size first adopted was 26 cubic yards; but it was subsequently found that blocks of 13 cubic yards remained undisturbed by the sea.

Two methods were adopted for placing these concrete blocks; in one case the blocks were manufactured on the site they were to occupy, and in the other case they were constructed on the bank, and were slid down slips to their destination.

The blocks made in place were formed by depositing the concrete in large bottomless timber frames, lined with tarred sacking. The frames were constructed on shore, and were towed to their position, where they were fixed by weighting them with little boxes filled with pig iron hung round the outside of the frame. The sacking was commenced at 1⅔ feet above the water-level, it was nailed to the sides of the frames, and spread loosely all over the bottom. The concrete was then poured into the loose sack lining the frames; and the sacking, whilst preserving the concrete from the wash of the sea, allowed it to adopt itself to the irregularities of the bottom. The frames were constructed so as to be readily taken to pieces, and they were removed from the concrete block ten or twelve days after its manufacture. M. Poirel considered the lining as the essential part of the system; the idea was suggested by the use the Italians made of concrete in bags for repairing hollows in masonry structures under water.

The blocks made on the bank were formed in boxes, resting on an incline down which the blocks were slid as soon as they were sufficiently hardened, and after the boxes had been removed. Hydraulic lime was used for these blocks which could set out of water; whereas puzzolanno, or Roman cement, was added in the manufacture of the immersed blocks.

In reconstructing the breakwater, a row of blocks in place were first made in sizes varying between 80 and 260 cubic yards. Boxes, having a capacity of from 13 to 65 cubic yards, were then placed on these blocks, and filled with concrete; and the blocks thus formed were thrown into the sea as soon as they had set. The space between these two rows of blocks was then filled with rubble stone of from 4 to 9 cubic yards; and behind this protecting mound, a trench was excavated down to 6½ feet below the water, and 10 feet wide, which was filled with concrete-in-mass. The work of reconstruction, extending over a distance of about 650 feet, was carried out in lengths each season; it occupied five years, and cost under £80,000.

Extension of the Algiers Northern Breakwater. After the work of reconstruction had been completed in 1838, the enlargement of the harbour was commenced by prolonging the northern breakwater. This extension was carried out by depositing a mound composed solely of concrete blocks of 13 cubic yards. These blocks were manufactured on land, and thrown at random into the sea after standing a month or two. Some of the blocks were pushed down inclined planes into the sea; whilst others, when partially immersed, were suspended by chains between two pontoons and floated into place. This double system was adopted to increase the rate of progress; and 280 lineal feet were executed by the middle of 1840, or in about a year and a half. The mound stood at a slope of 1 to 1 on the sea side, and ½ to 1 on the harbour side. It was proposed to raise a solid quay wall of concrete-in-mass, deposited within frames, on the top of this mound.

Breakwaters at Algiers. The harbour has been considerably enlarged since 1840. The northern breakwater was prolonged, curving round towards the east, to a length of 2300 feet, and terminated by a pierhead. Shelter has also been provided to the south and east by a breakwater starting from the shore and then turning at right angles parallel to the coast[1]. (Plate 5, Fig. 1.) An entrance, facing east, and 1115 feet wide, has been left between the northern and eastern breakwaters, in a depth of 12 fathoms. A smaller opening left in the eastern breakwater, 288 feet wide, has been recently closed, which will improve the tranquillity of the port.

An extension of the northern breakwater beyond the pierhead has been lately made, to reduce the swell entering the

[1] M. Ereves, one of the engineers of the port, supplied me with a plan and sections of the breakwaters.

harbour during north-easterly gales, owing to the waves turning round the end of the breakwater, and also to protect the entrance.

The harbour, being situated on the western side of a wide bay open to the north, is only exposed to the north and east. The area sheltered by the breakwaters is 220 acres; and quays have been formed in it in front of the town. (Plate 14, Fig. 25.)

Construction of Algiers Breakwaters. Some modifications have been introduced into the composition of the breakwaters from M. Poirel's first design. (Plate 6, Fig. 9.) The hearting of the mound is composed of small rubble with layers of larger stones surrounding it; and the protection of concrete blocks has been confined to the top and sea slope of the mound, with the exception of the portion of the northern breakwater within the pierhead where concrete blocks cover the harbour slope as well. This alteration has greatly reduced the number of concrete blocks, and has increased its compactness without diminishing its stability; for the slope of 1 to 1 on the sea side has been maintained, whilst the harbour slope has been made 1 to 1¼, instead of the 1 to 1 slope which was adopted for the concrete blocks on the harbour side of the inner northern breakwater. The section in Plate 6 Fig. 9, shows the construction of the recent northern breakwater extension, which has been carried into a depth of 18 fathoms, and has therefore been given a thicker layer of concrete blocks than the other breakwaters, which are less exposed and in shallower water. The concrete blocks employed are 11 feet long, 6½ feet broad, and 7⅓ feet high; the larger blocks of stone forming the upper layer contain not less than ⅔ cubic yard, weighing 1⅓ tons; the under layer consists of stones weighing from 2 cwt. to 1⅓ tons; and the rubble hearting is composed of stones from 40 lbs. to 2 cwt in weight. The blocks remain undisturbed and permanent under water, and do not require renewal.

To protect the sea slope close to the water-level, large blocks of from 40 to 60 cubic yards are deposited; and huge blocks are generally formed on the top of the mound, containing about 130 cubic yards each, providing a sort of parapet for the breakwater.

Comparison of the Algiers section with the sections of Plymouth and Portland. The section of mound adopted at Algiers is specially suited for the considerable depth of water in which it is situated; as a mere rubble mound would have absorbed a large quantity of material, besides needing some additional protection on account of the small size of stone that could be obtained. The saving in material effected by the Algiers system is readily indicated by a comparison of the section of Algiers breakwater with the sections of Plymouth and Portland. Thus the sectional area of the Algiers breakwater, in a depth of 105 feet, is 1600 square yards, little exceeding the sectional area at Plymouth of 1440 square yards in a depth of only 48 feet at high water; whilst at Portland, in a depth of only 68½ feet at high water, the sectional area of the breakwater is 2010 square yards, one-fourth more than that of Algiers. Now assuming that the breakwaters at Plymouth and Portland were situated in the same depth at high water as Algiers, the difference in the amount of material required by the two systems becomes comparable, and is still more marked. Thus the sectional area of the Portland breakwater in the same depth as Algiers would be 3320 square yards, or more than double that of Algiers; whilst the Plymouth section would amount to 4510 square yards, or not far from three times that of Algiers. Also the width at the base, in a similar depth, which is 235 feet at Algiers, would be 440 feet with the Portland section, and 590 feet in the case of Plymouth. The exposure at Algiers is, indeed, less than at Plymouth, but the depth is much greater; whilst, as compared with Portland, both the fetch in the exposed direction and the depth are greater.

State of Algiers Harbour. The harbour of Algiers is situated at the western extremity of a sandy bay; but though some shoaling appears to have taken place against the southern face of the south breakwater, and also to a small extent on the sea side of the northern breakwater, it is not likely to affect the entrance which is situated in deep water, and at a distance from the southern beach. The depth inside is good, even up to the quays along the shore.

The entrance is wide for the size of the harbour; and the formation of vertical quay walls on the site of a flat sandy beach, on which the waves used to expend their force, increased the agitation in the harbour, but this will no doubt be mitigated by the recent extension of the northern breakwater.

PORT SAID HARBOUR.

Description. A deep water entrance had to be provided to the Suez Canal, at its junction with the Mediterranean Sea at Port Said, across the sandy beach. This channel could be formed by dredging; but in order to maintain the channel, to afford protection for the dredging operations, and to provide shelter for vessels approaching or leaving the canal till they reached deep water in the open sea, it was necessary to construct a harbour.

The prevailing winds on the coast are from the north-west; and the currents follow the same course. Strong easterly winds, however, blow in winter; so that protection was needed on both sides of the entrance channel. The plan of converging breakwaters was adopted, in preference to parallel jetties which had been first suggested, on account of the larger area sheltered by this means. The western breakwater has also been made longer than the other, owing to the northerly direction of the western winds, and in order to protect the entrance from the easterly drift of alluvium from the Nile. (Plate 5, Fig. 7.) The breakwaters start at a distance of 4600 feet apart at the shore, and converge to an opening of 2300 feet at the extremity of the eastern breakwater. The western breakwater has a length of about 9800 feet, and the eastern one is 6233 feet long. Inner basins at the west side of the mouth of the canal afford accommodation for vessels waiting at Port Said, and for the local trade.

Construction of the Port Said Breakwaters Schemes were drawn up for constructing the breakwaters of mounds of rubble, or of concrete blocks. The rubble mound was abandoned on account of the difficulties in the way of transport from the quarries at Gebel-Geneffé, and the large quantity of stone that would have been required to form a permanent mound, entailing delay in execution. A mound of random concrete blocks was, accordingly, decided upon. Stone, however, was used for the hearting of the mound near the shore, as shown on the first section (Plate 6, Fig. 10); and also for the base of the extension of the western breakwater in deep water, after the opening of the canal, when impediments to the carriage of stone were removed. (Plate 6, Fig. 12.) The general section of the eastern and western breakwaters, consisting solely of concrete blocks, is shown in the second section. (Plate 6, Fig. 11.) A roadway and parapet of concrete-in-mass, on the top of the western breakwater, formed part of the original design, but was not executed.

The concrete blocks, composed of sand and Theil lime, each containing 13 cubic yards ($11\frac{1}{8}$ ft. × $6\frac{1}{2}$ ft. × 5 ft. high), and weighing about 20 tons, were formed in boxes, and left for six weeks or two months to set before being used.

Method of Depositing the Blocks at Port Said. A barge or pontoon, carrying three blocks on an inclined plane on its deck, and towed by a steam tug, conveyed the blocks to their destination. Where the mound was low enough for the tug and barge to pass over it, the barge was simply towed into position, and, as soon as it was in the line marked by two beacons, the catches retaining the blocks on the inclined plane were released, and the blocks slid into the water. When the mound had been raised too high for the vessels to pass safely over, the barge was cast off from the tug when nearing the mound, and then warped itself into position as close to the mound as practicable. A floating derrick was used for depositing the blocks above the water. The derrick, being moored near the breakwater, lifted the block from the barge, and hauling itself into position gently lowered the block into place, and the suspending chains were released by a strong pull. Calm weather was required for this operation.

Condition of Port Said Harbour. A detached piece of breakwater was constructed beyond the western breakwater, so that the waves entering through the gap might scour away any bar that might tend to form. (Plate 5, Fig. 7.) A channel has been dredged, averaging 4½ fathoms in depth, from the entrance of the canal to the 5-fathom line, and is maintained by dredging. The rest of the sheltered area has a depth of from 1 to 3 fathoms.

An apprehension was felt at the outset that the projection of a solid breakwater from a sandy shore, where a turbid littoral current from the west was known to exist, would result in an advance of the western foreshore, and eventually silt up the harbour. The shore line has, indeed, advanced along the outside of the western breakwater, and some of the sand, passing through the interstices of the concrete-block mound, has settled on the harbour side of the breakwater. (Plate 5, Fig. 7.) The apertures, however, in the mound have now become choked with sand; and the western foreshore appears to have reached a position of equilibrium, or at any rate the progression is now very slow, so that there is no prospect of the harbour being endangered by the accumulation. Dredging is being carried on at the extremity of the channel, and a more permanent depth is being gradually attained; but doubtless dredging will always be needed for maintenance.

BIARRITZ HARBOUR.

Breakwater. Works were commenced in 1863 for a refuge harbour at Biarritz, situated on the exposed stormy coast of the Bay of Biscay. The breakwater consists of a concrete-block mound interspersed with rubble. The blocks were mostly made of the size of 19½ cubic yards; and the interstices, above low water, were filled with rubble masonry below, and concrete at the top. (Plate 6, Fig. 13.) The mound rests upon a rocky foundation. The works were brought to a close by the war in 1870; and in 1873 the breakwater was partially destroyed. It has a length of about 400 feet, and the outer 160 feet were overturned. The inner 33 feet of this damaged portion were washed away to a depth of 6½ feet below low water. For the rest of the length, the breakwater is divided into large sections rising from 3 to 13 feet above low water. Though the breakwater is of little use, it has been decided to repair it, so that a work belonging to the State may not be left in ruins; and after the first breach, 33 feet wide, has been repaired, the remainder will be easily accomplished at a small cost. All idea of prolonging the breakwater has been abandoned.

Doubtless the failure of this breakwater must be attributed to the excessive exposure of the site, being open to the Atlantic, and in the Bay of Biscay noted for its storms and the height of its waves. The experience of French engineers in such works being limited to the Mediterranean and the Channel, due allowance was not made, in the size of the blocks, and the protection of the mound, for the much greater exposure of the locality.

ALEXANDRIA HARBOUR.

Description. The island of Pharos was situated a short distance off the coast of Egypt, in front of Lake Mareotis, to the west of the delta of the Nile. This island was joined to the mainland in the time of Alexander by a wide earthwork embankment; and it is upon the promontory thus formed that the modern town of Alexandria has been built. As the island projects on each side of the embankment parallel to the coast, a small bay was formed on each side of the embankment. (Plate 3, Fig. 1.) The bay on the north-eastern side is further protected by the point of Pharallon to the east, so that this bay is only open to the north; it is called the New Port.

The more recessed bay on the south-western side of the embankment forms the north-eastern extremity of a wide bay, extending about 6 miles along the coast, and nearly 2 miles deep, with a reef of rocks stretching across it through which there are three passes for vessels. This end of the bay is sheltered by the promontory of Pharos from north-east to north-west; but it was partially exposed to the west, and open to the south-west, the directions of the most prevalent winds. This portion of the bay is situated close to the town, and the Mahmoudieh Canal opens into it. In order to extend and improve its shelter, a detached breakwater was commenced in 1870, and completed in two years. Commencing near the south-western extremity of Pharos, called Eunostos Point, and extending its line, the breakwater eventually makes a bend towards the shore, leaving at its extremity an entrance 3340 feet wide, with a depth of 10 fathoms in the centre. This harbour has now a sheltered area of 1400 acres with a depth of 5 fathoms and upwards: its inner extremity has been further protected from any swell, coming in through the wide entrance, by a mole, run out from the shore opposite Eunostos Point and extending half-way across the harbour, thus forming a more sheltered inner harbour, and protecting a line of quays along the shore.

The breakwater, mole, quays, and other works cost nearly £2,000,000; but this outlay has provided the Egyptian Government with an excellent spacious harbour.

Construction of Alexandria Breakwater. The section of

the detached breakwater, 9675 feet in length, is shown on Plate 6, Fig. 14. Like its predecessor at Port Saïd, the breakwater is composed of concrete blocks with an admixture of rubble stone. In this instance, however, the concrete blocks form the mass of the outer portion of the mound; and the rubble has been deposited within their shelter, on the harbour side, to add width to the mound, and the harbour slope has been protected by large stones. The concrete blocks employed, weighing 20 tons ($11\frac{1}{2}$ ft. $\times 6\frac{1}{2}$ ft. $\times$ 5 ft.), were almost precisely the same dimensions as those of Port Said; but broken stone, as well as sand and lime, were used in their manufacture, stone being much more readily obtainable at Alexandria than at Port Said, some of the stone required for the Suez Canal having been procured from the same quarries at Mex, 4 miles only from Alexandria. The blocks were kept for three months to harden before being used.

Though the breakwater shelters an area where the depth in places reaches 10 fathoms, yet, by taking advantage of a line of shoals and sandbanks in front of the harbour, it has been possible to construct the breakwater in depths not exceeding 4 fathoms, and in many parts only from 1 to $2\frac{1}{2}$ fathoms.

Mode of Depositing the Blocks at Alexandria. The lower portion of the concrete-block mound was deposited, like at Port Said, by releasing the blocks from inclined planes on the decks of barges. The barges, however, at Alexandria carried five blocks each.

The blocks also near and above the water-level were deposited, as at Port Said, by a floating steam derrick which lifted the blocks from the barge and lowered them on to the mound[1]. (Fig. 15.) The suspending chains at Port

[1] The illustration of the derrick is from a photograph sent me by N. Lihou, master of the derrick at Alexandria in 1872, who also lent me a model of the slings employed.

Said were passed under the blocks, and kept in place by grooves in the blocks; and the removal of these chains from underneath the block, when in place, necessitated a strong

ALEXANDRIA HARBOUR.

Depositing blocks from Derrick.

Fig. 15.

pull by a rope worked by a steam windlass. This method of suspension, accordingly, proved inconvenient, and tended to weaken the block. Suspension by lewis bolts, which was at first adopted at Alexandria, had similar defects; but subsequently slings were contrived for lifting the blocks without impairing their strength, and they could be removed from the blocks without the slightest trouble.

The slings consisted of four angular pieces, like short lengths of angle iron, which fitted to the four side corners of the block, and were pressed tightly against the sides of the block by two chains passing through rings in the corner pieces. The ends of each of these chains were passed over two hinged hooks, one on each side of the block. The hooks, when holding the chains, were kept closed by a catch-loop, which could be easily released by a slight pull from a rope passing through a pulley above. The hooks were both suspended from a cross-piece held by the suspending chain from the derrick. Whilst the hooks bear the weight of the block, they keep the two chains tight which encircle the corner pieces, and thus grasp and sustain the block. The block is thus lifted, and when lowered into position the cords fastened to the catches of the hooks are pulled by a man on the derrick, the hinged half of the hook falls releasing one of the encircling chains, the grasp of both chains and corner pieces is thus relaxed, and the block is free from the slings; one of the chains with its two corner pieces still hangs on the hooks, and the other chain and corner pieces are retained by two accessory chains hanging from the ring of the derrick chain.

This system of slings enabled the blocks to be laid more rapidly, and with less injury, than by the previous methods. Thus, by aid of the slings, 40 to 50 blocks could be deposited in a day; whereas previously, 10 blocks per day was considered a good rate of work.

The rubble stone on the harbour side of the mound was deposited from hopper barges with trap-doors at the bottom and sides.

Comparison between Algiers, Port Said, and Alexandria Harbours. The harbour of Alexandria presents some resemblance to Algiers harbour, both in its origin and development; it was originally formed by connecting an adjacent island with the mainland, thus affording a moderate shelter which has since been extended by a rubble and concrete-block mound breakwater. Like Algiers, the entrance to Alexandria harbour is in deep water, and though situated in a part of a sandy bay, it seems likely to maintain its depth; and the port and harbour are combined in one. The breakwaters at Algiers have been constructed in a considerable depth of water, and were commenced when little experience had been gained in such works; whereas both Port Said and Alexandria breakwaters are situated in a moderate depth, and have been carried out with the advantages of the most modern engineering appliances.

The sites of both Algiers and Alexandria possessed natural advantages for the formation of harbours; and Alexandria was originally selected as the port of Egypt on account of its being to the west of the Nile delta, and therefore beyond the influence of the alluvial deposits of that river which are carried eastwards by the littoral current produced by the prevailing winds. Port Said, on the contrary, is situated on a straight flat sandy coast, in the path of a turbid current from the west, and therefore most unfavourably placed for a close harbour; it has, however, accomplished the object of providing a deep sheltered entrance to the Suez Canal; the advance of the foreshore to the west has not been sufficient to imperil the entrance, and dredging can maintain the channel inside.

The breakwaters at Algiers, Port Said, and Alexandria have successfully resisted the sea; and though Biarritz breakwater has been less fortunate, its failure may be attributed to a want of due consideration of the difference of exposure on the Atlantic and Mediterranean, and not to any inadequacy of the system, as indeed is sufficiently evinced by the intention to repair it.

The composition of the mound is somewhat different at the three Mediterranean harbours, occasioned by the different facilities of procuring stone, and the urgency of rapid execution. (Plate 6, Figs. 9–14.) The rate of construction has also varied considerably: the breakwaters at Algiers have been gradually carried out; the breakwaters at Port Said occupied about six years in construction, being a yearly advance of 2670 feet, 42,000 blocks being deposited on the average in a year; whilst the Alexandria breakwater was formed in about two years, giving the unprecedented rate of nearly a mile in a year. In comparing, however, the progress at Port Said and Alexandria, it must be remembered that Port Said harbour formed merely a portion of a gigantic undertaking, and that it would have been useless to complete the breakwaters a long time before the canal was opened for traffic, whilst some shelter was advantageous in the early stages of the work; whereas at Alexandria, it was desirable to complete the harbour as quickly as possible.

Advantages of the Concrete-Block Mound. The saving in material and in space by the employment of concrete blocks for forming a random mound breakwater has already been indicated by a comparison of the section of the Algiers breakwater with the sections of Plymouth and Portland. The benefit is specially marked in deep water and exposed situations.

The stability of the mound is greatly increased by the covering of large blocks, which can be made of a size suitable to the exposure; and the cost of maintenance is reduced.

Saving in time of execution naturally results from saving in material; and though the use of concrete blocks does not adapt itself to the running out of a breakwater from the shore, as in the case of a rubble mound, yet this is more than compensated for by the opportunity, when barges are used, of carrying on the work at various points simultaneously, and not merely at the extremity of the finished portion. In fact, the rate of progress is simply limited by the amount of plant which can be economically employed; and the rapidity of execution that can be achieved with ample plant and suitable mechanical contrivances has been forcibly illustrated in the case of the Alexandria breakwater.

CHAPTER XII.

MEDITERRANEAN HARBOURS PROTECTED BY SORTED RUBBLE AND CONCRETE-BLOCK MOUNDS WITH SLIGHT SUPERSTRUCTURES.

MARSEILLES. TRIESTE. BASTIA. ILE ROUSSE. ORAN.

Marseilles Harbour: Description: Breakwater, construction, cost; State; Extensions proposed. *Trieste Harbour:* Situation; Breakwater, construction; State. *Bastia Harbour:* Position; Extension; Breakwater for Old Harbour, construction, cost; Breakwater for New Harbour, construction. *Ile Rousse Harbour:* Description; Breakwater. *Oran Harbour:* Description; Breakwater. General Remarks.

SOME of the harbours in the Mediterranean Sea are protected by breakwaters very similar in construction to the Algiers breakwater, described in the previous chapter, in having a sorted rubble mound protected on the sea slope with concrete blocks, but differing from it in having both a flatter slope on the sea side and also a slight superstructure of masonry or concrete on the top of the mound. The flatter slope and the superstructure protect the top of the mound on the harbour side, and enable it to be used more or less as a quay according to the position of the breakwater and the state of the weather. In order to allow vessels to approach the inside of the breakwater, the harbour slope is in some cases terminated by a quay wall founded on the rubble mound below the water-level. The finest example of this type of breakwater has been constructed at Marseilles to shelter that port; and a similar breakwater has been completed at Trieste within the last few years.

MARSEILLES HARBOUR.

Description. The establishment of Marseilles as a port dates back to ancient times, owing to its excellent geographical position for communication between France and the Mediterranean ports, and on account of its possessing a creek forming a natural harbour. Up to 1844 this old harbour was the only accommodation it possessed for shipping; and the whole of the extensive modern port has been constructed since that period.

The old harbour has an area of 72 acres, a depth of between 3 and 4 fathoms in most parts, and a length of 8860 feet of quays; but the space on the quays for merchandise is very restricted.

The growth of trade necessitated an increased accommodation, whilst no other good natural shelter existed for the development of the port. The roadstead of Marseilles is somewhat embayed between two projecting rocky points, north and south of the town, about 3 miles apart; it is further protected to the south, in the line of greatest exposure, by a group of islands; and the general coast-line of the Gulf of Lyons is only a short distance off on the north. The coast, accordingly, in front of Marseilles was sheltered to some extent from all quarters except the west, in which direction, however, the greatest fetch is only the width of the Gulf of Lyons. The most prevalent wind at Marseilles is from the north-west (see Fig. 1, p. 10), in which direction the roadstead is sheltered by the coast.

As the object of an enlargement of the port was to provide for an increasing trade, it was important to adopt a design which, whilst affording the required shelter, should aid in giving additional quay space, and should also be capable of extension without interfering with the existing trade. The plan carried out combines all these requirements; it consists of a detached breakwater parallel to the coast-line, which protects a series of basins formed by jetties, or moles, projecting at right angles from the shore. (Plate 3, Fig. 4.) Quay walls have been constructed on each side of the jetties, and on the inside of the breakwater; and each basin has been completed by a quay along the shore. Communication between the shore and the breakwater quay has been provided by swing-bridges from the jetties to piers projecting from the breakwater, the intermediate opening serving for the passage of vessels from one basin to another, or to the outer harbour at each end.

Breakwater at Marseilles. The detached breakwater is continuous throughout, and consists of two straight portions joining at an angle in order to maintain the parallelism with the bending coast-line. The harbour is entered at either end by turning round one of the extremities of the breakwater; and the breakwater overlaps the outer jetty at each end, so as to form a sort of outer harbour, enabling vessels to take shelter without entering the basins, or to pass through the passage into them in still water. The breakwater was commenced at its southern end in 1845, and has been gradually extended as the requirements of trade necessitated the creation of fresh basins, without the slightest inconvenience to the shipping. Its total length is 11,930 feet.

By constructing the breakwater parallel to the shore, the basins have been made of a convenient shape; and the depth in which the breakwater is situated is more moderate than if it had stretched straight across from point to point, whilst the shelter provided in its unfinished state has been much better.

The Joliette Basin was commenced in 1845, and completed in 1852; the Lazaret, Arenc, and Maritime basins were constructed between 1857 and 1863; and the National Basin between 1866 and 1881. Thirty-six years have, accordingly, elapsed between the commencement and completion of the breakwater in its present condition; and as the interval between each extension has hitherto not exceeded five years, it is natural that a further enlargement of the port is under consideration.

Construction of Marseilles Breakwater. The breakwater was built in depths varying from 6½ to 12 fathoms. The portion sheltering the Joliette Basin, with a width of quay of 59 feet (Plate 7, Fig. 16), differs somewhat from the remainder, with a width of quay of 92 feet, sheltering the other basins[1] (Plate 7, Fig. 15). The principle of construction, however, is the same in both sections. The mass of the mound is formed in layers of sorted rubble stone, varying in weight from between 5 lbs. and 2 cwts. to 3¾ tons and upwards according to their situation in the mound, as shown on the sections. The sea slope is protected by a layer of random concrete blocks, containing 13¼ cubic yards (11 ft. × 6½ ft. × 5 ft.), and weighing nearly 22 tons. The paved quay is sheltered by a parapet wall about 22 feet high, and is bounded on the inside of the breakwater by a vertical concrete quay wall enabling vessels to lie alongside. Earthwork fills in the space at the back of the quay wall and under the quay. Larger stones have been used in the deeper breakwater; but whilst at the Joliette breakwater the layer of concrete blocks descends to the bottom, the blocks in the extension have only been carried up from 26 feet below the water-level, where they rest upon a bench of rubble stone, as it was unnecessary to extend the protection of concrete blocks down to a depth of from 7 to 12 fathoms. The outer slope of the Joliette breakwater is 1 to 1 throughout, with a bench of 13 feet in front of the parapet wall; the slope of the extended breakwater is the same up to the water-level, where it changes to a slope of 2 to 1; but the bench at the top is reduced to 6½ feet.

The benching and the flatter slope protect the parapet wall from the full force of the breaking wave; and the concrete

[1] M. Barret, engineer at Marseilles, sent me plans and sections with other information relating to Marseilles Harbour.

blocks do not merely serve to maintain the mound, but also, by the interstices and the rough slope they afford, tend to break up the wave before it reaches the summit of the mound.

The object of sorting the stones and depositing them in layers according to their sizes was that all the products of the quarries might be utilised, and each kind placed in the most suitable position in the mound, the smallest in the perfectly protected centre and base, and the largest on the more exposed slopes. Moreover, by not mingling the small rubble with the large, the proportion of interstices is increased, and a mound can thus be constructed of a given size with less material.

The benching, on which the concrete blocks rest in the deeper portion of the breakwater, appears not to have been made sufficiently wide to retain the blocks perfectly; so that during construction some of the blocks fell down to the bottom of the slope, and after completion some more were drawn down by the action of the waves and had to be replaced. M. Barret considers that, to ensure stability, the bench of rubble should have projected at least 10 feet beyond the face of the concrete-block slope.

Cost of Marseilles Breakwater. The portion of the breakwater sheltering the Joliette Basin, constructed in a depth of 40 feet, and about 2600 feet long, cost £75 11*s.* per lineal foot; the portion opposite the Lazaret and Arenc basins, 2050 feet long, in a depth of 55¾ feet, cost £109 14*s.* per lineal foot; and the portion in front of the Maritime Basin, in a depth of 52½ feet, and 1660 feet long, cost a similar amount.

State of Marseilles Harbour. In one of the designs for the enlargement of the port, an outer detached breakwater was proposed for protecting the roadstead outside the basins, in addition to the works actually carried out. This breakwater would have protected the approaches to the basins, and have formed a refuge harbour, but it would not have added to the accommodation for trade; and, as the depth along the proposed site is over 100 feet, it would have proved a very costly work, and, considering the fairly sheltered position of the roadstead, it cannot be regarded as essential. The entrances to the basins, however, are not quite as satisfactory as might be desired. The northern outer harbour is somewhat exposed to the prevalent north-westerly winds; and when the wind is very high a vessel experiences some difficulty in making the entrance into the National Basin. The southern entrance is also inconvenient of access under similar conditions, for though there is a width of 650 feet between the southern end of the breakwater and the shore, yet, owing to the restricted size of the outer harbour, and the great length of the large modern steamers, amounting to from 425 to 525 feet, it is impossible for these vessels to run in at full speed, and when going slowly they do not readily answer their helm in turning round the end of the breakwater, and are liable to be driven by the wind towards the rocks of Pharo Point.

The harbour is not exposed to any accumulation of silt, beyond the refuse discharged from the town sewers which still have their outlets into the port; and it affords depths of from 23 to 33 feet in the basins.

With the exceptions referred to, the harbour appears to be in a perfectly satisfactory condition; it has fulfilled all the requirements of trade; and the breakwater is secure, and adequately shelters the quays on the top as well as the basins beyond.

Proposed Extensions of the Port of Marseilles. The steady increase of the trade of Marseilles will before long necessitate an addition to the basins. The basins have now reached the northern limit originally assigned to them, as the breakwater extends almost as far as Cape Janet, the northern extremity of the bay in which the town stands. As, however, the line of the breakwater diverges a little from the actual shore line, it would be possible to extend the basins beyond Cape Janet without altering the line of the breakwater; and the bay of Madrague, which is already partly sheltered by the breakwater, could be easily converted into a regular basin. This bay has been suggested as the outlet for the proposed canal from the Rhone.

The objections to the extension of the port at the northern end are, that it would place the new basins at a considerable distance from the centre of the town, and that the railway communication would be restricted to a line along the quays by the projecting point of Pinéde. The Government has proposed an extension to the south of the present basin beyond Pharo Point. The scheme consists, like the existing port, of projecting jetties with quays protected by a detached breakwater. The length of breakwater would, however, be quite disproportionate to the extent of quays; so that, according to M. Barret, whereas the cost of the present northern basins was £165 per lineal yard of quay, the proposed southern port would cost four times as much. The two series of basins would, moreover, be quite distinct, and severed by Pharo Point.

M. Barret, on the other hand, proposes a plan for forming regular entrances with projecting converging piers at each end of the existing breakwater; and by placing the southern breakwater further out, and bringing it in front of Pharo Point, the basins would be all connected. This scheme appears to have the advantage of providing safer entrances for large vessels; of affording considerably more quay accommodation with a very similar length of breakwater; and of enabling vessels to pass within shelter through the whole series of basins.

The port of Marseilles combines both harbour and dock works together: the breakwater, the shelter it affords, and the entrances to the port appertain essentially to the harbour division; whilst the quays and other works belong to the docks, and will be further considered in Chapter XXVII.

TRIESTE HARBOUR.

Situation. The port of Trieste is very similarly situated to the port of Marseilles. The bay of Trieste, on which Trieste stands, is protected by the shores of the Gulf of Trieste from every quarter except the west; it is also more fully sheltered by its own curving coast-line from the north round by east to the south. The worst winds in that locality blow from the north-east and east; but, being land winds, they do not produce waves in the port, though they lower the water-level in the gulf, and, from their boisterous character, inconvenience the operations of trade: they prevail from November till March. Westerly winds prevail during the summer months. The stormy south-west (Libecchio) and south-east (Sirocco) winds are rare; and though the former raises a high sea, owing to the wider expanse over which it blows, it is of very short duration. The roadstead, accordingly, in front of the town needed protection from the west to south-west.

Proposals were made to transfer the port to the more sheltered Muggia Bay, south of Trieste, connecting it with the town by a railway, or canal, passing in tunnel under the sandstone ridge of San Giacomo which intervenes between the bay and Trieste. The cost, however, of the undertaking, and the distance from the town, caused the abandonment of this project.

After careful consideration of numerous schemes, the design of a detached breakwater, parallel to the coast, protecting a series of inner basins formed by projecting jetties, was adopted. (Plate 3, Fig 7.) It resembles precisely in its general outline the method adopted for the modern port of Marseilles, and exactly faces the entrance to the gulf a little to the south of west.

Breakwater at Trieste. The detached protecting breakwater, in front of the Trieste basins, is formed of a mound of sorted rubble stone, with the smaller material at the base and in the centre, and the larger rubble enveloping it, the largest stones being deposited on the sea slope. A parapet wall of masonry protects a paved quay, which is provided on the harbour side with a vertical quay wall composed of concrete blocks resting upon a levelled portion of the rubble mound, 20 feet below the water-level, and backed with earthwork. (Plate 7, Fig. 7.) The only difference between this section and the Marseilles breakwater (Plate 7, Figs. 15 and 16) consists in the omission of the concrete blocks on the sea slope; and, consequently, this slope has been given a flatter inclination of 1½ to 1 at Trieste The silty bottom at Trieste rendered it expedient to adopt a wider base in proportion to the height than at Marseilles, where the bottom is firm sand; and the sheltered site, aided by the flatter slope, renders the large rubble sufficiently stable.

Construction of Trieste Breakwater. The breakwater is situated in a depth of about 9 fathoms, which decreases towards the shore; the depth at the entrances is between 7 and 8 fathoms, and from 4 to 9 fathoms within the sheltered area. It curves slightly outwards at its extremities to facilitate the entrance of vessels, which can be effected at either end as at Marseilles. The total length of the breakwater is 3576 feet; it was commenced in 1868, and completed in 1874, having occupied 6¼ years in construction. The foundation is composed of such soft silt, that, in spite of the broad base of the rubble mound, it settled as much as 26 feet in the centre, increasing the width of the base, required to form the proposed width of quay of 39⅓ feet, from 200 feet to 330 feet, and necessitating a considerable increase in the mass of the mound. The breakwater in two years after its completion had settled 1½ feet more; and it is probable that after a certain time it will be necessary to raise the quay.

State of Trieste Harbour. The settlement of the jetties and quays, as well as the breakwater, upon the silty bottom forced up the material in front, and quite altered the original depths. Accordingly, the displaced material had to be dredged, which was rendered all the more difficult by its consisting of a mixture of silt, gravel, rubble, and large stones, rendering it necessary to resort to spoon dredgers and divers as well as bucket dredgers. The depth was first carried down to 20 feet throughout, and finally to 28 feet, at a distance of 23 feet from the face of the quays. The total amount dredged between 1872 and 1876 reached 707,000 cubic yards, out of which the gravel and stone, amounting to 260,000 cubic yards, were used for embankments, and 12,000 cubic yards of large stones were employed in raising the outer slope of the breakwater.

Two torrential streams, and the town sewer, which formerly discharged into the sea within the site of the new port, have been diverted, so as not to interfere with the quays and to prevent their pouring into the harbour the sediment and refuse they bring down. The harbour, however, being surrounded by accumulations of silt, will probably suffer from the introduction of wave-borne silt through its entrances, necessitating the maintenance of its depth by dredging.

There is ample space for the extension of the harbour to the north by the prolongation of the breakwater; but hitherto the accommodation provided appears to have outstripped the requirements of the trade, as the moles separating the basins have not been utilised for railways and sheds, and the quay on the breakwater has not been connected with the other quays by means of swing-bridges.

Bastia Harbour.

Position. The island of Corsica abounds in bays and creeks, several of which might be utilised for harbours if the trade of the island demanded it. Such an island, however, cannot develop a traffic like ports bordering large continents, such as Marseilles or Alexandria, and therefore cannot afford scope for very large works. Unfortunately the northern part of the east coast of the island, which is nearest to the mainland of Italy, is devoid of natural shelter, except what the small creek of Bastia affords. A good depth of water is found close inshore at Bastia, and there is little accumulation of sand, so that Bastia became a port in early times; and the creek was protected from the north-east by the construction of a mole in the 17th century, 500 feet long[1]. This ancient mole was repaired early in the present century; but the creek was still deficient in shelter from the east and south-east.

Extension of Bastia Harbour. The limited accommodation afforded by the creek having become quite inadequate for the increasing trade, it was decided, in 1845, to enlarge the port by enclosing the small bay of St. Nicholas, to the north of the town, by a breakwater projecting from the shore at the northern extremity of the bay and bending round so as to run parallel to the shore, leaving a southern entrance protected by the projecting point on the south side of the creek. (Plate 5, Fig. 6.)

During the delay which elapsed before the commencement of the St. Nicholas breakwater in 1864, the old harbour was improved by prolonging the mole 180 feet, and by the construction of a breakwater on the opposite side of the mouth of the creek to shelter the creek from south-easterly winds, so that it might be more serviceable for trade, and form a harbour for the vessels employed on the new breakwater.

Breakwater of the Dragon at Bastia. The breakwater on the south side of the creek is about 500 feet long, leaving an entrance 266 feet wide between its extremity and the mole: it was commenced in 1855, and completed in 1863. It consists of a mound of rubble stone, formed in sorted layers of three sizes of stone, protected on the sea slope by concrete blocks of 13 cubic yards, and surmounted by a high

[1] 'Annales des Ponts et Chaussées,' 4th Series, vol. xx. p. 175.

narrow parapet wall protecting a quay only 8⅓ feet wide. The parapet wall, rising 22½ feet above the sea-level, suffered in several places from the settlement of the mound, having been built too soon after the formation of the mound. The work was consolidated by placing concrete in bags in the hollows between the blocks, and by depositing additional concrete blocks on the sea slope.

The breakwater, situated in an average depth of 40 feet, cost £67 per lineal foot.

The shelter afforded by this breakwater allowed of the removal of a rock in the creek by blasting; for this rock, though greatly restricting the available area of the harbour, had been needed for shelter; and the quays could be extended without fear of increasing the swell in the harbour. The sheltered area was only 13½ acres.

St. Nicholas Breakwater at Bastia. As the old port was inadequate for the trade, and as it was desired to have a harbour capable of receiving a ship-of-war, the new harbour of St. Nicholas was commenced in 1864. The breakwater was designed to have a total length of 2790 feet, of which 1530 feet appertain to the outer arm parallel to the shore; and the estimated cost was £248,800. The completion of the last 330 feet of the outer arm has been delayed until the effect of the opening of 800 feet between the breakwater and the shore has been fully ascertained, as it is open to the south-east, and though winds from that quarter are less frequent at Bastia than north and north-easterly winds, they raise larger waves owing to the greater fetch of sea in that direction. It is considered possible that, to secure adequate tranquillity in the harbour, it may prove necessary to contract the entrance by a short jetty from the shore. When the works in progress are completed, the sheltered area will amount to about 53 acres, with depths reaching to a maximum of 11 fathoms.

The breakwater is composed of a rubble mound, with two

sizes of stone, protected on the sea slope, to a depth of 20¾ feet, and on the top by concrete blocks. (Plate 7, Fig. 4.) Owing to the good proportion of large stone obtained from the quarries, a quantity of large rubble was deposited with the smaller material in the base of the mound. The rubble was partly conveyed by land, and partly deposited from barges, the large blocks of stone being tipped from wagons run on to the barges. The concrete blocks, made with a large proportion of rubble stone, were similarly tipped on to the sea slope. The barge was capable of conveying four wagons, each loaded with a concrete block of 13 cubic yards; but three wagons formed the ordinary load. The top of the sea slope is protected by a layer of concrete-in-mass, surmounted by a superstructure of concrete blocks, formed in place, containing from 19 to 26 cubic yards each. The construction of a parapet wall within the shelter of these blocks, for protecting the quay, has been deferred till the settlement of the mound has ceased, to avoid the injuries experienced at the Dragon breakwater. Some shoaling has occurred along the shore to the north of the breakwater, owing to the drift caused by northerly storms.

ILE ROUSSE HARBOUR.

Description. The port of Ile Rousse is formed by a small bay situated on the north-western coast of Corsica, and from its position is the natural port of communication with France, as Bastia is for Italy. It is open to the north and north-east. Some shelter has been provided by a breakwater starting from the northern extremity of the bay and pointing east-south-east.

Breakwater at Ile Rousse. The breakwater had been carried out 280 feet in 1866, at a cost of £25,400; and an extension was commenced in 1867, to make the total length 600 feet. It is formed of a mound of rubble stone, with a slope of about 2½ to 1 on the sea side, which is protected by concrete blocks. The quay on the top is sheltered by a parapet wall, 26 feet high, which has been somewhat dislocated by the settlement of the mound. (Plate 7, Fig. 5.)

The breakwater extends into a depth of about 50 feet, but it would require further extension to shelter the bay effectually from north-east winds.

ORAN HARBOUR.

Description. The recess of the north coast of Africa, on which the town of Oran, the second port of Algeria, is situated, was exposed from north to east; so that vessels had to seek shelter, in storms from the north, within the adjacent bay of Marsa-el-Kébir which, however, is open to the east.

The port of Oran is now artificially sheltered by a northern breakwater, about 3280 feet long, and an eastern pier; the breakwater affords the main protection, and the pier has been formed to reduce the width of the opening between the breakwater and the shore. (Plate 5, Fig. 15.) The greatest exposure is from the north-east; so the breakwater has been carried out so as to overlap the entrance, and protect it from that quarter; and a short pier has been run out from the breakwater opposite the east pier, so that, whilst assisting in the contraction of the entrance to 500 feet, it may check any run of waves pivoting round the head of the breakwater. These works have provided an excellent shelter; and quays have been constructed in the inner part of the harbour. The maximum depth inside is 7 fathoms; and at the entrance, it is from 4 to 7 fathoms.

Oran Breakwater. The breakwater extends into a depth of 11 fathoms; it consists of a mound of sorted rubble stone, protected on the sea slope, to a depth of 30 feet, and on the top by concrete blocks; and it is surmounted by a low massive superstructure. (Plate 7, Fig. 12.) The work experienced some injury from storms during construction.

General Remarks on the preceding Breakwaters. The breakwaters just described are only a step removed from the ordinary rubble and concrete mound breakwaters previously described. As compared with Algiers or Alexandria, rather more material is employed in proportion to the depth, owing to the width required on the top for a quay; and the superstructure, or parapet wall, effects no saving in material, but merely serves as a protection for the quay. The economy in material, however, is notable when compared with the ordinary rubble mounds of Plymouth, Portland, and Table Bay, in consequence of the adoption of concrete blocks on the sea slope; and it is even greater than appears on the sections, owing to the large proportion of vacuities in the sorted rubble mounds and blocks, amounting sometimes to one-third of the whole. The sorting of the material involves labour; but the principle is correct, as the largest material is thus deposited where the force of disturbance is the greatest, instead of being placed in the interior where the smallest material is equally efficacious. The settlement of a sorted mound would naturally be greater than the more compact mound of large and small materials intermingled; but the outer layers of the sorted mound, consisting of stones of various shapes and somewhat different sizes, become consolidated under the action of the waves. Moreover, the large stones on the sea face, though less compact than the mingled mass at first, are eventually more firm, as the waves gradually wash away the smaller rubbish.

Experience shows that, in the Mediterranean, concrete blocks of 13 cubic yards are stable on a sea slope of 1 to 1; and it is not necessary to use this protection beyond the limit of 20 to 30 feet below the sea-level. The error committed at the extension of the Marseilles breakwater, of only making the benching of rubble just wide enough to support the base of the facing of random concrete blocks, whereby several blocks have become displaced, was avoided at Bastia and

Oran by causing the benching to project from 10 to 20 feet beyond the base of the concrete-block slope.

The possibility of placing a quay on the actual breakwater is due to the protection of the slope of concrete blocks, the high parapet wall, and the comparative shelter of the sites. Marseilles is situated in a fairly sheltered bay; Trieste stands in a bay within the Gulf of Venice; and the comparative want of shelter at Ile Rousse is compensated for by the flatness of the sea slope.

The inexpediency of building a high narrow parapet wall on a rubble mound soon after its formation has been amply demonstrated at Bastia and Ile Rousse.

All the harbours described in this chapter possess the advantage of a good depth near the shore, with little tendency to accumulation of silt, with the exception of Trieste; and quarries were within easy reach for an abundant supply of materials.

CHAPTER XIII.

HARBOURS PROTECTED BY A RUBBLE MOUND, AND A SUPERSTRUCTURE FOUNDED AT LOW WATER.

CHERBOURG. HOLYHEAD. PORTLAND. ST. CATHERINE'S.

Cherbourg Harbour: Site; First attempts to form a Breakwater; Rubble Mound; Completion of Breakwater with Superstructure; Cost; State of Harbour; Remarks on Breakwater. *Holyhead Harbour:* Situation; Original Design; Enlargement; Cost of Breakwater, Maintenance; Comparison between Holyhead and Cherbourg Breakwaters. *Portland Inner Breakwater:* Description; Construction. *St. Catherine's Breakwater:* Situation; Description; Construction; Comparison between Cherbourg and St. Catherine's Breakwaters. Remarks.

IN the harbours about to be described, the superstructure on the top of the rubble mound fulfils a much more important function than at Marseilles, Trieste, and other similar breakwaters. The superstructure, instead of being merely a parapet wall and sheltering the quay, serves also to protect or replace the weak parts of the rubble mound above low water, and forms an integral portion of the breakwater. By being founded at low water, it can be built without any costly diving operations, and can be laid in cement. Moreover, it enables the amount of material required for the mound to be considerably reduced, more especially in cases where the range of tide is considerable; but, on the other hand, it involves a considerably larger amount of skilled labour, and great care in construction.

CHERBOURG HARBOUR.

Site. No natural harbour of large dimensions, suitable for a naval station, exists on the southern coast of the English Channel; and the want of a harbour for refuge and repairs, bordering on the Channel, was seriously felt by the French in their naval engagements with England. Under these circumstances, attention was naturally directed to the bay of Cherbourg, so well situated from its projecting position in the Channel, and only exposed to northerly winds. The idea of converting this bay into a harbour was suggested as early as 1665; but the great width of the bay from Point Querqueville on the west to Pelée Island on the east, a distance of 4 miles 3 furlongs, necessitated such large works, that the project was only actually undertaken in 1783 by the commencement of a breakwater in front of the bay.

De Cessart's Scheme for a Breakwater at Cherbourg. The original scheme was precisely similar to the design actually carried out, as regards the nature and site of the sheltering works. From the first it was proposed to form a detached breakwater in a line between the two extreme points of the bay, leaving one opening to the west of about 1 mile 3½ furlongs, and another to the east of nearly 5 furlongs. (Plate 3, Fig. 5.) The method, however, of construction attempted in the first instance was quite different to the existing structure.

M. de Cessart proposed to construct the breakwater of a series of truncated timber cones filled with stone, 149 feet in diameter at the base, 64 feet at the top, and 64 feet high[1]. It was intended that these cones should touch at their bases throughout the whole length of the breakwater, and that they should be filled with large rubble from the bottom to low water, and with concrete faced with masonry from this level to the top. The first cone was floated out and sunk in June 1784, and the second close to it in the following month. A storm, however, occurred before the second cone had been filled up with stone; and its upper portion was destroyed. This accident, together with the time that the depositing of the cones would occupy, being only feasible at the top of spring tides, and the cost of this method, led to a modifica-

[1] 'Mémoire sur la Digue de Cherbourg,' J. M. F. Cachin, 1820, p. 4.

tion of the scheme; and it was determined to place the cones 1280 feet apart, and to fill the intervals between them with rubble stone. Eighteen cones were deposited in this manner at various times; but as they suffered considerable injury in their partially filled and isolated state, they were cut down to the level of low water in 1789, with the exception of the one marking the eastern extremity of the breakwater, which, however, fell down in 1799, though it had been protected by a coating of concrete.

Rubble Mound Breakwater at Cherbourg. After the failure of the cones, the formation of the rubble mound, which had been commenced to fill up the spaces, was carried on with great activity throughout the whole length of about 2¼ miles of the breakwater; and at the close of 1790 the rubble stone deposited in the mound amounted to 3,486,000 cubic yards. The mound was formed to slopes of 1 to 1 on the harbour side, and 3 to 1 on the sea side, and was carried up to low-water level. The action of the waves, however, soon modified the form of the mound, drawing some of the rubble down the sea slope by their recoil, and carrying some over the mound on to the harbour side. The mound was thus lowered below the lowest tides; the sea slope was flattened to an inclination of 10 to 1 down to a depth of about 15 feet below low water; but the lower portion of the sea slope, and the harbour slope, remained unaltered.

The lowering of the mound diminished the shelter afforded by the breakwater; and it was considered necessary that the breakwater should be raised 8 feet above high water, in order to protect the harbour adequately. It was useless to attempt to raise the mound with the small material hitherto employed; but as the sea slope of a portion of the mound, which had been covered with larger stones by way of an experiment, had stood well, it was decided to place larger blocks of stone on the mound for raising it, and also for forming and protecting the foundations of the batteries which were to be placed at the centre and ends of the breakwater.

In order to defend the harbour, it was considered necessary to have a battery in the centre of the breakwater, at the angle which had originally been designed with the object of commanding the entrances. The raising of the mound was, accordingly, commenced at this point in 1802, by placing large blocks on the sea slope. The projection thus formed arrested the travel of stones along the breakwater, which varied according to the direction of the wind; and therefore an accumulation took place on both sides. This mass was covered with large stones, and the mound which was raised with small stones was similarly protected. A battery was erected on the raised mound and equipped in 1803, which was subsequently enlarged and strengthened, and protected by large blocks of stone. It, however, was injured by storms on various occasions; and in 1811 it was determined to carry the foundations of the battery down to the level of the lowest tides, and to build it of masonry faced with granite, so as to present a solid mass to the waves. In the same year 17,500 cubic yards of the largest stone procurable were placed on the sea slope in front, to repair the losses of the previous winter. Before these works could be completed, they were stopped in 1813; and nothing was done to the breakwater for the next eleven years, when the battery was again repaired.

During the French wars, defence rather than the completion of the breakwater was considered; and it was not till 1830 that the continuation of the works was arranged, so that the breakwater remained submerged for about forty years.

Completion of the Cherbourg Breakwater. The first operation undertaken in 1830 was the deposit of rubble to raise the mound to the level of low water. As the new foundations of the central battery had proved stable at the level of low water, it was determined to raise the breakwater to the required height by a superstructure of masonry, resting upon a layer of concrete deposited on the mound at low-water level. (Plate 7, Fig. 1.) This is the system upon which the breakwater has been completed. A row of concrete blocks, formed in place, was laid in front of the foundation of the superstructure, to protect it during its construction. The sea slope was protected by a layer of large stones, extending down to 16 feet below the lowest tides, and carried up to the face of the superstructure, covering the row of concrete blocks and the footings of the superstructure. This layer, forming the face of the sea slope, was given an inclination of about 5 to 1.

The mound, having been deposited under varying conditions and at different periods, settled unequally in places under the weight of the superstructure; and breaches occasionally resulted during storms, owing to the dislocations thus produced, which were repaired as they occurred. To provide as far as possible against settlement, the foundations were laid in long lengths during the summer, when westerly winds generally prevail, and the sea is very calm at Cherbourg. The final raising of the top of the superstructure, and the construction of the parapet wall, were deferred for three or four years to allow the mound to consolidate. The ordinary superstructure of the breakwater was completed in 1846; and the reconstruction of the central fort, and the building of the pierheads at each extremity with forts surmounting them, were then undertaken. The work was finally completed in 1853[1]. The pierheads and the base of the central fort are protected with large concrete blocks, as the rubble would tend to travel round the pierheads under the influence of the winds; and the central fort projects in front of the general line of the breakwater.

Cost of Cherbourg Breakwater. The total sum expended

[1] 'Travaux d'Achèvement de la Digue de Cherbourg,' J. Bonnin.

on the works, from their commencement in 1783 to their completion in 1853, amounted to £2,674,500. This, however, includes the cost of M. de Cessart's cones, the works for forming the central battery between 1803 and 1814, and other expenses not properly incidental to the construction of the breakwater; and by deducting these M. Bonnin reduces the actual cost to £2,000,000, which, for the extreme length of breakwater he names of 12,178 feet, gives a cost of about £164 per lineal foot, for a detached breakwater built in an average depth of about 40 feet below low water. Taking, however, the actual distance between the pierheads of about 11,880 feet, the cost would amount to £168½ per lineal foot. The superstructure cost about £55 per lineal foot.

State of Cherbourg Harbour. The breakwater affords shelter to an anchorage of about 2000 acres. The harbour is readily accessible through its two wide entrances. The eastern entrance is 2730 feet wide at low water, with a depth in the centre of 28 feet. The western opening is divided in two by the rock upon which Fort Chavagnac has been built to guard the entrance, which is between the fort and the breakwater, 3750 feet wide, and situated in a depth of 36 feet.

The depth in the harbour has not been diminished by the construction of the breakwater, owing to the free flow of the currents through the entrances at each extremity of the bay. It has been proposed, for purposes of more perfect defence, to close the opening between Fort Chavagnac and the shore; in which case it would be also advisable to join Pelée Island to the coast, so that the currents through the entrances might be more equalised; and it is believed that these works would not lead to the silting up of the harbour, as the currents would be merely intensified, and the introduction of silt through the entrances is small.

The breakwater shelters the entrances to the Government docks on the north-west of the town, and to the commercial docks in the recess of the bay. (Plate 3, Fig. 5.) The dockyard was commenced at the beginning of the present century, and completed in 1866; the basins have been excavated out of the rock. The commercial docks are of much earlier origin, having been commenced in 1687. Cherbourg, however, is more a naval station than a port of commerce.

Remarks on Cherbourg Breakwater. It will be readily seen, from an inspection of the plans of harbours in Plates 1–5, that the breakwater at Cherbourg is one of the largest marine works ever undertaken, and has not been exceeded, as regards extent of area sheltered, by any more recent works. Alexandria breakwater, indeed, approaches it in length, whilst the mattress jetties at Charleston and Galveston are longer; but considering the solid nature of the work, the batteries which it supports, and the period of its construction, it is a monument of engineering perseverance and skill. It does not equal in section, or in depth, several other breakwaters; but it furnishes a perfect type of a breakwater formed of a rubble mound with a superstructure founded at low water. The superstructure supplies the place of the weak part in the rubble mound, and fulfils precisely the object for which it was designed; whilst the rubble mound is not raised higher than is just adequate for the protection of the foundation course of the superstructure.

The breakwater is not so exposed as the Plymouth breakwater on the opposite coast, as it only faces the English Channel, instead of the Atlantic Ocean. Moreover, it is sheltered from the most prevalent south-westerly winds. Northerly winds are rare, and the worst winds for the breakwater are north-east and north-west. North-west gales, though less frequent, raise the largest waves; and, consequently, the western arm was more difficult to build, and requires most care in maintenance. Though, during storms, some of the cannons placed on the quay of the breakwater have been overturned, and stones from the sea slope have been hurled over the breakwater into the harbour, the work of maintenance is not heavy. It consists in replacing concrete blocks in front of the pierheads and forts, and in depositing large blocks of stone on the foreshore to the extent of about $1\frac{1}{10}$ cubic yards per lineal yard of breakwater annually.

HOLYHEAD HARBOUR.

Situation. Holyhead Bay is situated near the extreme western point of the island of Anglesea, and is therefore the best place of departure for the shortest sea route between England and Dublin. This advantage, together with the importance of having a harbour of refuge on the west coast of Wales, led to Holyhead Bay being selected by the Government for the formation of a national refuge harbour.

The bay is of considerable extent; it is open to the north-west, but sheltered from west round by south to north. (Plate 4, Fig. 5.) Moreover, even in the north-western direction, its line of exposure is not very great, as the Irish Sea is hemmed in on the north-west by the coasts of Ireland, Scotland, and the Isle of Man.

An old harbour existed in a creek within the bay near the town of Holyhead, which has now been converted into a commercial harbour; but formerly it was inadequately sheltered, and too shallow and small. In order to form a refuge harbour, it was only necessary to shelter the bay from the north-west.

Original Design for Holyhead Harbour. When the harbour works were commenced in 1847, it was proposed to enclose a sort of secondary bay, to the west of the old harbour, by two breakwaters; a northern breakwater 5360 feet long, and an eastern breakwater of 2000 feet, sheltering an available area of 267 acres and a packet pier 1500 feet in length[1]. The northern breakwater was intended to provide the necessary shelter; the eastern breakwater, starting from Salt Island, was to cover some rocks in front; and the packet pier was to accommodate the mail service with Ireland.

The northern breakwater was commenced in accordance with the original design; but in 1854, when the rubble base of the breakwater had been extended about 4150 feet from the shore, the number of vessels seeking the shelter of the unfinished harbour was increasing so rapidly that it was considered advisable to extend the area of the proposed harbour.

Design of Holyhead Harbour as executed. To provide additional shelter, it was necessary to prolong the northern breakwater. The original easterly line was followed up to about 4860 feet from the shore, and it was then curved round to a north-easterly direction, and carried out to a total length of 7860 feet; the change in direction being made in order to facilitate the access to the enlarged harbour, and to shelter a more extensive area. (Plate 4, Fig. 5.) The eastern breakwater was abandoned; and provision was made for the mail service by the erection of a jetty at the old harbour. The Skinners and Platters rocks, which were to have been covered by the eastern breakwater, have been marked by buoys. The extension provides a sheltered roadstead of 400 acres, of which 200 acres have a depth of 5 fathoms and upwards, in addition to the 267 acres enclosed within the rocks. The creek within the old harbour has been deepened by dredging, and quays have been formed round it for the trade with Ireland carried on by the London and North Western Railway Company.

Holyhead Breakwater. An abundance of excellent stone was available for the construction of the breakwater from

[1] 'Holyhead New Harbour.' Final Report by John Hawkshaw, F.R.S., Superintending Engineer, 1873.

quarries at the foot of Holyhead mountain, only about a mile distant from the commencement of the work.

The design adopted and carried out consists of a rubble stone base rising up to high-water level, with a superstructure, on the harbour side of the mound, founded at low-water level, composed of a sea and harbour wall of masonry built with large blocks of stone set in lias lime mortar. The sea wall both provides an upper roadway, or promenade, on the top, 21¾ feet above high water, and also shelters the quay level, 11½ feet below the upper level. The quay is formed by the deposit of suitable small material on the mound above high-water level between the sea and harbour walls. (Plate 7, Fig. 2.) The breakwater has been built in an average depth of 40 feet, and extends into a maximum depth of 55 feet.

The slope of the mound, on the sea side, is about 12 to 1 between high and low water, 5 to 1 from low water to a depth of about 12 feet, and about 2 to 1 from thence to the bottom; on the harbour side it is about 1¼ to 1 throughout.

The breakwater is terminated by a wide head of ashlar masonry founded 20 feet below low water, 50 feet long in the direction of the breakwater, and 150 feet across. A square lighthouse has been built on the head, that form having been adopted for the sake of the lightkeeper's rooms.

Construction of Holyhead Breakwater. The breakwater was commenced in 1849. The stone was conveyed on a railway from the quarries, and tipped from timber staging, erected along the line of the breakwater, by means of special iron wagons carrying from 8 to 10 tons of stone. The staging, carrying five lines of railway, was supported on piles resting on the bottom and surrounded by rubble. (Plate 8, Figs. 10 and 11.) The rails were raised 20 feet above high water, so that the platform might be beyond the influence of the waves. The tipping wagon held the stone in a movable shoot hung on to the frame; the shoot was retained by a trigger which, when released, allowed it to fall and discharge its contents; and the shoot was so counterbalanced, that as soon as it was empty it returned to its original position[1]. The end staging was frequently washed away by storms, before it was buried some depth in the mound.

About 4000 tons of stone were deposited in a day when the works were in full operation, the maximum having attained 5220 tons. The largest amount of stone deposited in one year was 1,066,900 tons, in the year ending 31st March, 1856; the average annual deposit from 1850 to 1858 was 714,600 tons. The total amount of stone in the mound is about 7,000,000 tons; about 13½ cubic feet of the stone weigh a ton; but interstices occupy about one-third of the volume of the mound. The greatest advance of the mound in one year was 4190 feet in 1853-4.

The superstructure was not commenced till 1860, to allow the mound to consolidate thoroughly; but it was then built out rapidly, 6660 lineal feet having been constructed in four years. The works were finally completed in 1873.

Cost of Holyhead Breakwater. The total cost of the northern breakwater and accessory works was £1,285,000, which for a length of 7860 feet gives about £163 10*s.* as the cost per lineal foot of breakwater. The superstructure, exclusive of the head, cost £36 per lineal foot. The cost of a bay of 30 feet of the staging was £586 10*s.*

Maintenance of Holyhead Breakwater. One of the objects of the superstructure, in addition to sheltering a quay and affording more shelter to the harbour, is to prevent the smaller stones of the mound washing over into the harbour. As the mound rises to high-water level against the superstructure on the sea face, there is little recoil of the waves from the superstructure; and, moreover, the mound has been given a very easy slope down to low water. Accordingly, the maintenance of the mound has hitherto, for the most part, been very slight. At the extremity, however, the mound tends to

[1] 'Minutes of Proceedings, Inst. C. E.,' vol. xliv, plate 3.

travel round the head, and is not quite stable. It has, accordingly, been found necessary to protect the rubble slope for the last 200 feet on the sea side near low water. This was first attempted by throwing down concrete blocks; but as there was no adequate machinery for depositing the blocks, they got broken by their fall, and it would have been better to make the blocks in place within frames in fine weather at low water during spring tides. Eventually old chains were placed in long coils upon the foreshore: these chains keep the foreshore from shifting by their weight, amounting to 1000 tons, and at the same time do not offer a solid face to the blow of the waves[1].

A rubble mound like that at Holyhead, rising above low water, cannot be regarded as absolutely permanent: though the rubble is prevented from travelling over the top by the superstructure, and round the end to some extent by the chains, it is constantly triturated and rolled about by the waves. The duration of the stones on the surface of the slope depend on their size and hardness, and as they become worn they offer a greater surface in proportion to their weight; so that such a mound tends to deteriorate, and to become less stable; and sooner or later fresh deposits become necessary, unless the stones are of such size as to be immovable by the sea.

Comparison between Holyhead and Cherbourg Breakwaters. The breakwaters at Holyhead and Cherbourg resemble one another in having been constructed in a similar depth of water, in being subjected to the same rise of tide, and in facing north-westerly winds with a moderate degree of exposure, the greatest uninterrupted fetch of sea at Holyhead being about 80 miles, and the same at Cherbourg.

[1] A similar system was previously adopted, under my superintendence, at the Alderney breakwater, for weighting the lower face blocks of the superstructure, till the tide or weather permitted the upper courses to be laid upon them. In this instance a round coil of chain was placed upon each block, and kept it in place.

A comparison, however, of the sections of these breakwaters (Plate 7, Figs. 1 and 2) indicates a very marked difference in their construction. Both, indeed, have superstructures founded at low-water level; but whilst the Cherbourg superstructure forms a very important part of the breakwater, rising from near low water above the rubble mound, thus greatly saving the material of the rubble mound, and bearing the full force of the waves above low water, the Holyhead superstructure is buried in the mound up to high-water level, and serves to maintain, rather than to replace the rubble mound. The superstructure at Holyhead, rising higher than at Cherbourg, and protected by the high flat slope of the foreshore, shelters the quay more effectually, and enables a gravel coating to be employed for the roadway, instead of pitching as adopted at Cherbourg, but it does so at the expense of a very large additional mass of rubble. The long slope, moreover, between high and low water at Holyhead, has to be covered with large blocks of stone; for though, owing to the height of the mound, there is little recoil of the waves from the face of the superstructure, yet the slope has to withstand the main force of the sea, which is in a great measure borne at Cherbourg by the superstructure.

The Holyhead section affords little saving in the mass of the rubble mound; but the superstructure at Cherbourg, whilst reducing the mound, has to be made more compact, and needs more protection for its toe on the sea face. At Holyhead, however, there was an abundance of suitable stone, easily conveyed to the breakwater; whereas at Cherbourg, large stone was difficult to procure, the means of conveyance from the quarries were defective and slow, and, the breakwater being detached, the material had to be transhipped into barges.

The Cherbourg breakwater was at a disadvantage in comparison with Holyhead, both as regards its earlier period of execution before the advantages of steam for carriage were fully realised, and also from its position at a distance from the shore and the quarries. It is therefore to the credit of French engineers that, if M. Bonnin's estimates are correct, the Cherbourg breakwater has little exceeded the cost per lineal foot of the Holyhead breakwater, which can only be due to the great saving in the material of the mound by the more scientific section adopted for the former breakwater. The Holyhead breakwater, however, affords more efficient shelter for the quay and adjacent portion of the harbour; and the use of a large quantity of rubble stone was the result of the facility with which it was obtained; whilst necessity imposed a different course at Cherbourg, which involves rather more care and expense in maintenance.

PORTLAND INNER BREAKWATER.

Description. The harbour of Portland and its outer breakwater have been already described in Chapter X; but the short inner breakwater, connected with the shore, belongs to the type dealt with in the present chapter. (Plate 3, Fig. 11, and Plate 7, Fig. 11.)

The inner breakwater at Portland Harbour, extending in a straight line from the shore to the south-eastern entrance, faces south-east, and reaches a depth of 55 feet at its extremity. The type of design is almost identical with that of Holyhead breakwater, which was natural, as both were designed by Mr. Rendel, about the same time, and under very similar conditions; for both breakwaters are connected with the shore, and abundance of stone could be procured close at hand. (Compare Plate 7, Fig. 2, with Fig. 11.) The main difference between the two is due to the smaller rise of tide at Portland (6¾ feet as compared with 17 feet), which reduces considerably the length of flat slope between high and low water, and consequently the width of the mound, and also the height of the superstructure.

The chief object of the superstructure at Portland is to provide quay accommodation; but it also prevents the rubble being carried over into the harbour, and protects the top of the mound. Though the mound rises to high-water level against the superstructure, mean sea-level approaches much closer to the face of the superstructure than at Holyhead, owing to the slight range of tide, and therefore the superstructure is more exposed to the waves. This, however, is compensated for by the position of the breakwater, for south-easterly gales are rare in the Channel, and the wind generally veers round rapidly to south-west; whereas north-westerly winds, to which Holyhead breakwater is exposed, are more vehement, more frequent, and of longer duration. The exposure at Portland is moderate, as the greatest fetch of sea to the south-east is only 100 miles.

Construction of Portland Inner Breakwater. The method of depositing the rubble mound at Portland has been already described in Chapter X. The work was commenced, at the end of 1849, by depositing the mound at the shore end. Though the mound for the inner breakwater had been extended to its full length by 1852, the superstructure was not commenced till 1856, so as to give the rubble mound ample time to consolidate. The pierhead, however, at its extremity, being founded in a depth of 24 feet, was commenced in 1853, having a lower mound to rest upon, and was finished in 1856. The inner breakwater was practically completed in 1860.

ST. CATHERINE'S BREAKWATER.

Situation. The breakwater of St. Catherine's forms the northern arm of the unfinished harbour which was intended to be enclosed within St. Catherine's Bay, on the east coast of the island of Jersey. The harbour was commenced in 1847, with the object of providing protection for the Channel Islands, and it was to be sheltered by two converging breakwaters; but the works were discontinued in 1856, when the northern breakwater, about 2000 feet long, had alone been carried out. The harbour, therefore, is only partially sheltered; and sand has accumulated in it, so that it is of little value in its incomplete state. The position is one of slight exposure, as though the bay is open to the east, it is somewhat sheltered by the French coast only 13 miles distant.

Breakwater at St. Catherine's Bay. The breakwater consists of a rubble mound rising somewhat above low water, with a superstructure founded at low-water level and composed of a sea and harbour wall with an intermediate filling of loose rubble. (Plate 7, Fig. 10.) The harbour wall, and the sea wall up to high-water level, were built of masonry without mortar; but mortar was used for the upper part of the sea wall.

The breakwater extends into a depth of about 30 feet; but the actual height needed for the breakwater is considerably increased by the great rise of tide in the bay of St. Malo, amounting at St. Catherine's to 32 feet. This unusually large range of tide rendered it specially advisable to keep the mound at a low level, as a flat slope, like at Holyhead, between high and low water would have greatly added to the mass of the mound, and the supply of suitable stone in the immediate neighbourhood was not very plentiful.

Construction of St. Catherine's Breakwater. The breakwater, which was commenced in 1848, was completed in 1855. The mound was tipped out from the shore; the greatest amount of stone deposited in a year was 149,000 tons, and the total quantity was about 790,000 tons. The erection of the superstructure followed the extension of the mound, averaging about 260 lineal feet annually; and though some settlement occurred, it did not materially injure the uncemented masonry, but some cracks in the upper wall and in the head had to be repaired.

The total cost of the work was £197,000, and assuming that £26,000 of this amount may be reckoned as appertaining to the southern breakwater abandoned in 1849, the average cost per lineal foot would amount to about £81.

Comparison between St. Catherine's and Cherbourg Breakwaters. The section of St. Catherine's resembles in principle that of Cherbourg; but though the superstructure is larger in mass, it is much less solid in construction, being formed of two thin walls quite unconnected, and built without mortar, though standing in nearly twice the depth at high water. (Compare Plate 7, Fig. 10 with Fig. 1.) The breakwater, however, at Cherbourg was difficult to construct, and needs careful maintenance; whilst St. Catherine's breakwater has given no trouble, either in construction or maintenance. This difference can only be due to the difference in exposure and position, and illustrates the importance of duly weighing these conditions. St. Catherine's breakwater faces north-east, and is sheltered by the French coast from that quarter, the fetch of sea being only 13 miles in that direction; and north-east winds are neither so frequent nor severe as westerly winds in the English Channel. It has been already pointed out that Cherbourg is, in this respect, much less favourably situated.

Remarks on the Breakwaters described above. Though the breakwaters of Cherbourg, Holyhead, Portland, and St. Catherine's, all consist of a rubble mound with a superstructure founded at low water, they really form two distinct groups. Holyhead and Portland are rubble-mound breakwaters, with a sort of backbone of masonry inserted in them to keep the stones of the sea slope in place, and to protect the top of the mound, and are therefore somewhat undeveloped types of the mixed system, though approaching nearer to it than the Mediterranean harbours, described in the previous chapter, which possess merely the rudiments of a superstructure. Cherbourg and St. Catherine's are, on the contrary, distinct types of the mixed breakwater, where the superstructure replaces the rubble mound above low water. In the former, the rubble mound bears the main attack of the waves: in the latter, the superstructure opposes the principal resistance. In the former, the rubble mound must be strengthened by large blocks from high water to a little distance below low water, just like an ordinary mound breakwater: in the latter, the superstructure must be firm, and the slope at its toe protected. Mass is of most consequence for the first; compactness for the second. An abundance of stone, hard and of adequate size, is the principal requirement for the first; good cement, and careful workmanship for the second. The simple mixed system with a low mound, like Cherbourg and St. Catherine's, is most advantageous where the range of tide is large: the raised mound adds little comparatively to the mass where the rise is small, as at Portland. Holyhead and Cherbourg furnish instances where, under very similar conditions of site, the two different methods have been adopted, owing to unequal facilities for procuring materials.

The St. Catherine's section has stood securely in its well-sheltered site; but to adopt a similar design in more exposed situations is sure to lead to failure, as will be illustrated by the history of the first design for Alderney breakwater in a future chapter.

CHAPTER XIV.

HARBOURS PROTECTED BY SORTED RUBBLE AND CONCRETE-BLOCK MOUNDS, WITH SUPERSTRUCTURE FOUNDED AT LOW WATER.

GENOA. BOULOGNE. ST. JEAN-DE-LUZ. LEGHORN.

Genoa Harbour: Description; Western Breakwater, extension, cost. *Boulogne Harbour:* Description; Breakwaters; Construction of south-west Breakwater. *St. Jean-de-Luz Harbour:* Description; Early Works; Breakwaters, Socoa, Artha; Condition of Harbour. *Leghorn Harbour:* Description; Breakwaters, northern and outer, construction and cost. Remarks.

THE breakwaters about to be described differ from those in the previous chapter, like the Algiers and Alexandria breakwaters differ from the simple rubble mounds of Portland and Table Bay. A sorted rubble mound at Genoa, a sorted rubble and concrete-block mound at Boulogne and St. Jean-de-Luz, and a concrete-block mound at Leghorn, take the place of the simple rubble mound of Holyhead and Cherbourg. The concrete blocks reduce the amount of material required for the mound, and secure the toe of the superstructure, which has merely to be made strong enough to resist the impact of the waves.

GENOA HARBOUR.

Description. Genoa is situated on a small bay at the north of the Gulf of Genoa. The bay has the form of a semicircle, a little more than a mile in width, and seems formed by nature for a harbour, as it is only open towards the south; and it is not subject to the influences of alluvial deposits. Protection was only needed from the south to convert the bay into a sheltered harbour. The first work for this purpose consisted of a mole, commenced in 1134, carried nearly half way across the mouth of the bay from a projecting point on its eastern side, thus affording shelter from the south-east. (Plate 5, Fig. 3.)

In order to provide protection from the south-west, and to enlarge the sheltered area, another breakwater was built out from Cape St. Benigno, the south-western extremity of the bay, which had a length of 1640 feet at the beginning of the present century, and has been prolonged gradually at various periods, so that now it is 3280 feet long, and overlaps the old mole, leaving an entrance between the two of about 1760 feet facing east.

This breakwater is being now extended to the south and east, as shown by dotted lines on the chart, so as to form an outer sheltered roadstead, to the south-east of the bay, down to Point St. Giacomo; and it will afford additional protection to the existing harbour, where a swell enters at present during south-west and south-east winds.

The breakwaters shelter a series of inner basins, jetties, and quays, along the shores of the bay.

Western Breakwater Extension at Genoa. The quarries of Genoa afford an abundant supply of stone of large dimensions; and as they are situated close to the shore, it was only necessary to carry the stone by water to the breakwater, a distance of about two-thirds of a mile. The base of the breakwater, accordingly, has been formed of sorted rubble stone, the smaller material being placed in the centre of the mound, and the stones exceeding 20 tons in weight are deposited at the sides and slopes of the mound. (Plate 7, Fig. 3.) Blocks of stone up to 70 tons in weight have been used, being conveyed to their destination by a 400-ton barge. These masses were tipped into the sea by raising gradually the inclined plane upon which they rested.

The superstructure, founded at low-water level, has the great width of about 100 feet; it rests upon a layer of concrete, averaging 8 feet in thickness, placed on the top of the rubble mound. The actual quay, raised 10½ feet above the sea-level, has only a width of 11 feet, as the rest of the space is occupied by storehouses covered by arches, by the outer protecting wall, and an apron at its toe. Rubble concrete blocks have been used to protect the foreshore of the mound, as well as the large blocks of stone, merely to save the time required for procuring these large blocks in sufficient quantities.

The breakwater extends into a depth of 50 feet; but owing to its moderate exposure, a tidal rise of only 1⅗ feet, and the large rubble available, it has not attained the dimensions of the Holyhead breakwater, in spite of the accommodation for merchandise provided on its superstructure; and it approximates to the Cherbourg and Portland breakwaters. (See Plate 7.)

Cost of Extension of Genoa Breakwater. The last extension of the western breakwater, of 980 feet, previous to the commencement of the new works in 1878, cost about £128 per lineal foot, including the superstructure which cost £61 per lineal foot[1].

BOULOGNE HARBOUR.

Description. The new harbour at Boulogne, now in course of construction, is designed to be formed as shown on the plan[2]. (Plate 5, Fig. 8.) Two entrances are to be provided, on each side of a detached breakwater, facing north and west respectively, 820 and 490 feet wide; so that vessels will have a choice of entrance according to the direction of the wind. The new harbour will enclose the old jetty harbour, which provides merely for a tidal service of steamers and affords access to the docks. It is proposed to construct docks or quays within the area sheltered by the breakwaters.

[1] 'Annales des Ponts et Chaussées,' 4th Series, vol. xx, p. 168.

[2] The plan and section of the Boulogne breakwaters were taken from drawings supplied me by M. Guillain, Engineer-in-Chief of the Port.

The existence of a flat sandy beach in front of Boulogne has raised doubts as to the maintenance of a closed harbour in that locality. The foreshore outside the south-west breakwater has already advanced since the projection of this breakwater from the shore; but it is probable that the shore line will reach a fresh position of equilibrium before attaining the extremity of the breakwater, as has happened at Ymuiden Harbour. (Plate 5, Fig. 2.) This is the more probable as the breakwaters will form a projecting point from the coast, which will tend to produce scour in front of them, especially as a current now exists at that distance from the coast. The sea, moreover, is free from sandbanks in front of Boulogne, and the depth at the entrances will be 4 fathoms, so that it is unlikely that much wave-borne sand will find its way into the harbour; and any accumulation from this source could be easily removed by dredging in the sheltered water of the harbour.

Breakwaters at Boulogne Harbour. The three breakwaters are to be similar in construction, with a solid superstructure of masonry, resting, at low-water level, upon a mound of small rubble in the interior, enclosed by larger rubble, and protected by concrete blocks on the sea slope. (Plate 7, Fig. 8.) The concrete blocks have been dispensed with at the inner portion of the south-west breakwater, already constructed, as they will only be used for the deeper and more exposed portions of the breakwaters. The thickness also of the superstructure is varied according to its position. The estimated cost of the works is £1,280,000.

Construction of Boulogne South-West Breakwater. Up to 1883 the mound for the south-west breakwater was tipped from wagons, run out from the shore over the completed portion. A small haven, however, was formed that year in the southern angle of the harbour, where barges could lie in shelter and be loaded with stone, and then be towed out to deposit their load on the mound. It is hoped that by the employment of barges a more rapid rate of progress will be accomplished. Some of the stone is being obtained from the adjacent cliff; but the layers of stone are thin, and at several feet below the surface, so that additional stone has to be procured by railway from quarries some miles distant. The concrete blocks employed weigh 24 tons.

ST. JEAN-DE-LUZ HARBOUR.

Description. The bay of St. Jean-de-Luz, situated on the south-west coast of France, is open to the north and north-west, and therefore partly exposed to the vehement Atlantic storms of the Bay of Biscay. The worst seas come from the west and west-north-west, when the waves occasionally attain a height of about 23 feet, and a length of 400 to 600 feet outside the bay[1].

The bay is terminated on the west by the promontory of Socoa, and on the east by St. Barbe Point. The Artha Rock stands in the centre of the entrance to the bay; and the deepest passage, reaching a maximum of 8 fathoms, is between this rock and Socoa. (Plate 3, Fig. 9.) The bay was formerly very dangerous for vessels, on account of the exposure of the coast, the variety of the currents, and the surf on the Criquas off Socoa. The river Nivelle flows into the bay, and afforded the only safe anchorage for vessels; but its depth has been reduced by the formation of a bar at its mouth. The waves in a storm used to break upon the Artha Rock, and rendered the eastern entrance impassable.

Early Works at St. Jean-de-Luz. A breakwater was built out from Socoa Point early in the 17th century; and at the beginning of the following century the breakwater was repaired and extended. A scheme for enclosing the bay, by means of breakwaters from Socoa and St. Barbe Points, was started in 1783; but, owing to the disturbed state of France, the works

[1] 'Annales des Ponts et Chaussées,' 5th Series, vol. xi, p. 398.

were discontinued in 1787 for want of funds, when the breakwaters had been carried out 588 and 486 feet respectively; and the sea gradually demolished the greater portion of these structures.

Breakwaters at St. Jean-de-Luz. The project of converting the bay into a harbour was revived in 1864. The design consisted of a breakwater from Socoa Point, 1430 feet long, and a detached breakwater 820 feet long across the Artha Rock, leaving a passage to the west of 770 feet, and an opening on the east side to admit the flow of the currents and avoid silting. This plan has been somewhat modified during construction: the Socoa breakwater has been completed, and the Artha breakwater is in course of construction; but the width of opening between the breakwaters has been increased to 820 feet, by reducing the length of the Socoa breakwater 50 feet. Moreover, a breakwater has been commenced from St. Barbe Point, which is designed to have a length of 740 feet, and has already been carried out 560 feet; but it has been provisionally stopped in consequence of fears expressed as to its effect on the currents in the bay. (Plate 3, Fig. 9.)

The Socoa breakwater consists of a solid masonry superstructure, resting, about the sea-level, on a mound composed of a hearting of rubble stone and a covering of concrete blocks containing 26 cubic yards each. (Plate 7, Fig. 21.)

Artha Breakwater at St. Jean-de-Luz. The construction of the detached breakwater across the Artha Rock is progressing slowly. The section of this breakwater is similar to that of the Socoa breakwater; but concrete blocks enter more largely into the composition of the mound, the breakwater being more exposed[1]. (Plate 7, Fig. 22.) The rubble in the mound varies in size from 5 cubic feet to 1⅓ cubic yards; the concrete blocks contain 26 cubic yards. The mound was in a

[1] The section of the Artha breakwater has been taken from a drawing sent me by M. Pettit, the engineer in charge of the works, who also gave me some particulars about Biarritz breakwater.

forward state in 1883; the quantity of rubble deposited amounted to 16,200 cubic yards, and there were 108,300 cubic yards of concrete blocks, so that the proportion of concrete blocks to rubble in the mound is approximately as 7 to 1.

The superstructure was commenced in 1879, and only 230 feet had been completed in 1883. The works, however, have been delayed and injured by the violent storms to which they are exposed. The face-work on the sea side has been removed in several places; the small rubble, filling up the interstices between the large blocks, has been entirely washed away; and the parapet wall has been injured. It is proposed to strengthen the breakwater by forming an apron of masonry or concrete at the toe of the superstructure on the sea side, to increase the concrete blocks to 40 cubic yards, and to enlarge the parapet.

Condition of St. Jean-de-Luz Harbour. The projection of the Socoa breakwater has diverted the greatest force of the current in the bay to the east side, and shoaling has taken place on the west side; so that, in the first instance, whilst the shelter was being improved, the available anchorage area was being reduced. It appears, however, that the main cause of shoaling has been the erosion of the coast on the eastern side of the bay by the sea, the material thus removed being deposited on the sheltered western beach. When the inroads of the sea shall have been restricted by the Artha and St. Barbe breakwaters, the only sources of deposit will be the silt brought down by the river Nivelle, and whatever alluvium may enter the bay travelling along the coast from the mouth of the river Bidassoa. This deposit is estimated by M. Bouquet de la Grye to amount to 38,000 cubic yards in a year. A small tidal harbour at Socoa is thoroughly sheltered by the breakwater. The town of St. Jean-de-Luz, situated on the south-eastern side of the bay, had formerly to be protected from the encroachments of the sea; and the sea walls formed were thrice washed away. These encroachments will be effectually prevented by the breakwaters, which will, accordingly, both protect the town, and afford much needed shelter for shipping.

LEGHORN HARBOUR.

Description. The harbour of Leghorn, on the north-western coast of Italy, lies in a recess of a projecting part of the shore facing west. The first shelter for the harbour was provided by the wide mole of Medicis, 2575 feet long, projecting in a north-westerly direction from the southern point of the recess. This structure protected the harbour from the west, but left it open to the north-west. Two inner basins, however, furnished good shelter, in which the port was deficient in stormy weather.

The worst winds blow from north-west and south-west, so that the old harbour was exposed to one of the worst quarters, and was, moreover, inadequate in depth when the draught of vessels increased. Under the advice of M. Poirel, two detached breakwaters were undertaken by the Duke of Tuscany, and completed by the Italian Government. The straight smaller breakwater, facing north-west, shelters the old port, leaving an entrance of 420 feet between it and the old mole. The larger breakwater, in the form of a circular arc, shelters an outer roadstead extending into a depth of 5 fathoms; it also protects the entrance to the old harbour, and leaves an opening for vessels to the north and south. (Plate 3, Fig. 3.) The shape thus given to the harbour resembles that of Civita Vecchia, though on a more extended scale (Plate 3, Fig. 2), and also that of Cette. (Plate 5, Fig. 16.)

Dredging was begun in the inner harbour in 1866, with the object of giving it a uniform depth of 3½ fathoms, at an estimated cost of £73,300; but it would have been a troublesome operation owing to the existence of a stratum of rock, and it does not appear to have been carried out as the depth in the latest chart is only from ¾ to 2½ fathoms.

Northern Breakwater at Leghorn. The straight detached breakwater was commenced in 1854, and was terminated at its present limits in 1859, its final completion being deferred till the dredging shall have been executed. It is situated in the very moderate depths of from 1½ to 2½ fathoms, and its total length is 1640 feet. For a length of 330 feet nearest the shore, the breakwater was formed entirely of rubble stone; the remaining 1310 feet were constructed with a hearting of rubble enveloped in concrete blocks of 13 cubic yards.

The cost of the breakwater was £24,000, or £14⅗ per lineal foot.

Outer Detached Breakwater at Leghorn. The outer curved breakwater, convex to the sea, is 3610 feet long. It was commenced, in 1853, at the centre, and was completed in 1866; it is situated in depths of from 4 to 5¼ fathoms. The breakwater consists of a superstructure, founded at the sea-level, standing on a mound composed entirely of concrete blocks of 13 cubic yards, on account of the small-sized material furnished by the quarries. The superstructure is built upon a layer of concrete, formed on the top of the mound, 50 feet wide. It consists of a quay, 6½ feet above the water-level and 20 feet broad, and a wall rising 30 feet above the water, 16½ feet wide at the base and 7¾ feet at the top, protecting the quay, and protected itself at the base, on the sea side, by a row of blocks of 26 cubic yards. (Plate 7, Fig. 6.) Each end of the breakwater is terminated by a round pierhead built upon a carefully laid radial foundation of concrete blocks of 13 cubic yards.

The total cost of the breakwater was £424,000, which is equivalent to £117½ per lineal foot.

The outer breakwater shelters an area of about 153 acres, which is most affected by south-south-west winds; and the construction of a detached breakwater to the south-west has been suggested for the purpose of sheltering the outer harbour from these winds. The area of the old harbour is about 91 acres. Very little silt is brought into the harbour by the sea.

Remarks on Genoa, Boulogne, St. Jean-de-Luz, and Leghorn Breakwaters. In cases where breakwaters, having superstructures founded at low water, are situated in an almost tideless sea like the Mediterranean, the difference noticed in the last chapter between Cherbourg and Holyhead breakwaters, as regards the level to which their mounds are raised against the superstructure, ceases to exist, as high and low water are only a foot or two apart. Nevertheless, a difference in the mass of the mounds, quite as marked, may be noticed in the Genoa and Leghorn breakwaters. The large size of the Genoa mound is due partly to the great width given to the superstructure, and partly to the mound having been formed almost entirely of rubble; whereas the Leghorn mound, formed exclusively of concrete blocks, is as small as the width of the superstructure permits. The Genoa superstructure has been made exceptionally wide, for the sake of the storehouses it provides; but the width of apron and mound in front of the face of the superstructure appears unnecessarily large, considering the fairly moderate exposure of the site.

By comparing the sections of Boulogne, St. Jean-de-Luz, and Leghorn breakwaters with those of Cherbourg, Holyhead, and Portland (Plate 7), the saving in material by the use of concrete blocks is fully realised. For instance, the Boulogne section, in the same depth at high water as Holyhead, has little more than one-third of the width of the Holyhead breakwater at the base, and not one-sixth of the mass. The difference is less in the case of Cherbourg and Portland, but still very notable.

The Boulogne breakwater has a smaller section than Leghorn, though in deeper water at high tide; but this is partly due to the greater depth at Leghorn than the low-water depth at Boulogne, and partly to the Leghorn superstructure being furnished with a quay. The Leghorn and Socoa breakwaters are somewhat similar in section, being situated in similar mean depths; but it would appear from this that Leghorn is stronger than necessary, or Socoa too weak, as the exposure of the two is very different. It is natural that the Artha breakwater, situated in a somewhat similar depth at high water as Boulogne, should have a larger section, as the difference in exposure is very great, for Boulogne faces a narrow part of the English Channel, whilst Artha stands on perhaps the most exposed point of the whole French coast open to the Atlantic, and with great depths in front. There is no cause for surprise, under the circumstances, that the Artha breakwater has been found too weak; and it is probable that the injuries at St. Jean-de-Luz, and the disaster at Biarritz, are due to the fact that French engineers having mainly had experience of harbours in the Mediterranean, or on the Channel, or other somewhat sheltered places, did not at first fully realise the great exposure of the south-west coast of France. The rubble in the Artha mound appears to have been quite useless; and probably the only means of rendering Biarritz and Artha secure will be the employment of unusually large blocks of concrete.

CHAPTER XV.

HARBOURS PROTECTED BY A RUBBLE MOUND, AND A SUPERSTRUCTURE FOUNDED BELOW LOW WATER.

ALDERNEY. MANORA. MADRAS. COLOMBO. TYNEMOUTH. MORMUGAO.

Objects of the System. *Alderney Harbour:* Object; Design; Exposure; Breakwater, construction, modifications, maintenance, breaches; Cost of works; Repairs. *Manora Breakwater, Kurrachee:* Description; Progress; Damage; Cost. *Madras Harbour:* Designs; Description; Progress; Cost of works; Damage; Superstructure compared with other works; Methods of Reconstruction proposed, and cost; Progression of Foreshore. *Colombo Harbour:* Designs; Construction; Modifications; Cost of works; General Considerations; Remarks. Preceding Breakwaters compared; Superstructures; Rate of progress, and Cost. *Tynemouth Harbour:* Design; Section of Breakwaters; Modifications; Construction; Cost; Remarks. *Mormugao Harbour:* Description; Construction. Concluding Remarks.

THE harbours which it is proposed to describe in the present chapter, resemble those whose history has been given in the two preceding chapters, in being protected by breakwaters consisting of a superstructure placed on the top of a rubble mound. Their breakwaters, however, differ from those previously described in having the superstructure founded at a lower level: they are thus a stage further removed from the simple mound breakwater, and approach nearer to the upright-wall type of construction. This form of breakwater has been devised to unite the advantages, and avoid the defects, of the two types of breakwaters of which it is compounded, namely, the rubble mound, and the upright wall. The object in this type is to carry up the rubble base just to the limit of depth at which it is not liable to be disturbed by the waves; and from thence to erect an upright wall, solid enough to resist the unbroken, or only partially broken waves, and of a height sufficient to shelter the harbour it is designed to form. The smaller mass of rubble required for such a breakwater, as compared with the two types already described, renders it suitable for places where the depth is great, or the supply of stone limited; whilst the depth at which it is stopped makes its maintenance easier. The retention, moreover, of a rubble base for the lower portion of the breakwater tends to diminish the cost and delay frequently encountered in carrying up a vertical wall from the bottom; whilst the superstructure above is intended to afford all the advantages of an upright wall. The practical defects of this theoretically perfect type have been already mentioned in Chapter VI, and they will be fully illustrated by the history of the breakwaters selected as instances.

The successive changes in the Alderney breakwater, during its construction, mark the alterations from a dry rubble superstructure, founded at low-water level, to a solid superstructure commencing several feet below low water. Its history furnishes a most instructive lesson of the kind of injuries to which this type of breakwater is liable, when exposed to the waves of the open ocean intensified by rapid tidal currents, and situated in an unparalleled depth of water.

The Manora breakwater, at Kurrachee, illustrates a novel and rapid method of construction, by which it was endeavoured to avoid the most prominent defects of the Alderney system, whilst adhering to the same type of breakwater, and which proved perfectly successful on a site of moderate exposure.

The Madras breakwaters afford an example of precisely the same construction as at Kurrachee, applied to a situation of greater exposure, in which the weak points of the system have been rendered more evident by a severer test. This instance is the more valuable as showing how much more information may be gathered from failure than from success.

Colombo Harbour, though not yet completed, and only moderately exposed, has an interesting and instructive history, owing to the changes which experience has shown it desirable to introduce, during the progress of the works, both in the plan of the harbour and the designs of the breakwaters. Moreover, the breakwater in course of construction differs, in some respects, from the other three examples, though similar to them in type.

The Tynemouth piers resemble the Alderney breakwater in design, and in their subsequent modifications ; but they are situated in a moderate depth of water, and are less exposed.

ALDERNEY HARBOUR.

Object of the Harbour. The harbour works at Alderney were commenced in 1847, nearly simultaneously with several other Government harbours which had been designed with the object of refuge or defence. Alderney Harbour, however, together with St. Catherine's Harbour, Jersey, commenced at the same time, though classed with the other new harbours at Dover, Portland, and Holyhead, as refuge harbours, were in reality intended for the protection of the Channel Islands, and more especially as naval outposts to watch and to anticipate any movements of the French fleet near Cherbourg and the adjacent coast, which are within easy reach of those islands. Neither Alderney nor St. Catherine's harbours are suitably situated for refuge harbours, as they lie out of the course of passing vessels which naturally avoid the dangerous rocks, and rapid currents, with which the Channel Islands are surrounded. The application, accordingly, of the term *harbour of refuge* to Alderney Harbour, for which purpose it was avowedly unfitted, led to a popular depreciation of its real value, and hastened the final abandonment of the works in an incomplete state, and the discontinuance of the necessary works of maintenance for the outer portion. The history of the works, however, is none the less instructive in consequence of the harbour not having been fully completed, or owing to the outer half of the breakwater having been relegated to the mercy of the winds and waves.

Original Design for Alderney Harbour. The original design for a harbour at Alderney comprised an area of about 67 acres, having a depth of 3 fathoms and upwards, situated in Braye Bay on the north-western side of the island, and sheltered to the south and south-east by the island, and to the west and north-east by two breakwaters, one commencing at Grosnez Head, and the other at Roselle Point. (Plate 4, Fig. 1.) The estimate for this harbour was £620,000; and the western breakwater was commenced in 1847. This breakwater consisted of a mound of hard sandstone, obtained from a quarry about two miles from the works, and of a superstructure founded on the rubble base at the level of low-water spring tides. The superstructure was formed, in the first instance, of two dry rubble walls with loose rubble filling between; the sea wall having a width of 14 feet, and a batter on the face of 9 inches in a foot, and the harbour wall a width of 12 feet, and a batter of 4 inches in a foot. The quay level is 6 feet above high-water spring tides, or, as the rise of spring tides at Alderney is 17 feet, 23 feet above the foundations of the superstructure. A promenade wall, about 14 feet high, placed on the sea side of the superstructure, protects the quay; and a parapet wall, 4 feet high, runs for a short distance along the outside of the promenade level [1].

Exposure of Site of Alderney Harbour compared with St. Catherine's Harbour, Jersey. The section of the breakwater, as originally designed and commenced, resembles, indeed, very closely the section of St. Catherine's breakwater (Plate 7, Fig. 10), with the exception that the superstructure of the latter had to be made higher, owing to the much greater range of tide off Jersey, amounting to 32 feet at spring tides. Mr. James Walker, the foremost harbour engineer of that time,

[1] A section of this breakwater accompanies my Paper on Alderney Harbour. 'Minutes of Proceedings, Inst. C. E.,' vol. xxxvii, Plate 4, Fig. 1.

designed both harbours; and it shows how little the question of site was taken into consideration, at that period, in designing breakwaters, when we find similar sections proposed for two situations having such different exposure as the western side of Alderney and the east coast of Jersey; the Alderney breakwater having no sheltering land within about 3000 miles in the direction of the worst and most prevalent winds, whilst the breakwater at St. Catherine's is only exposed to the north-east, and is within 13 miles of the French coast.

The exposure of the site selected for Alderney Harbour was not long in manifesting itself as the breakwater progressed, and presented a most marked contrast to the favourable position of St. Catherine's Harbour. The breakwater, indeed, at St. Catherine's was carried out, to its full extent, according to the original design (Plate 7, Fig. 10); its cost of maintenance has been trifling; and some soundings taken by me round its head, in 1871, showed that the alterations in the form of the rubble mound had been very slight during a period of fifteen years. At Alderney, on the contrary, the experience of the storms of two winters was quite sufficient to show that the breakwater, in its original form, was inadequate to resist the waves; as the rubble base had been disturbed several times, and considerable damage had been done to the unfinished end of the superstructure.

Modification of Superstructure at Alderney. In 1849, when the breakwater had been only carried out 410 feet from the shore, steps were taken for strengthening the work; the face joints of the sea wall were filled with cement, and it was determined to modify the section of the superstructure. As it appeared that, at that distance out, the rubble mound was not disturbed at a depth of 12 feet below low water, the foundations of the superstructure were placed at that depth; and as, consequently, the mound could no longer be tipped out from the shore by wagons, hopper barges were employed for depositing the base. The courses of the sea and harbour walls of the superstructure, below low water, were laid by divers, and consisted of concrete blocks with face stones of dressed granite. The sea wall was increased in thickness to 23 feet at the base, and the harbour wall to 14 feet; and the batter of the face of the sea wall was reduced to 6 inches in a foot above low water, and 4 inches below. The walls above low water were built of the native Mannez stone set in cement, the face stones being dressed. The inner face of each wall was carried up vertically; and the sea wall, being brought inwards 9 feet at low-water level, left a space of 21 feet between it and the harbour wall, which was filled in with loose rubble to 700 feet from the shore, beyond which point concreted filling was substituted above low water. This section is shown on Plate 7, Fig. 13. The promenade wall was built solid, with Mannez stone set in mortar. Both the quay and promenade levels are paved with granite pitching; and a line of rails is laid on each level, the lower one being intended to facilitate the landing or embarking of troops, and the upper one for the conveyance of stone to maintain the foreshore near the sea face.

Enlargement of Alderney Harbour. The breakwater was constructed in accordance with the design just described, to a distance of 2700 feet from the shore, which point was reached in 1856. The original scheme for the harbour had, at an early period, been recognised as inadequate for an important naval station; and the design had been successively enlarged in 1854 and 1855, by an extension, in a straight line, of the western breakwater, which was to be curved round towards its extremity, and by shifting the position of the proposed eastern breakwater further out[1]. In 1856, however, a considerable extension of the harbour was decided upon, which necessitated an alteration in the line of the western breakwater, so that it might not

[1] A plan of the various schemes proposed is given in the last Plate illustrating my Paper on Alderney Harbour. 'Minutes of Proceedings, Inst. C. E.,' vol. xxxvii, Plate 7.

unduly narrow the harbour in being prolonged, and which, by carrying the breakwater into a much greater depth of water, rendered a modification of the superstructure advisable.

Alterations adopted for Outer Superstructure at Alderney. The final design adopted for the superstructure is shown on Plate 7, Fig. 14. The alterations adopted were: the substitution of rubble set in cement for the concreted hearting between the walls, thus rendering the superstructure solid from low water upwards; a reduction of the batter on the sea face to 4 inches in a foot; the lowering of the foundation of the harbour wall to the level of that of the sea wall; and a diminution of the width of the superstructure, making the quay 20 feet wide instead of 25 feet. This section was commenced at 2700 feet from the shore; and the breakwater was continued in a straight line to 2900 feet, where it was curved seawards for a distance of 520 feet, and from thence was carried straight out, in a north-easterly direction, to its termination. (Plate 4, Fig. 1.)

Design of Alderney Harbour as constructed. The scheme of 1856 was modified in 1858, to allow of a still further enlargement of the harbour, at a total estimated cost of £2,500,000 (Plate 4, Fig. 1); but this proposed extension was curtailed the following year to the points marked *A* and *B* on the plan; and the estimate of the scheme of 1855, amounting to £1,300,000, always appeared in the Parliamentary Returns since 1855 as the proposed cost. According to the design of 1858, the harbour would have possessed a well-sheltered area of 150 acres, having a depth of 3 fathoms and upwards, and a maximum depth of 20 fathoms. The western breakwater, however, was only carried out to the point *A*, as determined in 1859, giving it a length of 4680 feet; and the eastern breakwater was not constructed, so that the harbour is open to the north-east.

Method of Construction of Alderney Breakwater. The base of rubble stone, which, as already mentioned, was deposited from hopper barges, was carried forward considerably in advance of the superstructure, so as to allow it to consolidate, under the action of the sea, for three or four years before the foundations of the superstructure were erected upon it. The barges, capable of carrying loads of from 60 to 140 tons of stone, were towed out into their position by steam tugs.

The superstructure was built from staging, erected on the top of the mound, supported by round piles standing on, and secured to flat stones, and buried a few feet in the mound. The blocks of granite and concrete for the lower courses, being lowered from the staging, were laid by divers, on the levelled base, in lengths of about 60 feet. The diving work was carried on from May till the end of August: and each length was raised to low-water level in a fortnight, during favourable weather, so that the masons might set the lower courses of the masonry of the superstructure during spring tides. Directly the diving courses of the sea and harbour walls were in place, the space between them was filled up, as compactly as possible, with small rubble stone and sand; after which, the superstructure was built up from low water nearly to quay level. At the end of each length, a wall of concrete blocks and masonry was built across the superstructure, to protect it till the next length of wall could be completed; and at the point of termination of each season's work, a cross wall, 15 feet thick, was constructed to protect the end during the winter months. Large masses of stone were then tipped over each side of the superstructure to cover the lower courses up to low water. The only works that were carried on during the winter months were, the depositing of the base, and the building of the promenade wall, when the weather was favourable; and the preparation of materials for the next season's work. The laying of the copings, pitching, and lines of rails, on the quay and promenade levels, was deferred till the subsidence of the walls had ceased.

The average annual amount of rubble stone deposited in the mound, from 1849 to 1862, was 265,000 tons; and the greatest amount deposited in a single year was 608,000 tons.

The average annual progress of the superstructure, from 1849 to 1863 (omitting three years when the work was stopped), was 363 lineal feet; and the greatest advance in one season was 562 lineal feet. Some rocks in the harbour, which rose to within 5 fathoms of low water, were removed by blasting.

Head of the Alderney Breakwater. The termination, or head, of the breakwater was built in 1864. This very exposed portion of the structure, resting on the high rubble base situated in a depth of 130 feet at low water, was strengthened by placing the foundations, for the last 66 feet, at a lower level, and carrying them right across the superstructure. The outer 42 feet of the work was founded 24 feet below low-water level; special precautions were taken to connect together the face stones, both laterally and vertically; and the whole of the masonry above low water was set in cement, and it was carried up to the promenade level. The end was built square, as it made the workmanship simpler, and would facilitate the extension of the breakwater, if desired at some future time. Owing to the settlement of the high rubble mound, the bottom of the superstructure at the head eventually reached a depth of 30 feet below low water.

Maintenance of the Rubble Mound at Alderney. Theoretically, in a breakwater of the type to which the Alderney section was eventually assimilated, the cost of maintenance should be reduced to a minimum; for the superstructure, like a vertical wall, should be capable of resisting the attacks of the sea, and the rubble base is stopped at a level where it is supposed to be undisturbed by the waves. Practical experience has, however, too clearly demonstrated that this result was far from being attained at Alderney. The mound, indeed, was raised considerably above the assumed limit of disturbance, by stone being tipped over the sea wall in sufficient quantities to cover the face up to low-water level. This addition was useful for making up the deficiencies in the mound, occasioned by settlement and by its disturbance under the action of the waves, and eventually proved absolutely essential to protect the lower face blocks of the sea wall from being drawn out. Accordingly, the portion of the mound forming the sea foreshore of the breakwater above the limit of disturbance would, under ordinary conditions, require periodical renewal; though the average annual addition required might decrease as the slopes of the mound became flatter by the increase in the mass. The impact of the waves, however, against the high superstructure, and their subsequent recoil, exerted a far more powerful force than had been anticipated.

When a severe westerly gale of two or three days duration occurs during spring tides, a tremendous sea breaks upon the structure, at ebb tide, towards the close of the storm, the wind veering to the north-west and blowing with greater force. Each wave, dashing against the superstructure, rises to a great height above it (estimated to amount occasionally to 200 feet); a portion, driven by the wind, envelopes the superstructure in a sheet of water, occasionally damages the quay in its descent, and hurls large stones from the sea foreshore into the harbour; the remainder of the wave recoils on the sea slope, and, rushing down the mound, scours away the foreshore, till at length, on meeting the succeeding wave, it rebounds into the air about eighty feet from the superstructure. (Fig. 16.) The result of this almost irresistible recoil is, that the rubble mound is not secure from disturbance, on the sea side, except at a distance of 80 or 90 feet from the wall, and, where the depth in front is great, in a depth of 20 feet below low water at that distance out. It will be readily understood from the above facts that the mere maintenance of the mound would require a considerable annual deposit of stone on the sea slopes. The average annual amount of stone deposited, from the completion of the rubble base till 1872, was 50,400 tons; and I estimated that about 25,000 tons of stone would be required, each year, merely to maintain the sea foreshore at 4½ feet below low water against the superstructure; whilst a much larger amount would be necessary to raise and maintain the mound up to low water, as originally contemplated.

SEA STRIKING ALDERNEY BREAKWATER.

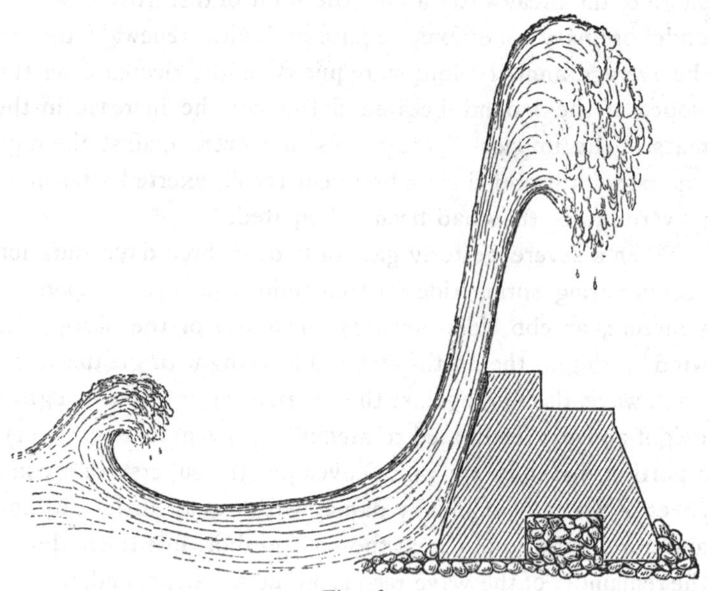

Fig. 16.

The fact of the disturbance of the mound to a depth of 20 feet below low water, at a distance of about 80 feet from the superstructure, is of considerable importance, as showing that the limit of 12 feet originally assumed was very far from correct; for if the waves can influence the mound down to 20 feet at that distance off, it is clear that this limit would be exceeded at the foot of the superstructure, where the scour due to the recoil is so much greater. The undermining, indeed, of the outer portion of the superstructure would have very soon occurred if the mound had not constantly been replenished, even though its base had been lowered by settlement to over 16 feet below low water. Probably about

25 feet would have been a safe depth for the foundations of the superstructure in the deeper and more exposed part of the breakwater; and the head is, therefore, the only part that might have been considered fairly secure from undermining without a constant maintenance of the sea slope of the rubble mound.

Damage to the Superstructure of Alderney Breakwater. The sea caused some injury to the superstructure during the progress of the works, which were repaired at a cost of £11,300. As however, with one exception, these damages occurred at the unfinished ends, or at places where they might be attributed to unequal settlement at the junction of a rocky

ALDERNEY BREAKWATER.

Section of Superstructure showing Breach. Scale $\frac{1}{500}$.

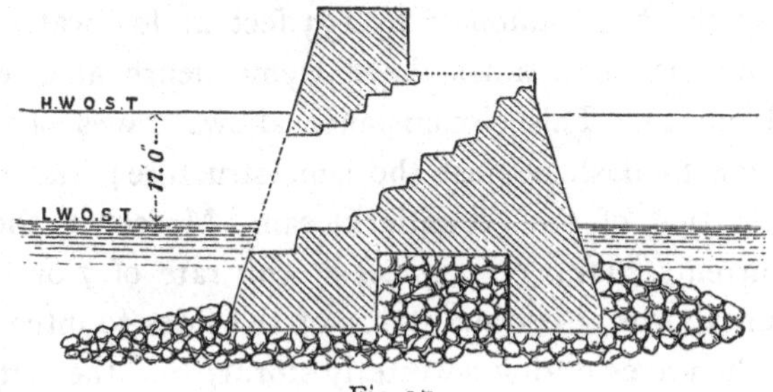

Fig. 17.

foundation with the yielding rubble base, no apprehensions were entertained for the maintenance of the superstructure up to the period of its completion in 1864. In January of the following year, however, two large breaches were made right through the superstructure, as well as some minor damages; and, from that time, hardly a winter passed without some injury being done to the superstructure. The repairs of the breaches and other damages, between 1865 and 1872, cost £44,225, or, on the average, £5500 per annum. The breaches commenced by the drawing out of face stones, usually about low-water level; then, if the storm continued, the hole was rapidly enlarged; and, finally, either a clean breach was made right through the superstructure, extending some feet below low water, or more commonly the opening extended upwards towards the quay level, and the waves, finding a vent through the quay, left the harbour face intact, and the promenade wall spanned the gap above. (Fig. 17.) This latter form of breach could sometimes be prevented from enlarging considerably, even when occurring early in the winter and consequently exposed to several storms, indicating that the vent thus provided for the waves diminished their effect on the structure at that part.

Causes of the Damages at Alderney Harbour. The breakwater at Alderney has been extended into an unprecedented depth of water, as may be readily seen by comparing its section, at 4200 feet from the shore (Plate 7, Fig. 14), with those of other breakwaters. (Plates 6 and 7.) In fact the depth at the head amounts to 133 feet at low-water spring tides, which is three times the extreme depth at Cherbourg and Plymouth. This circumstance allows waves of unusual magnitude to dash against the superstructure; and the exposure is that of the Atlantic Ocean. Moreover, the rapid tidal currents, running in places at the rate of 7 or 8 knots per hour, appear at certain states of the tides to intensify the fury of the waves during a westerly storm; and the settlement of the high rubble mound, under the weight of the superstructure and the action of the waves, averaging one-twentieth of the total height, tends to dislocate the superstructure.

Economical considerations prevented the modification of the original design of the superstructure to the full extent dictated by the exigencies of the situation; and the retention of an interval between the sea and harbour walls, as far as the foundations were concerned, led to the anomalous section of superstructure resting upon two feet, though solid above. (Plate 7, Figs. 13 and 14.) The settlement of the superstructure, accordingly, tended to be uneven; and longitudinal cracks were produced at the junction of the feet with the solid body, extending more or less up the superstructure; the rubble also was liable to be withdrawn from under the central portion during settlement, and thus fissures and hollows were formed which provided receptacles for compressed air during storms, and, when once the face stones were withdrawn, led to the rapid extension of the breaches. The high promenade wall also caused the whole force of the breaking wave to be concentrated against the superstructure, instead of allowing it to be partially expended in projecting a portion of the wave over into the harbour. The batter too of the sea face of the superstructure, though considerably reduced in the latest design, prevented the lower unconnected face stones deriving the full advantage from the weight of the superincumbent mass, and exposed them to the shock of the falling wave, which led to their more rapid withdrawal by the sea and the consequent commencement of breaches.

If the foundations of the superstructure had been carried level right across, and at a lower level; if the promenade wall had been dispensed with; and if the batter had been still further reduced, the stability of the superstructure would have been greatly increased. The value, indeed, of the two first measures was sufficiently demonstrated by the head, in the extreme depth and exposure, remaining uninjured till the works of maintenance were discontinued. Transverse dislocation of the superstructure would, however, still remain unprovided for, due to the effects of unequal settlement at the junction of each fortnight's work, and still more at the junction of the work of two seasons, which would be a source of danger to the maintenance of a superstructure on such a high mound.

Cost of the Works at Alderney Harbour. Up to 1872, when the works of maintenance were discontinued for a time, the total cost of the works amounted to £1,274,200, of which £57,200 were spent in repairing damages to the superstructure, and about £41,000 in maintaining the sea foreshore by the deposit of stone. The average cost of the superstructure per lineal foot amounted to £103⅓ for the first 1200 feet, £115 for the next 1500 feet, and £118⅓ for the outer 1800 feet up to the head, which cost altogether £20,000. The actual cost of the breakwater, up to the period of its completion in 1864, amounted to about £235 per lineal foot. The cost of maintenance for the eight years from 1864 to 1872 amounted to £85,000, or an average of £10,600 per annum.

Proposals for securing the Alderney Breakwater from Damage. Owing to the considerable expense incurred in repairing the frequent breaches, and in the maintenance of the foreshore, the Government, in 1870, asked Sir J. Hawkshaw and Sir A. Clarke to report on the best means of permanently securing the breakwater. They recommended that the promenade wall should be taken down, and that the superstructure and sea foreshore should be protected by a large mound of rubble stone or concrete blocks all along the sea face of the superstructure. This mound would have served to protect the rubble base from erosion by the recoil of the waves, and would have acted like the wave-breaker at Ymuiden and Mormugao (Plate 6, Fig. 19, and Plate 7, Fig. 20), in protecting the superstructure; and the principle of protection suggested was precisely similar to that which has been recently proposed for the outer arms of the Madras breakwaters. (Plate 7, Fig. 18.) Large concrete blocks would probably have proved the only satisfactory protection; as experience has shown that rubble stone, of the size obtainable from the quarry at Alderney, cannot remain permanently against the sea face of the superstructure. The cost however, of the proposed system of protection would have been considerable; and the Government did not consider that Alderney was of sufficient value as a military outpost, under the altered conditions of naval warfare, to justify this further expenditure.

Maintenance of the Alderney Breakwater since 1872. The breakwater is now being maintained by the Admiralty up to a distance of 2856 feet from the shore, where a temporary head, which served to protect the end of a season's work, forms a suitable termination. The sea foreshore is being kept up to about six feet above low-water mark, along this distance, by a regular deposit of stone, amounting on the average to about 20,000 tons annually. The repairs of breaches along this portion of the superstructure have cost approximately £10,000 during the ten years 1873–1883, or an average of £1000 a year. Taking the foreshore at its former price, the total annual cost of maintenance is about £3400.

Action of the Sea on the Abandoned Portion of the Alderney Superstructure. Since the discontinuance of the maintenance of the outer portion, the sea has gradually overthrown the superstructure from 2856 feet to 3100 feet, from 3200 feet to 3480 feet, and from 4220 feet to the end, thus leaving one short and one long length of superstructure standing disconnected from the rest[1]. (Plate 4, Fig. 1.) The débris of the damaged portions emerges about 3 feet at low water. It remains apparently undisturbed by the waves which break over it during storms; and the progress of the damage along the detached superstructure is slight.

MANORA BREAKWATER, KURRACHEE.

The harbour of Kurrachee, which is an instance of a jetty harbour similar in type to Dublin and Aberdeen harbours, has been already referred to in Chapter VIII, which deals with that class of harbours. The Manora breakwater, however, which was constructed for contracting, deepening, and protecting the channel across the bar at the entrance, belongs to the type of breakwater specially described in this chapter, and possesses the peculiar interest of being the first instance, on a large scale, of a superstructure built of sloping blocks

[1] The breaches in the superstructure, as they existed in 1883, are indicated on the plan by shading.

laid by an overhanging Titan[1] (Plate 8, Figs. 1 and 2), of which the Madras breakwaters were subsequent examples.

Description of the Manora Breakwater. The breakwater starts from Manora Point, which juts out into the Arabian Sea, on the west side of the entrance to Kurrachee harbour (Plate 2, Fig. 6.) It has a length of 1503 feet, and extends into a depth of 30 feet at low water. It is composed of a base of rubble stone, thrown from boats and levelled at 15 feet below low water, upon which a vertical superstructure, 24 feet wide and 24 feet high, was built with concrete blocks. (Plate 7, Fig. 19.) These blocks, each 12 feet by 8 feet by 4½ feet, and weighing 27 tons, were laid in two unconnected tiers of three blocks, each placed close together and having an inclination of ¼ to 1 towards the shore, so that each tier rested somewhat on the preceding one and was not liable to tip over forwards during the construction. This form of construction enables the block-setting Titan, running on rails on the completed portion of the superstructure, to approach near enough to the end to lay the successive rows of blocks without the aid of staging. The advantage of this method, in dispensing with staging so liable to injury from the sea, has been already pointed out in Chapter VI. The rubble base was given a width of about 100 feet at the level of the foundation of the superstructure, the central line of the superstructure being placed at one-third of this distance from the harbour side. The blocks were composed of 1 part of Portland cement, 4 of sand, 5¾ of shingle, and 3¼ of rubble stone, and were sometimes set a month after they had been made, though the upper blocks, upon which the Titan rested, required a longer interval. The Titan could set a block of 27 tons at a distance of 26½ feet out from the base of the outer wheels, and it could shift a block 12 feet across the breakwater.

Progress of the Manora Breakwater Works. The working season at Kurrachee extends from October to March. The

[1] 'Minutes of Proceedings, Inst. C. E.,' vol. xliii, p. 4.

south-west monsoon prevails from the middle of June to the middle of September, and raises a very heavy sea; and the weather at the setting in and close of the monsoon is unfavourable to sea-works. The rubble base was started in March 1869, and the superstructure was commenced towards the close of 1870. During that season the breakwater was extended to 270 feet from the shore: in the following season a distance of 793 feet was reached; and the breakwater was carried to its full length of 1503 feet by February 1873. The progress of the work was delayed by the slow rate of excavation of the base for the foundations of the superstructure, and by want of funds; whilst the setting capabilities of the Titan were never fully reached, even though the last 710 feet were accomplished in less than four months.

The superstructure along the outer half of the breakwater has settled from 3 to 4 feet on the mound; and the range of spring tides being about 8 feet 9 inches, the top of the superstructure has sunk below high-water level, and has been gradually raised by a layer of concrete on the top to facilitate access.

Damage to the Manora Superstructure. The large waves produced by the south-west monsoon dash over the breakwater, passing in an almost unbroken mass, about 18 feet high, above the top of the superstructure during high water, and even higher, but in a more broken state, at low water. Some damage was done to the superstructure, during the monsoon, each year from 1871 to 1874. Several of the top blocks on the harbour side were displaced, and in some places the second row was disturbed; a few also of the blocks on the sea side were overturned, and the central joint opened at some parts. The open joint was filled with concrete, and the blocks replaced; and since 1874 the work appears to have become sufficiently consolidated for the monsoon waves not to affect it.

Cost of the Manora Breakwater. The total cost of the breakwater, including a percentage of the engineering and other expenses belonging to the general harbour improvement works, amounted to £109,000, or £72 10*s.* per lineal foot. Out of this sum, £1300 was spent in repairing damages, which bears a small proportion to the total expenditure.

The results of the work have been already described when the harbour was dealt with as a whole (page 160).

MADRAS HARBOUR.

The sandy coast in front of Madras stretches in a straight line north and south, and is exposed to a continual surf which prevents vessels approaching near the shore, and renders the landing always difficult, and, at times, dangerous. The landing of passengers and goods has been effected by means of long surf boats, specially constructed for the purpose; and an iron screw-pile jetty was built out from the shore to facilitate the operation.

Winds at Madras, and their Action. The prevalent winds in that locality are the periodical winds known as the north-east and south-west monsoons. The north-east monsoon produces a chopping sea, in which the waves come in from the north-east, and, approaching the shore obliquely, cause a travel of sand towards the south; it blows from November to April. The south-west monsoon, blowing from May to October, raises a ground swell which proceeds in a north-westerly direction, and produces a travel of sand towards the north. Thus the wave action produced by the monsoons causes an alternating motion of sand along the coast; but the south-west monsoon has most effect in heaping up the sand against any solid projection from the beach. The exposed coast is occasionally visited by cyclones, of variable intensity, which raise a violent sea in front of Madras. Severe cyclones appear to occur most frequently in May and November, but they also take place occasionally in October and December. The height of the sea raised by these storms at Madras bears no definite relation to the force of the wind there, but is dependent on the vehemence, duration, and direction of the cyclone at a considerable distance from the coast. The shore in front of Madras shelves rapidly near the land, but the inclination becomes more gradual further out. The range of tide is between two and three feet.

Proposed Designs for a Harbour at Madras. The importance of Madras, and the absence of shelter along the adjacent coast, have for a long time caused the want of an artificial harbour to be severely felt. The exposure, however, of the coast, and more especially the apprehension that sand would rapidly accumulate under the influence of the south-west monsoon if a closed harbour was projected from the shore, prevented, for a considerable period, any definite steps being taken to provide the much needed protection. About fifty years ago, Sir Arthur Cotton proposed the construction of a detached breakwater parallel to the coast, which, whilst affording shelter and arresting the surf, would not interfere with the littoral drift of sand. This work was commenced in 1835 by the deposit of a mound of rubble stone, but it appears not to have been carried out beyond a length of 76 feet; and, as its site is not quite as far out as the end of the pile pier (Plate 4, Fig. 12), it is evident that an adequate extent of sheltered area would not have been provided between the breakwater and the coast.

Other schemes have been proposed for securing the necessary shelter without impeding the natural travel of sand. Thus Mr. Rymer-Jones suggested the construction of a second pile pier out from the shore, at some distance from the existing one; and he proposed to prolong both piers so as to serve as approaches to two solid breakwaters, which, together with a detached breakwater, beyond the others and parallel to the shore, would furnish the requisite harbour accommodation without causing an advance of the foreshore. Sir Andrew Clarke had conceived a similar idea, in which he proposed to employ open viaducts, resting on concrete cylinders, for getting beyond the influence of the littoral drift, instead of iron pile piers. Another proposal was to enclose an area, at some distance from the shore, between two curved breakwaters, leaving spaces for their entrances between their extremities, the inner breakwater being connected with the shore by a viaduct on piers.

Design of Harbour adopted at Madras. In all the designs already referred to, interference with the littoral drift was avoided, as much as possible, by placing the solid breakwaters beyond its influence, as it was considered that the advance or influx of sand would compromise the maintenance of the harbour. Mr. Parkes, however, in his design, laid aside all provision for leaving the travel of sand near the shore unimpeded, and decided to carry out solid breakwaters from the shore into deep water. He doubtless considered himself justified in neglecting the provisions contained in the previous designs, in consequence of the successful completion of a closed harbour at Ymuiden, on the sandy coast of the North Sea (Plate 5, Fig. 2), and owing to the depth of water into which he proposed to carry the ends of the breakwaters. The design adopted (Plate 4, Fig. 12) consists simply of two solid breakwaters, placed 3000 feet apart, and extending out at right angles to the shore for a distance of about 3000 feet; they then converge with return arms at angles of 70° to their original directions, and leave an opening between their extremities, 550 feet in width, parallel to the coast-line. The breakwaters, accordingly, form a perfectly symmetrical enclosed harbour, which would be a perfect square except for the slight extension seawards of the return arms, and which has an area of about 220 acres. Mr. Parkes naturally adopted the section for the breakwaters which he had carried out successfully and rapidly at Kurrachee; it is shown on Plate 7, Fig. 17, and has been already described (page 120). It consists, essentially, of a vertical concrete wall, resting upon a rubble base whose height varies with the depth of water, amounting to about 25 feet at the deepest part; so that the depth of the foundation of the superstructure remains constant throughout, at 22 feet below low water. The rubble base is composed of gneiss and laterite, the pieces varying from 5 lbs. to 2 cwt. in weight; it was deposited from steam hopper barges, and it is 78 feet wide at the top with slopes of 1 to 1. The top of the mound was levelled by divers to receive the foundation blocks of the superstructure. The vertical wall, composed of a double row of concrete blocks, 27 tons in weight, deposited by an overhanging crane (Plate 8, Figs. 1 and 2), is 28 feet wide at the base, 24 feet wide above, and 30 feet high. The Madras section differs only from the Manora section, in having four tiers of blocks instead of three, in being founded lower, and raised 3½ feet above high-tide level, and in each block having a tenon on its upper surface and a mortise at its base to unite the blocks in each tier together, in place of the stone joggles which were introduced at the Manora breakwater. A slight variation has also been introduced in the bottom blocks at Madras, which are wider at the base and shorter than the others. They are also square at the bottom, and not tapered, as was done at Manora in order that they might rest level upon the base. The blocks were made of 1 part of Portland cement, 2 of sand, 5 of broken stone, and 2¼ of large stone. The breakwaters have been carried into a depth of 45 feet at the extremities; and the superstructure was gradually stepped down, so that the foundations might reach a depth of 31½ feet below low water at the pierheads, where the width was also increased. The northern breakwater is designed to have a total length of 3,866 feet, and the southern breakwater 3,970 feet.

Progress of the Madras Harbour Works. The works were commenced in 1876. The northern breakwater was begun in December 1876, and the southern breakwater in November 1877. The projection of the solid breakwaters from the shore soon caused the anticipated accumulation of sand on their southern sides. As, however, the northern pier was begun first, the sand, in the first instance, accumulated on its southern side, within the site of the new harbour. Moreover, even for some time after the commencement of the southern pier, the travel of sand continued in front of it, and settled between the breakwaters. This movement of sand, indeed, impeded the progress of both breakwaters for a short distance from the shore, by causing an accumulation of sand on the rubble base in front of the work, which hindered the laying of the foundation blocks for the superstructure. This action, however, ceased when the breakwaters reached the depth of about 4 fathoms; and as soon as the southern pier was extended to this position, it arrested the sand travelling northwards. With the exception of the difficulty in laying the foundation courses of the superstructure, caused by the accumulation of sand on the rubble base near the shore, the work was carried on without any hindrance, and was approaching completion towards the close of 1881. The settlement of the superstructure on the mound was small, averaging less than 6 inches: on the laterite base it varied from about 1 to 4 feet, whilst on the gneiss it was scarcely perceptible.

Cost of the Madras Harbour Works. The original estimate for the harbour was £565,000, or 56½ lakhs of rupees; and up to the end of October, 1881, the expenditure on the works had amounted to little over 57 lakhs, according to the statement furnished me by Mr. W. Parkes, of which the details are given on the opposite page; and Mr. Parkes has informed me that if no injury had occurred to the breakwaters, the additional expenditure for completing the work, as designed, would have been covered by the sale of the stores in stock. The actual expenditure, reckoned in rupees, appears rather in excess of the estimated cost; but since the estimate was made, the rupee has depreciated in value from 2*s*. to 1*s*. 10*d*. in 1876, when the works were begun, and to 1*s*. 8*d*. in 1882, so that the cost, if reduced to pounds sterling, would appear less than the original estimate. The cost of the breakwaters per lineal foot averaged about £72. These figures, however, will be considerably increased by the expenditure that will be necessary for reconstructing and strengthening the outer arms of the breakwaters.

MADRAS HARBOUR.

Statement of Expenditure up to 31st October, 1881.

	Rupees.	*Rupees.*
PRELIMINARIES AND PLANT:		
Marine Survey of Site	6,506	
Railways	2,47,121	
Block Grounds	54,990	
Buildings	65,451	
Locomotive Engines	55,171	
Other Rolling Stock	26,038	
Floating Plant	1,85,288	
Stationary Engines, Cranes, and other Machinery...	2,95,685	
Office Furniture	4,030	
		9,40,280
NORTH PIER:		
Surf Bank, 869 feet long, 99,650 tons of rubble stone	1,99,300	
Rubble Base 387,281 tons	8,09,419	
Concrete Block Work, 3,866 lineal feet, 100,579 cubic yards	9,13,516	
Setting Blocks	2,09,394	
		21,31,629
SOUTH PIER:		
Surf Bank, 535 feet long, 32,295 tons of rubble stone	64,590	
Rubble Base 418,883 tons	8,53,580	
Concrete Block Work, 3,970 lineal feet, 101,764 cubic yards	9,24,279	
Setting Blocks	1,87,778	
		20,30,227
CONTINGENCIES:		
Protection of shore from erosion, north of harbour, 43,085 tons of rubble stone	91,009	
Intercepting Sewer along shore	33,855	
Sundry minor expenses	34,183	
		1,59,047
SUPERINTENDENCE		3,51,004
MATERIALS IN STOCK:		
Concrete Blocks, 4,435 cubic yards	40,281	
Purchased Stores (various)	65,669	
		1,05,950
Total		57,18,137

Damage to the Madras Breakwaters. After the works had been carried on successfully for nearly six years, and when they were on the eve of completion, a cyclone visited Madras, on the 12th November 1881, which raised an unusual sea, and inflicted considerable damage on the outer arms of the breakwaters. The cyclone, as observed at Madras, does not appear to have blown with any great amount of vehemence, as its greatest recorded velocity was only 33 miles an hour. It seems probable, however, that the storm had somewhat spent itself before reaching Madras, and that the waves that beat upon the breakwater were due to a greater cyclonic disturbance at a distance from the coast. This result, indeed, seems to be not unusual at Madras, as the most recent previous cyclone of importance, which visited that coast in May 1872, had a velocity there of only 57 miles an hour, though a very great sea broke upon the shore. Moreover, there is some conflict of opinion on which of these occasions the sea was the greatest, indicating that the force of the wind at Madras affords no certain criterion of the power of the waves on the coast. The height of waves being due to the continued action of the wind, as pointed out in Chapter II, it follows that waves would be greater at the end of the path of a cyclone than at the point of its greatest intensity; and, consequently, the waves at Madras, during the cyclones of 1872 and 1881, were in all probability larger than they would have been in the case of a storm of greater vehemence at Madras itself, but which had originated at a point nearer the coast.

The superstructures of both of the return outer arms of the harbour were more or less displaced or overthrown by the cyclone of 1881, throughout their length. The curved portions of the breakwaters, also, sustained considerable damage, the superstructure having been both undermined and partially thrown down; but the inner portions, at right angles to the shore, remained uninjured. The dotted lines on the plan (Plate 4, Fig. 12) indicate the extent of the damage. The inner tier of blocks suffered most along the straight portions of the outer arms; whilst, along the curved portions, the outer tier were more thoroughly overthrown.

Causes of the Failure of the Madras Breakwaters. The lines of the crests of the waves during the storm were nearly parallel to the northern return arm, so that the waves must have dashed with their whole force against that part of the structure, and with considerable, but somewhat less vehemence against the southern return arm; whilst they must have come very obliquely against the inner portion of the north breakwater, and cannot have struck at all against the corresponding portion of the south breakwater.

The injuries inflicted were entirely in accordance with what might have been anticipated from the direction of the waves. Thus the northern return arm suffered most, and the damage extended to the termination of the curve; whereas, on the southern side, the injury was somewhat less, and extended less far. The stability of the sides of the harbour is amply accounted for by the waves having run along their faces, instead of striking against them. The failure of the superstructure at the curves is mainly attributable to undermining; for it was found that the rubble base had been lowered there below the level of the foundation of the superstructure, the tier of blocks on the sea side had fallen outwards on to the sea slope, whilst in some places the harbour tier of blocks remained standing, though leaning seawards. It is evident that the oblique direction of the waves to the curved portions of the breakwaters caused the water to rush away to the sides, which produced an erosion of the base, and consequent overthrow of the sea tier of blocks. Along the straight outer arms, on the contrary, where the waves dashed almost direct against the superstructure, and the damage was more complete, the harbour tier of blocks suffered most. Owing to the lowness of the superstructure, a considerable portion of each wave, dashing against the sea face, rose over the superstructure and descended partly upon the top blocks of the harbour tier, and partly into the harbour. The tendency of this falling mass of water would be to throw off the harbour tier of blocks in detail, and possibly to scour away the harbour slope of the rubble base, and, eventually, to undermine the superstructure. Either of these results, or both combined, would have effected the overthrow of the harbour portion of the superstructure as actually experienced. Both Mr. Molesworth and Mr. Parkes, in their reports to the Government of India on the injury to the works, state that no signs of any lowering of the harbour slope of the rubble base could be detected. It is probable that the base was adequately protected, in the first instance, by the cushion of still water above it, and subsequently by the fallen blocks as well; and the failure of the superstructure must be therefore wholly attributed to the direct action of the sea. The waves, striking against the outer blocks of the superstructure, tend to force the inner blocks into the harbour; they also compress the air in the open joints, which aids in the displacement of the inner blocks. The portion of the wave, moreover, which rising over the superstructure falls down upon it, both opens the central joint of the superstructure, and drags off the inner blocks.

Comparison of the Madras Superstructure with other similar Constructions. Waves during severe storms tend to approach the shore in lines approximately parallel to the coast, owing to the influence of the shelving of the bottom. Accordingly, the outer arms of the Madras breakwaters will always be exposed to the principal shock of the sea; and, moreover, they are situated in deeper water than the other portions, and consequently encounter the largest waves. It is, therefore, somewhat unaccountable that the same section of breakwater should have been adopted throughout, instead of proportioning the strength to the forces to be resisted. Possibly it may have been assumed that, whilst the section possessed sufficient strength for the most exposed portions, it would have been inexpedient, in the interests of ease of construction, and for the sake of quay space, to reduce the width of the inner portions of the superstructure. A comparison, however, with other breakwaters would have indicated that the superstructure at Madras had less solidity than is usually given to structures in similar situations, which experience has, unfortunately, most clearly demonstrated. For instance, the Manora breakwater at Kurrachee, whilst similar in section, is less exposed to storms than Madras, and even its superstructure did not entirely escape injury; and the mortises and tenons, introduced into the blocks at Madras, were not adequate, in a more exposed situation, to cope with the sort of dislocation in detail to which this peculiar form of superstructure is liable. Indeed, in 1875, in a discussion on the Manora breakwater, I pointed out that the experience gained at that breakwater showed the advisability in future, in similar constructions, of connecting together the four upper blocks of each set [1], leaving each set unconnected, as before, across the breakwater to allow for unequal longitudinal settlement; but it was not till after the casualty of 1881 that any steps towards this end were proposed for Madras. Again, the superstructure at Madras, though founded at a lower level, was given a less width than either at Alderney or Colombo, being 24 feet, as compared with from 39 to 50 feet, and 36 feet, respectively, between high and low water. (Plate 7, Figs. 14, 17, and 23.) Doubtless Alderney breakwater is in a situation of extreme exposure, and of greater depth, even where the solid superstructure referred to was commenced, than is reached at Madras; but the difference in the width of the superstructure is considerable. Colombo breakwater, however, is more favourably situated, and in a less depth of water than the Madras piers, as will be seen in the sequel (page 305); and yet its superstructure has been made broader, and has been more solidly bonded together than at Madras. Moreover, the slight

[1] 'Minutes of Proceedings, Inst. C. E.,' vol. xliii, p. 44.

deflection of the sea wall at Colombo, in 1877 (page 308), when carried out unsupported in front of the harbour wall, though it was 24 feet wide at the base, might have served as a warning of the danger incurred in building such a narrow unbonded superstructure to protect the most exposed part of Madras Harbour. Lastly, comparing the Madras superstructure with the North Sea piers of the Amsterdam Canal, it will be seen that the views of the designers of the two harbours, as to the strength necessary for the breakwaters, were essentially at variance (Plate 6, Fig. 19, and Plate 7, Fig. 17); though the exposure of the North Sea piers, considering the limited extent of the North Sea, must be considerably less than at Madras, and the depth of water at the entrance of Ymuiden Harbour is only 29 feet, whilst it amounts to 45 feet at Madras. Thus Sir John Hawkshaw, after making the North Sea piers 30 feet wide at high-water level, increasing to about 37 feet at the base, considered it necessary, eventually, to protect the outer portions of the piers with a wave-breaker of large concrete blocks; whilst Mr. Parkes considered that a superstructure, 24 feet wide throughout, except at the foundation course which is 28 feet, was of sufficient strength without any protection.

There is one satisfactory conclusion to be drawn from the foregoing comparison, namely, that the damage at Madras need not be attributed, as suggested by Mr. Parkes in his report, to any novel phenomenon of cyclonic disturbance raising waves of unprecedented power. The exposure, or fetch, at Alderney and Wick is considerably greater than at Madras; whilst the depth of water at Alderney, and the rapid tidal currents at both Alderney and Wick, are not experienced at Madras. Accordingly, it may be reasonably anticipated that there will be no unusual difficulty in erecting a stable structure at the outer portion of Madras Harbour, provided that the blocks are bonded together transversely, and that the width of the superstructure is proportioned to what has been found expedient in other places similarly exposed.

Methods proposed for restoring the Outer Madras Breakwaters. To complete Madras Harbour, it will be necessary to rebuild the damaged superstructure; and as the original design has proved inadequate, some stronger construction must be adopted. Three different courses have been proposed. Mr. Parkes, in his report to the Indian Government in March, 1882, advocated rebuilding the superstructure on its original site by picking up and using again the fallen blocks which might prove sound, and making up deficiencies by new blocks. He, however, proposed to remedy the weak points of the original structure, by adding a row of concrete blocks on the base, along the toe of the sea face, to prevent undermining; by placing a third tier of blocks on the harbour side of the old superstructure, and sloping the top block down so as to withdraw it from the dragging away action of the descending wave; by filling the cross joints of the top blocks with cement so as to connect them longitudinally, and cramping the two top blocks of the outer tiers together to unite them transversely; by tying together the top blocks of the sea wall round the curves with a chain bedded in concrete; and by depositing additional rubble of large size on the base where scouring might occur.

The scheme suggested by Mr. Molesworth consisted in leaving the fallen blocks in their present position, and adding a few additional ones, so as to make the outer arms into mounds of random concrete blocks resting upon a rubble base.

The Joint Committee appointed to investigate the subject, consisting of Sir J. Hawkshaw, Sir J. Coode, and Professor Stokes, proposed a third course in their report of January, 1883. They advised, that the superstructure should be rebuilt, if possible, on its original site, or, in the event of the fallen blocks rendering this impracticable, that the superstructure should be rebuilt sixty feet further in, but parallel to the old lines; that the superstructure of each breakwater should be raised and bonded together, throughout its whole length, by a solid mass of concrete, carried up to 12 feet above high water, the top blocks having been previously joined together by cramps; that additional gneiss rubble of large size should be deposited on the sea slope of the mounds for a short distance from the shore; that from thence to the curves a wave-breaker of concrete blocks should be deposited against the sea face of the superstructure, with the necessary extension of the rubble mound in which gneiss only should be used; and that the curves and outer arms should be protected by a larger wave-breaker (Plate 7, Fig. 18); whilst two rows of concrete bags, laid on the harbour slope of the mound close to the superstructure, would secure the inner toe of the wall from undermining. Simple mounds of concrete blocks, for the outer portions, were estimated to be nearly as costly as the system proposed, if carried out on an adequate scale, and were considered unadvisable as not affording such effectual shelter, a point of great importance in such a limited harbour.

Cost of Restoration of the Madras Breakwaters. The design proposed by Mr. Parkes was estimated to cost from £135,600 to £159,700; that of Mr. Molesworth, £236,250; and that of the Joint Committee, £480,000 or £430,000 according as the superstructure could, or could not be built on its original site. In comparing these estimates, it must be noted that Mr. Molesworth proposed having a mound of random blocks which could not be regarded as large enough to resist effectually the forces to which it would be exposed. The advantages of the more costly of the two plans recommended by the Committee are, that the restored superstructure would rest upon the consolidated rubble base, and would therefore be less liable to settlement; and that the size of the harbour would not be reduced. The system of protection recommended is similar to that which was proposed for Alderney, and was adopted at Ymuiden. (Compare Plate 6, Fig. 19, with Plate 7, Fig. 18.)

Probable Total Cost of the Madras Breakwaters. The

plan of reconstruction and consolidation recommended by the Committee will in all probability be adopted by the Government, subject to such modifications as may be suggested by Mr. Parkes in the actual carrying out of the works. The cost will in any case be considerable; and assuming that the amount of the largest estimate is required, the total expenditure will be raised to £1,045,000, and the average cost per lineal foot of breakwater will be about £133, a large addition to the original cost, but not comparing unfavourably with other breakwaters in less exposed situations. If the superstructure had been given an adequate width, in the first instance, along the more exposed parts, and had been bonded together by cramping and a solid capping of concrete on the top, it is probable that, with a little protection of the base at the sea face of the superstructure, the breakwaters would have proved capable of resisting the attacks of the sea, and would have been carried out at a smaller cost.

Progression of the Foreshore at Madras. It will be remembered that the danger dreaded before the harbour works were commenced was not the violence of the waves, but the inroad of sand. As the principal travel of sand is from the south, it was feared that any solid pier projecting from the shore would act as a groyne, and cause an accumulation of sand on the southern side, which might eventually overlap the pier and endanger the maintenance of the harbour. Experience has shown that this anticipation was, to a certain extent, well founded. It will be seen, on referring to the plan of the harbour (Plate 4, Fig. 12), that an advance of the foreshore has taken place each year, to the south of the harbour, in the angle between the southern pier and the shore. The accumulation of sand at this part occurs during the continuance of the south-west monsoon; but as a considerable portion of the sand thus brought in is removed again by the sea during the period of the north-east monsoon, the actual yearly advance is thereby reduced. The annual advance of the foreshore is quite evident; but though the amount of yearly accumulation may not diminish, the actual yearly extent of progression will be reduced as deeper water is reached, and it is not likely to compromise the approach to the harbour for many years. Some sand finds its way into the harbour through the interstices between the blocks of the south pier, but it settles close to the harbour face of the breakwater, and can easily be removed by dredging. An oscillating movement of sand takes place on the Madras beach in about 4 fathoms depth of water, the sand moving inshore from February to September, and raising a bank in that depth, which was the cause of the difficulty in laying the foundations of the superstructure already referred to; this mound of sand is borne out to sea again during the other portion of the year.

The sand of the Madras beach does not appear to be disturbed much in depths exceeding from 4 to 5 fathoms; so that there is no chance of much sand being brought into the harbour through the entrance, which is in a depth of 7½ fathoms, even during storms. Slight changes, however, have occurred in the depths inside the harbour; the depth having increased a little between the 5-fathom line and the shore, whilst a corresponding accumulation of sand has taken place on the shore. This result appears to be due to the action of waves entering the harbour, and it varies with the monsoons.

The only agency, accordingly, which appears liable to be prejudicial, in the future, to the maintenance of Madras Harbour, is the progression of the foreshore on the southern side; but the advance is so gradual that many years must elapse before it will be necessary to take steps to cope with it. On the north side of the harbour, beyond the northern pier, erosion of the beach actually occurs; so that protective works are necessary at this part to prevent encroachment on valuable land in front of the town.

COLOMBO HARBOUR.

Colombo, the capital of Ceylon, has a large and rapidly increasing trade, the various products of the island being shipped at this port. It is situated on the west coast of the island; and its harbour, though lying in a sort of bay sheltered from the east and south, is open to the west and north. (Plate 4, Fig. 3.) Accordingly, the harbour was exposed to the monsoons, which blow from the south-west between May and November, and from the north-east during the other half of the year. Attention was directed by the colony, in 1871, to the serious difficulties experienced in the operations of shipping and landing goods, more especially during the prevalence of the south-west monsoon, and also during the early part of the north-east monsoon when along-shore winds prevail. The matter was taken into consideration by the colonial government, and in 1872 a report was obtained from Sir John Coode.

Original Scheme of Improvement at Colombo Harbour. The scheme of improvement proposed in Sir J. Coode's report consisted of a breakwater, starting from the western extremity of the bay, which, after running in a north-north-easterly direction for 1800 feet, turned eastwards to an east-north-east line, and terminated in a depth of 28 feet at low water, having a total length of 3600 feet. It was proposed to carry out two jetties from the shore near the commencement of the breakwater and within its shelter, each jetty being 1000 feet long and pointing to the north-east, so that vessels alongside would be lying in the direction of the wind during each of the monsoons. It was anticipated that the inner face of the breakwater would be also available as a quay for the purposes of trade; whilst the outer arm would possess the same advantages of direction as the jetties. The design for the breakwater consisted of a rubble mound, surmounted by a superstructure. The latter, founded 20 feet below low water, was composed of a sea and a harbour wall, 16 feet, and 13 feet wide at the base respectively, with rubble filling between, and strengthened by cross walls at intervals. The superstructure was 50 feet wide at the base; and its quay, at the top, was 35 feet in width, protected by a parapet wall, 10 feet 9 inches high and 7 feet 6 inches thick. The quay level was designed to be 9 feet above high water, or 11 feet above low water, the rise of tide being only 2 feet. It was intended to build the sea and harbour walls of concrete blocks not exceeding 12 tons in weight, so as to avoid the cost of heavy setting machinery. The estimate for this design was £630,000.

Construction of Western Breakwater at Colombo. The design above described was approved; but owing to further investigations on the site in 1873, it was decided to increase the proposed length of the breakwater to 4010 feet, raising the estimate to £718,800.

Preparatory works for the construction of the breakwater were commenced in June, 1874; a railway from the quarry at Mahara, connecting it with the main line to Colombo, having been previously made. The first stone of the sea wall was laid by the Prince of Wales in December, 1875. A large solid mass of concrete, measuring 250 cubic yards and weighing 437 tons, was formed at the root of the work by depositing concrete-in-mass within a timber casing, in 17 feet of water, on the irregular rocky bottom. This mass serves as an abutment from which the sea and harbour walls of the breakwater start.

The section of the superstructure of the breakwater was somewhat modified from the original design in actual construction: the principle of a sea and harbour wall with intermediate rubble filling was retained; but the sea wall was widened to 26 feet at the bottom and 24 feet at the top; the concrete blocks, also, were made from 16 to 33 tons in weight, and deposited by means of a steam Titan, on the sloping-block system, to allow for unequal settlement on the rubble mound, in this respect resembling the method introduced by Mr. Parkes at Kurrachee. The concrete blocks for the sea and harbour walls were composed of 6 parts of broken stone, 2 parts of sea sand, and 1 part of Portland cement. The interval of 14 feet left between the two walls was filled in with rubble stone protected at the quay level by a layer of concrete. It was intended that the quay should be eventually sheltered by a high promenade wall, or parapet, 10 feet wide, to be built when settlement had ceased, leaving a width of 40 feet for the quay.

The rubble mound at the commencement, or root of the breakwater, was deposited from side-tip wagons; but the rubble mound, forming the base of the regular breakwater, was deposited from a steam hopper barge conveying an average load of 43 tons. The rubble mound, tipped from the steam barge, was roughly levelled at the top by dropping material from hand barges, carrying about 6¾ tons, into any hollows that existed in the mound, previous to the final levelling by the divers for the foundations of the superstructure. The progress of the rubble mound, in each season, was limited to the length considered sufficient for the following season's extension of the sea wall of the superstructure. The amount deposited in 18 months, from January 1879 to June 1880, was 28,500 tons, increasing the length of the base by 450 feet.

The advance of the sea wall within the same period was 522 feet, and 784 feet of the harbour wall. The rate of progress, however, in each month varied considerably; the maximum advance being 154 feet, in January, 1880, for the sea wall, and 169 feet, in March, 1880, for the harbour wall. The work was entirely stopped during the prevalence of the south-west monsoon, from the middle of May till October. At the commencement, the sea wall was carried out in advance of the harbour wall, with the object of providing more shelter at an earlier period; but in 1878, during the south-west monsoon, when the sea wall was 700 feet in advance of the harbour wall, the outer portion of the sea wall, 450 feet in length, was deflected slightly inwards, the extreme end being moved 13½ inches towards the harbour. As this movement indicated that the sea wall was unable, by itself, to resist the continued shock of the waves, the harbour wall was pushed forward, and the other wall stopped, till the harbour wall was abreast of the sea wall. The laying of the foundations of the harbour wall was considerably impeded, in 1878 and 1879, by an accumulation of sand on the rubble base, which was heaped up to the extent of 12 feet during the monsoon.

Extension and Modification of the Colombo Harbour Works. When the rubble base of the breakwater was approaching the point where it had been proposed to curve the line of the pier inwards, the resident engineer, Mr. John Kyle, reported to the Colonial Government, in 1877, as to the advisability of extending the harbour, owing to the increase of trade, which had risen from 446,100 tons in 1869 to 1,145,300 tons in 1877. He proposed that the breakwater should be continued in a straight line for its whole length; and that, to afford protection from the north-east, a rubble-mound breakwater, 5600 feet in length, should be run out from the eastern shore of the harbour, opposite the end of the western breakwater and in a westerly direction, leaving an opening of 600 feet between the pierheads.

This suggestion was to a great extent endorsed in a design submitted by Sir John Coode in 1878. Experience had shown that the seas raised during the south-west monsoon were too great to allow of the breakwater being used as a quay at that period, as originally intended. The waves, indeed, are not unusually large, reaching only occasionally about 10 feet in height, but they break with great violence over the breakwater; so that even with the shelter of the parapet wall, not only would the quay be useless in stormy weather during the south-west monsoon, but access also to the proposed north-east arm would be prevented. It would have been possible, as mentioned by Sir J. Coode, to have obviated this latter defect by providing a covered way under a high parapet wall; but such a structure would have offered more resistance to the waves, and, moreover, would have interfered with the traffic along the quay during the north-east monsoon. As it would, consequently, have been useless to continue the construction of the breakwater with a width suitable for a quay, it was decided to widen the sea wall to 34 feet, so as to enable it, by itself, to resist the sea, and to omit the harbour wall and intermediate filling. The breakwater, accordingly, is now intended to serve only for a protection; and the outer portion consists of a solid structure suited to this special object. (Plate 7, Fig. 23.) As the original idea of using the breakwater as a quay had to be abandoned, it has become expedient to shelter the southern shore of the harbour, so that vessels may be able to lie undisturbed alongside jetties and quays, which will require to be constructed along the southern side. Accordingly, Sir J. Coode proposed to convert the bay into a close harbour by constructing a second breakwater, situated to the north of the bay, in order to protect the harbour from the waves, or swell, raised by the north-east monsoon. These waves are only from 2 to 5 feet high, but they continue for four-fifths of the period of the north-east monsoon, and therefore, though not in any way capable of endangering the safety of vessels in the harbour, they would impede the transfer of cargoes at the quays. The modified and approved design of the harbour is shown on the plan. (Plate 4, Fig. 3.) The western breakwater is to extend for a length of 4200 feet from the root, and is being constructed with a curve towards the east at its outer portion, so as to protect its inner face from the swell during the prevalence of the north east monsoon, at which period, by this arrangement, it can be used for a quay. The northern breakwater is to be detached. with an opening of 800 feet between its extremity and that of the western breakwater, and an interval of 1300 feet between its other end and the shore: it is to be constructed of a rubble mound deposited from barges, and capped with large concrete blocks put in place by floating shears. The opening of 800 feet between the ends of the breakwaters will be situated in a depth of about 40 feet, and will form the entrance to the harbour. The opening near the shore has been designed, partly to cause the circulation of the water along the shore and thus prevent stagnation in the harbour, and partly to reduce the length, and consequently the cost of the rubble breakwater, which was originally designed to be 2500 feet long, instead of 2000 feet as at present proposed. The width of entrance proposed has been enlarged from 600 feet to 800 feet, in deference to the wishes of the nautical authorities; and the change has the advantage of diminishing the length of the detached breakwater.

Cost of the Colombo Harbour Works. The additional cost, entailed by the enlargement and modification of the designs for the harbour, is the sole obstacle to the speedy completion of the works. The amended design, with the northern breakwater, and including two jetties, a depot wharf, and dredging in the harbour, was estimated at £1,173,000. This cost, however, was considered to exceed the amount that the probable revenue of the port could bear; and various modifications of the scheme were proposed, with a view to reducing the expenditure. Sir J. Coode pointed out that, owing to the peculiarities of the site, the alternative schemes would either unduly reduce the depth of water at the entrance, and cause the track of approaching vessels to pass over the Issaure rocky shoal, or would provide an inadequate shelter from the north. Accordingly, he proposed to meet the exigences of the case by reducing the cost of his final design, omitting, for the time, the less important portions of the works till funds could be obtained for their completion. By modifying the design of the northern breakwater, omitting the two jetties, and reducing the extent of the wharf, the estimated cost was reduced to £1,008,400. As, however, the limit of expenditure was fixed by the authorities in Colombo at £800,000, it became necessary to reduce the estimate still further. It was, therefore, eventually proposed to omit the concrete blocks from the top of the northern breakwater, and to reduce also the length of this breakwater, by which the estimate was lowered to £910,600. It was imagined that the resources of the colony would be able to bear the additional expenditure, especially as the tonnage of the vessels frequenting the port had increased from 446,100 tons in 1869 to 1,205,900 tons in 1879, or nearly trebled in ten years. If, however, the limit named above must be adhered to, the northern breakwater, estimated to cost £198,000, must be abandoned for the present, and the authorisation of its construction has been deferred.

General Considerations relating to Colombo Harbour. Objections were raised to the conversion of Colombo Bay into a close harbour, on account of some of the drainage of the town flowing into the bay. There appears, however, to be no impediment to the diversion of the drains beyond the harbour; and this course, moreover, would be necessary even if the northern breakwater is not constructed, as it would be objectionable, on sanitary grounds in a tropical climate, to allow any drainage to flow into the area more or less enclosed by the western breakwater. It will readily be seen from the plan (Plate 4, Fig. 3), that the northern breakwater is essentially necessary for the due protection of the harbour during the north-eastern monsoon; whilst the opening left inshore, together with the entrance, will prevent the water in the harbour from becoming stagnant.

The wide entrance will undoubtedly admit at times a certain amount of swell; but Sir J. Coode estimates that with waves of the exceptional height of 10 feet, there would only be an undulation of 18 inches where the large steamers lie, and only 1 foot with the more ordinary 6 feet waves, and that these undulations would die away near the shore; and, moreover, the width could be easily diminished by extending the detached breakwater if desirable. There does not appear to be any tendency towards the accumulation of deposit owing to the enclosing of the bay by the western breakwater; the construction of the northern breakwater could not affect this satisfactory result, and the water is too deep at the entrance for much wave-borne sand to be brought in and settle in the still water.

The harbour when completed, and deepened by dredging as proposed, will contain a well-sheltered area of 235 acres, having a depth of water exceeding 26 feet at low water; and the total area of water space at low water will amount to 502 acres.

Concluding Remarks on Colombo Harbour. The site is decidedly favourable for the formation of a harbour; and most effectual shelter can be provided by two breakwaters of moderate dimensions. The bottom shelves sufficiently to admit of the entrance being placed in a good depth of water; whilst the depth attained is nowhere so great as to be liable to affect the work prejudicially. Moreover, a moderate amount of dredging will enable a depth of 26 feet at low water to be obtained over nearly half the area of water enclosed, and the material dredged will be very serviceable for reclaiming land to form a wharf on the southern side of the bay. The western breakwater has to resist the seas raised by the south-west monsoon; but the storms do not appear to be particularly severe along that coast, and the outer portion is more favourably situated: the northern breakwater will be comparatively little exposed.

The works, if fully carried out, will possess the somewhat peculiar interest of being constructed according to different types. The western breakwater, formed of the composite type of rubble mound and superstructure founded below low water, exhibits two forms of superstructure, the first composed of two distinct walls with intermediate filling, and suitable for a quay; the second consisting of a solid structure to serve simply as a breakwater. Doubtless the latter design would have been adopted at the outset, if the force of the waves had been accurately known; but the inner portion, though too wide for a mere breakwater, may perhaps be useful as a quay during the north-east monsoon; and the deflection of the sea wall, previously referred to, was most valuable, in indicating the exact power of the sea, for the design of the proper thickness of the solid superstructure beyond. The northern breakwater is to be a detached rubble mound without any superstructure. The stone for the work has been procured without difficulty, and is of excellent quality, being very hard and consisting mainly of gneiss.

Colombo Harbour has escaped the misfortune, so common to harbours, of being designed on a small scale, and having to be subsequently extended in a different line. The harbour was, indeed, originally laid out to much smaller dimensions than the present plan; but the initial line of the western breakwater was consistent with considerable extension, and the whole scheme was carefully reconsidered before any irrevocable deviation was commenced. Accordingly, there exists no indication, in the works under execution, of any divergence of design; and since the larger scheme was brought forward, Sir J. Coode has steadily resisted all proposals to reduce it, in order to meet the limit of expenditure imposed, wisely preferring a perfectly satisfactory scheme, though left for a time in an incomplete state, to a design which might be completed at once, but which would be inadequate and somewhat objectionable, and which would bar for ever the carrying out of the best plan.

Considering that the island is 25,740 square miles in extent, that it is very fertile, and has a great variety of produce; that Colombo is its capital, and its principal port; and taking also into account the rapid increase in the trade of the port that has already taken place, which improved shelter is sure to develop still further, it may be reasonably anticipated that funds will eventually be forthcoming for the construction of the northern breakwater, and the formation of the necessary inner wharves and jetties. Moreover, even if at the outset a remunerative return could not be derived from the harbour dues, yet a perfectly sheltered harbour would furnish a valuable indirect return to the colonial government by increasing the traffic of the railways, and would prove an inestimable advantage to the whole colony.

Comparison between Alderney, Manora, Madras, and Colombo Breakwaters.

Depth of Foundation of Superstructures. The superstructures of all these breakwaters were founded on a rubble base some feet below low water; and experience has shown the expediency of carrying the foundations lower than was at first considered sufficient. Thus at Alderney, the depth was made 12 feet, though subsequently increased by subsidence: at Manora, 15 feet below low water was adopted: at Madras, 22 feet; and at Colombo, 20 feet. The depth of foundation at Alderney was quite inadequate; and undermining could only be prevented by constant deposit of rubble stone on the sea foreshore. At Manora, the depth appears to have sufficed, partly owing to the moderate exposure of the site, and partly owing to the lowness of the superstructure, which reduced the recoil of the wave by allowing a portion of it to pass over the breakwater. The depth of 22 feet at Madras appears to have been adequate, except at the curves; and the erosion there was in great measure due to the small size of the rubble, enabling the waves to act upon it at a greater depth. The foreshore at Colombo appears not to have been disturbed.

Comparison of Superstructures. The superstructures of Alderney and Colombo breakwaters present some resemblance. At Colombo, as at Alderney, the superstructure was commenced with a sea and harbour wall with filling between, and was intended to serve as a quay, and to be sheltered by a promenade wall. The design, however, was modified before the outer portion was reached at Colombo; the uselessness of an exposed quay was recognised; the promenade wall, which intensifies the recoil of the waves, and increases their shock against the superstructure, was abandoned; and the superstructure was made solid throughout. Moreover, the sloping-block principle was introduced, which obviates the danger of unequal settlement, so injurious at Alderney; whilst a certain amount of transverse bonding was attained, and longitudinal connection can be eventually secured by a layer of concrete on the top.

Manora and Madras breakwaters are essentially similar structures, designed by the same engineer, Mr. Parkes, and constructed in exactly the same manner. They possess the advantage of presenting no more obstacle to the waves than what is necessary to arrest their progress; but this gain is effected at some loss of stillness and shelter on the harbour side, which is of considerable importance in a small harbour like Madras; and the raising of the superstructure in the proposed reconstruction at Madras will completely alter this condition. The unbonded sloping-block system was expressly adopted by Mr. Parkes to provide against unequal settlement on the rubble base, resulting in dislocation and fissures. Mr. Parkes, however, carried the principle of absence of bond too far; and if the rigid connected system of the Alderney superstructure may be regarded as the Scylla of dislocation on a yielding mound, the Madras unconnected and unbonded system must be considered to be the Charybdis of displacement in detail. The disaster, moreover, at the exposed Madras breakwaters was foreshadowed by the damage to the comparatively sheltered Manora superstructure, where the harbour blocks were dragged over by the descending wave.

Rate of Progress and Cost of the preceding Breakwaters. It would be very difficult to draw any perfectly fair comparison between the rates of progress, or the expenditures, at different harbours, as the difference in depth and exposure, the distance from which the supply of materials has to be obtained and their quality, and also the price of labour, have a great influence upon results. It must, however, be admitted that the block-setting by movable steam Titans, inaugurated at Manora, and adopted at Madras and Colombo, constitutes a notable advance, both in speed of execution and economy, upon the system of fixed staging by which the Alderney superstructure and other early breakwaters were constructed.

Alderney breakwater, extending into a depth of 133 feet, cost £235 per lineal foot; and the greatest advance of its superstructure in one year was 562 lineal feet.

Colombo breakwater, reaching a depth of 40 feet, has been estimated to cost £170 per lineal foot; and its advance, in the early stages, averaged 430 feet in twelve months.

Manora breakwater, going into a depth of 30 feet, cost only £72 10*s.*; and 793 feet of superstructure were built in one season.

The Madras breakwaters, which terminate in a depth of 45 feet, were constructed in the first instance at a cost of £72 per lineal foot, or practically the same as at Manora; but this cost will be largely increased by the works of reconstruction. The southern breakwater was carried out at the rate of about

990 feet in a year; the average yearly advance of the northern breakwater being about 770 feet.

TYNEMOUTH HARBOUR.

The improvement of the entrance to the River Tyne formed the subject of various reports and schemes, from the first report of Mr. Rennie in 1813, down to Mr. James Walker's report in 1853.

Original Design. The design first recommended by Mr. Walker consisted of two piers, or breakwaters, starting from the shore to the north and south of the mouth of the river, at the points where the present breakwaters commence (Plate 4, Fig. 10), and converging to an entrance 1100 feet wide, in a depth of 13 feet at low water. The breakwaters were to have lengths of 2100 feet, and 4200 feet, on the north and south sides respectively; and their object was to induce scour across the bar, and to shelter vessels entering the river. As the worst winds blow from the north-east, the northern breakwater was the most important for shelter, and also the most exposed. The south breakwater, however, was also needed to stop the inroad of sand, brought in by the eddying current; to prevent vessels, entering the river during southerly winds, from being driven on to the rocks along the northern shore; and to concentrate the scouring current at the entrance.

Section of the Tynemouth Breakwaters. The section to which the breakwaters were built, at the commencement, resembled, in many respects, the earlier Alderney design, and the section of St. Catherine's breakwater, Jersey, carried out by the same engineer. (Plate 7, Fig. 10.) The superstructure of the north breakwater (Plate 7, Fig. 9)[1] has a promenade level 13 feet wide, protected by a parapet, and raised 10 feet above the quay level, which is 18 feet wide and 10 feet above high water; it was also founded at low-water level on a rubbie

[1] This figure representing the most recent section of the breakwater is from a drawing furnished me by Mr. P. J. Messent, the engineer of the works.

base, but the mound was raised against the sea face up to about half-tide level. The intermediate space between the sea and harbour walls was filled with loose rubble hearting. The south breakwater, being less exposed, was given a narrower section; and the promenade wall was dispensed with[1]. These works were commenced in 1856.

Modifications of the Sections of the Tynemouth Breakwaters. The experience of Alderney was repeated at Tynemouth. Though the North Pier is not nearly so much exposed as the Alderney breakwater, it is not land-locked like St. Catherine's. Accordingly, in 1863 it had become evident that the rubble mound was not stable, in that situation, near low-water level; and the foundations of the superstructure of the north breakwater were lowered to 12 feet below low water, the same depth as adopted at Alderney, which was then supposed to be the limit of the disturbing action of waves on rubble. With this modification, the works were continued up to December, 1867, when a violent storm washed away the rubble to such a depth as to undermine portions of both the north and south superstructures, causing the destruction of 240 feet of the sea wall of the North Pier, and damaging 250 feet of the harbour wall of the South Pier. Though the sea foreshore on the north side was protected with stones of from 5 to 10 tons, it was lowered to a depth of 17 feet below low water, at a distance of 100 feet from the face of the superstructure.

To prevent further injury, Mr. Ure and Mr. Messent, the engineers, determined to strengthen the superstructure by underpinning the old foundations, by rebuilding the damaged portions at a lower level, by placing the foundations at a lower level for the rest of the work, by substituting concrete hearting for loose rubble, and by protecting the sea foreshore of the North Pier, and both foreshores of the South Pier, with

[1] A section of this breakwater, as modified, is given in 'Rivers and Canals,' vol. ii, Plate 18, Fig. 3.

concrete blocks 36 tons in weight. The foundations of the northern superstructure have been laid from 16 to 21 feet below low water at the sea and harbour faces; but the concreted hearting has not been carried down as low. (Plate 7, Fig. 9.) The southern superstructure, which was previously founded at low-water level, has been laid on both faces at from 7 to 16½ feet below low water. (Fig. 18.) These alterations have proved adequate to protect the breakwaters from any further damage.

TYNEMOUTH HARBOUR.
Superstructure of South Breakwater.

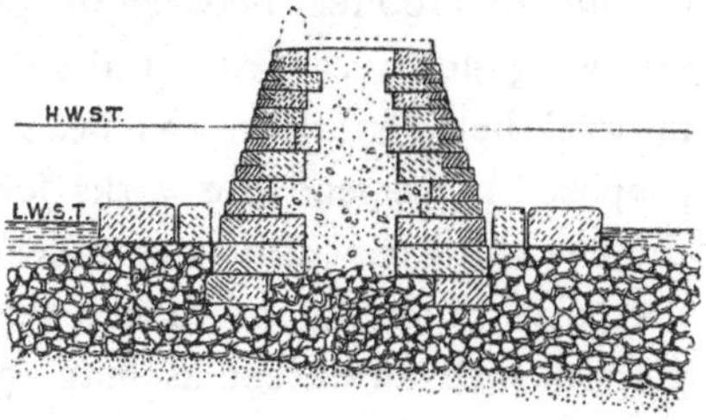

H.W.S.T.
L.W.S.T.

Scale $\frac{1}{600}$

Feet 10 5 0 10 20 30 40 50 Feet

Fig. 18.

It will be noticed that the sections (more especially the one of the North Pier) resemble the later section of the Alderney breakwater, in resting at different levels on the rubble mound. (Plate 7, Figs. 9 and 14.) As, however, the Tynemouth mounds are only a few feet thick below the base of the superstructure, there can be no fear of settlement; and, consequently, the arrangement is not objectionable in this special case.

The modifications of the foundations of the superstructures have considerably retarded the work; as the base, which had been deposited in advance at the higher level, has had to be excavated by divers for placing the foundations at a greater depth.

Alteration of the Design of Tynemouth Harbour. Owing to the strong recommendations of the Commissioners on Harbours of Refuge that Tynemouth should be one of the places at which a refuge harbour should be constructed, it was decided, in 1859, to carry out the piers into deeper water, so as to enclose a larger sheltered area, and obtain a greater depth at the entrance. The final design, adopted in 1862, is shown on Plate 4, Fig. 10, whereby the harbour was converted into a refuge harbour, the breakwaters being carried into a depth of 30 feet at low-water spring tides; the North Pier having a length of 2900 feet, and the South Pier 5400 feet, with an opening of 1100 feet between the pierheads. The breakwaters are being slowly carried out along these lines; but when the interval between them has been reduced to 1800 feet, it is proposed to suspend the works for a time, so as to observe the effect of this width of entrance on the harbour and river, and then gradually to reduce this width by prolonging the piers to such an extent as may appear advisable. A reduction of the width of entrance would increase the stillness in the harbour, and promote scour; but an entrance narrow enough to impede at all the free influx and efflux of the tide along the river, would be prejudicial to the maintenance of the river.

Construction of the Tynemouth Breakwaters. The rubble mounds, forming the base of the breakwaters, have been deposited from barges; and the apparently excessive width of the base (Plate 7, Fig. 9) is due, partly to its having been originally raised to a higher level, and partly to the action of the waves on the small stones of which it is composed in the comparatively shallow water.

The sea and harbour walls of the superstructure are built of concrete blocks faced with stone, backed below low water with concrete in bags; and the hearting of the outer portions consists of concrete-in-mass. The superstructures of both breakwaters had, up to 1883, been built from staging, erected over the site, and supported on piles resting on iron shoes buried in the rubble mound. Mr. Messent has however, designed a "Goliath" block-setting crane, which he has placed on the North Pier for completing that superstructure. (Plate 8, Fig. 8.) This Goliath will be able both to set the blocks of the superstructure, and also to deposit the concrete blocks on the foreshore for protecting the rubble mound.

Cost of the Tynemouth Breakwaters. The harbour works at Tynemouth have been so much modified, both in their general design, and in the sections of the breakwaters, and have been so much retarded in execution by the absence of available funds for the extended scheme of a refuge harbour, advocated by the Commission, but not aided by national funds, that their cost cannot be fairly compared with other harbour works executed under more favourable conditions. Up to 1879, the North Pier had cost about £160 per lineal foot, and the South Pier about £75. The difference in cost is due to the South Pier having been built in shallow water for some distance from the shore, and also to it being given a slighter section in consequence of its being less exposed.

Remarks on the Tynemouth Breakwaters. These works possess a special interest from the long period over which they have extended, and particularly from the modifications which it has been found necessary to adopt in the superstructures, and the new method of construction which is being carried out with the block-setting crane. The disturbing action of the sea on small rubble, near the face of a high superstructure, at a depth of 12 feet below low water, even in a site of moderate exposure, has been fully proved by the damage at Tynemouth; and the protection of large concrete blocks seems to have arrested this action. The harbour side of the South Pier has been exposed to injury from the same quarter as the North Pier, and has, therefore, had to be similarly protected; but it will cease to be exposed on that side when the works are completed.

Any definite conclusions as to the tranquillity of the harbour, and the effect of the piers on the river, must await the final completion of the works; but the tentative process for determining the width of the entrance ought to ensure satisfactory results. The sandy beach has advanced along the southern face of the South Pier (Plate 4, Fig. 10), but not to such an extent as to threaten the maintenance of the entrance.

MORMUGAO HARBOUR.

Description. A harbour is being formed on the west coast of India, at Mormugao in Portuguese territory, about 330 miles north of Bombay, with the object of affording a sheltered outlet for an important system of railway communication from the interior. The site selected is on the southern side of a sort of bay, forming the broad estuary of two or three rivers which converge to this part of the coast. The bay opens to the west, and is well protected by the land in other directions. Accordingly, a straight breakwater, carried out for 1800 feet in a northerly direction, will protect an area of 370 acres having a depth of from 18 to 26 feet of water. (Plate 5, Fig. 13.) Sheltered quays, extending into 24 feet of water, are to be formed at the same time along the coast, for a length of 1200 feet, to provide the necessary accommodation for trade; and the reclaimed area will be used for sidings and sheds[1].

The breakwater is being built in a fairly uniform depth of about 28 feet at low water. It is to a certain extent sheltered by the Mormugao headland from the south-west monsoon; and the waves raised by this wind will only come at an angle against the breakwater.

Construction of Mormugao Breakwater. The design of the breakwater has been already described. (Page 121.) It mainly resembles the type adopted at Manora and Madras, the superstructure being formed with sloping blocks in disconnected sections. (Fig. 7, page 120.) The blocks, however, in each section are bonded together, and the upper blocks are firmly connected. (Fig. 9, page 123, and Plate 7, Fig. 20.) The several sections are also to be finally connected longitudinally by a continuous capping of concrete-in-mass. The blocks, moreover, are made from 28½ to 37 tons in weight, thus exceeding those

[1] The plan and section of the breakwater have been taken from drawings lent me by Mr. H. Hayter, one of the consulting engineers for the harbour.

previously employed in the other similar types. The width of the superstructure is 30 feet, one-fourth more than that of the Madras superstructures. A wave-breaker of 20-ton concrete blocks has been added, as a further precaution against any chance of a recurrence of the Madras disaster. If the Madras superstructure, though slighter than usual, mainly failed from the absence of connection of the blocks, the Mormugao breakwater, which will be less exposed, and in less depth, must possess very ample strength.

The blocks are being laid by a Titan precisely similar to the one used at Manora and Madras, but somewhat stronger on account of the larger size of the blocks to be lifted. The wave-breaker, however, necessitates the employment of a crane capable of pivoting round, so as to lay the 20-ton blocks over the side of the breakwater, as the Titan does not revolve and can therefore only lay blocks along the line of the breakwater. Had a Goliath, like the one at Tynemouth, been adopted, the same machine might have accomplished both operations.

Concluding Remarks. Though three out of the five instances of constructed breakwaters, described in this chapter, have sustained serious damage, it does not appear that the injuries were in any case due to defects in the system, but rather to the designs not having been made of adequate strength for the exposure of the sites. The extreme exposure at Alderney seems not to have been understood at first, and a sort of tentative process of improvement was adopted as the work proceeded. The large increase, however, on the original estimates, due to the extension of the scheme and the great augmentation in depth, prevented full scope being given for an adequate change in design.

The changes also that have been effected in gunnery and naval construction, since the harbour was designed, have led to the harbour being regarded of less importance as a military outpost and naval station; and, consequently, it has not been considered advisable to undertake the expenditure necessary to render the breakwater permanently secure. The constant maintenance of the sea foreshore by deposits of stone is necessitated by the inadequate depth of the foundations of the superstructure, combined with their want of solidity, and by the recoil of the waves from the high sea face of the superstructure.

The object aimed at in the system upon which these breakwaters have been constructed, of dispensing with any maintenance of the rubble mound, appears to have been attained at Manora, and will be probably achieved at Colombo.

The foreshore at Madras seems to have been only affected at the curves; and probably the protection of the small-sized stone, of which the mound is composed, by large blocks at the top of the sea slope would have obviated this. The main failure of the superstructure at Madras was due to absence of bond, and insufficiency of width; and a design that was only just sufficient at Manora, which is beyond the limit of cyclones, could not be expected to stand at Madras which lies right in their path.

The injuries at the Tyne Piers were clearly occasioned by the undermining of the superstructures, under the recoil of the waves, and have been successfully provided against by lowering the foundations of the superstructures, and protecting the sea foreshore with large concrete blocks.

The chief principles which the history of these breakwaters enforce are: that the limit at which the rubble mound must be stopped, in order not to be disturbed by the sea, varies between 15 and 25 feet below low water, according to the exposure of the site; and that even at the greater depth, where the mound is composed of small stone, the top of the sea slope may need protection with large blocks; that a high parapet wall on the sea face of the superstructure is dangerous to the maintenance of both the superstructure and mound; that disconnections transversely obviate the injuries of unequal settlement on the mound; but that the different blocks of each section should be bonded together, and that the sections should be firmly connected as soon as settlement has ceased; and, lastly, that the force of waves in various localities is so different, that it is most unsafe to assume that a structure which has stood in one situation must be equally successful in another.

CHAPTER XVI.

HARBOURS SHELTERED BY UPRIGHT-WALL BREAKWATERS.

WHITEHAVEN. RAMSGATE. DOVER. YMUIDEN. ABERDEEN. NEWHAVEN. FRASERBURGH. BUCKIE. CHICAGO. OSWEGO. BUFFALO.

Whitehaven Harbour: Description; Extension; Construction of Piers; Condition. *Ramsgate Harbour:* Description; Alterations. *Dover Harbour:* Site; Designs; Commencement; Original Design of Pier, staging, diving bell, construction, modification, progress, cost, maintenance; Condition of Harbour. *Ymuiden Breakwaters:* Construction; Cost. *Aberdeen Breakwaters:* Construction; Progress; Cost. *Newhaven Harbour:* Description; Breakwater, construction; Other Works. *Fraserburgh Harbour:* Description. *Buckie Harbour:* Description. *Chicago Harbour:* Description; Construction of Breakwaters; Cost of Breakwaters. *Oswego Harbour:* Description; Construction of Breakwaters. *Buffalo Harbour:* Description; Construction of detached Breakwater. General Remarks on Upright-Wall Breakwaters.

THE number of harbours which have been formed by means of upright-wall breakwaters is comparatively limited, in the case of large harbours; the most noted instance being Dover, both on account of its importance, and the depth of water in which it is situated. The system, however, has been frequently applied to small tidal harbours, such as Whitehaven, Ramsgate, Scarborough, and others, where it is specially suitable, as the foundations are hardly below the water, and a quay wall occupying a small area and nearly vertical is required.

This method of construction possesses the merits of providing a quay, and of employing a small amount of material and taking up little space compared with the systems previously described.

The different modes of construction, and their respective advantages have been already mentioned in Chapter VII; and they will be further illustrated in the following descriptions of harbours formed with breakwaters of this type.

The cribwork breakwaters of the United States will also be referred to in the chapter, as they more closely resemble the upright-wall system than any other.

WHITEHAVEN HARBOUR.

Description. The town of Whitehaven is situated on a small sandy bay, on the west coast of Cumberland, protected from the south-west by a projecting cliff, but open from west to north. Quays and projecting jetties had been formed near the town in the south-western extremity of the bay, under shelter of the cliff, in the last century; and the new quay in front afforded protection from the north-west. The area, however, enclosed was very limited, and did not extend to low-water mark.

Extension of Whitehaven Harbour. As early as 1768, Smeaton proposed the enlargement of the harbour; but it was only in 1824 that the work was regularly undertaken by the commencement of the west pier, which, together with a northern pier, would shelter a larger space, and also place the entrance beyond the low-water line. (Plate 4, Fig. 11.) The west pier was completed in 1839, having been given a total length of 1015 feet, with two bends towards the north. A depth of 6 feet at low water had been reached at the first bend, where now a depth of only 2 feet exists outside the pier. During the progress of the work it became evident that silt was rapidly accumulating under shelter of the new pier, which arrested the south-westerly drift of sand. By 1833, the accumulation had reached 2 feet in depth at some places; and the north pier was commenced in the hope of diverting the deposit: it was extended to a length of 970 feet, and completed in 1841.

Construction of the Whitehaven Piers. The west pier consists of a sea and harbour wall of masonry, with curved batter and radiating courses; the space between them is filled with loose rubble. (Plate 6, Fig. 20.) Though the pier does not extend beyond low-water mark, it has been given a height of over 40 feet, partly owing to the range of the tides, amounting to 26 feet at springs, and partly in consequence of the greater depth when the pier was built.

The north pier has been also formed with a sea and harbour wall, and intermediate rubble filling. The construction, however, of the walls is somewhat different. They rest upon a concrete foundation placed on the sand, and consist of a masonry and concrete-pocket wall, almost precisely the same as the design adopted for the inner dock walls (Plate 14, Fig. 16), the sea wall being made rather thicker, with a foundation width of concrete of 19 feet.

The cost of the west pier was about £118 per lineal foot, and of the north pier £41 per lineal foot. This considerable difference in cost must be attributed to the greater exposure and depth of the west pier, and to the employment of concrete in the north pier.

Condition of Whitehaven Harbour. The two piers afford good shelter, but they have undoubtedly promoted the accumulation of silt, so that the low water in the harbour ebbs out to the end of the west pier, and the north pierhead is about 300 feet within low-water mark. The port is, therefore, only a tidal harbour, instead of possessing the depth of 10 feet at low water, within the piers, that was originally anticipated.

The accommodation of the port has been enlarged within the last few years by a diversion northwards of the line of the north wall leading to the north pier, thus widening the inner north harbour, and affording space for the construction of a small dock[1].

Though the bay of Whitehaven is open to the north-west, it is not very exposed, as it is situated at the entrance to the Solway Firth, and the opposite coast of Scotland is only about 20 miles distant. Moreover, the most exposed west

[1] 'Minutes of Proceedings, Inst. C. E.,' vol. lv, p. 39.

pier has only a depth of 3 feet of water near its sea face, and the depth only increases gradually further out; so that the method of construction of a sea and harbour wall with filling between has proved quite satisfactory. It is a system frequently adopted for tidal harbours, but quite unsuited for exposed situations.

RAMSGATE HARBOUR.

Description. The harbour of Ramsgate, situated to the north of Sandwich, is somewhat exposed to the influences of the sand-bearing currents along the coast, produced by the prevailing winds and direction of the waves, which have ruined Sandwich as a port. It is, however, fortunately on a projecting coast, outside the bay whose silting up has converted Sandwich into an inland village. The harbour, which is entirely artificial, has been formed by two breakwaters converging by successive angles; and the area enclosed is 46½ acres. (Plate 4, Fig. 4.) The eastern pier was constructed of solid masonry with vertical sides. The western pier was similarly built, with the exception of a portion near the shore, which was formed of open timberwork with the object of facilitating scour and affording an escape for the swell entering the harbour. As, however, this opening was found neither to keep down the silt, nor to tranquillise the harbour, it was ultimately made solid like the rest.

The harbour was commenced about 1750, and completed about twenty years later. Shortly after its completion, silt was found to be accumulating within the sheltered area much more rapidly than it could be removed. Accordingly, by Smeaton's advice, a wall was built across a portion of the harbour, separating off 12½ acres to form a tidal reservoir, into which the rising tide was admitted, and in which the water was penned up by gates till low water, when it was discharged through sluices into the outer harbour. This sluicing, combined with dredging, has sufficed to maintain the outer basin as a tidal harbour. More than half the area is dry at low water; and the depth at the entrance is about 4 feet. Artificial beaches were made in the outer harbour to reduce the swell. The entrance which originally faced south-east, in a direction at right angles to the flood and ebb currents, proved inconvenient for vessels entering or leaving the port, as the current tended to drive them against one of the pierheads. The east pier was, accordingly, lengthened in 1791, which turned the entrance to the south, and though reducing its width from 300 to 200 feet, protected it against easterly gales.

The chief points of interest about Ramsgate Harbour are: the early period at which it was constructed; its being the earliest instance of the adoption of a solid upright-wall breakwater, and of the employment of a diving bell for submarine foundations in the extension of the east pier; and a portion of an artificial harbour being turned into a sluicing basin for the maintenance of the remainder.

DOVER HARBOUR.

Site. The position of Dover, as the nearest port to the continent, and a place of great military importance, has naturally directed great attention to its natural capabilities, and the means by which they can be turned to the greatest advantage. Dover is situated in a slightly indented bay into which the old port opens, and which is protected from the west near the shore but is exposed from south to north-east. The natural shelter is small, but the exposure, being only that of the narrowest part of the English Channel, is moderate, as the sea is encircled by the neighbouring French coast from south to east. The depths within the actual bay are small, not exceeding 3 fathoms in the naturally sheltered area; but depths of about 7 fathoms are found further out. The bottom consists of chalk.

Designs for Dover Harbour. It was natural that when, about the year 1845, proposals were made for forming harbours of refuge, or defence, at various points of the English coast, Dover was one of the sites specially selected, and on the consideration of which most care was bestowed. Several engineers of eminence were requested to report to the Admiralty as to the position and section of breakwater they would recommend for the formation of a harbour of refuge in Dover Bay, about 520 acres in area, extending into a depth of 7 fathoms, and with the outer face placed in the general line of the current, that is approximately parallel to the shore. Eight reports were sent in. Most of the designs presented a great similarity in plan, owing doubtless to the instructions furnished, consisting of an oblong-shaped harbour protected by three breakwaters, two starting from the shore at each extremity, and a central detached breakwater separated from the other two by the width left for the entrances on the south and east faces of the harbour. They resembled, in fact, the finally approved design of Mr. Walker's. (Plate 4, Fig. 2.) Sir John Rennie, however, recommended a single detached breakwater, parallel to the shore in the centre and converging towards the coast at its extremities, based evidently on the design of the Plymouth breakwater; and Colonel Jones proposed a single entrance facing south.

The engineers differed considerably in the form of section advocated. Thus Sir J. Rennie, Mr. G. Rennie, and Mr. Cubitt, proposed to adopt the rubble-mound system, like Plymouth and Kingstown. Mr. Vignoles and Col. Jones selected the mixed system of a mound base surmounted by a vertical superstructure, the former proposing to introduce concrete blocks in the base, in imitation of the foreign system; whilst Mr. Walker, Mr. Rendel, and Capt. Denison preferred the upright-wall system.

The areas proposed to be enclosed, by the plans submitted, varied from the suggested limit of the Commissioners up to 1000 acres in Mr. Vignoles' design; and the estimates varied from £1,100,000 for a harbour of 688 acres, designed by Col. Jones, up to £5,000,000 for a harbour of 700 acres, designed by Mr. Cubitt.

The Commissioners appointed to investigate the subject reported, in 1846, in favour of Mr. Rendel's scheme, with two entrances facing south, which they proposed to reduce from 800 to 700 feet in width, and with an upright wall formed of masses of brickwork. The preference for southern entrances alone was caused by a fear of the introduction of silt through side entrances, and in consequence of the immunity of Kingstown harbour from silt with its entrance placed in the line of the tidal currents. The decision in favour of an upright wall was due to the weight of opinions given for this system, the concurrence of the Commissioners, the damages which had occurred to the sloping system at Cherbourg and Plymouth, and the lack of suitable stone for rubble mounds in the neighbourhood of Dover. Brickwork blocks were preferred to concrete blocks, merely owing to the small experience in concrete which existed at that time. More weight was given, in arriving at this decision, to the instance of a stable upright wall at Kilrush, near the mouth of the Shannon, than was quite admissible, as Kilrush is not exposed to the Atlantic, as suggested, being within a bay in the estuary of the Shannon, and therefore somewhat sheltered, besides only extending into a depth of 7 feet at low water. Nevertheless, though the Commissioners acknowledged that the upright-wall system was an experiment, there is no doubt that, with a chalk bottom, an absence of suitable building materials, and a moderate depth, the upright wall was the best system to adopt.

Eventually Mr. Walker was entrusted with the construction of the harbour, according to the approved design of 1847 (Plate 4, Fig. 2); but though nearly forty years have elapsed since the commencement of the work, the western breakwater alone has been partiallv constructed; and the positions of the entrances, size of harbour, and ultimate cost, still remain open for further consideration. Dover, the most important of all the harbours started by the Government about 1847, remains in the most unfinished state.

Commencement of Dover Harbour. Works were commenced, in 1847, for the construction of 800 feet of the western breakwater, starting from Cheesman's Head at the extreme western end of the bay, at an estimated cost of £245,000. This breakwater, extending out almost at right angles to the shore, and facing the most exposed quarter of south-west beyond the shelter of the bay, constitutes the most difficult portion of the work. It was naturally selected as the first work to be undertaken, for besides serving to shelter the entrance to the port from south-westerly gales, it forms a convenient pier for the passenger and goods traffic with the continent, as it was easily connected with the adjacent railway. Moreover, as in the unfinished state of the harbour, it faces also the second most exposed quarter of north-east, the steamers coming alongside obtain shelter from the worst winds by going to one or other side of the pier, according to the direction of the wind.

Original Design of Dover Pier. The first object in view at the commencement of the harbour works was to provide, as soon as possible, a pier suitable for the accommodation of the continental traffic. Accordingly, a timber jetty was begun, with its platform 10 feet above high water to serve as a staging for the work. Stone was packed behind the piles of the jetty; and masonry walls were brought up in front on each side, with a batter of 1 in 3 on the face; and a parapet was raised 15 feet above the platform to protect the quay[1]. Though fair progress had been made with the work by 1850, when the end of the foundations had reached

[1] Mr. Druce, the engineer in charge of the harbour works, has furnished me with several particulars relating to the construction of the Dover Pier.

650 feet from the shore, storms towards the end of that year inflicted considerable damage on the staging: the first storm carried away 78 feet of staging with the diving bells and machinery; and the unfinished end of the pier was somewhat injured. In the early part of the following year some stones were displaced, and staging carried away.

Staging at Dover Pier. As it had become evident that the staging could not be considered secure from damage, unless its platform was raised beyond the reach of the waves, a new system of staging was adopted. The piles of the new staging consisted of two round timbers, scarfed together at their butt ends so as to form a mast, about 100 feet long, 21 inches in diameter in the centre, and 16 inches at the ends; and they were driven about 7 feet into the chalk. Each support was composed of three piles, the two outer ones being placed clear of the pier; and the distance apart of each support was 30 feet. The piles were moored by chains to concrete blocks; and a single horizontal tie connected the moorings. A platform was erected on these piles, from which the diving bells were worked, and the blocks lowered for the construction of the pier. The round piles offered little surface to the waves, and the close platform was raised above the waves, so that this staging suffered little from storms, except on rare occasions, as for instance during a south-westerly gale in December, 1869, when it was washed away.

Diving Bells at Dover. Diving bells have been exclusively used at Dover for levelling the chalk bottom, and laying the lower courses of the pier. Mr. Druce, the harbour engineer, has a decided preference for diving bells over helmet divers in such a site as Dover, where the water is turbid, the current strong, and where it is desired to work through the winter. Helmet divers cannot see properly in turbid water; they cannot stand well in a strong current; and they cannot work in the cold and rough water during the winter months. Turbidity of the water is of no consequence in a diving bell; the men in it are out of reach of the current; and they complain generally of heat rather than cold, so that working in the winter is no hardship to them. Moreover, when the sea is not very rough, the diving bell can be used, as it can be rapidly lowered through the rough water, and the men are able to work below, unimpeded by the motion above. The use of the diving bell, however, limits the sizes of blocks that can be employed.

Construction of Dover Pier. The inner portion of Dover Pier was constructed as a solid upright wall, with a batter of 1 in 3 on both faces, the mass of the wall being composed of concrete blocks, containing about 5 cubic yards, and it was faced throughout with granite. The soft material overlying the chalk, which had an average thickness of 3 feet, was excavated by aid of the diving bells; and the bed for the foundation course was most accurately levelled, before the granite facework and the 8-ton concrete blocks were set. The excavation and levelling of the foundations, and the setting of the foundation course, formed, Mr. Druce states, the most tedious portion of the work; but it was most important, as any inaccuracy in level led to difficulty in laying the succeeding courses, and could not be rectified except by the very troublesome process of trimming off the beds. This section, with concrete-in-mass for the hearting above high-water level, and with the width of quay and protecting parapet as shown on Plate 6, Fig. 21, was continued up to the point where the pier curves round to the east, only 300 feet from its present extremity. (Plate 4, Fig. 2.)

Modification of Section of Dover Pier. When the pier had reached its then authorised limit, of 1800 feet, in 1865, the question of its method of termination or further extension was considered. As the end of the pier, in its existing position, was scoured by a strong tidal current, it was feared that the soft chalk at its base might be undermined, and endanger the stability of the pier. It was also urged, that an extension of the pier in an easterly direction would increase the sheltered area, and was a work of national importance. On the other hand, some naval men feared that a considerable extension of the pier towards the east would hasten the process of silting, which had been already noticed within the shelter of the pier, whilst they doubted the necessity of additional shelter for purposes of refuge with the natural shelter of the Downs so close. Eventually it was decided, in 1866, that the extension should only be carried far enough to the east to make the pier terminate facing eastward, so as to secure its end from scour.

As this extension would face south, instead of south-west, and would not be required as a landing-pier, its section was modified, and economies introduced. This extension, as executed, consists of an upright wall, with a batter on the sea face of 1 in 6, and on the harbour face of 1 in 12, which, as the width at quay level has been kept the same as before, considerably reduced the section of the pier near the base. (Plate 6, Fig. 21.) Moreover, in this portion, the granite facing has been dispensed with up to low-water mark; whilst, in the foundation course, a sort of apron of concrete blocks has been introduced to protect the toe of the wall from scour. Mr. Druce is of opinion that even this reduced amount of granite facing would hardly be justifiable at the present time, on account of the expense, amounting to five times that of concrete; for the concrete blocks, which form the entire facing of the end of the pier, exhibit no more signs of wear than the granite along the western face. The pier has been terminated with a square head, upon which a fort has been erected.

Progress of Dover Pier. The rate of progress of the pier has been comparatively slow, the greatest advance of the foundations in one year, except just at the commencement, having been 147 feet in 1861; and the average yearly advance, from 1847 till its completion in 1871, only amounted to 91 feet. This slow progress, as compared with most other breakwaters of modern date, must be attributed, partly to the great care required in levelling the bottom and laying the foundation course, and partly to the greater depth than usual to which the regularly bonded work was carried down below low water, and which, unlike a rubble or concrete mound, can only be extended from its outer end.

Cost of Dover Pier. The work was carried out under three distinct contracts; the first contract for 800 feet, in 1847; the second for 1000 feet, in 1854; and the final extension of 300 feet, in 1867; at estimated costs of £245,000, £405,000, and £75,000. The total amount paid to the contractors up to 1874 was £679,368; and taking this sum as a basis, the average cost per lineal foot of the pier would be £323 10*s.*; but this sum does not include the cost of land, or salaries, and other disbursements made by the Government. The actual cost, however, per lineal foot, varied considerably according to the construction; for Mr. Druce informs me that, whereas at the commencement, when stone was largely used, the cost in only 12 feet at low water amounted to £293, in the extension, in an average depth of 45 feet at low water, the cost was reduced to £250 per lineal foot by the exclusive use of concrete blocks below water.

Maintenance of Dover Pier. The main portion of the pier has withstood the waves perfectly since its completion; though the work was occasionally disturbed to a small extent during its progress. Face stones have sometimes shown a tendency to be withdrawn on the more battered portion of the sea face; whereas no injury of this kind has ever been observed on the more upright face of the extension. This immunity in deeper water may be partly attributed to the advantage possessed by the steeper batter of throwing the weight of the upper courses more thoroughly on the lower face stones, thus keeping them in place, and partly to the less exposed position of the outer portion of the pier owing to its change in direction.

The parapet which protects the quay has not been equally fortunate in escaping injury. During a storm from the west, at the height of spring tides on the first day of the year 1877, the whole of the parapet wall carrying the promenade roadway between the two curves of the pier, for a distance of about 1050 feet, was carried away. This portion exactly faced the direction of the storm, and was therefore most exposed to the waves, which accounts for the injury being confined to this particular length of parapet. The original design provided a thickness of 13½ feet for this wall, and it was commenced with this thickness. The remaining width, however, on the top of the pier available for a quay, amounting to 31½ feet, proved insufficient for the requirements of the increasing railway service; so it was decided to reduce the parapet wall to about 5½ feet, thus gaining about 8 feet for additional width of platforms. The wall, being thus reduced to less than half its intended width, was unable to bear the strain due to an exceptionally severe storm beating direct upon it; but where the stroke of the waves was somewhat oblique it was able to withstand their force.

The Dover Pier, accordingly, in spite of this accident, due entirely to an excessive reduction of strength, may be regarded as a very stable breakwater, requiring little maintenance, and affording excellent shelter as far as it extends. Scour, however, round the head requires to be guarded against.

Condition of Dover Harbour. The western breakwater, in its present unfinished condition, serves rather as a landing pier than as a regular breakwater. With this object, five timber berthing stages, or landing places, have been provided alongside the pier, two on the outside and three on the harbour side, the three inner ones being used for the mail service, and the two in deeper water for vessels of the Royal Navy.

One objection raised against the construction of a closed harbour in Dover Bay was that the shingle, carried along the coast by the prevailing westerly winds, would be arrested by the western breakwater, and accumulate to the detriment of the harbour. The Commissioners, however, did not anticipate inconvenience from this source; and though it has been stated that some advance of the shingle is apparent since the commencement of the works, it is certain that the travel of shingle has been in a great measure arrested to the westward by other causes, so that, according to Mr. Druce, no shingle reaches the pier from that side; and the pier protects the beach in Dover Bay, preventing its travelling eastwards as it otherwise would do.

Mr. Druce considers that the present unfinished state of the pier is actually injurious to the anchorage in the bay, as an eddy is formed by the tide, after rounding the pierhead, which scours out the soft chalk from the shallow portions of the bay and deposits it alongside the pier, and that this eddy would be checked by the extension of the breakwater. It is quite certain that, whether this view is correct, or the opposite opinion as to the injurious effect of an extension expressed by naval authorities in 1866, the sheltered area near the breakwater must be silting up; for whereas the engineers in their report to the Admiralty, in 1864, stated that there was a depth of 40 feet at low water alongside the outer landing jetties, the most recent chart gives depths of only 23 feet and 35 feet, respectively, inside and outside the pier at these places. Capt. Calver also, in a comparison of surveys taken in 1859 and 1865, came to the conclusion that silting up to the extent of 4 feet had occurred during the interval in the most important portion of the sheltered area.

The breakwater in its present state, 2100 feet long, affords protection to the entrance of the port, and shelters about 35 acres from westerly gales; but the bay remains quite open to the east. If the harbour is to serve the purpose for which it was originally designed, and not merely as a landing pier, extensive works will be needed. The most exposed portion has, however, been already completed; and the experience gained will lead to greater economy in construction, especially as the reliable character of well-made concrete blocks is now so fully established. The closing of the harbour, and placing the entrances in deep water, would in all probability arrest the deposit of silt, whatever might be the result of merely completing the western portion of the breakwaters; and even if the depth in the harbour had to be maintained, it could be accomplished at a small cost with the improved dredging appliances now in use at several ports.

YMUIDEN BREAKWATERS.

Construction of the Breakwaters. The form of Ymuiden Harbour, which provides a communication between the North Sea and the Amsterdam Ship-Canal, is shown on Plate 5, Fig. 2; and the harbour has already been briefly described at page 181.

The site at Ymuiden was even more destitute than Dover in suitable material for a breakwater; for whilst shingle and sand were readily procured at Dover, very fine sand was the only substance near at hand at Ymuiden. It was therefore necessary to construct the breakwaters with as little material as possible; and concrete blocks forming an upright wall were naturally adopted. The bottom, however, consisted of very fine unstable sand, so that when the blocks were deposited on it, the eddy produced round the blocks by their interference with the currents caused a scour, which removed the sand and undermined the blocks. An artificial foundation of basalt rubble, 3¼ feet thick, and extending out to about three times the width of the breakwater, was accordingly formed; and when scour occurred, the loose rubble fell into the hollow, till the mound attained a stable slope, leaving a level portion in the centre on which the breakwater was built. (Plate 6, Fig. 19.)

The two breakwaters are symmetrical in shape, and identical in construction. They have been built entirely of concrete blocks, from 6 to 12 tons in weight, placed in bonded courses and laid by steam cranes. The blocks in the four upper courses were connected together by iron cramps; and the structure was capped at the top by concrete-in-mass. The inner half of each breakwater consists simply of an upright wall, 20 feet wide at the top, with a batter of 1 in 8 on each face. The outer portion has been protected by a wave-breaker of from 10 to 20-ton concrete blocks, forming a random mound on the sea side; and the wall was gradually increased, as it reached deeper water, to the width shown on the section.

The worst gales are from the south-west, veering round to north-west, so that the two outer bends are most exposed; but the depth of water at the pierheads is only about 4½ fathoms, and the bottom shelves gently, whilst the exposure being only that of the North Sea is not excessive; and the upright breakwaters protected by the wave-breaker have withstood the waves without injury.

Cost of Ymuiden Breakwaters. The breakwaters are each about 5085 feet long, and cost altogether £1,082,200, which amounts to £106 8*s.* per lineal foot. The upright portion of the structure cost £803,750, out of the total amount; whilst the remaining £278,450 were expended on the wave-breaker, which did not form part of the original design, but was subsequently added to ensure stability.

Aberdeen Breakwaters.

Construction of the Breakwaters. The harbour of Aberdeen, and the methods of construction adopted for the breakwaters have been already fully described in Chapters VII and IX. (Plate 1, Figs. 11 and 12, and Plate 6, Figs. 15 and 16.)

The conditions of the site differed from those of Dover and

Ymuiden in the presence of stone in the neighbourhood. Granite, however, the stone of the district, is so costly to work that concrete was adopted as in the two cases just described. The bottom resembled Dover in being firm, and therefore suitable for an upright-wall breakwater, and the depth of water is moderate. As however at some places the granite rock crops up, a level foundation, like at Dover, would have proved very costly to prepare. This led Mr. Cay to adopt concrete in bags for levelling the foundations (Plate 6, Fig. 15), and it proved such a satisfactory and expeditious system that concrete blocks were eventually dispensed with for the work under water, and bags deposited from hopper barges were employed instead. (Plate 6, Fig. 16.) The whole of the work above low water, in each breakwater, was constructed of concrete-in-mass, instead of partially with blocks in courses as at Dover and Ymuiden.

Progress of the Aberdeen Breakwaters. The south breakwater occupied about four years in construction, being completed in 1873. This would give an average rate of progress of about 260 feet a year. The progress, however, when the work was in full swing, was actually 300 feet in a year; and on one occasion in the summer, 87 feet were accomplished in four weeks.

The progress of the northern breakwater in prolongation of the North Pier was even more rapid, as the extension of 1000 feet was completed in two years, giving a rate of 500 feet annually.

This rapid progress must be attributed partly to the moderate amount of work below low water, as the extreme depth into which the breakwaters have been carried is 22 feet, but more especially to the use of concrete in bags below water, and the extensive adoption of concrete-in-mass above. Neither of these methods of depositing concrete are actually novel, as concrete in bags was used by Italian engineers for repairing foundations below water before M. Poirel deposited concrete-in-mass, protected by sacking, under water at Algiers in 1834; and concrete-in-mass above water had been previously partially adopted at Dover and Ymuiden.

The extension, however, of the system of bags to uneven foundations for an upright-wall breakwater, and their subsequent exclusive use below low water, and the method of depositing the bags from a hopper barge, were initiated for the first time at Aberdeen, and together with the depositing of concrete-in-mass within framing from low water upwards, largely contributed to the rapid progress achieved.

Cost of the Aberdeen Breakwaters. The total cost of the south breakwater was £68,100, which gives for a length of 1050 feet, as actually constructed, a cost of £64 17*s.* per lineal foot, extending from the shore into a depth of 22¼ feet at low water.

The northern breakwater, in a depth of about 17 feet at low water, cost about £46,000, which is equivalent to a cost of £46 per lineal foot. This considerable reduction in cost, as compared with the earlier work, is partly due to the smaller depth of water, and consequent reduced section, and in great measure also to the economy effected by the extended use of bag work. (Plate 6, Figs. 15 and 16.)

NEWHAVEN HARBOUR.

Description. The harbour of Newhaven is the nearest port on the English Channel to London, and provides a very direct route from that city to Paris by way of Dieppe. It is situated at the mouth of the river Ouse, which flows into Seaford Bay. The outlet of the river was frequently impeded or choked by shingle and sand brought along the coast by the prevailing westerly and south-westerly winds; and the accumulation was increased by silt brought down by the river. Before 1835, the entrance to the harbour was maintained by two parallel wooden jetties, 100 feet apart; and at that period the eastern jetty was prolonged to the same distance as the western one to concentrate the current, and to keep out the shingle which used to drift round the west jetty. The advance also of the shingle from the west into the channel was checked by a groyne on the western beach; but eventually the shingle overlapped the groyne. Though the harbour was improved at various times by deepening the river, removing shoals, widening the entrance channel, and making a straight cut, the entrance was still impeded by shoals, so that vessels had frequently to lie off from the harbour till the tide rose high enough.

Breakwater at Newhaven. Till recently the harbour at Newhaven was strictly a jetty harbour, maintained by the tidal and fresh waters in the river Ouse, and guided by jetties across the beach, but subject to the inroad of shingle, just as the French jetty harbours are to that of sand. A breakwater, however, is now in progress, which, starting from some rocks to the west of the entrance, will gradually curve round till it covers the approach to the river[1]. (Plate 4, Fig. 7.) The breakwater will, accordingly, protect the entrance from west round to south, and will arrest the travel of shingle and thus prevent the choking up of the mouth of the river. The new harbour will thus afford shelter from the most frequent and stormy westerly winds, but it will be exposed to the east and south-east unless another breakwater is constructed on the east side. Newhaven is being converted by these works from a jetty, or river harbour, into a partially closed sea harbour. The idea is not quite a new one, as Seaford Bay was proposed as a favourable site when the establishment of harbours of refuge along the English coast was being considered, about forty years ago; and it was reckoned by some persons to be second only in importance to Dover and Portland as a Channel refuge harbour. The present works, however, are being

[1] Plans of the harbour were furnished me by Mr. Banister the engineer, and by Mr. Carey the resident engineer of the harbour.

undertaken for the improvement of the traffic of the port with Dieppe.

Construction of Newhaven Breakwater. The breakwater is being constructed entirely of concrete, for which an abundance of materials exists in the locality. It is being built on the upright-wall system, on the same principle as the north breakwater at Aberdeen, with concrete in bags deposited from hopper barges[1] on the chalk bottom up to low water, and concrete-in-mass above, thus forming one solid mass of concrete. (Plate 6, Fig. 17.)

The only differences introduced are, that the bags are larger than those used at Aberdeen, as they each contain 104 tons of concrete, and the framing above low water is fastened together on the outside, instead of by through bolts, thus saving the bolts, and rendering the framing easy to remove. An arched parapet wall protects the quay, and affords access to the breakwater when the sea is breaking over it. The breakwater will extend into a depth of 18 feet at low water, and is to be 3000 feet long.

The concrete mixer consists of a screw working in an inclined cylinder, and is continuous in its action, the materials being introduced at one end, water being added during the transit, and thoroughly mixed concrete coming out at the further end.

Other Works at Newhaven Harbour. To provide for the increased trade which it is expected will result from the good shelter afforded by the breakwater, various other works are in progress or contemplated. A concrete sea wall has been built in front of the shore, between the breakwater and the west jetty, reclaiming 7 acres of land: the west jetty has been rebuilt, and the line of the east jetty has been altered, widening the entrance from 150 to 200 feet; and the channel is to be dredged to a depth of 12 feet at low water. It is proposed to

[1] A section of the hopper barge will be found at page 134.

build a dock of 24 acres on the east side, on the site of the reservoir which was designed to utilise the tidal flow.

FRASERBURGH HARBOUR.

Description. The harbour of Fraserburgh is situated at the western side of an extensive bay, open to the north, on the north-eastern coast of Scotland. The small fishing harbour, established in a small inlet, has been enlarged, and sheltered recently by a small breakwater, serving as a protection from the north and north-east. (Plate 4, Fig. 14.) The breakwater, about 850 feet long, extends from low water into a depth of 3 fathoms, and both shelters a sort of outer harbour, and protects the entrances to the two inner harbours. It has been constructed of concrete, on the upright-wall system; the same principle of concrete bags, and the same plant, having been employed as at the northern breakwater at Aberdeen.

Fraserburgh Bay is a site endowed by nature with larger capabilities than have been hitherto utilised; but the whole bay is somewhat too wide, and too little recessed, to enable a refuge harbour to be formed, enclosing the whole bay, at a moderate cost.

BUCKIE HARBOUR.

Description. The small harbour of Buckie was carried out, by private enterprise, for the protection of the fishing boats, on the coast of Banffshire in Scotland. The breakwater and west pier, sheltering the harbour, extend into a depth of 10 and 7 feet at low water respectively[1]. They are constructed entirely of concrete deposited on the rocky uneven bottom, within framing lined with jute cloth, and resemble the Aberdeen northern breakwater and Newhaven breakwater in section. (Plate 6, Fig. 18.) The framing was tied together with

[1] 'Minutes of Proceedings, Inst. C. E.,' vol. lxx. p. 350.

bolts, which were protected with timber covers so that they could be drawn out after the wall had been built. The method of construction is very similar to that adopted by M. Poirel, at Algiers, for forming huge concrete blocks in the sea, about forty years previously.

CHICAGO HARBOUR.

Description. The first improvement for navigation at Chicago consisted in training the river Chicago by jetties, at its outlet into Lake Michigan, and dredging the straightened channel.

In order to provide more accommodation and a better depth of water, an outer harbour, sheltered by cribwork breakwaters, was designed in 1870. This has been carried out by the construction of two detached breakwaters, as shown on the plan[1] (Plate 5, Fig. 5), with an entrance of about 800 feet between them.

The east breakwater, parallel to the shore, is 4000 feet long, and the south-eastern breakwater, placed at an angle, is 3000 feet long: they shelter altogether an area of 455 acres, of which 185 acres are to be reserved for piers and slips; the remaining 270 acres have been deepened by dredging to a minimum depth of 16 feet. An exterior detached breakwater is being constructed, 5432 feet long, in a depth of from 24 to 31 feet of water, about a mile to the north-east of the harbour, for the purpose of protecting the entrance to the outer harbour and forming a refuge harbour. It was commenced, in 1881, at the centre, and is being extended at each extremity.

Construction of Breakwaters at Chicago. All the breakwaters at Chicago are constructed of strong timber cribwork, floated into place, in lengths of about 100 feet, and sunk and filled with stones. The cribs for the eastern breakwater, commenced in 1870, and completed in 1875, were sunk upon

[1] A plan of the harbour and drawings of the cribs were sent me by Major Lydecker, the engineer in charge of the works.

the natural bottom, in the usual manner, in a depth of about 20 feet. As, however, the cribs were liable to settle irregularly upon the soft foundation, the cribs of the south-eastern breakwater were deposited on a pile-work foundation, which gave satisfactory results.

The pile-work foundation was not attempted for the exterior breakwater, owing to the difficulty and expense of driving and cutting off the piles to an exact level in deeper water and in an exposed situation. To afford increased stability, and provide against unequal settlement, the side walls of the cribs were not carried down within about 8 feet of

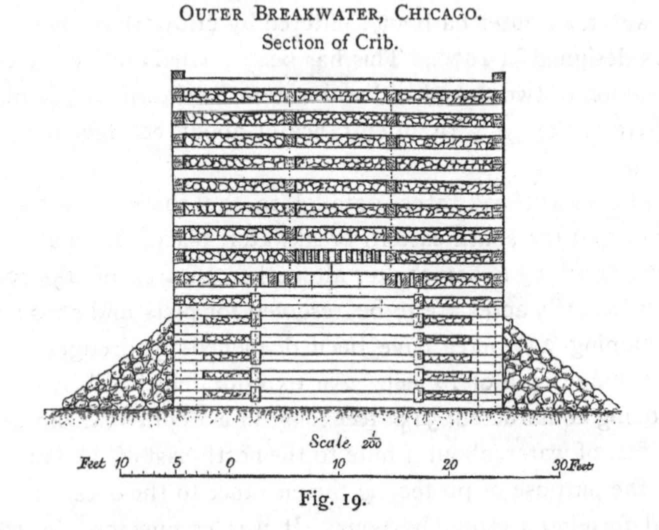

Fig. 19.

the bottom, and the interval was filled with a mound of rubble stone. (Fig. 19.)

Cost of the Chicago Breakwaters. The total cost of the whole of the works is estimated at about £160,000. The cost of the exterior detached breakwater is estimated at £22 18*s.* per lineal foot. This appears a very small cost for a detached breakwater in from 24 to 31 feet of water. The materials, however, are cheap; and the site, though in the open lake, is not exposed like in a sea or the open ocean. Moreover, the system can only be adopted where timber is abundant, and where the teredo does not exist; and the breakwaters, though cheap to construct, will require frequent renewal near and above the water-level, where the timbers are sure to decay rapidly.

OSWEGO HARBOUR.

Description. The harbour of Oswego on Lake Ontario, situated at the mouth of the Oswego River, has been improved lately, like Chicago Harbour, by the construction of an artificial harbour along the coast sheltered by cribwork breakwaters. It consists of two portions, to the north-east and south-west of the river respectively. (Plate 5, Fig. 11.)

The south-western portion has been completed; but the north-eastern breakwater is in course of construction, and is separated from the other by an entrance, 350 feet wide, for the exit of the river and the approach to the harbour. The south-west breakwater has a length of 6025 feet; and the north-east breakwater is designed to be 2700 feet long.

Construction of Oswego Breakwaters. The south-western breakwater has been formed of cribs, about 35 feet square, filled with stone and sunk in a depth of from about 18 to 21 feet of water on a sandy bottom overlying rock. When the cribs had settled down, the superstructure was erected upon them. Instead of raising the cribs throughout to a uniform height of 8 feet above the water-level, the outer portion has been raised, at the exposed corner, to a height of 12 feet above the water-level, whilst the inner portion has been correspondingly lowered three or four feet. The corners have been strengthened by adopting oak, in place of pine, from 6 feet below to 5 feet above the water-line, and by being plated with iron for 12 feet on each side of the angle for a height of 9 feet. This breakwater cost £25 8*s.* per lineal foot.

The north-eastern breakwater is similar in construction, and cost £38 15*s.* per lineal foot.

BUFFALO HARBOUR.

Description. The river Buffalo has been trained into Lake Erie by parallel jetties, similarly to the rivers Chicago and Oswego; and, like the harbours at those places, Buffalo Harbour is being enlarged by the formation of a harbour along the coast protected by two cribwork breakwaters. (Plate 5, Fig. 12.) The harbour has been designed on a more extensive scale than Oswego, and even larger than Chicago. The detached breakwater parallel to the shore, in course of construction, is to have a length of 7600 feet. It was begun in 1868.

Construction of detached Breakwater at Buffalo. The breakwater is being formed of cribwork filled with stone, each crib being about 50 feet long, and averaging 36 feet in width. The soft material, overlying a hard bottom, has been removed to a depth of from 10 to 20 feet by a dipper dredger before the cribs are deposited. The cost of the cribwork has varied considerably in different contracts, having reached £43 per lineal foot, and been as low as £33 12*s.*; whilst it was estimated that another portion could be constructed for £27 2*s.* per lineal foot.

The cribs have been raised 8 feet above the ordinary level of the lake; but as the lake is raised during storms, the waves are liable to break over the cribwork, and produce a commotion in the harbour sufficient to make vessels occasionally drag their anchors in the soft bottom. Accordingly, it is proposed to adopt the plan followed at Oswego, of raising the outer portion of the superstructure of the crib 12 feet above the water-level, and lowering the harbour side.

General Remarks on Upright-Wall Breakwaters. The number of large harbours sheltered by breakwaters of this type is very limited, so that some small harbours have been cited in illustration as well.

The laborious process of levelling the bottom, as carried out at Dover, has been superseded by the simpler, more expeditious, and cheaper process of concrete deposited in bags, as practised at Aberdeen, Newhaven, and Fraserburgh. The method of depositing concrete-in-mass within timber framing, as accomplished at Buckie, is only suitable for small depths of water and on a hard bottom.

The general employment of concrete, either as blocks or in mass, for these constructions is very marked ; for out of all the instances given, Whitehaven alone was built of masonry, and concrete was introduced in the later design ; whilst at Dover, granite was only used for the facework, and eventually was replaced by concrete below low water.

All the more recent breakwaters described, namely, Ymuiden, Aberdeen, Newhaven, Fraserburgh, and Buckie are composed entirely of Portland cement concrete. The increasing preference for concrete in this type of breakwater is undoubtedly due, not merely to the absence of suitable stone which dictates the choice of this kind of breakwater, but also to the facility with which it is manufactured to any suitable size, and its cheapness; whilst in bags and in mass, it fulfils the main qualifications of an upright-wall breakwater, of being readily adapted to an uneven bottom, and being perfectly solid throughout. Where the bottom is firm and hard, with no tendency to settlement or scour, concrete in bags below low water and concrete-in-mass above form a monolithic structure, which is the object aimed at in all solid portions of breakwaters, whether as superstructures or upright walls.

Though the Dover Pier has realised the object of forming a stable breakwater with a small amount of material, as may be readily seen by comparing it with other types, and more especially with the rubble mound, yet this result has only been accomplished at a considerable cost, contrasting unfavourably in this respect with most other breakwaters. This has naturally led to suggestions of cheaper expedients, deduced from constructions in more sheltered situations. Thus rubble stone within timber framing was proposed, on the model of the Port of Blyth piers situated in shallow water in a fairly sheltered position, due account not having been taken of the considerably greater exposure and depth at Dover. It may be concluded that the storm which overthrew a long length of solid concrete parapet would have completely demolished a timber framing filled with loose stone, even if the timber could have resisted gradual decay near the water-level and the attack of the teredo. Another proposition consisted in forming an outer casing of stone or concrete blocks, bound together by an iron framework, and filling up the interior with loose rubble. This system, though more substantial than the other, had only been carried out at the piers of Greenock dock lying in a well-sheltered portion of the Firth of Clyde, and had not been tested in the open sea. The iron, forming an essential part of the structure, would be certain to rust and perish; and if the sea once pierced the casing, and reached the hearting, the whole structure would be exposed to rapid destruction. Fortunately neither of these systems was adopted; and Dover Pier, though costly, is at least stable. The employment of concrete bags below the water, and the saving of the costly levelling of the bottom, would considerably reduce the price of such a breakwater. The only advantage an upright wall possesses over the mixed system, of a rubble mound with a superstructure founded about 20 feet below low water, consists in the saving in material, and the freedom from settlement when the bottom is firm.

The harbours of Chicago, Oswego, and Buffalo are very similar, both in general plan and in the construction of their breakwaters. Being lake harbours in a moderate depth of water, the breakwaters are not severely tried; and the cheapness and rapidity of construction, furnishing in a short space of time a large extension of shelter, compensate for the expense of considerable maintenance which has already been required, and is sure to continue. These structures, though resembling in principle the Port of Blyth jetties, are much more strongly built; as the cheapness and abundance of timber in America enable a free use to be made of this material.

CHAPTER XVII.

HARBOURS ON SANDY COASTS.

Difficulties in forming Harbours on Sandy Coasts. Ostia Harbour. Methods of forming these Harbours. Parallel Jetties. Converging Jetties. Closed Harbour. Breakwater connected with the shore by an Open Viaduct. Detached Breakwater. General Remarks.

VARIOUS kinds of harbours have been described in the preceding pages, executed under various conditions; and it has been seen how some designs or methods of construction have been most suitable for deep water or great exposure, whilst other designs have proved adequate for shallow harbours in sheltered sites. The nature of the design has to be adapted to the special circumstances; but generally it suffices to make an adequately strong and connected structure to attain a successful result.

In the case, however, of harbours on sandy coasts, the difficulties to be contended with are so impalpable, the results of works so uncertain, and the best designs so liable to be marred by unforeseen or unpreventible occurrences, that these harbours constitute the most difficult type of design with which a harbour engineer has to deal. It is possible, with large mounds protected by huge blocks or massive superstructures, or with compactly built and well-connected upright breakwaters, to resist almost any force of waves which is liable to be encountered under ordinary conditions. Sometimes, indeed, engineers have erred in underrating the power of the sea in certain localities; but it has been only necessary to strengthen the structures, to arrive at a successful issue: and though in the case of Alderney and Biarritz the conflict with the waves has not been maintained, this is rather due to the abandonment of the works as unnecessary or too costly, and not with any notion that it would be impossible to erect a stable breakwater in those most unfavourable and exposed situations. When, however, it is attempted to form a harbour on a flat sandy or shingly beach, where every projection from the coast is followed by a progression of the foreshore, or where a gradual change of the coast is taking place, the problem presented is far more difficult, and the forces to be overcome or evaded, though less visible, and apparently less powerful than the action of the waves, are more insidious and unchangeable.

Ostia Harbour. The harbour of Ostia is probably the earliest instance of an attempt to construct a harbour on a sandy coast; it was built, by order of the Emperor Claudius, to secure a deeper entrance from the sea for the trade of Rome than the bar at the mouth of the Tiber afforded. The harbour was sheltered by two piers carried out at right angles to the shore, and terminating with return arms parallel to the coast. The wide space between the pierheads was partly occupied by a central detached breakwater, parallel to the shore, and leaving an opening at each end which formed the entrances to the harbour. The piers seem to have been constructed, like a viaduct, with openings below and a continuous roadway above. The upper part served to check the undulations; whilst the apertures left a vent for the sand-bearing littoral currents. The harbour, however, was situated within the influence of the alluvial drift from the Tiber, and, being devoid of means of scour, it rapidly silted up, and its site is now two miles from the coast.

Methods of forming Harbours on Sandy Coasts. Various methods have been resorted to, or proposed, for constructing harbours on sandy coasts within recent times; namely, (1) Parallel Jetties, such as those of the foreign North Sea ports described in Chapter VIII: (2) Converging Jetties, of which several instances are given in Chapter IX: (3) Closed Harbour, like Madras and Boulogne: (4) Breakwater connected with the shore by an open viaduct, as adopted at Rosslare, Wexford, and proposed for Port Elizabeth: and (5) Detached Breakwater.

Parallel Jetties. The system of forming the entrances to harbours on sandy coasts with partly open parallel jetties has been given a long and frequent trial; but it cannot be regarded as successful in itself, for the foreshore always progresses to the extremities of the jetties, and a bar tends to form just beyond the pierheads. The extension of the jetties has been abandoned as useless; and a fair channel can only be maintained inside the jetties by scour, and outside by sand-pump dredgers. The jetties merely serve to train the channel across the beach, and to direct the sluicing current; they do not allow the littoral current to pass unchecked, and so far as they are made open for this purpose they are unsuitable for shelter.

Satisfactory results have been obtained with solid jetties when an extensive area inside is available as a reservoir for tidal scour, as at Malamocco and Kurrachee, and when fairly deep water can be reached.

Converging Jetties. The method of solid converging jetties has proved more uniformly successful than the parallel jetty system. It interferes less abruptly with the general line of coast; it enlarges the capacity for tidal scour; it concentrates the scour at the entrance; and it affords area for shelter and dredging, as well as for the passage of vessels. The permanence of the depth at the south entrance to Sunderland, and at Ymuiden, and the improvements effected at Dublin and Charleston, all attest the general efficacy of the system. It is true that at all these places an advance of the shore has occurred; but its progress seems to have ceased at Ymuiden and Sunderland, it is very gradual at Dublin, and will probably prove equally so in other instances. Even at Port Said, which it was feared might be imperilled by the alluvial drift from the delta of the Nile, and which is undoubtedly exposed to a turbid current from the west, the advance of the western foreshore has become very slight; and though no tidal scour exists for maintaining its outlet, dredging has been adopted as a substitute.

Closed Harbour on a Sandy Coast. The projection of solid breakwaters at right angles to a shore where a littoral drift of sand is known to exist might appear a somewhat unwarrantably rash proceeding; and it is therefore not wonderful that Mr. Parkes' scheme at Madras was strongly opposed, and that MM. Stœcklin and Laroche drew up an elaborate report in support of their scheme for a large harbour at Boulogne, now in course of construction[1]. Such a plan can, indeed, only be justified under certain conditions. If the proposed site is in the recess of a bay, if deep water cannot be reached at the entrance, and if the drift along the shore is considerable, the formation of a closed harbour can only result in failure, the silting up of the bay would be hastened, and sand would come in through the entrance. If, however, the solid breakwaters can be carried far enough out to place the entrance in deep water, if a scouring current exists at that distance out, or if the harbour forms a sort of projecting point, it is possible that the harbour may prove successful. In such a case sand does not come in through the entrance, as waves do not stir up sand in deep water; the current round the extremity assists in maintaining the depth; and, owing to the depth which the advancing foreshore has to fill up, and the projection from the shore which it has to assume, it may either take a long period in reaching the end of the breakwater opposed to the littoral drift, or previously attain a position of equilibrium.

The projecting position of the new harbour at Boulogne,

[1] 'Ports Maritimes.' MM. Stœcklin et Laroche, 1879, p. 14, and plate 2.

and the current which is found along the site of the outer breakwater, are relied upon by the engineers for securing the maintenance of the harbour. The sand has already progressed along the outer face of the west breakwater, in course of construction; but it is believed that the advance will be slow, and that the depth at the entrance will be maintained.

The Madras breakwaters extend into a depth of 45 feet, which prevents wave-driven sand from entering the harbour, as waves do not stir up the bottom at that depth; and the increasing depth will check the advance of the sandy foreshore. Judging, however, from the accumulation of sand which has already occurred on the south side of the harbour (Plate 4, Fig. 12), it seems probable that the advance of the shore line to the extremity of the south breakwater will be only a question of time, and that at some future period the breakwater will have to be extended to maintain the harbour.

This system, accordingly, of constructing harbours on sandy coasts with breakwaters built straight out from the shore, can only be adopted when the conditions of the site enable the advance of the foreshore to be disregarded, and when they are not favourable to the introduction of sand through the entrance.

Breakwater connected with the Shore by an Open Viaduct. On coasts where a littoral drift exists, it has been seen that any solid structure, interfering with the currents, produces an advance of the shore line. The interference with the littoral currents may be much reduced by constructing an open viaduct from the shore, and placing the protecting breakwater parallel to the coast. This method of forming a harbour on a sandy coast, where an accumulation of sand was apprehended, has been partially carried out at Rosslare near Wexford (Fig. 20), and has been proposed, on a more extended scale, for a harbour at Port Elizabeth in Africa. The viaduct, however, does check the littoral drift to a certain extent; and, accordingly, some advance of the beach has occurred at Rosslare. Moreover, the breakwater, if simply constructed parallel to the shore, does not afford ample shelter, except from on-shore winds; and the area thus sheltered is comparatively small. Accordingly, in this type of harbour, tranquillity is somewhat

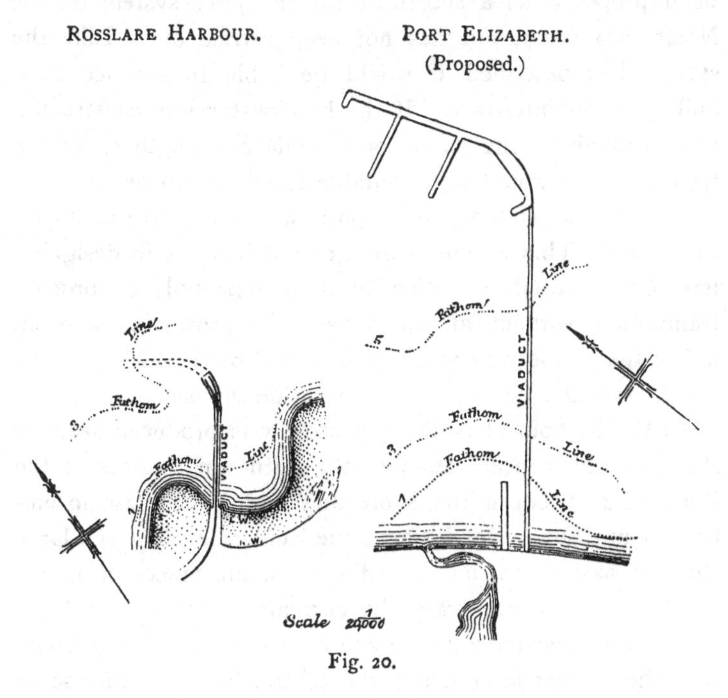

Fig. 20.

sacrificed for the sake of maintenance. This kind of harbour is not well adapted for commercial purposes, unless a supplementary inner harbour, or port, is provided, as the breakwater would be useless as a quay, on an exposed coast during a storm, with waves striking directly against its face; and with winds from other quarters, there would be an absence of adequate shelter, except in a land-locked bay.

Detached Breakwater along a Sandy Coast. The proposal of a breakwater parallel to, and at some distance from the shore, for affording shelter on a sandy coast without interfering with the littoral currents, resembles the design just described, with the exception of the absence of a connecting viaduct from the shore. The omission of the viaduct eliminates that source of interference with the currents, but it prevents access to the breakwater from the shore. The plan has been proposed as a substitute for the jetty system on the North Sea coast, but has not been carried out. Like the system just described, it would be liable to sacrifice tranquillity to maintenance. If the breakwater was constructed close enough to the shore, or of sufficient length to ensure tranquillity, it would be favourable to silting up, as the sand would have a tendency to deposit where not stirred up by the waves. This is one of the great difficulties in designing harbours exposed to silting, as it is impossible to procure tranquillity without at the same time promoting deposit, unless the silt-bearing waters are excluded, in which case the silt accumulates outside instead of inside the harbour.

At the harbour of Niewediep, scour was produced near the shore by the concentration of the outflowing waters of the Zuider Zee between the shore and a detached fascine embankment; but in this instance the Zuider Zee acts as a large sluicing basin, and, under ordinary circumstances, it is not practicable to concentrate the currents along the coast between a breakwater and the shore; and the tendency would be rather for the intervening channel to silt up, than for scour to be produced.

Proposals had been made in the case of both Calais and Madras to form a harbour entirely in deep water, sheltered by one or two breakwaters enclosing a circular or oval area, and connected with the shore by a viaduct. Such a structure would interfere as little as practicable with the littoral drift, and would furnish a more efficient shelter than a single breakwater parallel to the coast; but, as the harbour would have to be surrounded on all sides by breakwaters situated in deep water, it would necessarily be very costly in proportion to the sheltered area.

General Remarks. The parallel jetty system was very suitable when large receptacles for tidal water existed inland, as in the 17th century at Calais and Ostend. These tide-covered marsh lands, however, silt up and become useless for scour in process of time. The artificial sluicing basins cannot be given adequate capacity to maintain the lengthened channel, and dredging has to be resorted to outside. The only two instances named, where a parallel jetty channel is maintained by scour, are Malamocco and Kurrachee, where extensive lagoons still exist. Solid converging jetties are superior, in many respects, to the narrow parallel jetty channels of the North Sea coast, and appear to be the system best adapted for coping with the difficulties of such a position. This type somewhat approximates to closed harbours like Madras and Boulogne; but its piers present less abrupt obstruction to the littoral drift, and concentrate the scouring current at the entrance. Closed square harbours enclose a greater space than the converging jetties, with a similar length of breakwater; but, as they favour the progression of the foreshore, they can be only introduced in special cases. The Rosslare type of breakwater (Fig. 20) is placed beyond the influence of the littoral drift, but it is deficient in providing adequate shelter. Detached breakwaters at the entrances to bays afford excellent shelter without interfering with the currents in cases like Plymouth and Cherbourg; but along the straight coast of the North Sea, they would require to be long to provide shelter from various quarters, and the wave-driven sand, with which the sea is charged in these situations, readily deposits its burden on reaching still water.

CHAPTER XVIII.

LIGHTHOUSES; BEACONS; BUOYS; AND REMOVAL OF SUNKEN ROCKS.

Lighthouses: Materials; Elevation. Rock Lighthouses; Form of Towers; Height, Foundation, Methods of building; Rate of Progress and Foundations; Internal Construction; Cost; Illumination; Fog Signals. Screw-Pile Lighthouses. *Beacons:* Wooden; Iron; Masonry and Concrete. *Buoys:* Ordinary; Bell; Whistling; Light-giving. *Removal of Sunken Rocks:* Instances; Blasting operations at Hell Gate; Explosives; Compressed-air Diving Bell.

LIGHTHOUSES.

IN order to mark at night the sites of outlying rocks, of projecting headlands, and the entrances of harbours, as well as to guide the course of the mariner and indicate his position, it is necessary to provide erections from which a light can be exhibited from sunset to sunrise. These lighthouses are built in the shape of a tower, generally conical in form, with a lantern at the top containing the lamps. Timber, stone, iron, and of recent years concrete, have been employed for their construction; but the hardest procurable stone is preferred for such structures, where durability rather than economy has to be considered, especially when exposed to the sea. Timber is liable to decay, and to catch fire, as actually occurred at Rudyard's Eddystone lighthouse in 1755; and iron corrodes under the influence of salt water. Many lighthouses are built on the mainland, and therefore present no features of interest to the marine engineer; and, frequently, from the elevated site which they occupy, the tower is given a very moderate height. When they have to be placed on low land a considerable elevation is sometimes given to the tower in order that the light may be visible at a greater distance. One of the highest of these towers is situated on the low coast at Barfleur, to the east of Cherbourg; another stands on the northern point of Ré Island, off the west coast of France, having a height of 162 feet from the ground to the lantern gallery; and the iron lighthouse erected on the Roches-Douvres reef, to the west of Jersey, is 167 feet high from the base of the tower to the lantern gallery.

Some of the lighthouses erected upon islands above the sea level, of which two instances have just been given, present little more difficulties in construction than those erected on the mainland, beyond such as may be caused by the absence of suitable materials and appliances which have to be landed on the island. Distance from the mainland, and difficulties in landing are the main obstacles in such cases, and were the causes of the preference for iron being given in the construction of the Roches-Douvres lighthouse.

Rock Lighthouses. The really serious difficulties in lighthouse construction arise when a lighthouse has to be built on an isolated rock in the sea, at a distance from the mainland, where no shelter for the workmen can be provided, where the materials have to be landed on the rock as they are required, and more especially when the rock is covered by the tide, and the approach is difficult, uncertain, and even dangerous at times.

The important considerations for these lighthouses are the level of the rock in relation to the sea level, the extent of the rock and its hardness, the exposure of the site, and its distance from the nearest suitable port for the conveyance of materials.

In many cases the rock forming the foundation for the lighthouse is below high water, of which all the illustrations given are instances. (Plate 9, Figs. 1 to 9.) In some places the rock rises very little above low-water level, and is restricted in area, as at Bell Rock, Minot's Ledge, Des Barges, and Ar-men; whilst at Spectacle Reef in Lake Huron, where there is no rise of tide, the rock is several feet below the surface of the water.

Unless the rock is hard, compact, and without fissures, it is inexpedient to erect a lighthouse upon it, as the tower increases the surf on the rock, and by its weight on a small base, and the shocks which it transmits to its foundations, would soon impair any unsound portions of the rock. The strokes of the sea against Smeaton's Eddystone lighthouse caused the shaking and undermining of the hard gneiss rock upon which it was built.

The exposure of the site frequently diminishes the advantage which the height of the rock above the sea level would otherwise confer. Thus the Dhu Heartach and Skerryvore lighthouses, though founded on rocks above the sea level, are so open to the Atlantic that the waves frequently break over the rocks, and render them almost as difficult of access as if the rocks were at a lower level in a more sheltered position. Occasionally, the initial landing on the rock has proved one of the most difficult of the operations; for at Tillamook lighthouse, in the Pacific Ocean, the rock, though rising 120 feet above the sea, is so precipitous and surrounded by a belt of surf, that a life was lost in the first attempt to reach the rock, and special apparatus was provided for effecting the first landings after a rope had been secured to the rock by the aid of a surf boat. Frequently considerable exposure is combined with other disadvantages, as at Bishop Rock, Wolf Rock, Eddystone, and Ar-men lighthouses.

The distance of the depôt where the materials are prepared, and from whence they have to be conveyed, is frequently considerable. Thus the Eddystone is about 13 miles distant from the place on the Lairn within Plymouth Sound, selected for the workyard; the Wolf Rock is 17 miles from Penzance; the Dhu Heartach is 14 miles from Earraid on the coast of Mull;

the Bell Rock is 11 miles from Arbroath; the Skerryvore is 12 miles from Hynish on the island of Tyree; whilst Stannard's Rock lighthouse, in Lake Superior, is 51 miles from the place which served for its depôt in Huron Bay; and the Great Basses lighthouse is 80 miles from Galle Harbour. These long distances between the site of the work and the workyard enhance the difficulties and period of the construction, as they necessitate a longer continuance of favourable weather for the transport, as well as the landing and erection of the materials, and therefore considerably reduce the available opportunities for actual work.

Form of Rock Lighthouse Towers. The shape given to rock lighthouses is generally that of a conical frustum, with a curved instead of a straight batter. The circular section has the advantage of opposing the least resistance to the sea and the wind; whilst the increased curvature towards the bottom distributes the weight on a broader base, gives additional strength to the structure where it is most exposed to the shock of the waves, and increases the stability of the tower. The batter adopted in different cases includes four varieties of conical sections. For instance, an elliptical batter has been adopted at the Wolf Rock and the new Eddystone; a hyperbolic batter at Chicken's Rock and Skerryvore; a parabolic batter at Dhu Heartach; whilst straight batters have been given to Ar-men and Minot's Ledge. The curved outline, however, possesses the disadvantage of tending to make the waves run up the structure, occasionally enveloping the lantern in spray and eclipsing the light, as at the Bell Rock lighthouse; whilst at Smeaton's Eddystone, where the height was less and the exposure greater, the waves in a storm frequently ran up to the cornice of the lantern gallery, where, breaking into spray, they rose above the top of the tower and descended in a heavy shower on the lantern. To obviate this inconvenience, Sir J. Douglass, whilst retaining a curved batter for the new Eddystone lighthouse, has placed the tower upon a cylindrical base raised to a level of $2\frac{1}{2}$ feet above high water. The waves are divided and checked by this base; and the blow, though heavier than when the wave runs up the tower, mainly falls upon the portion of the structure which is best able to bear it.

The straight batter was adopted at Minot's Ledge, on account of the small area available for the foundations, which precluded the enlargement of the base produced by a curved batter. At Ar-men, the preference for a similar batter was probably due to the difficulties attending the foundations, and the expediency of ensuring the stability of the small stones of which the base was composed by placing the weight of the structure directly upon them. Huge packs of ice are the chief forces to be resisted at Spectacle Reef; and they are best resisted by a straight batter, as a curved batter would tend to pile up the ice.

Height of Rock Lighthouses. It will be observed by a reference to Plate 9, where the low-water line has been drawn to a uniform level, that the heights given to some of the most important lighthouses vary considerably. A high tower cannot be erected on a limited base without impairing the stability of the structure, which must in any case be absolutely secured. The stability varies with the exposure; but, generally, a high light is most needed in exposed situations at a distance from the coast.

Sometimes a lighthouse is only required to mark a dangerous reef, which lies near the course of passing vessels, not far from the coast; and for this purpose a moderate elevation suffices, of which Spectacle Reef, Minot's Ledge, and Des Barges lighthouses are instances. Where, however, a lighthouse stands upon an outlying rock, at a distance from the coast, or away from the range of other lighthouses, it serves not merely as a warning to vessels of a particular danger, but also to inform them of their position and to guide their course; and in such cases a high light that is visible from a long distance is very advantageous. Bishop Rock, Wolf Rock,

Eddystone, Bell Rock, and Ar-men lighthouses answer this double purpose, and have been, accordingly, given a greater height. The new Eddystone is the highest of the immersed rock lighthouses, having its focal plane 133 feet above high water, as compared with 72 feet in Smeaton's structure. (Plate 9, Figs. 5 and 6.) There are, however, rock lighthouses which have been raised even higher above the sea level, such as the Dhu Heartach which has the centre of its lantern 145 feet above high water, of which height, however, about 30 feet is due to the elevation of the rock on which it stands, and Skerryvore lighthouse whose tower is about the same height as the new Eddystone, but whose light is exhibited at a height of about 150 feet above high water owing to the greater elevation of the rock on which it is founded. In these latter instances, however, there would have been no difficulty in raising the tower to any desired height, owing to the extent of rock available for foundations; whereas when a rock is immersed at high water, the area that the foundations can be laid upon is generally very limited, and any extension of the base involves considerable increase of expenditure, and an addition to the most troublesome portion of the whole work.

Foundations of Rock Lighthouses. The rocks upon which lighthouses are founded naturally consist of the hardest kinds of rock, as they are the portions of strata which have longest resisted the action of the waves. Thus the Eddystone rock is gneiss, and the Skerryvore and Ar-men rocks are of the same formation; the Dhu Heartach rock is hard trap; Minot's rocks are granite with seams of trap; Wolf Rock is porphyry; the Bell Rock and Great Basses rocks are hard red sandstone, and Spectacle Reef consists of limestone. The hardness of the rock is of the utmost importance in securing the stability of the lighthouse erected upon it, but it adds to the difficulties encountered in preparing the foundations, more especially as blasting has to be avoided for fear of its shaking the adjacent portions of the rock.

The first operations, after the exact site for a lighthouse has been determined, consist in removing unsound rock, and in forming level benchings to receive the foundation course. Holes are also bored in the rock, in which bolts or dowels are inserted to secure the bottom stones to the rock, which are then laid in cement. The stones of the lower courses of the tower are dovetailed together horizontally and vertically in most of the more recent lighthouses, such as the Hanois, Wolf Rock, Great Basses, and new Eddystone, so as to fasten each stone into the structure as tightly as possible. Joggles were also employed at the Great Basses for connecting the courses, and bolts at the Wolf Rock. At Ar-men, owing to the peculiar difficulties of the site, rubble stone set in cement, built round a series of iron bolts sunk vertically in the rock, was adopted for the foundations; and the work above was connected together by bolts and iron ties.

The diameter of the base depends partly upon the available area, and partly on the height or design of the tower. The Great Basses and Bréhat lighthouses have both been given wide bases, 32 feet and 45 feet respectively at the bottom, and rising about 50 feet above high water, upon which slighter towers have been erected. In most cases, however, the tower rises straight from the foundations. The diameter of the base of the Bell Rock tower is 42 feet; and the Skerryvore, the Chicken's Rock, and Wolf Rock lighthouses have the same base. The base of the Dhu Heartach lighthouse is 36 feet in diameter, of the Bishop Rock 34 feet, of Spectacle Reef 32 feet, of Minot's Ledge 30 feet, and of Ar-men only 23½ feet; whilst the lower tower of Barges has been given a diameter of 39 feet at the base. The cylindrical foundation of the new Eddystone has a diameter of 44 feet, in place of the 26 feet of Smeaton's tower.

The greatest difficulties of the whole work are necessarily encountered in preparing the foundations, and laying the lower courses. The difficulties in landing, the unfrequent combination of the favourable conditions of a low tide and a calm sea, and the short periods that, in any case, can be spent upon a rock which is covered at high tide and is exposed to almost constant surf, render the earlier operations uncertain and slow. The greatest care, moreover, has to be taken to execute the work with the utmost precision; but this is facilitated, after the foundations have been prepared, by each stone being dressed to the exact shape at the workyard, and its position in the work indicated by a number.

Methods of building Rock Lighthouses. When there is a sufficient space on the rock, and more especially when the rock is above high-water level, a temporary barrack has been sometimes erected near the site of the proposed lighthouse to serve as a place of shelter for the workmen, from which they can easily go to their work on every favourable opportunity, and in which they can take refuge when the sea drives them from the rock. This expedient was adopted for the construction of the Bell Rock lighthouse, where, though the rock was covered at high water, a beacon was erected, on open timber staging, out of reach of the sea, and served as a barrack for the men, being placed near enough the site of the tower for a temporary bridge to furnish a means of communication between the barrack and the structure as soon as the tower had been raised nearly up to the level of the barrack. A similar plan was subsequently employed for facilitating the construction of the Skerryvore and Dhu Heartach lighthouses, where the rock in each case rises above the level of high water.

An attendant steamer is needed to bring the materials from the workyard, and this steamer sometimes serves also as a floating barrack for the workmen, being moored as near the rock as is consistent with safety. At the Great Basses lighthouse, two steamers were employed for supplying the materials, which had to be brought 80 miles from Galle; and a light-ship provided both accommodation for the workmen, and also a light to warn passing vessels. In constructing the Wolf Rock lighthouse, a steam tug and five barges were employed for bringing materials to the rock, and another vessel was specially fitted up as a barrack for the men. The building of the new Eddystone was accomplished with the aid of one of the steamers which had been built for constructing the Great Basses lighthouse.

Where there is available space on the rock, a landing platform is sometimes built out from one side of the lighthouse, on which a derrick is hoisted for landing the materials; and this stage facilitates the access to the rock, both during construction, and after the completion of the lighthouse. This method was adopted at the Great Basses, Wolf Rock, and Ar-men lighthouses. When the available area of the rock is very limited, the derrick has to be erected in the centre of the structure itself, a plan followed at the new Eddystone lighthouse, where the central traversing iron jib received the stones, landed by a wooden jib, and deposited them in place. The stones were landed from the vessel by steam power in about three minutes.

Rate of Progress and Foundations of Lighthouses. The progress achieved in building rock lighthouses is reckoned by the number of landings effected in a season, and the number of hours of work accomplished. This necessarily depends upon the level of the rock, and the exposure of the site. The opportunities for work are generally very limited at first; but, as the tower rises, the progress becomes more rapid, till at last the workmen can remain continuously on the tower.

At Ar-men, where the rock is very low, and the exposure very great, being in the Bay of Biscay and open to the Atlantic, only 8 hours work were accomplished, in seven landings, during the first season in 1867; and although the available time increased during the following seasons, the hours of work in 1873 were reduced to 15½, in only six landings; and it was not till 1875 that the hours of work exceeded one hundred in a single season; and the work was only completed in 1880. The work was, however, one of unusual difficulty; and the first holes in the rock were bored by fishermen, who worked with life belts round their waists owing to the almost incessant surf. (Plate 9, Fig. 9.)

Two seasons were occupied at the Bell Rock in preparing the foundations and raising the masonry to the level of the rock, and two more sufficed to finish the whole tower, the work having been commenced in 1807, and completed in 1810. (Plate 9, Fig. 4.)

The Skerryvore lighthouse, which is considerably higher and more exposed than the Bell Rock, though founded above high-water level, was only actually commenced in 1840, two previous seasons having been occupied in preparations on the rock, and it was completed in 1843.

Bishop Rock lighthouse was commenced in 1852, and completed in 1858. (Plate 9, Fig. 7.)

During the first season of the construction of the Wolf Rock lighthouse, in 1862, 83 hours of work had been effected in twenty-two landings, which during the next season were increased to 206 hours in thirty-nine landings, much more satisfactory work than at Ar-men; but, nevertheless, it was not till 1864 that the first stone of the tower was laid. The tower was completed in 1869. (Plate 9, Fig. 3.) The greatest number of hours of work accomplished at the Wolf Rock, in a single season, were 313 with forty landings, in 1867; as compared with the best working season at Ar-men of thirty landings, and 261 hours work, in 1877.

The great advantage of building on a rock raised several feet above high water is fully exemplified by the case of the Dhu Heartach lighthouse, even where the site is very much exposed, the landing difficult, and the workyard some miles distant. The base of the Dhu Heartach tower stands 32 feet above high-water mark; sixty-two landings were made upon the rock during the first season's work in 1870, and the tower was raised 48 feet above the rock. The masonry of the tower was completed in 1871; so that only two seasons were occupied in constructing the tower, though the lantern and optical apparatus were not put up till 1872.

The Great Basses lighthouse, like that at Skerryvore, is founded on the rock a few feet above high-water level; but the preparations of the foundations were impeded by the constant wash of the sea over the rock. This was provided against by the construction of a brick dam set in Medina cement, 2 feet thick and 3 feet high, which protected the workmen from the surf. The lighthouse was commenced in 1870, and constructed in two and a half seasons, as many as 679 hours of work having been accomplished in one season.

Minot's Ledge lighthouse, which stands on a rock of small area, and close to low-water level, was commenced in 1855; and three seasons were employed in preparing the foundations. The space was too limited, and the sea too rough, to admit of the construction of a cofferdam; though small portions of the area were protected for a time with dams formed of sand bags. An iron scaffold was erected across the rock, which afforded protection to the men. From 130 to 157 hours of work were done upon the rock in the first three seasons; and six courses of stone were laid during the fourth season, when the working hours rose to 208. The working time amounted to 377 hours in 1859; and the work was completed in 1860. (Plate 9, Fig. 2.)

Spectacle Reef lighthouse, which was commenced in 1870 and completed in 1874, had to be executed under peculiar conditions, for the surface of the rock was not less than 7 feet under water, and the actual foundation of the lighthouse was 12 feet below the water-level. A square cribwork protection was first formed round the site of the lighthouse, 22 feet wide, enclosing an area of 48 feet square. This served for a landing stage, and a site for quarters for the workmen, as well as for sheltering the space inside. A cofferdam, having an external diameter of 36 feet, was erected within this sheltered area; the water was pumped out, and the bottom levelled; and the tower was then built up.

A similar cribwork protecting pier was adopted at the Stannard's Rock lighthouse in Lake Superior, where the rock was about 12 feet under water at the site selected. In this instance, however, a wrought-iron casing was placed round the site of the foundations, 62½ feet in diameter, and was fitted closely to the bottom, and made watertight by aid of a canvas pipe filled with oakum under its bottom edge. The water was then pumped out; the rock was cleaned; and concrete was deposited within the casing, to form a foundation for the tower, to a height of 35 feet. The protecting pier was sunk in its place in 1878; the iron casing was fixed in 1879; and a portion of the concrete was deposited daily during about ten weeks. The cylindrical concrete base was completed in 1880; and the tower, 78 feet in height, was built in two months in 1881; and the lantern was erected, and the light exhibited in 1882. Ice drifts were the chief impediments to the work; they rose 18 feet above the concrete base in 1882, or 41 feet above the water-level of the lake, and were only cleared away near the end of May.

This is probably the first instance of a concrete foundation being adopted for a rock lighthouse; though concrete-in-mass had been previously used as the material for the Corbière lighthouse, above high-water level, close to the south-west coast of Jersey, and has also been employed for the lighthouse on Newhaven pier.

The new Eddystone lighthouse affords an instance of very rapid progress in the construction of a submerged rock lighthouse. The former lighthouse tower was built in three years, a rapid rate considering the early period (1756–1759) at which it was executed; but the rock at that site rose entirely above low water, and a portion emerged at high water. At the new site, a large portion of the foundations has had to be laid below low-water level; the base is 44 feet in diameter, as compared with 26 feet in Smeaton's lighthouse, and the height of the tower has been doubled. (Plate 9, Figs. 5 and 6.) The work was commenced in 1878; the first stone was laid in 1879, about a year after the preliminary works had been begun; and during this first complete season of work, as many as 518 hours work were accomplished in one hundred and thirty-one landings. The cylindrical base was completed in the following year; and the tower was completed in 1881, within three years after the first landing on the rock. The lantern was erected the same year; and the optical apparatus was set up, and the light exhibited in 1882. This rapidity in construction, considering the low level of the foundations, their extent, and the size of the tower on an exposed site, must be attributed, firstly, to the erection of a platform over the site, 10 feet above low water, on which the workmen could land in readiness to commence work directly the tide had fallen low enough; secondly, to the formation of a dam of bricks laid in Roman cement, 7 feet high, round the site of the foundations, out of which enclosure the water was pumped by steam power from the attendant steamer, enabling the men to work for a much longer period during each tide; thirdly, to the use of rock-drills worked by compressed air for preparing the foundations; and lastly, to the rapid manner in which the stones were landed by aid of the steam engine on the steamer[1].

Internal Construction of Lighthouses. When lighthouses are built out of reach of the waves, the buildings for the light-keepers and stores are erected near the tower; and the tower serves merely as a pedestal for the light, and a means of access to it by an internal winding staircase. In the case, however, of rock lighthouses, founded within reach of the

[1] 'Minutes of Proceedings, Inst. C. E.,' vol. lxxv, p. 29.

waves, it is necessary to construct the store-rooms and dwellings for the light-keepers in the tower itself. To ensure stability in exposed situations, the tower is carried up solid for some distance above high water; and the rooms are then built one over the other up to the lantern, with a staircase connecting them. The height to which the solid portion is raised varies according to circumstances, as may be seen by a reference to Plate 9. The old Eddystone was carried up solid to 10¼ feet above high water; Great Basses 12 feet; Bell Rock 14 feet; the new Eddystone and Ar-men 15 feet; Wolf Rock 16 feet; Chicken Rock 21 feet; Bishop Rock 23 feet; Skerryvore 30½ feet; Minot's Ledge 32 feet; and Dhu Heartach 64⅓ feet.

The rooms are generally given the same diameter; so that the walls are gradually reduced in thickness, as they ascend, by the batter on the outer face. Sometimes, however, the lowest rooms are made smaller, affording an additional width to the walls, as at the Wolf Rock and new Eddystone. The winding staircase is usually put inside each room next the wall; for though the separate shaft, adopted at Ar-men and Bréhat, shuts off each room, it reduces considerably the size of the rooms.

Cost of Lighthouses. The expense incurred in erecting lighthouses necessarily varies considerably with the conditions of the site, and the distance from which materials have to be conveyed; but the following table, drawn up by Sir J. Douglass, gives the actual and relative costs of some of the most important rock lighthouses:—

Lighthouse.	Total Cost.	Cost per Cubic Foot.		
	£	£	s.	d.
Eddystone (Smeaton)	40,000	2	19	11
Bell Rock	55,620	1	19	0
Skerryvore	72,200	1	4	8
Bishop Rock	34,560	0	19	7
Smalls	50,125	1	1	7
Hanois	25,296	1	0	7
Wolf Rock	62,726	1	1	3

Lighthouse.	Total Cost.	Cost per Cubic Foot.		
	£	£	s.	d.
Dhu Heartach	72,585	1	14	6
Longships	43,870	0	18	5
Eddystone (Douglass)	59,255	0	18	2

to which may be added :—

Great Basses	63,560	1	6	7
Minot's Ledge	62,500	1	17	2
Spectacle Reef	78,125	1	16	2
Ar-men	36,000	1	2	1

Lighthouse Illumination. Four sources of light are employed for lighthouses; namely, electricity, gas, mineral oil, and colza oil. The electric light possesses great intensity; but in foggy weather, owing to the comparatively large proportion of blue and violet rays in its composition, it does not possess as much penetration per candle unit as the other lights; though its power is so great as to more than compensate for this deficiency. It is not, however, applicable to rock lighthouses, owing to the machinery and coal necessary for its production.

Gas gives a very good and cheap light, but its manufacture cannot be undertaken in rock lighthouses. Colza and mineral oil are about equally efficient as illuminants, whilst the latter is the cheapest. Mineral oil is, accordingly, employed for lighthouses on shore where oil is adopted; but colza oil is used for all rock lighthouses, on account of its greater safety.

The light from the lamp is directed in a series of parallel rays towards the horizon, by carefully adjusted tiers of prismatic lenses encircling the light. The progress that has been made in lighthouse illumination is well indicated by a comparison of the lights exhibited at the old and new Eddystone. The first light at the old Eddystone consisted of twenty-four tallow candles, unaided by any optical apparatus, and having a total power of 67 candle-units. Latterly, the lamp in the old lighthouse gave a light having an intensity of about 6850 candles; whilst the light exhibited from the new lighthouse can be raised in foggy weather to an intensity of 159,600 candles, the ordinary light being 37,800 candles; and the cost of this light is only ½*d.* more per hour than the estimated cost of Smeaton's light.

Formerly, different lighthouses were distinguished, either by simple revolving lights having a distinct period of revolution, or by introducing a red beam of light between the white, or, in the case of the Casquets off the island of Alderney, by having three distinct lights on the same rock. At the present time the distinction is more simply and effectively made by producing a definite number of flashes in quick succession, followed by a certain period of obscuration; and by varying both the number and rate of the flashes, and the interval between the groups, each light can be furnished with a distinctive, but simple character, which enables the sailor to recognise the particular light.

Fog Signals at Lighthouses. When foggy weather obscures the light of a lighthouse, it becomes necessary to substitute an audible signal for warning sailors and indicating the position of the lighthouse. Fog signals are made, either by ringing a bell, or blowing a horn, a whistle, or a trumpet. Fortunately, the atmosphere is most favourable for the transmission of sound when it is foggy, and consequently when audible signals are most needed. Windy weather is unfavourable for the transmission of sound; whereas fogs more commonly occur with a still atmosphere. Also fine and hot weather renders the atmosphere opaque to sounds, owing to the air being made an unhomogeneous medium by the variable evaporation from the sea.

The range of sounds is, however, somewhat limited, as it is reduced much more rapidly than in the ratio of the inverse square which applies to light; and the practical limit is about 6 nautical miles for the most powerful sounds, as the power needed to extend this limit increases very rapidly. The range of a sound depends upon its intensity, its pitch, the state of the atmosphere, and the force and direction of the wind. M. Allard considers that the intensity of a sound may be considered as proportional to the power producing it; and some idea may be formed of the expenditure of power needed to augment the range of sounds, from the fact that, in M. Allard's investigations, a small bell, which was rung by a power of 2⅛ foot-pounds, had a range of 1·17 miles; whereas the blast of a siren trumpet, produced by the expenditure of 8680 foot-pounds, had a range of only 5·87 miles[1].

Bells have generally been placed on English rock lighthouses, as for instance at the Bell Rock, which was provided with two bells, at Skerryvore, Dhu Heartach, and Wolf Rock; whilst the new Eddystone has two 2-ton bells, on opposite sides of the cornice, so that the windward bell may be sounded during a fog. A siren trumpet has been adopted at Ar-men. Two steam sirens were erected at Tillamook Rock lighthouse; but in this case the area and height of the rock enabled suitable buildings for their reception to be erected at the base of the tower; whereas deficiency of space is usually the obstacle to their introduction in the ordinary rock lighthouses. They are the most efficient of all fog signals, but, as has been pointed out, necessitate a large expenditure of power to attain their maximum efficiency.

A distinctive character can be given to the sonorous signals of each station by arranging the series of strokes of the bell, or blasts of the siren, and the interval between each set of sounds, just as in the flashing combinations.

The variations in the sounds can also be multiplied by adopting two notes of different pitch, which are preferable to a longer blast, owing to the less expenditure of power.

Screw-Pile Lighthouses. When a light is required on a sandbank, where no foundation for a solid structure can be obtained, Mr. Mitchell's system of iron screw piles has been advantageously adopted.

[1] 'Annales des Ponts et Chaussées,' 6th Series, vol. v, p. 593.

The piles are screwed down till they reach a hard bottom, or till they have been sunk a sufficient depth in the sand to support the superstructure for the light. The first instance of the adoption of this system was for the Maplin Sands lighthouse, in 1838, in the Thames estuary; and a similar lighthouse has been placed in a more exposed situation, on the sandy beach to the north-east of Calais, near low-water mark, which is known as the Walde lighthouse. (Plate 9, Fig. 11.)

The first screw-pile lighthouse off the American coast was built in 1847–50, on a shoal near the entrance to Delaware Bay, in a depth of 6 feet at low water. Owing, however, to its exposure to drifting ice, it has been necessary to encircle it with a separate screw-pile barrier acting as an ice-breaker.

This system has been extensively adopted in the United States for marking shoals in the great bays and entrances to harbours off the southern coast: over fifty such structures have been erected, and some exhibit lights of the first order.

The slender round piles present little obstruction to the waves; and the structures are well adapted to sandy shoals in moderately sheltered situations. In the open sea, however, moored light-ships have to be resorted to for warning vessels off dangerous shoals.

BEACONS.

Secondary shoals are sometimes indicated at night by a beam from an adjacent lighthouse. This method has been adopted for the Hard Deeps shoal, 3½ miles from the Eddystone lighthouse, which is marked by a beam from a low fixed light in the tower. Submerged rocks in the course of vessels, or near the entrances to harbours, not sufficiently important to be marked by a light, are indicated during the day by a beacon.

Wooden Beacons. The simplest form of beacon is a

wooden pole, supported at the base by a mound of stone, and having a large ball at the top to mark it more clearly. These beacons are used to indicate the sides of the channel in an estuary, and possess the advantage over buoys of not being shifted in position by the rise and fall of the tide and the change in direction of the current. As the navigable channel in an estuary is liable to shift, it is necessary to have simple marks which can be readily moved.

Where a more permanent structure is required, wooden staging, pyramidal in forms, somewhat resembling wooden dolphins, form a cheap sort of beacon; but these wooden erections are only suitable for sheltered situations.

Iron Beacons. A structure consisting of iron tubes or bars sunk into the rock, and well braced and tied above, and exhibiting a painted ball at the top, forms a strong stable and fairly durable beacon, resembling a screw-pile lighthouse in presenting little surface to the waves. (Plate 9, Fig. 10.) The illustration represents a beacon erected on the Antioche Rock, to the north of Oleron Island, off the west coast of France.

Masonry and Concrete Beacons. The most durable form of beacon consists of a conical frustum of masonry, brickwork, or concrete, securely bedded in the rock. A common form of masonry beacon is shown in Plate 9, Fig. 13.

The base of the Lavezzi beacon, erected on a submerged rock in mid-channel between Sardinia and Corsica, is composed of cement concrete deposited in mass on the rock within timber framing, and further secured to the rock by twelve iron dowels sunk into the rock and embedded in the concrete. (Plate 9, Fig. 12.) The actual beacon consists of masonry set in cement.

Access to a beacon is provided by an iron ladder fixed in the masonry; and the beacon is frequently painted on the face, sometimes with bands of different colours, to make it more distinctly visible. These beacons, from their smaller elevation, offer less resistance to the waves than lighthouses; their base is larger in proportion to their height, and they are generally near the coast. Nevertheless, beacons are sometimes placed in positions of great exposure, as exemplified by the Charpentier beacon, at the mouth of the Loire in the Bay of Biscay, which, as mentioned in Chapter II, was injured by a storm, though constructed with the greatest care.

BUOYS.

Buoys are made solid or hollow of wood, or hollow of iron; but solid wooden buoys are liable to get waterlogged, and hollow ones decay, so that iron buoys are now generally preferred. The ordinary type of iron beacon buoy adopted in France is shown in Plate 9, Fig. 14; and a similar type of buoy, formed of wooden staves hooped with iron, is shown in Plate 9, Fig. 22. A large iron buoy, placed at the mouth of the Loire for mooring large steamers, is shown in Plate 9, Fig. 19. A cask buoy (Plate 9, Fig. 21), made either of wood or iron, is frequently employed for the moorings of vessels; whilst a can buoy (Plate 9, Fig. 20) is commonly used for indicating the fairway of a channel, or the limits of the channel on either side, in which case the buoys on one side are generally painted one colour, and on the other side another colour, being further distinguished by numbers on the top.

Bell Buoys. In order to indicate the position of buoys at night, or in foggy weather, different plans have been contrived. The oldest and simplest method is that of hanging a bell near the top of the buoy, so that the oscillation of the buoy, when moved by the waves, causes the ringing of the bell. The ordinary type of iron bell buoy used in France is shown in Plate 9, Fig. 16; another form is the boat bell buoy, which is more easily moored and more readily shifted, but is more expensive than the common type. (Plate 9, Fig. 18.) The objections to these buoys are that they require a considerable agitation of the sea to make the bell ring, and that they are useless during the night in calm weather.

Whistling Buoy. Another contrivance by which the motion of the waves causes a buoy to emit a sound, indicating its position, is the whistling buoy. (Plate 9, Fig. 17.) A pipe passes through the centre of the buoy, and dips deep enough in the water for the water-level in the pipe not to be affected by the motion of the waves, and the pipe is closed by a diaphragm a little above the water-line. When the buoy rises on a wave, air enters the pipe, through two small pipes, to fill the void produced in the pipe between the diaphragm and the constant water-level. As the buoy descends into the trough of the wave the air in the pipe is compressed, and, as the exit through the small pipes by which it entered is closed by valves, it is forced into a third small pipe terminated by a whistle which the air sounds in issuing. Thus a whistle is blown each time that the buoy descends on a wave. The strength of the sound emitted depends on the weight of the buoy, and the air space below the diaphragm. During some observations made on the French coast, the whistle was audible three times out of four at a distance of $2\frac{7}{8}$ miles, and twice out of four times $4\frac{1}{4}$ miles off. This type of buoy has been in use for some time on the coasts of the United States, and some have been stationed on the French coast. The cost of the buoy is £150.

Light-giving Buoy. Self-acting bell and whistling buoys have the disadvantage of emitting no sound in calm weather, and of having a very limited range in very windy weather. To remedy these defects, endeavours have been made to exhibit a small light on the buoy. The best result has been obtained by filling a reservoir in the buoy with compressed mineral oil gas, which is burnt in a lamp at the top of the buoy. (Plate 9, Fig. 15.)

One of these light-giving buoys was tested off Havre in 1881. The reservoir in the buoy, having a capacity of 13 cubic yards, was filled with gas under a pressure of 6 atmospheres, which supplied the lamp for nearly four months; and the light emitted was visible to a distance of 5 miles. The light was very rarely invisible at distances of from 1¼ to 2½ miles, and once invisible out of every three observations 2½ miles off. The buoy costs about £400. This system might be equally well applied to the illumination of beacons.

REMOVAL OF SUNKEN ROCKS.

Sometimes the capacity of a harbour, or the safety of its entrance channel, is impaired by the existence of a rock in the harbour or channel. Thus at the old port of Bastia, a partially submerged rock existed, which, though somewhat sheltering the inner harbour, rendered the approach dangerous, and prevented an extension of the harbour. Wells were sunk in the rock and holes bored, into which charges of gunpowder were placed and fired. Divers completed the work of removing the displaced or shaken rock, 5216 cubic yards of rock being thus blasted and removed between 1858 and 1868, at an average cost of £1 0*s*. 9*d*. per cubic yard. This operation, whilst rendering the port safe, enabled it also to be considerably enlarged.

Rocks have been lowered by blasting in Holyhead, Alderney, and other harbours, to improve the capacity and safety of the sheltered areas, the holes for the charges being bored by divers when below the water.

Blasting Operations at Hell Gate. The most extensive work for the removal of sunken rocks has been proceeding during several years for improving the channel of the East River opposite New York, where several extensive and dangerous obstructions impeded the navigation of the river. Not only were the sunken rocks dangerous in themselves, but, by narrowing the waterway, they gave rise to irregular and rapid currents which increased the difficulties of navigation. Some of these shoals had a considerable area needing lowering; thus Hallett's Point had an area under water of 3 acres, and Middle Reef, standing in mid-channel at Hell Gate, with a small backbone rising above high water, had an area of 9 acres.

The smaller isolated rocks have been lowered by aid of diving bells in which the miners worked.

Hallett's Point and Middle Reef are being lowered by sinking shafts, under shelter of a dam, down into the rock, and then driving parallel horizontal galleries in one direction, and another series at right angles to the first, thus forming a network of galleries, or mines, under the reef, which, after a set have been completed, are simultaneously exploded by electricity. The work was begun by mining under Hallett's Point in 1869; and a series of galleries having been completed in 1876, the mines, over 3600 in number, were fired by electricity, and the reef shattered; the only effect apparent at the surface of the water being an upheaval of the water about 3 feet over the site of the explosion, with here and there waterspouts rising about 20 feet in the air, and the water covered with foam. Though the disintegration of the rock was instantaneous, producing a distinct improvement in the channel, the actual removal of the débris has been a long operation, as dredging and grappling the loose rock was continued up to 1882. A depth of 26 feet at low water has been attained over the larger portion of the reef; and the works are in progress for lowering the whole area to the same level.

The boring of the galleries under the Middle Reef are being pushed forward rapidly by aid of rock drills; and it was anticipated that the mines would be ready for the explosion in 1884.

The smaller detached rocks are being bored by means of a steam drill boat, the rock being then blasted, and the pieces removed by grapple dredgers.

The total expenditure on this work, up to the middle of

1883, was £653,500; and the estimated additional cost for completing the work amounted to £409,000.

Explosives used for Blasting. Formerly gunpowder was the only explosive used for blasting under water; but now dynamite is largely employed, owing to its greater power; and by using a small quantity of fulminating mercury for the initial detonation, the whole charge can be exploded with certainty, even when the dynamite is in a frozen condition. By the employment of dynamite, the holes do not need to be bored so deep as for gunpowder, and very little tamping is necessary.

Compressed gun-cotton appears very suitable for blasting under water, as by aid of a detonating capsule it can be exploded when immersed in water, and, like nitro-glycerine or dynamite, its explosive force is manifested when merely in contact with the rock, so that drilling might be dispensed with, though confinement adds to the effect. Some uncertainties, however, with regard to its manufacture, and the fear of its spontaneous combustion, have checked its employment on a large scale.

Gunpowder and dynamite were both employed for blasting Hallett's Point reef.

Compressed-air Diving Bell. A new form of diving bell was designed by M. Hersent for lowering the Rose Rock in Brest harbour. It consists of a large iron case, 33 feet long, 26¼ feet wide, and 23 feet high; it is divided into two parts by a horizontal watertight diaphragm, the lower section forming the working chamber, and the upper portion giving flotation to the whole when it is desired to shift its position. A central shaft rises from the top of the working chamber, and passes through the watertight compartment up to a level higher than the ordinary high-water level, and is provided with a platform at the top to receive the workmen with their tools. The air-locks are situated at the bottom of the shaft; and a staircase inside leads from the lock to the platform.

Two smaller shafts, going from the working chamber to the platform, with air-locks on the top, furnish a passage for the excavated material. The bell is weighted with 200 tons of masonry, built at the sides of the working chamber and between the girders of its roof, and with 30 tons of pig iron placed in the working chamber, so that it may float in an upright position. The sinking of the bell is accomplished by letting water into the air compartment, above the working chamber, through valves in the side. The bell is raised by admitting compressed air into the upper compartment through valves in the bottom, which drives out the water.

The working chamber is like the ordinary type for compressed-air foundations, 6½ feet high; the water is expelled from it by compressed air, and it is lighted by electricity. When a charge is fired, the men retire into the two large air-locks, which can together hold thirty men. A small air-lock gives access to the chamber, without passing through the large air-locks which are usually kept open to the working chamber.

The first trials at Brest showed that a price of £1 18*s.* 3*d.* per cubic yard would enable the work to be executed profitably. The work at Brest, completed in 1880. consisted in excavating about 23,500 cubic yards of rock.

A similar bell is being used at Cherbourg for deepening the approach to the naval dockyard where the bottom is rock.

PART II. DOCKS.

CHAPTER XIX.

SITES AND PRELIMINARY WORKS FOR DOCKS.

Sites for Docks and Quays. Estuaries and Sea-coast. National Dockyards. Commercial Docks. Tidal Ports. Tideless Ports. Quays along Rivers. Situations suitable for Docks; low-lying lands; river bends; foreshores. Docks in old river course. General Considerations. *Preliminary Works for Docks.* Designs, varieties; jetties; outer basin. Borings. Cofferdams. Dams. Pumping Excavation, ordinary methods, excavators, dredging.

SITES FOR DOCKS AND QUAYS.

DOCKS are generally placed in well-sheltered situations. Low-lying lands along the tidal portions of rivers or estuaries are the most common sites for docks; for whilst natural protection is thereby secured, and the merchandise can be brought or shipped near important centres of commerce situated on the banks of rivers, easy access is also afforded to the sea. London, Liverpool, Hull, Bristol, Cardiff, Leith, Belfast, and Dublin (Plates 10 and 11), furnish examples of docks so situated in the United Kingdom; whilst Havre, Honfleur, St. Nazaire, Antwerp, Flushing, and Rotterdam (Plates 12 and 13), serve as similar instances of foreign works. Occasionally, however, docks are formed on the sea-coast, and small artificial harbours have to be constructed to provide a sheltered approach to the inner docks, as for instance at Hartlepool, Sunderland, Fleetwood, and Maryport.

National Dockyards. Naval dockyards may be formed wherever good natural shelter with an ample depth of water exists, like at Portsmouth, Chatham, Queenstown, Toulon, Brest, and Spezzia (Plate 12, Figs. 3, 4, and 5), or where the natural capabilities have been improved, as at Plymouth and Cherbourg (Plate 3, Figs. 5 and 8). The best sites in fact can be selected, provided they are fairly accessible, and afford space enough in some cases for the evolutions of men-of-war, quite irrespective of commercial considerations. Portsmouth with the Solent, Cork Harbour, Brest, Toulon, Spezzia, Cronstadt, and Sebastopol, are some of the finest natural harbours in the world, and, with the partly artificial harbours of Plymouth and Cherbourg, are the most important naval stations. Dockyards form necessary artificial adjuncts to these harbours, for the construction and repair of men-of-war. Commercial harbours are not generally found in close proximity to naval stations, as the imperative requirements of naval tactics interfere with the development of trade; though small commercial harbours exist at Plymouth, Cherbourg, and Brest. (Plate 3, Figs. 5, 6, and 8.)

Commercial Docks. Commercial docks cannot merely be constructed at suitable sites, but must be formed where trade demands them, though they assist greatly in the development of an existing commerce. As large towns have generally grown up by the banks of rivers, docks have generally to be provided in the same vicinity. Docks furnish that constant level of water, that freedom from motion and from currents, and that proximity to roads, railways, and warehouses, so essential to the rapid lading and discharging of vessels, and which cannot generally be provided by a tidal river. Owing to the growth of trade, and the increase in the size of vessels, docks have been added to many tidal harbours, where formerly the vessels entering the port used to lie on the bottom of the basins, as the smaller craft still do in the outer tidal basins of these harbours. Most of the jetty harbours of the North Sea have been provided with inner docks, which are approached through the jetty channel and old tidal harbour, and are thus effectually sheltered from the sea. Thus Ostend, Calais, and Dunkirk possess docks for the accommodation of trade: at Calais the docks are being considerably extended; whilst Dunkirk is developing into a very important commercial port. (Plate 1, Figs. 5, 7, and 10.) The jetty harbours also of Boulogne and Dieppe, on the English Channel, are furnished with docks. (Plate 1, Fig. 1, and Plate 5, Fig. 8.)

Tidal Ports. Where there is a tidal rise, vessels can only enter the docks near the time of high water, owing to the insufficiency of depth of water in the entrance channel at other states of the tide; though the tide is prevented from ebbing out of the docks by gates, which are closed soon after the turn of the tide. Accordingly, under these conditions, it is a great advantage for vessels, reaching the port at a time when they cannot enter the docks, to have some sheltered place in which to remain till they can be admitted into the docks; and this convenience is afforded by rivers possessing a fair depth at low water, like the Thames and the Humber, or by deep sheltered estuaries like Southampton Water and the Firth of Clyde.

Tideless Ports. In tideless seas, the deltas formed by rivers flowing into them bar these rivers, in most cases, from admitting vessels of any size, as for instance the deltas of the Rhone and the Nile in the Mediterranean Sea. Rivers therefore, which in a tidal state afford such convenient access for vessels into the interior of a country, are useless for communication with tideless seas, unless special means are adopted for the improvement of their outlets, such as have been successfully carried out at the mouths of the Danube and the Mississippi[1]. Other sites for docks have, accordingly, to be found; and the system adopted on the Mediterranean consists in forming basins and quays, sheltered by breakwaters, in suitable and somewhat protected places on the coast. In the absence of tidal oscillations, gates can be dispensed with; and the only requisites are, tranquillity, an adequate area and depth of water, and the accommodation of quays. These are provided at Marseilles and Trieste by detached breakwaters,

[1] 'Rivers and Canals,' L. F. Vernon-Harcourt, p. 310 and p. 315.

parallel to the shore, protecting an area separated into basins, surrounded with quays, by moles or jetties projecting at right angles from the coast. (Plate 3, Figs. 4 and 7.) They combine, in fact, the protection of a harbour with the accommodation of docks. At other ports on the Mediterranean, harbours of the more ordinary form have been constructed; and under their shelter, quays, and occasionally small basins, have been constructed along the shore. Genoa, Leghorn, Bastia, Cette, Barcelona, Malaga, Tarragona, Alexandria, Algiers, and Oran are examples of ports where quays sheltered by artificial works provide suitable accommodation for trade. (Plate 3, Figs. 1 and 3, and Plate 5.)

Quays along Rivers. Sometimes, when the tidal range on rivers is small, quays are formed alongside the river; and the lading and discharging of goods are carried on in the river, instead of in docks. This method has been exclusively adopted at New York, where the tidal range averages $4\frac{3}{4}$ feet; at Rouen, where the tidal range is $6\frac{1}{2}$ feet; and also to a large extent at Dublin, with a rise of tide of 13 feet. Occasionally the system of river quays is used to supplement docks, where there is a good depth of water in the river, even when there is a fair rise of tide, as for instance at Antwerp where quays are being formed all along the river in front of the town, and also at Belfast and Bordeaux. By this arrangement the vessels are saved the trouble of entering docks, a long length of quay is readily formed, and the river regulated at the same time; but a considerable range of tide is a source of inconvenience.

Situations suitable for Docks. It is expedient, if possible, to form docks on low-lying land, for the amount of excavation is thereby reduced, and part of the excavated material can be utilised in raising the land round the docks for forming quays. The land enclosed by the bend of a river furnishes a very suitable site for a dock, for, being generally alluvial, its surface is low; and where the bend is sharp, it is possible to provide an entrance at each end of the dock by carrying the dock right across the bend. The double entrance enables the two kinds of traffic to be separated, the sea-going vessels using the down-stream entrance, whilst the river craft is accommodated by the up-stream entrance. The bends of the river Thames below London, which form a circuitous route to the sea, offer conveniences for docks which have not been lost sight of. Thus the Surrey Commercial Docks have occupied one of the upper bends; and the West India, Victoria, and Albert docks have secured the advantages of a double entrance, as well as a position on low-lying land. (Plate 10, Fig. 1.) The new Tilbury Docks are also being placed on marsh land near a bend of the river.

Sometimes the site of a dock is actually reclaimed from a river, whereby the excavation is still further lessened, and the area required obtained at a small cost. The Albert Dock at Hull on the Humber, and the new docks at Havre on the Seine are examples of docks formed on sites reclaimed from a river. (Plate 11, Fig. 4, and Plate 12, Fig. 2.)

SEA EMBANKMENT.
Maryport Docks.

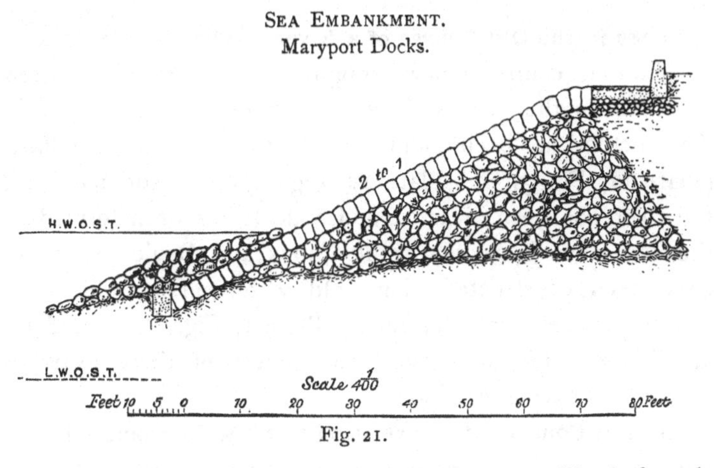

Fig. 21.

Occasionally the low sea foreshore is utilised for the formation of a dock, as for instance at Maryport, in Cumberland, where an embankment (Fig. 21) has been formed for excluding the sea and enclosing a dock.

The embankment in a river, for enclosing the site of a dock, can be also made to serve as a river quay for small craft by erecting timber wharfing upon it. A portion of the river embankment on the Humber for the Albert Dock was constructed so as to provide such a quay. (Fig. 22.)

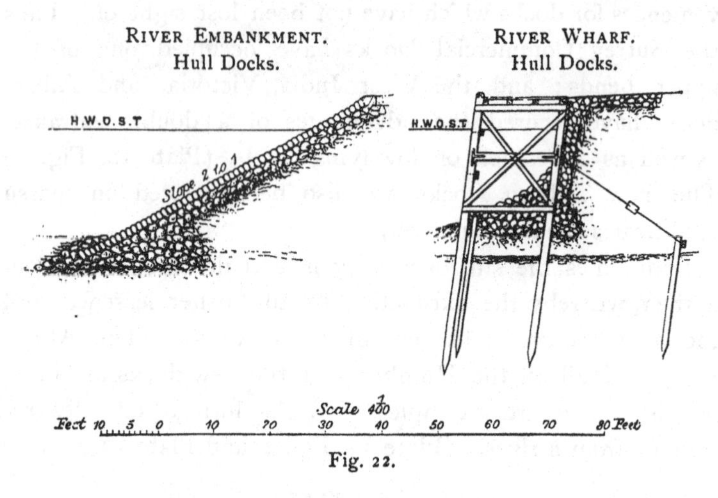

Fig. 22.

Docks in the Old Course of a River. Where a river follows a circuitous course, a new straight cut has sometimes been provided for the river, and the old channel is used for a dock. This method has been adopted at Bristol, and also at Belfast. (Plate 11, Fig. 3, and Plate 10, Fig. 4.) At Aberdeen too, the river has been diverted, and a dock has been formed in the old river bed; whilst at Calais and Boulogne, docks have been constructed along old watercourses, which accounts for their irregular line. (Plate 1, Figs. 5, 11, and 12, and Plate 5, Fig. 8.) The main object of these arrangements is to save excavation.

General Considerations relating to Sites for Docks. Docks have to be formed where a trade already exists, as, though they foster and develop a trade, they cannot be relied upon to create it. The course of trade, moreover, is very capricious, being regulated partly by the inclinations of shipowners, and partly by the resources of the adjacent district, and also by the most suitable lines for through traffic, as well as the capabilities of the particular port. Thus Dunkirk, though unfavourably situated by nature as regards the depth of its approach channel, appears to be in a good line of communication; and by dredging at its entrance, and the enlargement of its docks, it is becoming a port of considerable importance, and has far outstripped the more favourably situated ports of Calais and Boulogne. Again, as railway communication renders the distance of a dock from the centre of commerce of minor importance, the tendency of docks in the present day is to get nearer the sea-coast, thus enabling sea-going vessels to reach their destination sooner, to be less impeded by the tidal current in a river, to avoid the upper portions crowded with smaller craft, and to secure a greater draught of water. Thus every dock extension on the Thames has been carried further down the river. Nevertheless Flushing, though situated at the mouth of the Scheldt, and fostered by the Dutch Government, has been unable to wrest the trade from the rival Belgian port of Antwerp, 47 miles up the same river; and whilst the quays of Flushing are almost deserted, Antwerp can hardly extend its docks and quays fast enough to provide adequately for its rapidly increasing trade.

Though in the case of a large port like London the systems of docks may be widely scattered without inconvenience, in a small port it is desirable that the docks should be concentrated near the town. It is therefore important, when docks are designed, that the prospect of future extension should be considered, and the docks so placed and constructed that an enlargement or extension can be arranged without hindrance to the existing works, and without great expenditure in the purchase of land.

PRELIMINARY WORKS FOR DOCKS.

Designs. The length given to a dock must depend upon the available site; and the width should be regulated by the size of the vessels which may be expected to use the dock. There must be room for vessels alongside the quay on each side, and a clear passage down the centre of the dock for vessels entering or leaving the docks; and it is also advantageous to have space as well for vessels to moor in the centre of the dock. It used to be considered advisable to have a sufficient width for vessels to be able to turn round in the dock, and this was sometimes effected by making the dock wider near its entrance, as at the Albert Dock at Hull. (Plate 11, Fig. 4.) By this means a vessel can turn, and start with its bow foremost, without a greater width than necessary being adopted in the rest of the dock.

There is great diversity in the designs of docks, as may be readily observed by reference to Plates 10, 11, 12, and 13, the figures in which, with the exception of those in Plate 12, are all drawn to the same scale. The main object generally is to obtain a large amount of quay space, along which vessels can lie to discharge and take in cargoes. This is best effected by a number of moderate sized oblong docks, like the Liverpool Docks (Plate 11, Fig. 1), or a narrow dock like the Albert Dock, London (Plate 10, Fig. 1); whilst a greater water space, for the same length of quays, is furnished by wide docks, like Chatham, Portsmouth, and Queenstown (Plate 12, Figs. 3, 4, and 5), or almost square docks, like the Cavendish Dock at Barrow. (Plate 11, Fig. 2.) A considerable water space may be advantageous in Government docks, where vessels may have to be stationed for some time in the docks; but it is undesirable in commercial docks, where vessels should not lie idle.

Where the docks are wide, additional quay space can be provided by extending moles, or jetties, out into the dock, as at the London Docks, the West India Docks, and the Victoria Docks (Plate 10, Fig. 1), and also at Marseilles and Trieste. (Plate 3, Figs. 4 and 7.)

The precise shape of the docks must depend upon the conditions of the site, as well as the nature of the traffic, and should also be arranged with a view to possible future extensions. Though the oblong shape is the best, various other forms are sometimes adopted to suit special circumstances, as at the St. Catherine's Docks, the Surrey Commercial Docks, Dunkirk, and Aberdeen. (Plate 10, Fig. 1, and Plate 1, Figs. 7 and 12.)

The depth of a dock must depend upon the water-level at neap tides, the depth of the approach channel or river below the dock, and the greatest draught of the vessels that may be expected to frequent the dock. The height of the quays must be regulated by the level of the highest spring tides, and the average height above the ordinary water-level most suitable for vessels.

Occasionally it is found expedient to provide an outer basin between the entrance lock and the main dock, whose water-level can be lowered before high water so as to be level with the tidal channel, and can also be kept open for some time after the turn of the tide, thus enabling vessels to enter and leave the port some time before, and some time after high water, without the trouble and delay of locking. This intermediate, or half-tide basin serves as a sort of huge lock, which is only drawn down before high water, and then filled again from the dock after the tide has somewhat fallen, enabling vessels to pass out of it and into it straight from outside, and preventing undue fluctuations in the water-level of the principal dock. Instances of these basins exist at the London Docks, the East and West India Docks, the Albert Dock (London), Hartlepool, Cardiff, Penarth, Sunderland, and Antwerp. (Plate 10, Figs. 1 and 3; Plate 11, Fig. 6; Plate 12, Fig. 6; and Plate 13, Fig. 3.)

Borings. In addition to the regular surveys of the proposed site for a dock, and levels of the ground, it is essential that borings should be taken to ascertain both the nature of the soil which has to be excavated, and more especially the strata upon which the dock walls and locks will have to be founded. Sometimes the excavations furnish sand or gravel suitable for mortar or concrete, which materially reduces the cost of the work; and on the nature of the strata upon which the walls have to be built, depends the depth to which the foundations require to be carried, and the kind of foundations necessary. As the depth and thickness of the different strata are liable to vary considerably, even within a comparatively small area in the case of alluvial deposits on which docks are so frequently constructed, it is desirable to obtain as many borings as possible consistently with economy. A moderate sum expended on borings is fully repaid by the more exact knowledge acquired as to the precise conditions under which the work has to be carried out. Great care must be taken in the examination of the materials brought up by the boring tools, and due allowance must be made for alterations in their consistency from the disturbance by the action of boring. Borings are not such satisfactory indications of the exact nature of the strata traversed as trial pits; but a series of trial pits in alluvial soil, where water soon infiltrates, would require pumping to keep them dry, and, in many cases, could not be carried down far without timbering as well; so that whilst one or two trial pits are advantageous to supplement the borings, and to afford more certainty as to the strata, they could not, owing to their cost, take the place of borings which, even though occasionally not absolutely reliable, furnish very valuable information for practical guidance. By plotting the results of the borings to the proper levels, indicating each change of stratum, and joining the lines of separation in the various borings, a good geological section is obtained of the ground to be traversed; and the depths and nature of the foundations can be determined.

Cofferdams. As it is necessary to shut off the access of water from the site of a proposed dock, for the purpose of carrying on the excavations and building the walls and locks, an embankment or a dam has to be left or constructed between the future outlet and the works. When a dock is formed on land out of reach of the tide, it is only necessary to protect the excavations where they approach close to the proposed entrance channel. This may be accomplished by leaving for a time a portion of the existing bank between the river channel and the excavations. As, however, the entrance lock is generally constructed close to the outlet, and a bank of earth with its slopes occupies a considerable width, it is generally found expedient to form a cofferdam projecting into the approach channel beyond the site assigned to the lock.

A cofferdam is usually composed of two rows of piles driven into the ground to an adequate depth, and separated by an interval of about three to five feet, which is filled up with clay puddle. It is desirable to dredge out the loose or silty materials at the surface between the piles, so as to carry down the clay to as impermeable a stratum as possible; and the clay should be brought up in layers, and well punned, so as to prevent percolation of water. The piles are prevented from yielding to the pressure of the clay inside by a series of horizontal walings, on the outside of each row of piles, which are connected together by tie bolts passing through the more deeply driven guide piles, on each side, placed a few feet apart. When the cofferdam is straight, balks of timber are used for the walings; but when a cofferdam is made segmental, for the purpose of gaining additional space inside, and securing an increase in strength, laminated walings, composed of three or four planks bent to the curve of the dam, are adopted. As the strain exerted on the tie bolts is sometimes considerable, hard wood washers are interposed between the iron washers of the bolts and the walings, so as to distribute the pressure. The number of rows of walings placed along the dam must depend on the height of the cofferdam; but generally from two to four rows are formed on each side, placed eight to ten feet apart. A cofferdam can be further strengthened by struts on the inside, and by a bank of earth against the inner face. It should always be raised a foot or two above the highest water-level, so that there may be no danger of the water overtopping it. A cofferdam is, in fact, a watertight wall of clay puddle, supported by sheet piling on each side; it occupies very little width, it is easily closed against the tide, and little difficulty is generally experienced in removing it when the work is completed.

Occasionally, when the head of water to be kept out is moderate, a cofferdam is formed by a single row of piles calked and strutted at the back; but unless the soil is soft and homogeneous, it is impossible to drive the piles close together with sufficient accuracy. The same result, however, can be attained by driving piles close together below the surface, and then placing planks against piles projecting at intervals to the top of the dam, the covering of planks being carefully calked: this method was adopted by Mr. Brunlees at the Avonmouth Dock, where the head of water was 44 feet.

Dams. Concrete or masonry dams have sometimes been employed for protecting foundations, but they are costly to construct, and somewhat difficult to remove. They are more suitable for excluding the water from existing works, for the purposes of repair, where sills or aprons do not admit of driving piles, and where the space available is limited.

When a large area has to be surrounded by a dam, as for instance when the site for a dock is to be reclaimed from the foreshore, earthwork embankments are formed enclosing the site; except just in front of the entrance, where an ordinary cofferdam is usually inserted to economise space, and to facilitate the excavation and construction of the foundations extending towards the outlet channel. Thus the sea was shut out from the site of the existing Maryport Dock by an embankment, pitched with stone on the sea face and protected by a rubble mound at its toe. (Fig. 21, p. 391.) The same system was adopted, at an earlier period, for constructing the Albert Dock along the foreshore of the Humber. (Plate 11, Fig. 4.) The embankment was formed by depositing a small rubble mound of chalk along the line of the outer toe of the bank, and then tipping the bank with material obtained from the excavations. (Fig. 22, p. 392.) The area was enclosed in separate sections by forming cross banks, at intervals, from the main bank right up to the shore, thus enabling the work to be proceeded with without waiting for the final completion of the main bank. Wharfing was subsequently erected over the sea slope, as shown in Fig. 22 (p. 392). A cofferdam was employed for completing the enclosure, opposite the entrance lock at the east end of the dock. These embankments, besides serving the temporary purpose of a dam, form permanent quays for the dock.

Sometimes cofferdams are dispensed with, even opposite the outlet channel, and embankment dams used instead. Thus an embankment was formed for constructing the lock of the Whitehaven Dock; and an embankment 1100 feet long, composed of material from the excavations, was tipped into the Medway, in front of the recently completed Chatham Dockyard extension works (Plate 12, Fig. 3), to exclude the river during their construction. Generally the water area enclosed by these embankments is small; otherwise, with a considerable tidal range, the final closing of such a bank would be a very difficult undertaking, owing to the rush of water in and out of the small opening left in the bank just before the closing is accomplished, tending to scour away the ends of the bank on each side. At Chatham, a large number of wagons filled with earth were collected near the embankment; as soon as the tide had fallen sufficiently the tipping was commenced, and the bank across the opening was raised in advance of the rising tide, and the dam completed.

The objection to temporary earthwork dams is that, though cheaper in the first instance, their removal is a long and somewhat costly operation, owing to the large amount of material contained in them; whereas cofferdams are more readily taken away, and the timber, if originally sound, can be employed again.

Pumping. As the levels of the foundations of dock works are considerably lower than the ordinary level at which water is found in the excavated soil, it is necessary to resort to pumping to lower the water after the excavations have been carried down a certain depth. The borings indicate the general dip of the strata, and therefore the best positions for the sump holes, so that the water may flow most freely to the pumps. The sumps have to be carried below the level of the lowest foundations, in order that these foundations may be drained; but, in special cases, where a small extent of foundations is exceptionally low it may be drained by a small additional pump, raising the water to the general drainage level.

As the lift frequently exceeds 30 feet, the ordinary suction pump is not generally suitable for the drainage of dock works, unless a double lift is adopted; but chain pumps, centrifugal pumps, and others are commonly employed.

Excavation. The excavation of the earthwork of a dock generally forms an important item of the cost. The usual method of removing the principal part of the earthwork consists in throwing the earth into ordinary tip wagons, and drawing them by horses, or, if the place of deposit is at some distance, by locomotives to the site to be raised. The earthwork near the slopes at the sides, and for the foundations of the walls, is frequently removed by navvies who, with the load on their barrows, are pulled up the slope, on an inclined plane of planks, by a horse at the top dragging the chain attached to the barrow. When the walls have been built, the remaining earthwork near them can be filled into large buckets or skips, and raised to quay level by means of a steam crane. This furnishes a convenient method of completing the backing up of the walls.

The lower part of the backing to the walls can be filled in by barrows from the surplus earth deposited for the purpose on the quay level.

The backing should be brought up in layers, and well punned, and should consist of the driest material available. No part of the backing should ever be performed by tip wagons, as the shock from the tip is very liable to force the wall forward. It is, moreover, advisable to keep the tip roads at a distance from the back of the walls, as the shaking produced by the tipping is unfavourable to the stability of the wall.

Mechanical appliances are sometimes substituted for manual labour in carrying out extensive excavations. A steam excavator, with a chain of buckets like a bucket dredger, moving on rails along the top of a slope, excavates the material along the slope and raises it to the top, depositing it direct into wagons running on a line alongside.

A steam navvy running on wheels, resembling a dipper dredger, with an excavating bucket at the end of a long beam, has been found to give satisfactory results.

These excavators, besides effecting a saving in time and labour, appear to afford important reductions in the cost of excavation where the amount of earthwork is considerable. They also have the advantage of rendering the work independent, to a great extent, of the supply of labour and its uncertainties.

Dredging provides a cheap method of excavation; but though it is employed for forming and deepening the outlet channel, and for maintaining the depth in docks, it is not generally available for the excavation of docks. It would be possible to apply dredging to the excavation of sites reclaimed from an estuary before the site is completely enclosed; but the silting up within an area protected by banks is so rapid, till the tide is finally excluded, that it would be unadvisable to leave such a site open for the entrance and exit of dredgers. Such sites should be enclosed as quickly as possible; otherwise the additional excavation, necessitated by the accumulation of silt within the sheltered area, becomes considerable, and a portion of the advantage gained in forming a dock along a low foreshore is lost.

CHAPTER XX.

DOCK WALLS; PITCHED SLOPES; AND JETTIES.

Dock and Quay Walls. Importance, Objects; Pilework Foundations; Well Foundations; Concrete-block Foundations; Compressed-Air Foundations; Pressure on Walls; Width in proportion to Height; Batter; Ordinary Foundations; Materials for Walls, Masonry or Brickwork, Concrete; Coping; Cost; Failures. *Pitched Slopes:* Objects; Instances. *Jetties:* Projecting Jetties; Wharfing; New York Quays; Rouen Quays. Concluding Remarks.

DOCK AND QUAY WALLS.

OF all the works which appertain to the construction of docks, the most important are the walls which surround them. Not only do these walls constitute the most material item in the cost of the docks, but it is in their design, their foundations, and the materials of which they are formed, that the most important modifications in the total expenditure can be effected. The amount of the earthwork is readily ascertained; and though economies may be realised by skill in laying out the work, and the method by which the excavations are performed, an estimate of the approximate cost can be arrived at; and any gain falls to the contractor who is entrusted with the work. The entrance locks and subsidiary works entail a considerable expenditure within a small area; but there is little scope for economy in this part of the work, where the utmost solidity and stability have to be ensured. A small modification, however, of the section of the wall, of the nature of its foundations, or an alteration in the materials forming it, may produce a considerable change in the total cost of extensive dock works, owing to the long lengths which are affected thereby.

Objects of Dock and Quay Walls. In order to utilise the whole area of a dock, and to render it convenient for vessels, it is necessary to surround it with a wall, nearly vertical on the face, so that whilst retaining the earthwork forming the quays, it may enable vessels to lie alongside. This arrangement prevents any waste of space, and facilitates the loading and discharging of vessels by bringing them close up to the quays. Quay walls also, along a river, convert its banks into continuous wharves, thus enlarging its capabilities for trade.

Pilework Foundations for Dock Walls. The nature of the foundations adopted for dock walls must depend upon the nature of the soil to be built upon. Where a firm stratum can be reached, it is worth while to increase the depth of the excavations and the foundations in order to build upon it. When, however, soft or silty soil extends for some distance below the level proposed for the foundations, it becomes necessary to adopt some other system than the mere deepening of the foundations.

Bearing piles, driven some depth into or through the silty stratum, and connected together at the top by walings or a layer of concrete, are frequently employed to support a wall on soil too soft to bear its weight without settlement. The piles sustain the wall, either by resting on the hard stratum below, or by the adherence of their sides; and the walings or concrete distribute the load. The supporting power of a soft soil is augmented by enclosing it within sheet-piling, thus preventing its displacement when the weight of the wall comes upon it; but this provision adds considerably to the cost of the foundations. Instances of pilework foundations will be found in Plate 14, Figs. 9, 14, 17, and 19, and in Plate 15, Fig. 15.

Well Foundations. Sometimes cylindrical well foundations are sunk, by excavating in the central hollow, and weighting the cylinders. This method has been adopted at some quays on the Clyde, and at the Surrey Commercial Docks. It has been carried out on the largest scale at the Penhouët Dock of the port of St. Nazaire, where the foundations of the walls had to be taken down through a thick bed of alluvium on to a sloping stratum of rock. Where the thickness of the mud exceeded 13 feet, square masonry wells were sunk through it on to the rock, which had to be excavated under their base to provide a level foundation. These wells, being placed at intervals apart, and filled in solid with masonry, form piers for arches which span the intermediate spaces, and support the continuous dock wall above. (Plate 15, Fig. 13.) In some places the slope of the rock was so great that it was found expedient to sink two smaller wells in line to form the pier, in order to reduce the amount of rock excavation, filling in the intermediate space with masonry; but this proved a difficult operation, as the smaller wells were more troublesome to sink, and as the back well was drawn towards the front one in the process of sinking.

Similar wells have been adopted in forming the walls for the new dock at Havre, in course of construction, on the alluvial foreshore of the Seine estuary, previous to its reclamation by an embankment along the river. (Plate 14, Fig. 23.) These oblong wells are sunk at intervals apart, and their hollows, as well as the spaces between them, are filled up with concrete. The wells are carried up to about mean tide level; and a continuous masonry wall is built on them. The advantages of this system are, that no previous excavation, pumping, or exclusion of the tide is necessary, and that the wells can be carried down to any depth according to the varying nature of the bottom.

Concrete-Block Foundations. Where circumstances render it advisable to build quay walls actually in the water, concrete blocks are sometimes used for raising the wall above the water-level. Thus at the Marseilles quays (Plate 3, Fig. 4), the portion of the wall below water round the Arenc and National basins, consists of tiers of large superposed concrete blocks, resting upon a levelled rubble base from 19½ to 23 feet below the water-level (Plate 15, Fig. 14); whilst the im ersed part of the Joliette quay walls is composed of concrete-in-mass founded 13 feet under water. (Plate 7, Fig. 16.) The quays surrounding the Trieste basins (Plate 3, Fig. 7) are similarly formed of four tiers of concrete blocks. (Plate 7, Fig. 7.)

The most important application of this system has been designed by Mr. Stoney for the construction of some deep quay walls at Dublin. (Plate 14, Fig. 15.) These walls have been founded at a depth of about 28 feet below low water, on a bed previously excavated and levelled by men in a diving bell. The portion of the wall from the bottom to a little above low water consists of a row of huge blocks, composed of rubble stone and concrete, weighing about 360 tons, placed side by side, with a groove left in each side face which is filled with concrete after the blocks are in place, thus forming a sort of connecting dowel between the blocks. The blocks are built on a timber wharf; two cast-iron girders are laid on the platform, and built into the block, forming washers for the four suspending bolts, 5 inches in diameter, which are inserted in vertical tubular holes left in the block. The block is lifted, conveyed to its destination, and deposited, by means of floating shears specially designed for the purpose[1]. (Plate 8, Figs. 11 and 12.) The barge carrying the shears is brought end on to the wharf, the suspending bars are connected with the shears by chains which are hauled in by winches on the barge, and water is at the same time pumped into the tank at the stern of the barge. The pull on the chains in front causes the barge to dip down more at the bow than the water in the tank lowers it at the stern; the winches are then stopped, and the water, continuing to flow into the tank, gradually raises

[1] 'Minutes of Proceedings, Inst. C. E.,' vol. xxxvii, p. 335.

the bow as it lowers the stern, and the block is thus gently lifted from the wharf. The barge is then towed to the site of the wall, and it is moored in position; the block is lowered alongside the end block, and its exact position is readily adjusted.

Part of the quay wall at Greenore was constructed below low water in a similar manner, with blocks of concrete, weighing 100 tons, which were built upon a tide-covered site just above low-water level. The block was lifted by two barges, near the time of high water as the tide rose, which conveyed it away, and deposited it in place by the aid of winches and chains. Each block, 21½ feet high, was founded at that depth below low water, and its length was 10 feet. The less exposed portions of the quay wall were built up to low water with blocks in front, and concrete in bags at the back, the intermediate space being filled with concrete deposited from skips.

The Dublin system involves a considerable expenditure on plant, amounting in that instance to about £34,000; so that it is only suitable for works where a long extent of quay has to be constructed under water. The simpler method employed by Mr. Barton at Greenore requires little special plant, and may be advantageously adopted where the amount of work is small.

Compressed-Air Foundations for Quay Walls. The quay works in course of construction along the right bank of the Scheldt, in front of Antwerp (Plate 13, Fig. 3), are being executed under somewhat exceptional conditions. The object aimed at is not merely to substitute a vertical quay wall for the natural sloping banks of the river, but also to regulate the width and depth of the river at that part. Accordingly, the line of quay is being placed at variable distances from the banks, and consequently in different depths of water. Moreover, as a considerable portion of the wall has to be founded in the alluvial bed of the river, the depth below the surface to which the excavations have to be carried, to secure a solid base, is variable; and the wall has to be both very stably founded, and very solidly built, to support the large mass of backing required to fill up the reclaimed river bed at the back.

The compressed-air system of foundations has been adopted by the contractors, MM. Couvreux and Hersent, to meet the special requirements of the site. The depth of water amounts in some places to about 40 feet at low water; the rise of tide is $15\frac{1}{6}$ feet at springs, and the velocity of the stream is sometimes 6 feet per second.

The plant for laying the foundations of the wall, in lengths of 82 feet, and building the wall from $26\frac{1}{4}$ feet below low water up to 2 feet above low water, consists of three parts; namely, a wrought iron caisson which is filled with concrete, and remains in place forming the foundation; a superstructure of wrought iron plates bolted round the top of the sides of the caisson, serving as a movable cofferdam within which the wall is built; and, lastly, a floating stage for floating out and depositing the caisson, for adjusting and lifting the cofferdam, and for carrying the air-compressing machines and the materials for the wall. (Plate 8, Figs. 5 and 6.)

The caissons are all 82 feet long and $29\frac{1}{2}$ feet wide, but they vary in height from $8\frac{1}{2}$ feet to $19\frac{1}{2}$ feet according to the position they are to occupy in the river. The lower part is bottomless, and constitutes the working chamber into which the compressed air is introduced; it is $6\frac{1}{4}$ feet high, and is covered by a roof strengthened with girders for supporting the weight of the wall. The upper portion is enclosed by plate-iron sides, varying in height according to the depth of the foundations, and it is surmounted by angle irons to which the plate-iron cofferdam is bolted. The roof of the working chamber is pierced with seven holes; a large central one for the admission of the workmen, four smaller ones for the introduction of concrete, and two still smaller for the discharge of the excavated material.

The floating stage is composed of two barges, 85 feet long, 17 feet wide, and placed 33 feet apart. The barges are connected together by six iron upright girders on each barge, the end pairs being braced together along their whole height; whilst the four central ones are only connected overhead, so as to leave a free space for raising and lowering the movable cofferdam. The uprights also support two stages over the deck of each barge.

When a length of wall is to be built, the floating stage carrying the movable cofferdam is taken out, and moored over the site of the wall; the caisson which has been built on shore is then launched and towed out to the spot, and during slack tide is placed between the barges, under the cofferdam which is then bolted on to it. The upper portion of the caisson, above the roof of the working chamber, is then filled with concrete. The wall is next commenced on the top of the concrete, hollows being left for the vertical shafts communicating with the working chamber; and the unequal balance, produced by the unsymmetrical section of the wall, is adjusted by loading the space in front of the face with sand. The building of the wall is continued till the caisson touches the ground at high water, which generally occurs when the wall has been raised from 11½ to 13 feet in height. The wall is then adjusted to its exact line, on being slightly lifted by aid of the suspending chains and the introduction of compressed air below, and is finally lowered into place. The masonry is proceeded with till its weight is sufficient to keep down the caisson, at high tide, when the working chamber is filled with compressed air. The men then enter the working chamber, and excavate the soil till a solid foundation is reached. The earthwork is thrown into a box where it is mixed with water, and, on opening a cock, it is forced by the pressure of the compressed air through a pipe communicating with the exterior. The construction of the wall is continued whilst the caisson is being sunk; and as soon as the caisson has reached a solid base, and the wall is completed to the required level, the working chamber is filled with concrete, and the cofferdam is unbolted, raised, and removed to another length of foundation. The whole operation occupies about twenty-five days; so that the average progress of the foundations is $3\frac{1}{4}$ feet per day.

The two adjoining caissons are fairly close together, but there is a space between each length of wall. Panels are placed against the face and back of the wall to close the openings, and the interval is filled with concrete, three grooves left at the side of each length rendering the connection more complete. The shafts are also filled with concrete; and the wall is built continuous on the top in the ordinary manner. (Plate 14, Fig. 21.)

Pressure on Quay Walls. The quay walls which surround a dock, or embank a river, act as retaining walls, for they retain the earth at the back, which is raised to the level of the coping for the purpose of forming a quay. They have to withstand a horizontal thrust at the back, equal to the weight of half the prism contained between the back of the wall and the natural slope of the ground. This pressure tends either to overturn the wall by causing it to pivot on its outer edge which sinks into the soil in the process, or to make the wall slide forwards on its base. The pressure depends upon the nature of the soil, as the natural slope varies with different soils, and also upon its condition, as certain clays, which can stand almost vertical when dry, are totally altered by exposure to the weather and to water, and tend to swell and assume a semi-fluid state, producing a great pressure at the back of a wall. The pressure is always increased by the presence of water at the back of the wall; and in the case of silt, and some kinds of clays in the presence of water, a sort of fluid pressure must be allowed for, like in the case of water, but increased in proportion to the density of the mixture.

Width of Dock Walls. The pressure varies with the depth, and this is provided for by augmenting the width of the wall towards the bottom, by means of a batter on the face and steps at the back. The best examples in this respect, given in Plate 14, are the Liverpool, Portsmouth, Barrow, and Dublin walls. (Plate 14, Figs. 1, 2, 9, 13, and 15.) The average width of the dock walls at the level of dock bottom, in the instances selected for illustration, is exactly half the height of the wall above that level. The Albert Dock (London), Millwall, Dublin, Belfast, and Havre dock walls are precisely this width (Plate 14, Figs. 7, 8, 15, 17, and 22); whilst the Bute, Portsmouth, Barrow, Dieppe, Antwerp, Havre extension, Dunkirk, Bristol, and Queenstown walls approximate to it. (Plate 14, Figs. 4, 9, 14, 19, 20, 21, 23, 24, and 26; and Plate 15, Figs. 11 and 12.) The West India Dock, Whitehaven, and Marseilles walls have a width at dock bottom much below the average, amounting in the case of Marseilles to barely more than a third of the height (Plate 14, Figs. 3 and 16, and Plate 15, Fig. 14); whereas the walls at Liverpool new Hornby Dock, Hull, Chatham, Penarth, Barrow tidal basin, and more particularly at St. Nazaire, considerably exceed the average width. (Plate 14, Figs. 2, 6, 10, 12, and 13; and Plate 15, Fig. 13.) The cause of the small width in the case of the West India, Whitehaven, and Marseilles dock walls must be attributed to the design not admitting of much enlargement at the base; whereas the excess of width at Hull and Penarth is due to the arched form of wall, and was adopted for stability at Chatham, Barrow, and St. Nazaire, on account of the walls being placed on reclaimed land.

The vertical concrete-pocket wall, adopted at the South Dock (West India Docks), and at Whitehaven (Plate 14, Figs. 3 and 16), is not well adapted for a theoretically correct section of wall, as the size of the pocket, being the same from top to bottom, leads to too much material being placed at the top and too little at the bottom.

The arched wall, built in the first instance at Penarth and Hull (Plate 14, Figs. 6 and 12), has a form of great stability against overturning from the width of its base and the weight of filling over the arch; but its broad foundations entail expense in construction, and take up space; and, moreover, though the amount of masonry is somewhat less than in a solid wall of the same height, yet the amount of facework, the dressing of the arch stones, and the turning of the arches make the arched wall more costly to construct than the ordinary solid wall. Accordingly, a solid wall was substituted in the subsequent extensions of these docks. (Plate 14, Figs. 5 and 11.)

In designing retaining walls, the width at the centre of their height in proportion to the height is frequently employed as a measure of their stability. Adopting this standard of comparison, and starting from dock bottom as before, for below that level the pressure is equalised on both sides, it appears that the average width half-way up in the walls referred to slightly exceeds one-third of the height, being actually eleven-thirtieths. As before, however, there are cases of divergence from this mean width. The walls which approximately coincide in width with this average are, West India Dock, Bute, Millwall, Chatham, Penarth extension, Barrow tidal basin, Dieppe, Antwerp, Havre (Eure Dock), Dunkirk, Bristol, Queenstown, and Marseilles. (Plates 14 and 15.) The Herculaneum (Liverpool) and the Portsmouth walls are greatly under the average, and so to a smaller extent are the Liverpool new dock, Hull extension, Albert Dock (London), Dublin, and Belfast walls; whilst the Hull and Penarth arched walls, and the Havre tidal extension wall have widths at the centre exceeding half the height from dock bottom, and the St. Nazaire wall nearly attains this proportion of width. The excess is readily accounted for by the arched form of the Hull and Penarth walls, and by the well foundation of the Havre wall; and the St.

Nazaire wall belongs to quite an exceptional type. The Dublin quay wall, though somewhat slighter in the centre than the average section, is composed of massive blocks, as already described, reaching above mid-height; and the Portsmouth and Albert dock walls are strengthened by counterforts.

The peculiar slimness of the Herculaneum dock wall at Liverpool (Plate 14, Fig. 1) is justified by its situation, for its foundations have been partially excavated in the red sandstone upon which it rests, and therefore it has not to sustain the same weight of backing as ordinary dock walls. Some portions, indeed, of the quay walls of this dock consist merely of a facing of ashlar tied by keys of masonry dovetailed into the natural rock, as shown in Fig. 23, thus in some measure compensating for the additional cost of the rock excavations.

MASONRY FACE TO ROCK.
Herculaneum Dock, Liverpool.

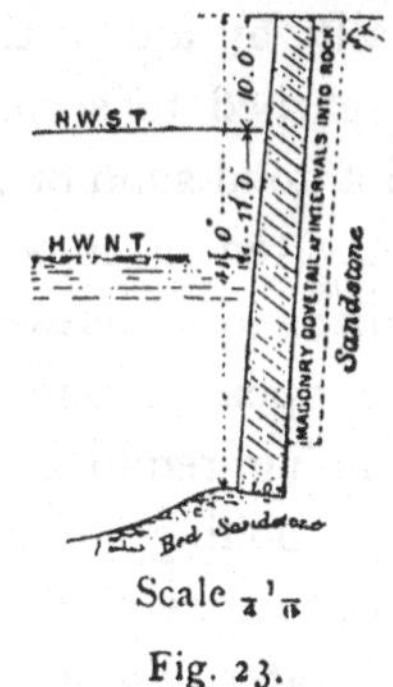

Scale

Fig. 23.

Batter of Dock Walls. A batter, or slightly sloping face, is almost invariably given to dock walls (Plates 14 and 15); though exceptions to this rule exist at the quays of Marseilles and Trieste. Theoretically a curved batter should be adopted, as in masonry dams, owing to the increase in pressure due to the depth. Slightly curved batters have been given to the walls at Cardiff, Hull, Penarth, Barrow, Dublin, and Bristol; and a greater curvature has been used at St. Nazaire and the new Hornby Dock (Liverpool), the good curve at the base in the latter example making up for the somewhat slight width higher up.

A straight batter is generally preferred to a curved one, on account of the straight line usually given now to the sides of vessels, which prevents their approaching close to quays with a curved face. A vertical wall would possess an advantage in this respect; but the superior stability of a battered wall, and the unsightliness of an overhanging wall, should the slightest movement occur, render the retention of a batter on the face advisable.

Laying Ordinary Foundations for Dock Walls. The methods of constructing the more difficult foundations have been already described. Where, however, a firm stratum is found near dock bottom, the ground is sloped back behind the site of the walls, and the excavations are carried down vertically, by aid of timbering and props, till a sufficient depth is reached to secure the wall from being pushed forward, or till a good solid bed is attained. The foundations are generally excavated in convenient lengths for the construction of the wall, as it is undesirable to have long portions of deep excavation open at one time, both on account of the additional pumping required to keep out the water, and also to avoid slips, boiling up of sand, and the disintegration of certain soils, such as clay, which do not bear exposure to the weather and water. Sometimes, in sandy or silty strata, it is necessary to enclose the foundations between two rows of sheet piling, which are strutted across as the excavations are carried down.

Directly a length of foundation has been excavated, the foundation courses should be deposited without delay, so as to relieve the pumping and struts, and protect the bed from the weather or disturbance. Concrete is very frequently adopted for the base of the wall, as it adapts itself to irregularities in the bottom, serves to fill up pot holes, spreads over the whole of the excavations, and, moreover, if formed with cement or hydraulic lime, it can block up small springs, can be deposited with care in water, and, by being carried gradually along, it drives the water before it and excludes it from the foundations as it progresses. When the concrete is brought up in layers, care should be taken that each fresh layer is deposited on the clean, rough, and, if necessary, moistened surface of the previous layer. When concrete is deposited in water, and especially where some current exists, or is to be stopped, a larger proportion of cement must be employed to allow for waste and to ensure more rapid setting.

The bottom of the wall is generally widened out by a broad base on a toe of concrete, or by footings if the masonry or brickwork rests directly on the ground. The broad foundation distributes the weight of the wall over a larger area, and the extension in front of the face adds materially to the stability of the wall. The various methods adopted for foundations are amply illustrated in the sections of walls given in Plates 14 and 15.

Masonry or Brickwork Dock Walls. Formerly all dock walls were built with stone or with brick, according to the local supply. Masonry was adopted when quarries of good stone existed near at hand. The employment, however, of large stones necessitates the erection of staging and gantries; whilst scaffold poles and planking suffice for brickwork. Some of the old dock and lock walls near London were built entirely of brickwork. There are numerous examples of masonry walls, or of masonry walls resting on concrete foundations: the Herculaneum (Liverpool), Cardiff, Belfast, Queenstown, and St. Nazaire walls belong to the former; and the Hull, Bristol, and Havre walls to the latter. (Plates 14 and 15.) When the stone is of inferior quality, brickwork is employed for the facing of the upper portion, which is exposed to the shocks and rubbing of vessels, as for instance at Havre and Dieppe; whereas the Antwerp brick quay wall is faced with hard Belgian limestone from a little below low-water level to the top.

Concrete in Dock Walls. The employment of concrete in dock walls has been developed considerably within recent years. Concrete was extensively used by the French at Algiers, more than forty years ago; and they subsequently adopted it for the Marseilles quays and at other places. It was employed as filling in the lower portion of the Southampton dock walls in 1841; and it served as backing to a wharf, faced with iron piles and plates, erected about the same period at Blackwall, the same system being subsequently adopted on a more extended scale at the Victoria Docks, where the beds of gravel in the excavations afforded excellent and ample materials for concrete. The large supply of gravel in the Isle of Dogs led to its being largely used, between 1860 and 1870, in the construction of the dock walls at Millwall and the West India Docks. (Plate 14, Figs. 3 and 8.) The facework in both these cases consisted of brickwork; and walls of a similar composition were commenced at the Portsmouth and Chatham dockyards extension works in 1868. Up to that period Portland cement concrete had been little tried, and lime concrete was not considered strong enough for facework.

Two systems of introducing the concrete into the mass of the wall were adopted, one consisting of pockets of concrete with face, cross, and back walls of brickwork or masonry, of which the West India and Whitehaven dock walls are types (Plate 14, Figs. 3 and 16); and the other, in which the concrete was brought up in layers, with horizontal bands of brickwork introduced between the layers and connected with a face wall, as illustrated by the examples of Millwall and Portsmouth. (Plate 14, Figs. 8 and 9.) The pocket system has the disadvantage of forming a vertical wall too uniform in thickness, too wide at the top and correspondingly narrow at the base. It also forms a somewhat unconnected wall, as the concrete is not tied at all to the brickwork or masonry, and adds rather to the weight merely of the wall than to its tenacity. The system of horizontal layers forms a stronger wall against rupture; and the facework is tied to the concrete backing by the horizontal bands of brickwork. The concrete layers, however, require time to settle before the horizontal bands of brickwork resting upon them are laid on, otherwise these bands, by settling, may break away from the facework and leave it disconnected from the rest of the wall. Hoop-iron bond is useful in these types of walls for securing the connection between the facework and cross pieces.

Concrete made with hydraulic lime was used for the face, as well as the backing of quay walls, several years ago abroad, as for instance at Algiers, and up to low-water level at Marseilles (Plate 14, Fig. 25, and Plate 15, Fig. 14); but it was not till the excellence of Portland cement concrete was fully established that concrete dock walls were adopted in England.

At the Chatham Dockyard extension works, a wall exactly similar to the Portsmouth wall (Plate 14, Fig. 9) was commenced. Owing, however, to a difficulty in obtaining a sufficient supply of lias lime of good quality, Mr. Bernays substituted Portland cement, and as it was dearer than lime he reduced the proportion of cement in the concrete to 1 in 12. This mixture gave such satisfactory results that the brickwork was abandoned, except for the upper facework; and eventually Mr. Bernays made the wall entirely of concrete, with a double proportion of cement and more carefully manufactured concrete for the face and copings[1]. (Plate 14, Fig. 10.)

Walls composed entirely of cement concrete have been erected at the Albert Dock (London), with 1 of cement to 7 of gravel and sand (Plate 14, Fig. 7); at the new northern Hornby Dock at Liverpool (Plate 14, Fig. 2); and at Maryport Docks, with the same proportion as Chatham. A dock wall composed of lime concrete, resting upon a base of cement concrete, has been built at Barrow. (Plate 14, Fig. 13.) The exclusive use of concrete possesses the advantages of requiring neither skilled labour nor scaffolding; concrete is also very cheap where sand and gravel are abundant, and it can be rapidly deposited. Special care, however, is needed in the measuring and mixing where small proportions of cement are adopted; and the

[1] 'Minutes of Proceedings, Inst. C. E.,' vol. lxii, p. 94.

stronger concrete for the facework must be most carefully prepared and incorporated with the rest.

Concrete is not always used for the facework where it forms the mass of the wall. Thus brick facing has been introduced at Dieppe (Plate 14, Fig. 20), and is being used at the Tilbury Dock works; whilst at Dunkirk, there is a facing of stone and a back of brickwork, with occasional connecting cross walls (Plate 14, Fig. 26); and a stone face has been built into the concrete rubble blocks of the Dublin Quays. (Plate 14, Fig. 15.)

The walls of the Surrey Commercial Docks contain specimens of all the various stages in the introduction of concrete. Brick walls with concrete pockets constituted the earliest stage; then concrete blocks were laid below, and brickwork above; next concrete with a facing of brick; and, lastly, a concrete wall has been built.

Concrete is a peculiarly suitable material for dock walls in the neighbourhood of London, as gravel of the best quality is invariably found in large quantities in the alluvial beds of the Thames basin.

Coping of Dock Walls. Granite, from its hardness, has been frequently used for coping, especially at locks where the traffic and the wear and tear are particularly great. Its cost is the sole impediment to its more universal adoption; but it is commonly used at Government dockyards, where durability rather than expense is considered. The Portsmouth dock wall is coped with it; and it was at first employed at Chatham, but subsequently discarded for the cheaper and almost equally hard concrete manufactured with the utmost care. Granite coping has been used for a new dock wall at Dunkirk; but cheaper stone is now generally selected for commercial docks. At the Antwerp docks and quays, a hard Belgian limestone forms an excellent coping.

The concrete coping which caps the Chatham dock wall appears likely to be durable; but the concrete coping at the Albert Dock (London), whether from defects in manufacture or unusually hard usage, has not stood well in places. At the concrete walls of Liverpool, Barrow, Algiers, and Marseilles, stone coping has been inserted, which is a wise precaution, as concrete, unless exceptionally good, cannot stand the wear to which coping is exposed.

Dowels of slate or cement are inserted in the vertical joints of the coping stones, so that a blow may not displace single stones. Angular grooves are dressed in the sides of each stone for receiving the dowel, which is grouted in with cement; and it is preferable not to run the groove right through the stones, so that the dowel may not show at the top.

Cost of Dock Walls. It would be difficult to form a fair comparison between the costs of different dock walls, owing to the differences in the conditions of construction, and the variable cost of materials according to the locality.

Two standards of comparison may be adopted; namely, the cost per unit of length, and the cost per unit of contents; the former indicates more clearly the actual cost of a wall, but is materially affected by differences in height and thickness, whereas the latter gives the cost of the materials in a wall independently of the section of the wall. The former depends upon the dimensions of the wall, the latter upon the cost of the materials put in place.

The cost of the Hull dock wall was £19 9*s*. per lineal foot, including excavation below dock bottom; of the South West India Dock wall, £12 10*s*. with excavation from dock bottom, and £12 2*s*. without; of the Penarth extension, without excavation, £26 1*s*.; of the Portsmouth extension wall, £64 3*s*. with excavation, and £41 10*s*. without; and of the Dublin quay wall, £40 per lineal foot. (Plate 14.) The cost of the Southampton dock walls appears to have been remarkably small; for the first wall, 38 feet high, built in 1841, and composed of brickwork and concrete pockets up to low-water level, and of rubble masonry faced with granite ashlar above, and a granite coping, cost only £13 6*s.* 8*d.* per lineal foot; whilst the wall of the second dock, built in 1851, with a rubble face connected to concrete backing by hoop-iron bond, and 35 feet high, was executed for £6 10*s.* per lineal foot, owing to the large adoption and cheapness of the concrete[1].

The cost of the West India Dock wall per cubic yard was 12*s.* 6*d.*; of the Penarth extension wall, 17*s.*; of the Portsmouth wall, £1 2*s.*; and of the Chatham wall, 7*s.* 10*d.*

It is sufficiently evident from the above instances that the cost of dock walls varies between rather wide limits, partly in consequence of differences in section, owing to varieties in height and pressure, and partly in consequence of differences in the cost of the materials employed.

Failures in Dock Walls. When the pressure at the back of a wall is considerable, the wall is liable, if the bottom is soft, to settle in front and turn over on its toe, or, if the ground is slippery, to slide forward bodily. The Southampton dock wall and the Belfast dock wall furnish instances of the first kind of failure, and the West India Dock wall and the Avonmouth Dock wall of the second.

The wall at Southampton was built upon a sandy and silty bottom, with a row of piles under its toe. A portion of the wall was exposed to a severe strain, owing to slips of the ground at the back and the pressure of the water behind at low water; and the wall, settling on its toe, fell forward into the dock. Another portion showed a tendency to come forward at the base, owing to the pressure behind and the small resistance the soft soil offered to its motion, as the foundations were not carried much below dock bottom. The portion which fell was rebuilt in a more massive form, with a broader and deeper base; and the unstable portions of the wall were relieved of a part of the backing, for which a timber platform was substituted at quay level. The Belfast wall (Plate 14, Fig. 17) had to be built upon such a soft bottom, that

[1] 'Minutes of Proceedings, Inst. C. E.,' vol. xvii, pp. 542 and 545.

it was placed upon round bearing piles from 10 to 7 inches in diameter and 15 feet long. The material available for backing was so silty that it occasioned a considerable thrust at the back of the wall; and the wall soon showed signs of instability, owing to the slightness of the piling in the bad foundations, and in consequence of the base of the wall having been carried very little below dock bottom. The wall was, accordingly, strengthened by bolts tying it to piles driven down at some distance behind the wall, and by footings of timber and masonry against the toe of the wall in front. A length, however, of 70 feet fell bodily over into the dock entrance two years after its erection, and had to be rebuilt[1].

This sort of failure in silty foundations can only be avoided by providing a stronger foundation of bearing piles, adequate to resist settlement; by carrying the wall down to a greater depth; and by using the very best available material for backing.

The South West India Dock wall (Plate 14, Fig. 3) was founded on the London clay, or on the gravel and hard natural conglomerate overlying the blue clay; and there was not the slightest tendency to settlement. Water, however, tended to accumulate at the back; and when the wall was being backed up, a portion suddenly moved forward with hardly any alteration of level or of batter. The wall exhibited a vertical crack at the point of greatest motion, and at each end of the disturbed portion. It was discovered on examination of the strata below the wall, when it had been taken down to be rebuilt, that the slip had occurred some two or three feet below the base of the wall, between two surfaces of blue clay which probably had been separated originally by a vein of sand washed out by the pumping. Another small slip occurred later on near the same place. The damaged walls were rebuilt with deeper foundations; and the foundations for the remainder of the walls, built subse-

[1] 'Minutes of Proceedings, Inst. C. E.,' vol. lv, p. 33.

quently to the slip, were also carried down to a lower level, and proved perfectly stable, having a depth of about 8½ feet of earth against the toe of the wall to resist any tendency to slide. Pipes also through the wall at frequent intervals relieved the pressure of water at the back.

The wall which failed at the Avonmouth Dock resembled the South West India Dock wall in being a pocket wall, and in being founded upon a bed of clay. The foundations for this wall were only carried 2½ feet below dock bottom; and 90 lineal yards of the wall slipped forward upon a soft layer of sand and clay, intervening between the bed of clay on which the wall was built and an underlying bed of hard sand. The disturbed portion of the wall maintained its batter, but sank 4½ feet at the centre of the slip where it was broken from top to bottom. This length of wall was rebuilt, and its foundations, as well as those of the subsequently executed walls, were carried down to the bed of sand. Numerous drainage holes were also formed through the wall for the escape of the water at the back. The backing consisted of heavy clay, and the drainage was imperfect, so that the clay became saturated. In consequence, another slip occurred just before the water was admitted, though the bottom of the foundations at that part averaged 9 feet below dock bottom. To avoid the chance of further failures, the wall was left standing, a fresh face was formed, the cracks were filled up with grout, counterforts were placed at the back, the wall which had sunk 7½ feet at the centre of the slip was raised to quay level, and the water was admitted at the earliest opportunity[1].

The remedies for this kind of failure are, a strong wall of ample dimensions, wide at the base and carried down to as good a foundation as possible several feet below dock bottom, any space between the wall and the solid earth in front being filled in with concrete; an efficient system of

[1] 'Minutes of Proceedings, Inst. C. E.,' vol. lv, p. 17.

drainage at the back of the wall, by placing loose rubble at the base of the slope, and building in pipes through the wall at a low level; and, lastly, the employment of the driest and lightest material for the backing, which should be brought up in thin layers. With very bad foundations and unsuitable backing, a row of sheet piling in front of the toe, and counterforts at the back assist in maintaining the wall.

Sometimes the foundations of the wall are sloped down towards the back (Plate 14, Figs. 2, 7, 9, 10, 20, and 22, and Plate 15, Fig. 11), which helps equally to secure the wall against slipping forwards, with a less amount of excavation and material than if the foundations were carried level throughout to the lowest depth. This expedient is advantageous where the bottom is firm; but where the bed is soft, such an arrangement is liable to promote settlement under the toe.

The worst feature in the failure of dock walls is that they generally occur when the works are far advanced, and when therefore it is too late to introduce modifications in the greater part of the walls. Moreover, slips frequently take place without the slightest warning. Accordingly, when any want of stability manifests itself in the walls at a late period of the works, it is impossible to tell, when the walls are built throughout under similar conditions, how far the mischief may spread, and where it may next appear. The only plan, under such circumstances, is to reserve the final backing up till the dock is filled with water, placing the material on the quays at some distance from the wall, and to let in the water as soon as possible to relieve the pressure on the walls. Moreover, when indications of instability have appeared during the construction of a dock, the water should never be lowered in the dock to any considerable extent after it has once been admitted, for the water accumulates at the back of the wall as soon as the outlets for drainage are stopped up, and a greater pressure would thus be exerted against the back of the wall, if the dock was emptied, than it had had to sustain even before the admission of the water.

PITCHED SLOPES.

In the earlier docks, the ground was often left at its natural slope at the sides, and vessels were reached by means of timber jetties. Slopes are sometimes now formed at the sides of a dock, instead of quay walls, for the sake of economy, or to allow a future extension to be more readily constructed. These slopes, however, are generally pitched so as to secure them from injury, and in order to make them steeper and thus economise space. Occasionally the length of slope is reduced by placing vertical sheet-piling against its toe, which enlarges the available water area, and enables vessels to approach nearer the quays. This method was adopted at the Hull Victoria Dock.

A pitched slope was adopted on the south side of the Penarth Dock (Plate 11, Fig. 6), along which the projecting coal tips are placed, and where consequently a continuous quay is not required. For the same reason, a pitched slope was formed along a portion of the south side of the Hull Albert Dock (Plate 11, Fig. 4); and a pitched slope was also placed at its western extremity, where an extension was contemplated. The outer slope of the river embankment of the Hull Albert Dock was also similarly protected. (See Fig. 22, p. 392.)

JETTIES.

Projecting Jetties. Where the sides of a dock are sloping, vessels cannot approach the quays, and it is necessary to erect jetties projecting into the dock, at the ends of which vessels lying in deep water can discharge and take in their cargoes. These jetties are generally constructed of timber, but when they have to carry coal tips they are built of masonry or brickwork.

Sometimes timber jetties are built out at right angles to the walls of a dock, for the purpose of enabling vessels to discharge and load with their bows end on to the quay. Each jetty admits of two vessels lying alongside; and the length of quays, and consequently the accommodation for trade, is thereby augmented. This arrangement necessitates a tolerably wide dock; it has been adopted at the East and West India Docks and the Victoria Docks on the Thames. (Plate 10, Fig. 1.) At Great Grimsby, two long jetties convey the trucks nearly to the centre of the Royal Dock; whilst three long jetties for coal drops occupy the end of the new dock. (Plate 11, Fig. 7.)

A long wide jetty divides in half the greater portion of the London Dock; and a similar arrangement, on a somewhat larger scale, has been adopted at the Edinburgh Dock, Leith, for extending the accommodation of the quays. (Plate 10, Figs. 1 and 2.)

At Marseilles and Trieste, a considerable part of the quay space is obtained by moles projecting from the shore; but these works resemble rather extensions of the quay walls than ordinary jetties. (Plate 3, Figs. 4 and 7.)

Wharfing. Owing to the great increase in the number of steamers that has taken place, in substitution for sailing vessels, the projecting timber jetties, which used to suffice for the loading and unloading of vessels, are found inconvenient, on account of the different arrangement of the hatchways and the requirements of rapid discharge. Accordingly, it has been found necessary, at the Surrey Commercial Docks, to provide a continuous wharf, in place of the old timber jetties standing on the natural slope, in order to accommodate the steamers engaged in the timber trade. To build a long length of quay wall would have been a costly and slow operation, and, accordingly, Mr. McConnochie has constructed a cheap and easily executed timber wharfing, in the line of the top of the slope, along the sides of the

Russia Dock (Plate 10, Fig. 1), and dredged away the slope, to a depth of 19 feet below high water, up to a few feet from the face of the wharf.

The wharfing consists of whole timber guide piles, placed 8 feet apart, with intermediate sheeting supported by walings. The guide piles are tied by bolts to piles driven into the solid ground at the back, 20 feet from the face of the wharf. A serviceable quay has thus been provided, at a very moderate cost, which will last for some years, and the upper portion of which may eventually be gradually replaced by a concrete wall.

New York Quays. As the tidal range is small at most of the important American ports, varying from 9 feet 8 inches at Boston to 1 foot 8 inches at Baltimore and Galveston, and as they are situated within land-locked harbours or well sheltered estuaries, no protected basins shut off from the tide are required. The chief works at these ports, accordingly, consist in the improvement of the approach channels, and the erection of bulkheads, and piers, to serve as quays for vessels. The docks, or river quays, of New York, are formed by continuous bulkheads, or quay walls, and a great number of timber jetties, projecting into the water at right angles to the walls along the banks of the Hudson and East rivers.

The soft alluvial character of the river bed, and the abundance of timber and rubble stone, have led to the adoption of pilework wharfing, enclosed in a mound of rubble, with a facing or support of masonry at the top[1] Sections of the most recent of these bulkheads are given in Plate 14, Fig. 18, and Plate 15, Fig. 16. Six different plans have been successively adopted for the North, or Hudson River walls. In the first design, a granite wall rested upon a rubble base carried down to a firm bottom after dredging out the mud; but it appears that only a short length of this wall was

[1] Sections and various particulars about the quays of New York were sent me by Mr. T. J. Long, engineer of the Department of Docks in that city.

built. Subsequently a wall, 24 feet high, was adopted, composed of concrete blocks below, with granite facing and concrete backing above, resting upon a foundation of piles with stone filling between, which cost about £83 6*s.* per lineal foot. A modification was introduced in 1874 with a view to economy, concrete deposited in mass under water being substituted for concrete blocks, which reduced the cost of the wall to £64 16*s.* per lineal foot. Unfortunately this concrete did not set properly, so the wall had to be strengthened by building a platform behind it with a mound of rubble stone to keep back the silt. A portion of the quay wall was then built like the earlier system, but with the addition of a relieving platform behind; and another portion was strengthened with a row of bracing piles in the foundations, and the pilework was connected together at the top by a double floor of 4-inch planking upon which the wall was erected. The final design, which has been in course of construction since 1876, in a part where the soft stratum of sand, clay and mud overlying the rock attain a considerable depth, so that the piles cannot traverse it, is shown in Plate 14, Fig. 18.

The beds of both rivers afford good holding for piles; and on the East River and a portion of the North River, the piles can reach a hard stratum, and thus provide a good support for the walls. On a part of the North River, however, where the latest section of wall has been built, no hard stratum was reached by the trial borings, the mud exceeding 200 feet in thickness, so that it was doubted whether a stable wall could be built on piles which were only supported by the adhesion at their sides. A series of careful experiments were accordingly made, and it was found that, though a 100-foot pile, driven its full length into the mud by a ram weighing 18 cwts. and given a fall of 10 feet, went down some inches at the last blow, and could be pulled up again directly afterwards with a strain of only 3 or 4 tons, yet if the pile was left for some hours, it was impossible to pull it out, and when left for several days, several blows of the ram only injured the head of the pile without driving it down. Moreover, these piles, when weighted with 22 tons, showed no more indications of settlement than that which occurred in the rest of the work.

The construction of the wall is commenced by dredging out the mud, along the line, to the width of the base, either down to a solid stratum, or, where the mud is deep, to 30 feet below low water. The trench thus formed is at once filled in with cobble stones, not exceeding 6 inches in diameter, on the site of the piles, and with small rubble beyond on each side. The piles are then driven as shown in Plate 14, Fig. 18, and are connected by timber framing below the wall, and with timber capping and planking on the top at mean low-water level. The 70-ton concrete blocks for the base of the quay wall are next laid by a floating derrick; the wall is then carried up; the rubble mound is raised behind the quay wall; and the filling at the back, composed of earth and ashes, it brought up to quay level. The cost of the wall has averaged £49 16*s*. per lineal foot. Where the piles reach a hard stratum, the settlement of the wall has not exceeded 4 or 5 inches; but where the mud is very thick, the settlement amounts, in the earliest built walls, to 18 inches, and the wall appears to be coming to rest. The piles do not sink in the mud, but the piles and mud descend together under the compression of the wall and the backing.

In this design, the rubble mound forms the retaining wall for keeping up the embankment constituting the quay, and the small granite-faced concrete wall serves for vessels to lie alongside.

Rouen Quays. The quay walls which have been recently built along both banks of the Seine at Rouen present some resemblance to the New York quays. They are founded upon long piles driven down to a solid stratum through the alluvial mud, and rest upon a timber platform at the top of the piles. (Plate 15, Fig. 15.) The walls consist of a concrete base, rubble backing, a brick face, and granite coping; and they are relieved from pressure behind by a wide layer of rubble. The lower part of the wall has been built within watertight timber frames sunk on to the platform on the top of the piles; and the intervals between the frames are spanned by arches, or by oak platforms, so as to support a continuous wall above. The wall is strengthened by occasional tie bars of iron, fastened to blocks of masonry at the back. The system, in fact, consists of a timber jetty supporting a dwarf quay wall, supplying all the requirements of trade, and widening the quays at a moderate cost, and with little difficulty in construction in spite of the depth at which the hard stratum is situated, and where a solid quay wall of the ordinary type would have involved the erection of a structure almost as massive as the Antwerp quay wall.

Concluding Remarks. The varieties in the designs adopted for dock walls and quays are amply indicated by the examples selected. (Plates 14 and 15.) The differences are due, partly to the nature of the foundations, and partly to the materials most readily available. One of the most prominent features in their construction is the increased employment of concrete, owing mainly to the great improvements effected in the manufacture of Portland cement, and the consequent far greater confidence in its strength and durability. The experience at Chatham shows that a good concrete wall can be constructed with as small a proportion of cement as 1 in 12; but unless very careful supervision is exercised in the mixing of the concrete, it would be safer to adopt a somewhat larger proportion of cement. Extreme care is necessary to render a stronger concrete suitable for facework, and it is owing to this that a brick or stone face is frequently preferred. The economy effected by the use of a concrete wall is occasioned, not merely by the cheapness of the materials, but also by the facility of construction and the saving of plant.

The most uncommon types of walls are those of Havre and St. Nazaire with well foundations (Plate 14, Fig. 23, and Plate 15, Fig. 13); the huge block foundations of the Dublin Quays (Plate 14, Fig. 15); the Antwerp quay wall founded by compressed air (Plate 14, Fig. 21); and the mixed system of New York and Rouen. (Plate 14, Fig. 18, and Plate 15, Fig. 15.)

The well system of foundations is not novel; but it is an extension of the system to build the mass of a dock wall in this manner, as at Havre, where the object was to save time by sinking the walls in the silt of the foreshore, without waiting for the reclamation of the site and the excavation of the foundations in the ordinary manner. The method, which was simply expedient at Havre, was essential at St. Nazaire, where the well formed the actual foundation as in previous instances, but on a greatly enlarged scale. The section, indeed, of the St. Nazaire wall is so colossal that much work of this nature for a dock would prohibit its accomplishment by private enterprise.

The objects in view at Antwerp, Dublin, New York, and Rouen were practically identical, namely, the construction of quays along the sides of rivers, on a silty foreshore, below water; but the methods by which this result has been accomplished are very different. A firm stratum was attainable at Antwerp, Dublin, and Rouen; whilst at places in the North River at New York, the layer of silt was too thick to be traversed. Though the plans adopted for laying the foundations of the Antwerp and Dublin quays were quite distinct, (Plate 8, Figs. 5 and 6, and Figs. 11 and 12) yet the walls in both cases rest upon a firm bed, and are solid and very durable. These walls in fact are quite as stable and well built as if they had been constructed out of water within the shelter of cofferdams.

The quay walls at New York and Rouen are of a lighter construction. The trade of Rouen is not large enough to justify such an expenditure upon solid quay walls as at Antwerp; and the length of banks to be provided with quays was considerable. (Plate 13, Fig. 2.) The quay walls at Rouen have been expeditiously and cheaply executed, but they appear very slight. and their durability will depend on the life of the timber on which the actual walls rest. The walls stand in a well-sheltered site with little range of tide, and near the upper extremity of the tidal Seine, thus differing considerably in position from the Antwerp quay walls.

The quay walls at New York, whilst built upon a similar principle to those at Rouen, have a foundation of rubble stone in addition, as the piles in this instance do not reach solid ground, and therefore they have to be consolidated, and the base enlarged, by a mound of stones. It is satisfactory to find that a stable quay wall can be built, with only a moderate amount of settlement, on a bed of silt; the success of the operation being due partly to the adherence of the piles, partly to the consolidation of the silt round the piles by the small cobble stones, and partly to the broad base of the larger rubble mound.

CHAPTER XXI.

ENTRANCES AND LOCKS; DOCK-GATES; CAISSONS; GRAVING DOCKS; MOVABLE BRIDGES.

Entrances to tideless and tidal ports. *Entrances and Locks:* Objects and Advantages, Instances, Widths, Depths; Reverse Gates; Foundations; Floor of Lock; Side Walls; Sluiceways; Sluice-gates. *Dock-Gates:* Description, Strains, Rise; Construction, wood and iron; Forms, straight, Gothic arched, segmental; Methods of Support, buoyancy, roller, self-adjusting flotation; Methods of moving; Cost. *Caissons:* Uses; Floating Caisson, method of working, advantages and disadvantages; Sliding Caissons, method of working, advantages and disadvantages, with rollers. *Graving Docks:* Object, Construction, Sizes, Cost; Gridirons; Hydraulic Lift; Depositing Dry Dock; Slipways; Remarks. *Movable Bridges:* Swing-bridges, different types; Traversing Bridges; Bascule Bridges.

In ports on tideless seas, the basins and quays have direct access with the sea through entrances, which are made narrow for the purpose of excluding the swell; and they are protected from the waves by outlying breakwaters. The basins at Marseilles and Trieste are protected by detached breakwaters, and are approached through narrow entrances, at the sides, sheltered by the projecting ends of the breakwaters. (Plate 3, Figs. 4 and 7.) In other Mediterranean ports, quays are formed along the shore within the shelter of the outer harbour, as at Alexandria, Civita Vecchia, Leghorn, Algiers, Genoa, and Bastia (Plate 3, Figs. 1, 2, and 3, and Plate 5, Figs. 1, 3, and 6); and the entrance to the harbour furnishes the entrance to the port.

At some ports also on tidal rivers, where the range of tide is small, quays are formed along the river banks; and additional space is sometimes provided by basins at the side opening direct into the river. Thus Rouen and New York have river quays (Plate 13, Fig. 2, and Plate 14, Fig. 18); whilst

Belfast, Rotterdam, and Hamburg have both quays and open basins. (Plate 10, Fig. 4. and Plate 13, Figs. 1 and 5.)

Where, on the contrary, there is a considerable range of tide, it would be troublesome to load and discharge vessels constantly rising or falling with the tide, and it would be also costly to excavate and maintain open basins of sufficient depth to enable vessels to remain afloat at low water. Accordingly in ports where the entrance channel is exposed to a large tidal range, it is customary to retain the water in the inner basins, or docks, whilst the tide is low, so as to maintain a tolerably constant water-level, and a good depth of water inside, at all states of the tide.

ENTRANCES AND LOCKS.

The narrow passages through which vessels enter docks are closed by gates or caissons when the tide is falling, and thus the emptying of the basins is prevented.

An entrance is provided with only a single pair of gates; whereas a lock possesses two pairs of gates containing the lock-chamber between them.

Relative Advantages of Entrances and Locks. An entrance, with its single pair of gates, serves to retain the water in the basins; it occupies much less space than a lock, and consequently leaves more room for the docks inside, and is also considerably less costly than a lock. The gates, however, of an entrance can only be opened when the water-level outside is the same as that in the dock; and therefore only a short time, at high water, is available for the entrance and exit of vessels, unless the water inside is drawn down before high water, or allowed to flow out for a certain time after the tide has turned.

A lock, on the contrary, enables vessels to be passed in or out of a dock, at any state of the tide, without any appreciable alteration of the water-level of the dock, and could even admit a vessel at low water provided there was depth enough of water for it to get up to the lock, which would be quite impracticable in the case of an entrance. Locks also possess the advantage of not making the safety of the vessels in a dock depend on a single pair of gates.

Instances of Entrances to Docks. Numerous instances of entrances exist at the Liverpool Docks (Plate 11, Fig. 1), where indeed they form the rule, and locks the exception. The distinction, however, in this case is more apparent than real, as the entrances from the river Mersey open into half-tide docks connected with inner docks by other entrances; so that these outer docks can be lowered to half-tide level without affecting the level of the docks opening into them, and thus they practically serve as huge locks for enabling vessels to pass into the regular docks.

A similar arrangement of a half-tide basin between two entrances has been adopted for affording access to the Victoria Dock at Hull (Plate 11, Fig. 4); to the northern end of the Hudson Dock at Sunderland (Plate 12, Fig. 6); to the Devonshire Dock at Barrow (Plate 11, Fig. 2); to the dock at Dieppe (Plate 1, Fig. 1); and to the Kattendyk Dock at Antwerp. (Plate 13, Fig. 3.)

At Penarth, and at the Bute West Dock, a lock is interposed between the half-tide basin and the dock, whilst there is only an entrance at the outer end. (Plate 11, Fig. 6.) There is also a similar mode of access to the Ramsden Dock at Barrow. (Plate 11, Fig. 2.)

The docks at Havre are only provided with entrances, with the single exception of the lock to the Citadel Dock. (Plate 12, Fig. 2.) The Steam Basin and Rigging Basin at Portsmouth Dockyard, and the dock on Haulbowline Island at Queenstown are closed by a single caisson across the entrance. (Plate 12, Figs. 4 and 5.)

Locks at Docks. In the port of London (Plate 10, Fig. 1), locks always form the entrances to docks. These locks, however, are in most cases supplemented by basins, which provide a freer passage for vessels near high water without lowering the water-level in the inner docks.

The length of the lock-chamber between the gates is regulated by the size of the longest vessels ordinarily trading with the port; but it is evident that an exceptionally long vessel might be admitted, near high water, by lowering the water-level of the basin or dock to the outside water-level, and opening both pairs of gates. The lock at the entrance to the Ramsden Dock at Barrow has the unusual length of 700 feet; whilst the largest lock on the Thames, at the entrance to the Albert Dock, has a length of 550 feet. The largest lock at Liverpool, admitting to the Canada Dock, has a length of 498 feet; the Penhouet Lock at St. Nazaire is 492 feet long; the South Lock at Sunderland 482 feet; the large locks at Portsmouth are 458 feet, and at Chatham 438 and 436 feet long. The Millwall lock on the Thames is 450 feet long; the lock at Dunkirk 423 feet; two locks at Birkenhead are 398 feet; locks at London, Leith, Cardiff, and Bristol 350 feet, at the Surrey Commercial Docks and Hull 320 feet, and at the West India Docks, London Docks, and Grimsby 300 feet long[1] The lock-chamber is sometimes made capable of subdivision, by an intermediate pair of gates, so as to allow the size of the chamber to be adjusted to the varying length of vessels, and thus save time and water in locking. Instances of this arrangement are found on the Thames at the London, Surrey Commercial, Millwall, and Albert Docks (Plate 10, Fig. 1); and also at Birkenhead and Dunkirk. (Plate 1, Fig. 7, and Plate 11, Fig. 8.) By making the two divisions of unequal length, three sizes of locks are obtained.

Width of Locks and Entrances. The greatest width adopted for locks or entrances is 100 feet, of which there are examples

[1] A list of the length, width, and depth, of most of the principal locks and entrances at various docks is given in Appendix IV. The lengths given above, as well as in the Appendix, all refer to the lock-chamber between the gates.

at Liverpool, Birkenhead, Barrow, and at the Eure Dock entrance at Havre. The large locks at Chatham Dockyard are 94½ and 84½ feet wide, and at Portsmouth 82 and 80 feet. The latter width of 80 feet is frequently adopted for large locks or entrances, as for instance at the Victoria and Albert Docks, at Millwall, Liverpool, Hull, Cardiff, Belfast, Antwerp, and St. Nazaire. The most ordinary width is about 60 feet, of which width instances exist at London, Hull, Barrow, Leith, Belfast, Penarth, and Hartlepool. Intermediate widths of 65 feet have been constructed at Liverpool, 70 feet at Grimsby and Avonmouth, 72 feet at Bordeaux, and 69 feet at Havre and Dunkirk; whilst the West India Docks and Cardiff possess important locks only 55 feet wide.

The width of 100 feet was adopted at Liverpool and Havre to accommodate the wide paddle steamers formerly built; but as screw steamers have superseded them, the recent entrances at Liverpool have been built only 65 feet wide.

Depth of Water on Sills. There is naturally considerable variety in the depth at which the sills of locks and entrances are placed, depending upon the date at which the works were constructed, the tidal range, and the draught of vessels frequenting the port. Many sills of old locks are not low enough to accommodate the newer class of vessels with their increased draught, so that, for instance, the more recent entrances at Liverpool have sills from 4 to 6 feet lower than the sills of the older docks. The range of tide has a very material bearing upon the depth advisable for lock sills; for though, with a large range, an ample depth is secured at spring tides, yet at neap tides, owing to the much smaller rise, the available depth is greatly reduced; whereas with a small range, the difference between spring and neap tides is comparatively insignificant. Thus whilst the depth on the lowest lock sill at Portsmouth is 41 feet 8 inches at spring tides, and 38 feet 11 inches at neaps, making a difference of only 2¾ feet, the corresponding depths on the sill of the Alexandra Dock at Newport are 37 and 27 feet, and at Bristol 33 feet and 21 feet, or differences of 10 and 12 feet respectively. Accordingly, whilst a large range of tide affords exceptional facilities, at spring tides, for admitting vessels having a great draught, it has the disadvantage of causing large variations in the available depth.

A port should be accessible at high water whatever may be the state of the tide, and the depth over its lock sill at neap tides is the measure of its capability from this point of view. It is customary, however, to reckon the depth over the sill at high-water spring tides, thus giving the greatest average available depth for vessels, and not the depth which is always attainable. The depths both at springs and neaps are accordingly inserted in the dimensions of the various locks and entrances given in Appendix IV

The lock sills of important docks are placed from about 26 to 35 feet below high water of spring tides. Thus the depth over the sill at Grimsby is 26 feet; at the South West India Dock, and the South Lock, Sunderland, 27 feet; at the Shadwell Basin, Millwall Dock, and Victoria Dock on the Thames, 28 feet; at the Albert Dock, Hull, 28½ feet; at the Albert Dock, London, 30 feet; at Liverpool, 31 feet; at Barrow, 31½ feet; at Swansea, 32 feet; at Cardiff, 35⅔ feet. The depth at Penarth is 32 feet, and at the Eure Basin at Havre 35 feet; but these depths are reduced to 25⅔ and 29½ feet respectively at neap tides. The depth at the new Tilbury Docks is to be 45 feet. The Government Dockyards in England have the deepest sills, so as to admit readily the largest ironclads; thus the depth at Chatham is 32 feet, at Queenstown 32⅔ feet, and at Portsmouth it attains the unusual extent of 41⅔ feet, but this even will be exceeded at Tilbury.

Reverse Gates at Entrances, Locks, and Passages. Where the entrance is exposed to swell or waves, it is protected by an outer pair of gates pointing outwards, which prevent the waves beating against the lock-gates. Similar reverse gates are sometimes added to entrances or locks when high tides have to be excluded, and also to passages between docks and basins for the purpose of enabling the water-level to be varied in one without affecting the other. Examples of outer sea gates exist at Dieppe, Ostend, Liverpool, Birkenhead, Penarth, Grimsby, Sunderland, Honfleur, Bordeaux, and Antwerp. (Plates 1, 11, 12, and 13) The sea gates at Honfleur are only carried down a few feet below high-water level, so as to arrest the surface swell without entailing the expense of regular gates.

The entrance lock at Ostend, and the inner lock at St. Nazaire possess the peculiarity of having a double pair of reverse gates. This arrangement enables a high tide to be excluded, whilst permitting the passage of vessels through the lock.

Reverse gates are frequently erected at passages between adjacent docks; they have been very generally adopted at Antwerp; they may be seen at Havre, Honfleur, and Dieppe; and there are numerous examples of their use in English docks, as, for instance, at the West India Docks, Liverpool, Birkenhead, Bristol, Barrow, Grimsby, and other places.

Foundations of Entrances and Locks. Special care has to be given to the foundations of entrances and locks, owing to the variations which occur in the water-level. It is desirable to carry the foundations down to a watertight stratum where feasible, so as to prevent infiltrations due to the head of water, which are liable to wash out the lighter particles of the stratum on which the foundations rest, and thus undermine the structure. When the bed is not watertight, it is expedient to enclose the site with sheet-piling, so as to arrest the flow of currents underneath. The foundations under the sills are specially protected by being carried down to a greater depth than the rest, or by rows of sheet-piling right across, as the pressure of the water is concentrated at this part.

Floor of Lock. The floor of a lock consists of an invert in the middle forming the bottom of the lock-chamber, a sill at each end of the chamber against which the gates close, a gate-floor over which the gates turn, and an apron at each extremity of the lock.

The invert is constructed of masonry or brickwork in the form of an inverted arch, so as to withstand the upward pressure of water when the lock-chamber is empty, and to support the side walls. (Plate 15, Figs. 4, 6, 8, and 10.)

The sill is raised about two feet higher than the general level of the floor, so as to form a projection against which the gates may close. (Plate 15, Fig. 4.) The inner sill is situated between the invert and the gate-floor (Plate 15, Figs. 3, 5, and 7); and the outer sill is situated between the gate-floor and the outer apron. (Plate 15, Figs. 1, 2, and 9.) The face of the sill, against which the gates abut, is generally made of granite ashlar, set in cement, and most carefully dressed to an exact line, so that no leakage may occur between it and the sills of the gates. The rest of the sill is paved with masonry, in thicker courses than the rest of the floor. The sill is generally flat; but sometimes the sill stones are radiated to increase their resistance to upward pressure, and occasionally the sill is formed like an invert.

The gate-floor is paved with masonry, and cast-iron bed plates are fixed upon it to serve as roller-paths for the rollers of the gates. (Plate 15, Figs. 1, 2, 3, 5, and 7.) No roller-path has been put on the gate-floor of the Dunkirk Lock, as rollers have been dispensed with in the gates at Dunkirk. (Plate 15, Fig. 9.) The gate-floor, besides supporting the weight of the gate, is exposed to scour, as the inlet sluices are situated at the sides of the gate-floor.

The paving of the apron at each extremity of the lock protects the floor from scour, caused, at the upper end, by the flow into the upper sluices, and at the lower end by the violent rush of water out of the lower sluices.

Side Walls of Locks. The walls at the sides of a lock fulfil the same purpose as a dock wall, and for the most part are similar to it in section. (Plate 15, Figs. 4, 6, and 8.) Beyond the inner gates, however, they are subjected to the changing pressure due to the fall of the water in the lock-chamber and outside; but this varying strain is fully compensated for by the solid toe in front, consisting of the invert and floors, which renders any sliding forward impossible.

The walls are made thicker at the back of the gates, both in order to sustain their thrust, and also in order to allow of the introduction of sluices at the bottom, chain-passages in the centre, and the foundations for the gate-machines and the anchors for the gates at the top. (Plate 15, Figs. 5, 7, and 9.)

Recesses are formed for the gates at the sides of the gate-floors, so that the gates, when open, may lie within the lines of the walls, and not impede the passage of vessels. The depth of the recess depends upon the curvature of the gate; and allowance must be made for clearance, and for the space occupied at the back by the roller and adjustment rod.

Hollow quoins of granite are built into the wall at one extremity of the gate-recess, forming a hollow vertical semi-cylindrical recess into which the heelpost of the gate fits, and in which it revolves. The hollow quoins are laid and dressed with the utmost precision, so that no leakage may occur round the heelpost. The bottom hollow quoin, and the heelpost of the gate rest upon a large granite stone, called the heelpost stone, whose top is level with the gate-floor.

The walls have a batter on the face, like ordinary dock walls, except in the gate-recesses, where they are necessarily built vertical. (Plate 15, Figs. 6, 8, and 10.)

Sluiceways in Lock Walls. In the earlier and smaller locks, sluices in the gates formed the only means of filling and emptying the lock-chamber, and of scouring the entrance channel. When, however, larger locks were constructed, it became necessary to provide a more rapid method of adjusting the water-level in the lock, and a more efficient scouring current for the deeper approach channel. Sluiceways are consequently constructed in the walls of all large modern locks. These sluiceways consist of large culverts in the walls at the back of each gate, having their bottom usually level with the gate-floor, paved with masonry, and lined with masonry or blue bricks. (Plate 15, Figs. 6 and 8.) The entrance to the sluiceways is formed by several small openings in the gate-recesses on each gate-floor; the outlets of the inner sluiceways open on to the invert in the lock-chamber; and the outlets of the outer sluiceways generally emerge on to the outer apron of the lock at various points, so that the rush of water through them may keep the apron clear of silt, as well as maintain the approach channel. (Plate 15, Figs. 5 and 7.) By placing the inlets to the sluiceways at the sides of the gate-floors, these floors are kept free from an accumulation of sand and silt.

The sluiceways at Dunkirk, egg-shaped in section, have been carried through the whole length of the walls on each side, and have their end outlets in the wing walls beyond the outer apron. (Plate 15, Figs. 9 and 10.) This arrangement enables the water for sluicing the entrance channel to be drawn straight from the dock without passing through any part of the lock, and renders the filling and emptying of either subdivision of the lock, formed by closing the intermediate pair of gates, independent of the other. It however entails the cost of wider side walls throughout, besides the expense of forming longer lengths of culverts.

The large sluiceways in the side walls of the Langton Dock entrances at Liverpool, for scouring the Canada Basin, likewise extend right through the walls; but they are quite independent of the ordinary sluices behind the gates of the lock and entrance. They merely serve for drawing the large supply of water from the dock needed for feeding the numerous branch sluices along the outer walls and under the apron of the basin, which are capable of discharging $1\frac{3}{4}$ million cubic feet of water in a minute, and thus effectually scour away the deposit from the river.

The above sluicing arrangements at Liverpool are the most elaborate and effective that have been designed; but the ordinary sluices are used at the entrances of most docks, in conjunction with the sluices in the gates, for scouring the outer channel at low water of spring tides. Sluicing can only be advantageously carried on during the lowest tides, foı besides the reduction of the head of water at neap tides, the scouring power of the issuing current is greatly reduced by the mass of inert water covering the bed of the channel.

Sluice-Gates at Locks. The gates for opening or closing the sluiceways in the side walls of locks are formed of wood or iron, and generally are made to slide up and down in grooves at the sides of the culverts, being wedge-shaped in section, and are raised or lowered by machinery at the top of the wall. When the gate is down, it fits into a groove at the bottom of the sluiceway, and being pressed by the head of water on the upper face against the accurately dressed sides of the grooves, it closes the sluice. Two gates are generally placed side by side in the main sluiceway, so that a supplementary gate may always be in readiness in case of accident or during repairs. (Plate 15, Figs. 5 and 7.)

Peculiar forms of sluice-gates have been constructed at Dunkirk (Plate 15, Fig. 9), turning upon a central vertical axis, like the revolving sluice-gates at the outlets of sluicing basins, which have been previously described (p. 70, Fig. 6). The sluice-gates closing the side outlets are made in pairs, revolving upon central pivots, and meeting in the centre when closed, and coupled together. The axis has been placed in the centre of the gate, as these gates, unlike the sluicing basin gates, have to be closed against a flowing current, and therefore must have the pressure equal upon each half of the gate.

The gates closing the ends of the main sluiceways are fan-shaped, having two leaves at right angles, with the vertical axis in the angle. The leaf forming the actual gate is slightly smaller than the other, so that when, by opening a small revolving gate in a small sluiceway at the side, the larger leaf is free to turn, the greater pressure on its larger surface makes it revolve into a recess at the side, and, in doing so, removes the other leaf from the sluiceway and leaves the passage open. In order to close the sluiceway, the small revolving gate is closed, a valve in the larger leaf is opened, which adjusts the pressure and affords a passage for the water, and the gate is pulled round by a chain. As it would be inconvenient and costly to provide a supplementary set of gates of this type, grooves are formed in the sluiceway near the gate, down which an ordinary gate can be lowered if necessary.

DOCK-GATES.

The gates which close the passage through a lock or entrance leading to docks, with the object of maintaining the water-level inside, consist essentially of a pair of gates turning upon vertical heelposts placed in the hollow-quoin recesses of the side walls, meeting at an angle in the centre of the opening, and shutting, at the same time, against the pointing sill at the bottom. (Plate 15, Figs. 1, 2, 3, 5, 7, and 9.) It might be supposed that such gates, having merely to resist the very regular and invariable pressure of water, would be constructed according to some very definite type which experience had shown to be the best and most economical; but a glance at the illustrations selected shows that a considerable diversity exists in their design.

Strains on Dock-Gates. The strains to which a dock-gate is exposed, when closed against a head of water and meeting another gate at an angle, arise from two sources; firstly, the direct water-pressure on the gate itself, increasing with the depth and varying as the length, which produces a transverse strain resembling the strain on a uniformly loaded girder; and, secondly, the pressure transmitted at the meeting-post by the other gate. This second pressure produces a compressive strain, in the direction of the length of the gate, which is equivalent to half the water-pressure on the opposite gate multiplied by the tangent of half the angle at which the two gates are inclined when they are closed. This angle evidently depends on the inclination of the faces of the sill, or the projection of the point of the sill in proportion to the span of the lock. This proportion is commonly called the rise of the gates. When the gates are curved to such an extent that when closed they form a continuous circular arc, they partake of the nature of an arch, the transverse strain disappears, and the whole strain is converted into a compressive strain in the direction of the length of the gate, and is the same throughout each horizontal section, whilst it increases with the depth owing to the increased increment of water-pressure. This strain is equal to the pressure on a unit of surface multiplied by the radius of curvature.

Rise of Dock-Gates. The proportionate projection of the meeting point of dock-gates to the span of the lock or entrance, which is known as the rise of the gates, causes important variations in the compressive strain on the gates, since this strain varies inversely as the rise. A small rise, accordingly, increases the strain on the gates, but the length of the gates is at the same time reduced; whilst a large rise, though reducing the strain, increases the length of the gates. Taking both these variations into account, it has been found that the lightest gate can be formed with a rise of about a third of the span. Other considerations, however, have to be taken into account besides the least cost of the gate. A large rise increases the length of the side walls of the lock or entrance, and, in the case of curved gates, necessitates deep gate-recesses which are inconvenient for the passage of vessels, without affording any counterbalancing advantage beyond economy in the gates. Consequently in practice, a rise of one-third of the span is not generally adopted.

The rise of the Penarth gates is only slightly less than one-third of the span. (Plate 15, Fig. 7.) A rise of one-fourth is frequently adopted, of which the gates at the Victoria Docks, the South West India Dock, the Albert Dock, Hull (Plate 15, Fig. 5), and the Avonmouth Dock may be cited as instances. The Bristol dock-gates have a rise of between a fourth and a fifth (Plate 15, Fig. 3); whilst this latter proportion has been adopted at the Bute dock-gates (Plate 15, Fig. 1) and Whitehaven. Examples of a rise of one-sixth are found at Liverpool, Dunkirk (Plate 15, Figs. 2 and 9), St. Nazaire, and Sunderland; whilst at the Eure entrance at Havre, a somewhat larger rise has been employed.

Construction of Dock-Gates. Wood or iron are employed for the construction of dock-gates. Iron appears to be best suited for large gates, and it can be adapted to any special form. Wooden gates, however, possess more elasticity, and therefore are better able to withstand shocks Iron is liable to rapid corrosion in salt water, but this can be in a great measure prevented by careful and thorough painting. Wood, on the other hand, is attacked by the teredo in tidal waters; and the only remedies are, either to protect the wood by studding it over with large-headed nails, as practised abroad, or the exclusive use of greenheart timber which is invulnerable but costly.

The balance of advantages is in favour of iron for large gates; it is used for the dock-gates in the port of London, at Hull, Cardiff, Penarth, Bristol, and Barrow. Iron also has been adopted in preference to wood for the more recent gates at Dunkirk, St. Nazaire, and Antwerp, where wooden gates were formerly constructed.

Wood, on the contrary, is invariably used for the dock-gates at Liverpool, which are made of greenheart. At Havre also, wooden gates studded with nails are employed; and the gates at Grimsby, at Avonmouth, and at the recently constructed South Lock at Sunderland are formed of wood.

Iron gates are constructed with a double skin of plates fastened to a framework of horizontal and vertical ribs formed of plates and angle irons. (Plate 15, Figs. 1, 3, 5, 7, and 9; and Plate 16, Figs. 3, 4, 5, 6, and 7.) The increased pressure towards the bottom is provided for by placing the horizontal ribs closer together, making them stronger, and increasing the thickness of the skin plates. The heelpost, meeting-post, and sill-piece are generally made of greenheart.

Wooden gates are composed of a framework of horizontal ribs, connected with the heelpost and meeting-post at each end, to which a sheeting of planks is fastened forming the skin of the gate and retaining the water. (Plate 15, Fig. 2, and Plate 16, Figs. 1 and 2.) The joints of the planking are calked, so as to make the skin watertight; and the beams forming the ribs are placed closer together towards the bottom.

Greenheart is the best wood for gates exposed to salt water, as not only is it generally unassailable by the teredo, but it is also considerably stronger than oak. Some greenheart used in iron dock-gates at Boulogne was injured after several years by the teredo, which is at variance with experience in England; but its life, even in that case, was much longer than unstudded oak at the same port. Teak has been sometimes used for dock-gates, but like greenheart it is costly, and though it possesses considerable power of resistance to strains, it is not durable like greenheart in sea water. Oak is very suitable for dock-gates, and has been largely employed; but the difficulty of procuring pieces of sufficient size for large gates has led to the introduction, in some cases, of creosoted pine.

Occasionally composite gates of iron and wood have been constructed, as for instance at Fecamp, where iron plates were used for strengthening the wooden ribs, and at Cardiff, where a wooden skin was fixed to a wrought iron framework.

Forms of Dock-Gates. Three forms have been adopted for dock-gates, namely, (1) straight gates meeting at an angle, of which the Dunkirk gates are a recent type (Plate 15, Fig. 9); (2) curved gates meeting at an angle and assuming in plan, when closed, the form of a Gothic arch, a more frequently adopted form than the first, and exemplified by the Bute and Penarth gates (Plate 15, Figs. 1 and 7); and (3) segmental gates which are so curved that, when closed, they form together a continuous circular arc, of which the Liverpool, Bristol, and Hull gates are instances. (Plate 15, Figs. 2, 3, and 5.)

The chief advantages of straight gates are, that they admit of less deep recesses in the side walls than curved gates; that they fit a straight sill without requiring any special projecting piece at the bottom; and that they are more simply and cheaply constructed when made of wood.

Gates in the form of a Gothic arch have less curvature than segmental gates with the same rise, and therefore do not necessitate such deep recesses in the side walls. Moreover, the widening out in the centre of the gate, which the transverse strain demands, is advantageous in the moderate-sized iron gates for which this form is adopted, as it leaves more space for the manholes by which access to the inside of the gate for painting and repairs is secured. The advantages also afforded by the segmental form in the reduction of the strains cannot be fully attained in the smaller gates, as the thickness of the plates cannot be reduced in these gates to the full extent indicated by theory under a mere compressive strain, owing to the greater proportionate deterioration by corrosion of thin plates, and the necessity of giving them adequate strength to resist shocks. Accordingly, in most iron dock-gates up to 60 feet span, the Gothic arch form is preferred.

Segmental gates, or cylindrical gates as they are sometimes termed, should theoretically have concentric sides, or skins, in the case of iron gates; and the curve of resultant pressures should lie midway between them. A little divergence, however, at the meeting-posts, from wear, an obstacle getting between the posts when closing, or any other cause, produces a deflection in the curve, and throws the main pressure on to one or other skin. To provide for this deflection, and also to give more space inside, iron gates are usually made thicker towards the centre by giving the skin on the outer face of the gate a larger curvature than the other; so that whilst the curve on the inner face is a continuous segment, the curves on the outer face next the sill form a Gothic arch. The segmental form is very suitable and economical for gates of large span, where iron is employed; but it is not so well adapted for large wooden gates. Nevertheless, the most recent gates at Liverpool, with a span of 65 feet, have been constructed on this type (Plate 15, Fig. 2); but the Avonmouth and Sunderland wooden gates are Gothic arched in form.

The sill is sometimes curved to suit the curve of the outer face of the gate, as for instance at Bristol and Hull (Plate 15, Figs. 3 and 5); but more commonly the sill is made straight, and a straight projecting sill-piece is fastened to the bottom of the curved gate. (Plate 15, Figs. 1, 2, and 7; and Plate 16, Figs. 3 and 4.)

Sluices are made near the bottom of the gates, which are closed by sliding doors raised or lowered by a rod fastened to each door, terminating in a screw which is turned from the top of the gate.

Methods of Supporting Dock-Gates. The heelpost of a dock-gate is supported, and turns, at the bottom, on a steel pivot let into the heelpost stone; and it is held against the hollow quoins, at the top, by an anchor, encircling its head, built into the wall and further secured by long tie bars at the back. These constitute the sort of hinge upon which the gate revolves. As there can be no intermediate support between the top and bottom of the heelpost, owing to the necessity of maintaining a watertight joint between the hollow quoins and the heelpost, a large and heavy gate would tend to drop at the outer end, unless very efficiently tied back at the top, and would make the heelpost leave the hollow quoins at the top and crush against them at the bottom. This tendency is, however, to a great extent counteracted by the natural buoyancy of the gates. Wooden gates indeed, when constructed of light wood, are sometimes too buoyant at first, and have to be weighted to keep them down when fully immersed; but after a time they get water-logged, and the ballast can be removed. The air also inside watertight iron gates with double skins affords an excess of buoyancy, which has to be provided against by the admission of some water as ballast.

The aforesaid considerations might lead to the conclusion that a dock-gate should be easily supported, and swing round on its heelpost as a well-balanced door on its hinges. Wooden gates, however, when constructed of the better and denser class of timber, together with their iron fastenings and adjuncts, and their platforms and other portions above water, have a greater weight than the water they displace, even at the highest tides; and this is still more the case when they are moved at a lower state of the tide. Iron gates, also, have to be over-ballasted to provide for the highest water-level, and therefore have a considerable excess of weight when the water is some feet lower, and this weight is liable to be increased by leakage. Accordingly, the theoretically possible perfect state of equilibrium is not generally attained; and consequently where there are considerable variations of water-level, the common practice is to provide an additional support for the gate in the form of a roller near its outer end, running on a curved roller-path fixed on the gate-floor. (Plate 15, Figs. 1, 2, 3, 5, and 7, and Plate 16, Fig. 4.) The centre of the roller is placed so as to lie in the same straight line as the centre of gravity of the gate and the centre of the heelpost, which in curved gates falls outside the gate. The roller is made slightly conical, for rolling along its curved path; and it is capable of adjustment, by means of a rod reaching to the top of the gate, so that it may always be made to bear upon its path. The use of a roller adds to the friction in moving a gate, which is much increased by any irregularity in the roller-path. This circumstance, and the fact that occasionally rollers have been found not to bear upon their path, thereby merely adding to the weight of the gate, has led to proposals for their abandonment. They are unnecessary for small gates, and even for large gates, where, as in Holland, there is little variation in the water-level. Iron gates can, moreover, be designed so that their flotation is self-adjusting with the rise and fall of the water, as accomplished at South Shields and St. Nazaire; and their counterbalancing weight of water ballast can be placed close to the heelpost, which reduces the strain upon the anchor straps. By this latter arrangement, the iron gates at Dunkirk, across a lock having a width of 69 feet, have been constructed without rollers. (Plate 15, Fig. 9.) Care, however, must be taken in such cases to remove the water that may leak into the air compartments. Wooden gates are not capable of similar adjustment.

Methods of moving Dock-Gates. Chains are generally used for opening and closing dock-gates, a separate chain being employed for each operation, so that four chains are required for a pair of gates. The chains are wound round drums, which are turned by capstans or by hydraulic engines. On account of the resistance offered by the roller, it is advisable to attach the chains more than half way down the gate, near the line of support of the roller, in order to facilitate the motion, and to avoid straining the gate. This necessitates the construction of two chain-passages in the side walls, for the opening and closing chain of each gate, with horizontal and vertical rollers for guiding them.

At Bordeaux, where the first application in France of hydraulic power for working gates has been recently made, the four chains of one pair of lock-gates are worked by a single hydraulic engine, on one of the side walls, with two pistons, one serving for opening and the other for closing the gates.

The chains fastened to the pistons are so connected together, directed by pulleys, and counterpoised on the opposite side wall, that the gates are opened or closed simultaneously[1].

Chains are liable to get in the way of vessels passing through; and the closing chains, lying on the top of the sill, diminish slightly the available depth, especially where the two chains cross in the centre. When gates are constructed without rollers, chains can be dispensed with; and a single bar attached to the top of the gate, and worked by a rack and pinion on the quay, serves at Antwerp for opening or closing each gate. It has been proposed to adapt hydraulic power to this method of moving dock-gates by attaching the piston of a hydraulic engine to the back of each gate, which would be a great improvement, both in simplicity and rapidity, on the ordinary system with chains.

Cost of Dock-Gates. Though the security and facility of access for vessels frequenting docks depends greatly upon the strength, durability, and ready movement of the gates, yet the gates form a comparatively small item in the cost of dock works. Accordingly, in designing dock-gates, durability and facility in manipulation are of more consequence than a saving in the first cost. The cost of iron gates varies considerably at different times, owing to fluctuations in the price of iron; so that whilst wood was substituted, in place of the first design in iron, for the Avonmouth dock-gates, owing to the high price of iron in 1873, some old wooden gates at St. Nazaire have been replaced recently by iron gates at nearly the same cost.

The iron gates erected at the South West India Dock in 1869 cost £4000 per pair. They are cellular iron gates of a Gothic-arched form, spanning an opening of 55 feet, with a rise of one-fourth, and resting against a sill which is 27 feet below high water of ordinary spring tides.

The Avonmouth wooden gates, having a similar form and

[1] 'Annales des Ponts et Chaussées,' 6th Series, vol. i, p. 540.

rise, across an opening of 70 feet, and with a depth of water of 36 feet on the sill at high water, cost £5586 per pair.

The gates across the large entrance to the Eure Dock at Havre, 100 feet wide, are made of wood, and cost £16,700 per pair. The depth of water on the sill at high water is 35 feet.

At St. Nazaire, a pair of gates, made of pitch pine and studded, which were placed across the entrance, 82 feet wide, in 1858, cost £8800, and required renewal in 1879. As the average cost of the four pairs of iron gates at the Penhouët Lock, of the same width, only amounted to £10,000 per pair, it was decided to construct the new entrance gates of iron, as the difference in cost is far more than compensated for by the much longer life of the iron gates. The depth of water on the sill at high water is 29 feet.

The cost of dock-gates has sometimes been estimated per square foot, at prices varying from £1 5*s*. to £2; but it is evident that such a method of calculation can only furnish a very rough approximation, as the cost of each square foot of the gate depends not merely upon its span, as in the case of bridges, but also upon the head of water which it may have to sustain. Thus a gate at Liverpool or Cardiff (Plate 16, Figs. 2 and 4) has to be made considerably stronger than a gate at Hull or Dunkirk (Plate 16, Figs. 6 and 7) of the same span, owing to the increased pressure on the lower portion from the great rise of tide in the Mersey and the Severn estuaries. The probable life of a gate should also enter into any comparative estimate of cost. Taking, however, a superficial foot of surface of the gate as the unit of cost in the examples given above, the cost is as follows: South West India Dock, iron gates, £1 19*s*.; Avonmouth, wooden gates, £1 6*s*.; Havre, wooden gates, £4 10*s*.; St. Nazaire, wooden gates, £2 18*s*., and iron gates, £3 6*s*. The large price of the Havre gates may be accounted for by their large span, combined with a considerable head of water and their straight form, together with the additional expense of providing struts for the outer gates against the swell entering the harbour, and the studding as a protection against the teredo.

Timber is undoubtedly the cheaper material for small gates; but iron is more economical for large gates, especially if its greater durability, and the power of adjusting the flotation of iron gates and, consequently, facilitating their working, are taking into account.

CAISSONS.

Wrought-iron caissons are generally employed for closing the entrances to graving, or dry docks, in preference to the ordinary form of dock-gate; and at Government dockyards, they are frequently adopted also at entrances and locks, instead of the dock-gates so invariably used at commercial ports. There are two forms of caissons; namely, Floating Caissons, and Sliding Caissons, the former being generally used at graving docks and entrances, and the latter at locks. (Plate 16, Figs. 8 and 9.)

Floating Caissons. Floating caissons are constructed somewhat in the shape of a vessel in cross section, and in consequence are sometimes called ship-caissons[1]. (Plate 16, Fig. 9.) They are formed with plate-iron sides stiffened with angle irons, diagonal braces and horizontal decks. The decks and bulkheads divide them into several watertight compartments. Some of these compartments form air-chambers which provide the necessary flotation, whilst others are ballasted. The caisson is, in this way, so adjusted that it can be floated in or out of position. When the entrance is to be closed, the caisson is floated into its place against a projection on each side of the side walls of the entrance, water is admitted into the air-chambers, and the caisson sinks into a recess at the bottom in which its keel fits. When the water

[1] The illustration given in the Plate is a section of a caisson at Devonport, of which a drawing was furnished me by Col. Percy Smith, R. E., Director of Works to the Admiralty.

is lowered on one side of the caisson, either by the fall of the tide in the case of an entrance, or by the pumping out of the water from a graving dock, the head of water on the other side presses the caisson against the sill and the projections of the side walls, the keel of the caisson and the sides in contact with the masonry being fitted with wooden pieces to prevent leakage. The caisson is removed by pumping the water out of the air-chambers; and as soon as the water is level on both sides, the caisson rises, and can be towed away, leaving the passage open. By battering the side walls and the ends of the caisson, the rising of the caisson, on removal of the water, frees it from the recesses in which it fits when sunk in position.

A floating caisson is cheap to construct, and dispenses with the costly accessories of gate-floors, roller-paths, gate-recesses, chain-passages, and gate-machines. Its upper deck also forms a very strong wide bridge, so that no necessity arises for the erection of a swing-bridge. The size, however, of a floating caisson is necessarily large to secure stability when floating and consequently involves more material than dock-gates, as may be readily seen by a comparison of their sections. (Plate 16, Figs. 3, 4, and 9.) Moreover a caisson occupies much more time in being floated in and out of position than the shutting and opening of dock-gates, a very important point in entrances to ports, but of little consequence for graving docks where the opening and closing have to be performed comparatively seldom. Another objection to its employment at entrances to crowded docks is the space which has to be provided for it when out of position. These objections have prevented the adoption of floating caissons for commercial wet docks, though for graving docks their use is very common.

As the traffic through the entrances to the basins of Government dockyards is much less than for commercial ports, and as strong bridges are very useful for the passage of the heavy loads required in naval equipments, floating caissons are frequently used for closing the entrances to these docks, as exemplified at Portsmouth and Chatham.

It is proposed to utilise the floating caissons at the entrances to the graving docks at Dunkirk, if necessary, for the repairs of the new lock at that port (Plate 15, Fig. 9), a recess having been formed at each end of the lock against which the ordinary caissons can be placed, and the lock pumped dry. This will render the operation of enclosing the lock for repairs, and its re-opening, much simpler and more rapid than the usual method provided by two parallel grooves at each end for temporary cofferdams.

Floating caissons are sometimes made box-shaped of iron or of wood.

Sliding Caissons. A ship-caisson is unsuitable for closing locks, on account of the impediment to the traffic which it offers in being moved, besides the time which its adjustment or removal occupies. Accordingly, caissons have been made, with parallel vertical sides (Plate 16, Fig. 8), sliding along ways in grooves across the lock, or into long chambers formed for them in one of the side walls, thus closing or opening the lock. This type of caisson has watertight compartments, like a ship-caisson, and it is ballasted to adjust its flotation, and fixed, when in place, by the addition of water ballast. Its meeting faces are likewise lined with timber. It differs, however, from a ship-caisson in having a chamber provided for its reception, into or out of which it is readily hauled by chains, the operation being thus rapidly effected, and quite independently of the state of the weather, which sometimes seriously impedes the movements of a ship-caisson. The top deck of the sliding caisson, which forms a bridge, is made capable of being lowered, so as to slide under the roadway, which is supported on girders, and covers the top of the chamber. If repairs are needed, the caisson can be shut into its chamber by a temporary dam at the end, and the recess pumped dry.

The sliding caisson possesses the same advantages as the ship-caisson, of taking up little length in the lock, and of providing a very strong movable bridge. It is however costly to construct, with its large chamber which needs a good deal of space at the side, and requires powerful machinery for moving it. When there are two adjacent locks, the intermediate space can be utilised for the recesses; and the positions of the caissons at the two entrance locks to the Fitting-out Basin of the Chatham Dockyard extension works, have been so arranged that the chambers of all the four sliding caissons, which close the locks, have been placed in the pier between the two locks, where the hydraulic machinery has been also erected. (Plate 12, Fig. 3.)

At Portsmouth Dockyard, there are seven sliding caissons in the extension works, two at the outer ends of the double locks leading to the Repairing Basin, one at the entrance to the adjacent deep graving dock, and four across the entrances to the other basins and the passages between them. (Plate 12, Fig. 5.) There is also a sliding caisson at Haulbowline Island Dock in Queenstown Harbour (Plate 12, Fig. 4), and across the lock and a graving dock of the docks at Milford Haven; and one of the earliest examples of a sliding caisson was placed across a graving dock at Malta several years ago. The inner caissons of the Portsmouth locks are ship-caissons.

Provision has to be made at these dockyards for admitting a vessel at various states of the tide, and also that the locks may be used, in an emergency, as graving docks. Accordingly, if ordinary dock-gates had been adopted, it would have been essential to have placed reverse gates at each end, like at the St. Nazaire lock; and consequently each sliding caisson in a lock or at an entrance performs the functions of two pair of dock-gates, as well as providing a movable bridge; and these advantages ought to be taken into consideration in comparing the cost of the two methods.

Rolling caissons differ from sliding caissons merely in having rollers attached to the bottom of the caisson, or in sliding upon rollers fixed on the floor of the entrance, which guide and facilitate the motion. Several of these caissons have been constructed for closing graving docks.

GRAVING DOCKS.

In order that the lower portions of the outer shell of a vessel may be cleaned or repaired, it is necessary for the vessel to be placed high and dry, where it can be adequately supported, where there is space and light for the repairs to be properly executed, and from whence it can be readily floated out again. These conditions are fulfilled by graving docks, or dry docks as they are sometimes termed in contradistinction to the ordinary floating or wet docks.

An ordinary graving dock is a chamber sufficiently large and deep to receive the largest sized vessel which generally frequents the particular port. It communicates with a wet dock, or has access to deep water outside; and its entrance is closed by a caisson or a pair of dock-gates[1]. (Plate 16, Figs. 10 to 14.) The sides of a graving dock are formed somewhat to the shape of a vessel, and are constructed with steps, or altars, for receiving the timber props which support the vessel in an upright position. Keel-blocks along the bottom sustain the keel of the ship. Slides are formed, at intervals, down the side walls for lowering beams and other materials for repairs; and flights of steps are also provided to afford access to the bottom. Culverts are constructed, leading from the bottom of the dock, which convey the water standing in the dock, or draining into it, to a well from which it is removed by pumping.

When the graving dock opens on to a tidal river, the vessel is introduced at high water, the entrance is closed, and the

[1] Drawings of the Devonport and Chatham graving docks were given me by Colonel Percy Smith, R. E., from which the illustrations at the end of Plate 16 have been made.

water is run off through sluices at low water; and only the water remaining below low water, or draining into the dock, has to be pumped out. The vessel is propped up on each side as the water lowers, and is then ready for inspection and repairs. When the repairs have been completed the water is re-admitted, and the entrance is opened and the vessel floated out at high tide.

Where graving docks are placed in tideless harbours, or at the sides of wet docks, the admission of the vessel is easier than when exposed to the currents of a tidal river, and can be accomplished at any time. The water, however, in these cases has to be pumped out of the dock, for which purpose special pumps are provided near the dock.

Solid foundations have to be formed for graving docks, as unequal settlement caused by the weight of the vessel inside would injure it, and also tend to produce cracks and leakage. The amount of leakage depends upon the nature of the soil, and the material of the side walls. Thus the Herculaneum graving docks at the southern end of the Liverpool Docks, excavated in the red sandstone rock, are much less watertight than the concrete graving docks at the northern end, owing to the percolation which takes place through the porous red sandstone.

Materials for Graving Docks. Dry docks are lined with timber, brickwork, concrete, or masonry. The earlier and smaller docks in this country were lined with timber, and provided with a brick or concrete floor, and a brick or masonry entrance. As vessels increased in size, larger docks became necessary; and the most important dry docks are lined with masonry, granite being largely used for the altars, slides, steps, and copings of the largest docks in the Government dockyards, where great durability and strength are required. Within recent years, concrete has been introduced for the entire lining and side walls of graving docks, of which the dry docks at the Albert Dock, London, are examples. (Fig. 24.)

The face concrete, however, requires to be made with special care to withstand the rough usage to which the edges of the altars are exposed.

CONCRETE GRAVING DOCK.
Albert Dock, London.

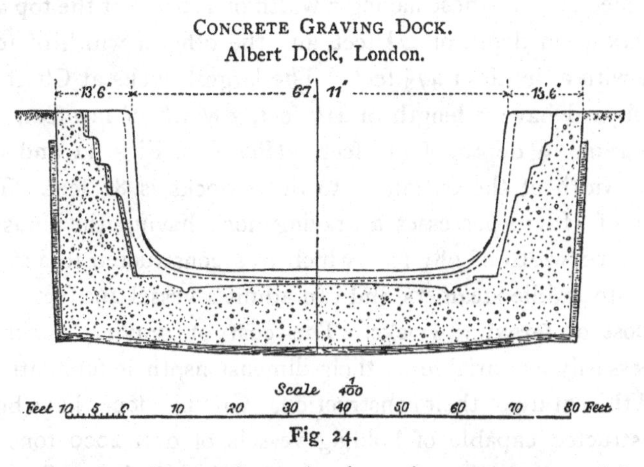

Fig. 24.

Though dry docks of granite have been constructed in the United States, the preference is now being given to timber docks, not merely on account of their cheapness, but also because they can be more rapidly constructed, are less injured by frost, are drier, and are lighter and more accessible with their narrow altars and gently sloping sides. The two largest dry docks in the United States, situated at Brooklyn, have been constructed of timber up to high water; they are 510 and 610 feet long, and 111 and 124 feet wide, respectively, and capable of docking the largest ocean-going steamers. The same system has also been favourably entertained for dry docks in Canada. In North America, however, timber is very cheap, and the frosts severe; and though the repairs of the timber docks are stated to be small, portions would need renewal after a period of twenty years.

Sizes of Graving Docks. The dimensions of graving docks are very variable, being suited for various classes of vessels. according to the period of their construction, the requirements of the port, and the special type of vessels for which they are designed. Some of the biggest graving docks have been built at Government dockyards to accommodate the largest ironclads. There are three at Portsmouth having a length of 491¼ feet, two of these having a width of 110 feet at the top and a maximum depth of 42½ feet, and the other a width of 101½ feet with a depth of 47⅜ feet. The largest docks at Chatham Dockyard have a length of 410 feet, a width of 110 feet, and a maximum depth of 41½ feet. (Plate 16, Figs. 11 and 14.) The width of the entrances to these docks is 82 feet. The port of Havre possesses a graving dock having the unusual entrance width of 98¾ feet, which was constructed to accommodate paddle steamers, and can admit a vessel 490 feet long.

Cost of Graving Docks. The cost of graving docks is necessarily as variable as their dimensions, their foundations, and the nature of their construction. Graving docks have been constructed capable of holding vessels of over 2000 tons for less than £20,000; whereas some graving docks at Spezzia cost £78,000 each; the Somerset Dock at Malta, 468 feet long, cost £150,000; the dock at Table Bay, 500 feet long at the bottom, cost £156,700; and the large dock at Havre cost £175,000. Reckoning, however, by the more comparable method of cost per unit of contents, it has been found that their cost varies from £1 to over £4 per cubic yard, the cost of the most recent docks at Portsmouth averaging about £2.

Gridirons. A framework of crossed timber beams, supported on piles or other solid foundations, is sometimes laid upon the foreshore of tidal rivers, where the range of tide is considerable, somewhat above the level of low water. The vessel to be inspected is floated over the gridiron at high water, and moored in position; and as the tide falls, the vessel settles upon the gridiron, and is out of water at low tide. The vessel can then be very easily examined, its bottom cleaned, and small repairs executed, during low water, without the trouble of docking it. A gridiron thus forms a very useful appendage to a port; and moderate-sized ports, such as Newport on the Usk, and Garston on the Mersey, are provided with them, being suitably situated for the purpose.

Floating Dry Docks. Where the absence of suitable foundations, or other considerations, render the construction of a graving dock inexpedient, floating docks have been adopted, consisting of an open box-shaped pontoon with one movable side. By ballasting the pontoon with water, it is sufficiently lowered for the vessel to enter by the open side. The vessel is then secured, the movable side is put in place, and as the water is pumped out the vessel rises out of water on the floor of the pontoon. Sometimes the caissons at the ends are dispensed with where sufficient stability is ensured by a broad rectangular base.

Sir F. Bramwell designed a floating dock for St. Thomas' with open lattice girder sides, thus affording ample light and air. The flotation was provided by the pontoon at the bottom, and adjusted by movable floats at the sides.

These docks were formerly constructed of wood, one of which is in existence at Havre; but the recent and larger types have been formed of iron. They can be constructed in any suitable locality, and either sent to their destination in separate pieces, or towed out, as in the case of the well-known Bermuda Dock which was launched in the Thames and towed out to Bermuda.

Hydraulic Lift Graving Dock. Another system of raising a vessel out of water for repairs was adopted by Mr. E. Clarke, at the Victoria graving docks, London [1]. Two rows of cast-iron columns were sunk, on each side of the entrance to the series of graving docks, 60 feet apart; each column contains a hydraulic ram which has a lift of 25 feet, and these rams can raise a series of girders, stretching across the entrance, from the bottom of the lift pit up to high water. When a vessel is to be docked, a pontoon, about 58 feet wide and of a suitable length, is sunk on the girders at the bottom of the lift pit,

[1] 'Minutes of Proceedings, Inst. C. E.,' vol. xxv, p. 291.

with keel-blocks and side-blocks upon it; and the vessel is then floated into the lift pit, and moored. The hydraulic power is then admitted to the presses, and the girders rise, lifting the pontoon with the vessel upon it clear of the water, when the water flows out of the open valves of the pontoon, which are then closed. The girders are then lowered, but the pontoon with the vessel upon it remains floating on the water, and, drawing only from 4 to 6 feet, can be hauled into one of the shallow docks opening out of the graving-dock bas n. The lift is only 310 feet long, but it can receive a vessel of 350 feet; it cost £25,500 with the machinery, and can raise a weight of 5,780 tons. The only additional expenses required, to provide for the docking of as many vessels as the lift can serve, are the pontoons, the shortest of which is 134 feet and cost £3,600, and the longest 321 feet costing £10,850. To render the system economical, it is necessary that several pontoons should be frequently in use.

Depositing Dry Dock. Another form of repairing dock has been invented by Messrs. Clark and Stanfield, resembling a floating dry dock in being provided with pontoons, which being brought under a vessel can raise it out of water; but it differs from that system in simply serving to convey the vessel to a suitable gridiron jetty, on which it is deposited, leaving the dock free to raise another vessel.

This dock consists of a central oblong box pontoon, carrying the machinery, to one side of which are fastened a series of parallel projecting pontoons, forming a row of brackets on which the vessel can rest. On the other side of the central pontoon, a hinged flat floating outrigger is attached, which remains on the surface of the water whatever may be the immersion of the pontoons, and whose function is to maintain the equilibrium of the dock during the raising or lowering of the vessel on the projecting pontoons[1].

[1] An illustration of this depositing dock is given in 'Engineering,' vol. xxxv, p. 512.

A long timber staging is erected near the shore, with narrow projecting piers, 5 feet wide, placed 15 feet apart so that the projecting pontoons of the dock can be inserted in the intervals between the piers.

When a vessel is to be docked, the projecting pontoons are brought underneath it; the water is pumped out of the pontoons, and the vessel rests on the series of pontoons; the vessel is then conveyed, on the dock, over the staging near the shore, and when it is in place, water is admitted into the pontoons, which settle between the piers and deposit the vessel gently on the staging. A reverse operation lifts the vessel off again, and floats it in deep water when its repairs have been completed. The advantage of the system is that the same dock serves for a number of vessels, the only additional requisite being a sufficient length of staging which can be easily and cheaply provided.

Three docks of this kind have been constructed. The first was sent to Nicolaieff on the Black Sea, and was subsequently transferred to Sebastopol; the second was made for Vladivostok, a Russian port on the Sea of Japan; and the third has been erected at Barrow. This latter dock has been formed in two equal parts, each capable of lifting a vessel of about 1500 tons, and together able to raise a vessel of about 3200 tons; and the construction of a third length has been arranged, so that vessels of 5000 tons may be docked.

Slipways. Vessels are sometimes hauled up slipways for repairs, instead of being placed in a graving dock. The vessel is deposited on a cradle, and the hauling is effected by hydraulic power. The advantages of such slipways for moderate-sized vessels are, the better light and ventilation afforded on a slipway than in a graving dock, less cost of construction in certain cases, and freedom from the pumping needed for draining a dock. Against these must be set the disadvantages of the greater space occupied by the long slope of a slip, the length of the foundations, and the difficulty and risk in landing a vessel in an exposed situation. Shipowners would hesitate to entrust a large vessel to a slipway; and if the foundations settle irregularly, the vessel is exposed to strains, so that graving docks are generally preferred. Slipways are not suitable on tideless shores, as at such sites the slipway has to be constructed for a considerable distance below the water-level to reach a sufficient depth of water with its gentle inclination.

Remarks on Graving Docks. Every port of any consequence should be provided with facilities for repairing vessels, as the absence of such a provision is injurious to the prospects of a port. Accordingly, though graving docks have not always proved profitable undertakings for dock companies who do not undertake the repairs, they are under any circumstances indirectly beneficial by enhancing the attractions of a port. They are specially valuable at isolated harbours, such as Table Bay, where the first graving dock was opened in 1882.

The dry docks of the port of London are mainly in the hands of private owners; and till the Victoria Dock Company constructed the system of eight hydraulic graving docks referred to above, the London dock companies did not possess any graving docks. Two concrete dry docks were constructed at the same time as the Albert Dock, and in communication with it (Fig. 24, p. 459), and one has been built at the Millwall Docks; so that 11 docks constitute the accommodation provided by the London dock companies. Where, however, ample facilities exist, as on the banks of the Thames for repairing vessels, it is merely a matter of greater convenience for the dry docks to adjoin the wet docks; but in most docks of importance, graving docks form an essential part of the system. Thus there are 24 dry docks at Liverpool, the three at the Herculaneum Dock being from 753 to 768 feet long, and consequently capable of admitting two fairly large vessels together. The Birkenhead Docks possess three long docks, the largest being 930 feet in length. There are eight dry docks at Cardiff; six at

Bristol and Antwerp; four at Leith, Marseilles, and Havre; three at Hull, Belfast, Hartlepool, and St. Nazaire; two at Grimsby and Sunderland; and one at Barrow and Newport.

Graving docks are so extensively used at Government dockyards, both for repairs and construction, that the large dockyards are very amply supplied with them. For instance, at Portsmouth there are fourteen docks, and three slips; at Chatham, there are eight docks, and seven slips; and at Cherbourg, there are nineteen docks, two of which, excavated in rock, are being enlarged.

The Devonport graving dock, shown in Plate 16, Figs. 10, 12, and 13, has been specially constructed with a roadway on arches, on each side, over the upper altars, so that the travelling cranes, conveying the armour plates for ironclads, may lower their loads close to the side of the vessel. This system, however, whilst facilitating the work, reduces the amount of light and air at the sides of the dock.

The ordinary graving dock, opening out of a sheltered basin, possesses the important qualification of absolute security, which has undoubtedly led to its very general adoption in preference to floating docks, pontoons, and slipways, except under peculiar conditions, as these other methods are not equally safe, and are liable to be affected by the state of the weather.

MOVABLE BRIDGES.

As direct means of communication has to be provided to the various quays of docks, and as locks and passages between docks frequently traverse public roadways, movable bridges generally form essential accessories to dock works. Moreover, it is important that neither the traffic over these bridges, nor the passage of vessels, should be impeded longer than is absolutely necessary, and therefore the bridges have to be constructed so as to be readily opened or closed.

The type of movable bridge generally adopted at docks is the swing-bridge, which is supported and revolves horizontally on a pivot. Bascule and traversing bridges are also occasionally employed.

Swing-Bridges. The most common form of swing-bridge, which used to be placed across locks, consists of two segments, meeting in the centre of the span, formed with cast-iron arched ribs supporting the roadway. Each segment revolves on a pivot and a ring of rollers on the side walls of the lock, the swing portion being balanced by a prolongation of the bridge, on the shore side of the pivot, called the tail-end. The segments act rather as cantilevers than as an arch, but they afford each other mutual support when meeting in the centre; and the rise towards the centre enables low craft to pass under the bridge, without the necessity of opening it. The bridge is moved by hand, by aid of gearing connected with each segment. This type of bridge was convenient for roadway traffic across small spans; it has been superseded by more modern forms, but several old specimens are still in existence.

When railways had to be carried across, and the spans were increased, it became necessary to modify the type; and the ordinary form, at the present time, is a wrought-iron girder bridge spanning the whole opening and resting at one side on a hydraulic press. The swing portion, spanning the lock, is balanced by a tail-end which, being made shorter than the other, is weighted with cast-iron kentledge. This arrangement necessitates a heavier bridge than if it was divided into two halves, but it forms a stronger bridge, and, being level, is suitable for a railway. Moreover the machinery is concentrated at one side, and the motion controlled by one man. When it is in position across the lock, it rests upon bed plates near the edge of the walls on each side. The bridge is turned by admitting water to the press, which raises the bridge off its bearings, and it is then swung round on its water centre by means of a chain encircling a grooved drum fastened under the bridge over its pivot. The tail-end is always made slightly heavier than the other, so as to prevent any chance of the swing portion overbalancing; and two rollers are placed at the extremity of the tail-end, on which it rests lightly when turning.

By aid of hydraulic power, these bridges are rapidly swung round; so that a swing-bridge at the Marseilles docks, spanning an opening of 92 feet and weighing 684 tons, is opened by one man in three minutes.

Occasionally, when there is not adequate available length of quay for the bridge to rest along, when open, if formed in one length, two equal segments are constructed turning, like the early type of bridges, on each side wall. This arrangement has been adopted for a swing-bridge at Barrow, crossing the short passage, 80 feet wide, between the Buccleuch and Ramsden docks.

Traversing Bridges. When a bridge crosses a short passage where the quay space is limited, a traversing bridge is sometimes adopted. This type of bridge rolls backwards and forwards, away from and over the opening which it spans, occupying no space along the quay beyond its own width. A traversing bridge has been erected across the passage between the outer and inner Millwall Docks, 80 feet in width. The bridge has a counterbalancing tail-end like a swing-bridge, and it is lifted and moved by hydraulic power. In order to open the passage, the bridge is first lifted by a press, so that its underside is just above the level of the approach roadway, and then it is drawn back by means of chains, worked by hydraulic pistons, and runs on rollers let into the roadway under the path of each main girder. Similar bridges have been erected over the Penhouët Lock at St. Nazaire with a span of 82 feet, and over the entrance to the Kattendyk Dock at Antwerp, having a span of 88½ feet. This type of bridge occupies a little more time in opening than a swing-bridge, and requires a larger expenditure of power.

Bascule Bridges. A bascule bridge is balanced on a horizontal axis, round which it revolves. It is generally constructed in two halves, meeting in the centre of the span, and when it is opened, the portion forming the bridge rises with its end in the air, and the tail-end descends into a pit in the abutment. Instances of this type of bridge exist at the Surrey Commercial Docks and at Havre. A bascule bridge, like a traversing bridge, takes up no quay space beyond its own width, and moreover, when open, occupies no portion of the roadway beyond, so that it interferes less than any other form of movable bridge with the quay or its approaches. Its high end, however, when open, is liable to catch in the rigging of vessels passing through the lock; it exposes a large surface to the wind, and it is not suitable for large spans [1].

The use of steel which, owing to its superior strength over wrought-iron, has proved so useful in enabling the weights of large bridges to be reduced, would be specially advantageous for movable bridges, as by reducing the weight of the actual bridge, the counterbalancing weight of the tail-end could be equally diminished, and thus a double saving effected.

[1] A more detailed account of the various forms of movable bridges, with illustrations, will be found in 'Rivers and Canals,' pp. 145-152, and plate 8.

CHAPTER XXII.

VARIOUS WORKS AND APPLIANCES FOR DOCKS.

Quays: Arrangement; Proportion of length to water area; Long lines; Increase by Jetties; Width; Position of Sheds and Sidings. *Mooring Posts:* Ordinary construction; Position; Modifications. *Capstans:* Position. *Buoys:* Construction. *Cranes:* Varieties; Advantages of Hydraulic Cranes. *Coal Tips:* Balance Tip; Hydraulic Tip. *Sheds:* Construction. *Warehouses:* Construction; Provisions against Fire; Cost. *Granaries:* Distribution of Grain. *Hydraulic Machinery:* Advantages; Supply of Power; Accumulator. *Compressed-Air System:* compared with Hydraulic System; Disadvantages. *Maintenance of Docks:* Ingress of Silt, removal, prevention.

THE space on the quays, the arrangement of sidings, the machinery provided for unloading and loading vessels, the shelter and storage room afforded for goods, and the provisions made for their distribution, form almost as important elements in facilitating the trade of a port as the accommodation afforded for vessels by the locks, basins, and docks. The capacity of a port depends, not merely upon the actual area of its docks, but also upon the rapidity with which vessels can be discharged and sent to sea again; and this is regulated by the length of quays, the power and number of the mechanical appliances, and the facilities for removing the merchandise, as soon as it is landed, either into warehouses, or to its final destination.

Quays. Two distinct points have to be considered in the laying out of quays, namely, the actual length of quays for a given area of docks, and the width, or superficial area that can be provided. The first point is determined by the form of the dock; the second is regulated partly by the form of the dock, and partly by its position in relation to the available area. The length of quay fixes the number of vessels, of any given size, that can be discharging or taking in cargoes at the same time; whilst the area and form of the quay space regulate the extent and arrangement of the sidings, sheds, and warehouses. The length of quay, for any given enclosed water area, is in a great measure under the control of the engineer in designing the form of a dock; but the form and extent of the quay space, though they may be modified by the position given to the dock, are very dependent upon the amount and shape of the land that can be acquired. Moreover, the arrangement of the sidings and sheds must be suited to the special requirements of the particular trade of the port, and the position of the main line of railway.

The proportion of the length of quays to water area varies considerably in different docks, and it is naturally greatest where the docks have a small average area, as at Liverpool (Plate 11, Fig. 1), or where, though large, they are narrow, as at Hull and Penarth. (Plate 11, Figs. 4 and 6.) On the contrary, where the docks are wide in relation to their length, the proportionate length of quays is small, as for instance at the Surrey Commercial Docks and Barrow. (Plate 10, Fig. 1, and Plate 11, Fig. 2.) Taking the number of lineal yards of quays per acre of water area, it appears that there are between 100 and 120 lineal yards of quays per acre of dock at the London and St. Katherine Docks, Liverpool, Hull, Leith, and Belfast; between 80 and 100 yards at Millwall, Birkenhead, Grimsby, Hartlepool, Sunderland, Penarth, and Antwerp; between 60 and 80 yards at the Albert and Victoria Docks, the East and West India Docks, Cardiff, Portsmouth, and Havre; whilst there are only between 50 and 60 lineal yards of quays per acre of dock at the Surrey Commercial Docks, Barrow, Chatham, and St. Nazaire. This proportion, however, must not be accepted as an absolute criterion of the comparative value of each acre of water area; for a port possessing a favourable proportion of quays might be cut up into a series of inconveniently small docks, entailing a large cost for passages between them, and awkward for the movement of vessels unless provided with numerous entrances; or long docks might be too narrow for the convenient transit of vessels. The Albert Dock, London, for instance, might have had its quay length increased by being divided into a series of basins; but the dock would have been thereby rendered less convenient, and its cost would have been considerably increased.

A long continuous line of quay is very convenient for traffic, and is adapted for all sizes of vessels, so that no space need be lost; and these advantages more than compensate for the loss of quay length in proportion to water area. The Albert Dock furnishes the finest example in the world of this simple arrangement (Plate 10, Fig. 1); and the West India Docks (Plate 10, Fig. 1), the Hull Albert and West Docks (Plate 11, Fig. 4), and the Cardiff and Penarth Docks (Plate 11, Fig. 6) possess very conveniently formed quays.

Where a long length is not attainable, a large increase of quay length can be gained, in a wide dock, by projecting jetties which afford great facilities for traffic, of which the Victoria Dock, London, the Alexandra Dock, Liverpool, the Edinburgh Dock, Leith, and the Freycinet Docks in course of construction at Dunkirk are good instances. (Plate 1, Fig. 7, Plate 10, Figs. 1 and 2, and Plate 11, Fig. 1.) The Tilbury Docks on the Thames are also being constructed on this principle. (Fig. 26, p. 500.) The Marseilles and Trieste Docks are examples of a similar system applied to tideless ports. (Plate 3, Figs. 4 and 7.)

The large number of steamers now employed in the shipping trade, which within recent years have replaced to a great extent the sailing vessels of former days, require a wide extent of quay. In order that these steamers may stay as short a time as possible in the dock, it is necessary to provide sufficient accommodation on the quay both for the cargo which a steamer discharges, and that which it has to take on board. On a well-arranged quay, the part next the dock, for a width of about 30 feet, should be devoted to the mooring posts, a line of way for the travelling hydraulic cranes, and an ordinary line of way for trucks beyond, so that the cranes may unload direct into the trucks or load from them. A covered space, of over a hundred feet in width, should be provided alongside these inner lines for the sorting and inspection of general merchandise; and on the further side of these sheds, more lines of way and a cart road should be also laid down, taking up altogether a width of quay of from about 250 to 300 feet. This would be the suitable arrangement for cargoes despatched at once from the docks. Where, however, goods have to be stored for a time at the docks, warehouses may be placed directly behind the sheds; or in some cases where the quay is devoted to a special trade, such as grain, the warehouses may be placed closer to the dock, so that the cranes may discharge direct into the warehouse. On quays set apart to the timber trade, it is only necessary to provide sufficient vacant space in front for the vessel's cargo, and an ample area of stacking ground behind, covered over in some instances by high sheds.

The lines of way on the quays should be connected together, as well as to the main line, to facilitate the interchange of wagons. This can be effected with the least amount of space by means of turntables, which are very convenient but costly. Moreover, as the wagons travel very slowly in the docks, very sharp curves can be adopted in laying down the roads, which facilitates the access to various parts of the quays and the connections between the lines. The railways along the quays of the Penhouët Dock at St. Nazaire have been provided with a very complete system of intercommunication by turntables (Plate 13, Fig. 4); whilst the sidings at Cardiff and Grimsby are connected by sharp curves. (Plate 11, Figs. 6 and 7.)

Mooring Posts. A series of mooring posts are fixed at intervals along the quays, to which vessels can make fast whilst lying alongside. These posts, made of stone or cast-iron, are placed usually a little distance behind the quay wall, and used to be merely buried partially in the ground, and further secured by iron tie-rods. A better system, adopted at the South West India Dock, consisted in forming a large block of concrete-in-mass round the lower part of the post, as shown in Fig. 25. Old cannons are frequently used for the purpose at Government dockyards.

MOORING POSTS.
Scale 6 Feet = 1 Inch.

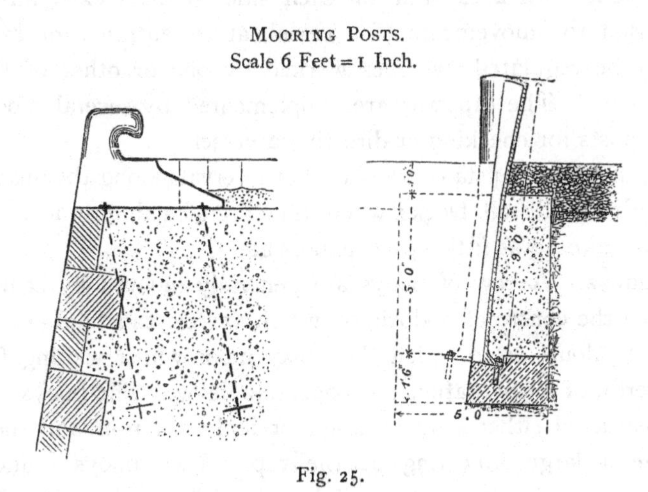

Fig. 25.

The position of the mooring posts, a little distance back from the walls, somewhat encumbers the quays with ropes; so that now the mooring posts are generally built into the wall, a system adopted at the new Hornby Dock at Liverpool (Plate 14, Fig. 2), at the Portsmouth and Chatham extension works (Plate 14, Fig. 9), and on the Antwerp Quays. (Plate 14, Fig. 21.)

At several recent dock works, low cast-iron hooked posts have been let into the coping, in place of the ordinary mooring posts. (Fig. 25.) Instances of this arrangement exist at the Albert Dock, London, at the Havre extension works, and at Dieppe. (Plate 14, Fig. 20.) This plan removes the mooring ropes from the quays, which is a great improvement for the traffic, though it throws more strain upon the coping.

Mooring rings, let into the face of the wall, are sometimes provided; but they are more commonly used along breakwaters than in docks, where they would be somewhat difficult to reach.

Capstans. At entrances, locks, and passages, the haulage and control of vessels is effected by means of capstans, which are usually turned by hydraulic machinery. Every lock is provided with a capstan, on each side, at both extremities, so that the movements of a vessel, at its entrance or exit, may be regulated by ropes worked by one or other of the capstans. The capstans are supplemented by several mooring posts for checking or directing a vessel.

Sometimes capstans are placed at intervals along the quays, as at the Millwall Docks, where they are of a dumpy form so as to take up as little space as possible.

Buoys. A row of buoys are generally moored by chains along the centre of a dock, by which vessels can haul themselves along, or to which they may moor whilst waiting for a berth, or when stationed alongside a jetty. The buoys are constructed either solid of wood, or hollow of iron, and they have a large iron ring at the top. Iron buoys require frequent painting, and careful supervision to ensure the speedy stopping of leaks; but they are preferable to wooden buoys, as these latter are liable in time to get water-logged, when they sink and become useless. A series of square wooden buoys which were placed in the South West India Dock at the time of its opening, in 1870, became gradually water-logged, and were all replaced by round hollow iron buoys within a period of twelve years.

Cranes. Formerly the quays of docks were provided with a few fixed cranes, to which the vessels had to be specially brought, and which were only used for raising heavy weights; and the ordinary work of discharging and lading a vessel was performed by hand, by the aid of the simplest tackle, as is still done in many ports in France.

Movable steam cranes were next introduced, to a limited extent; but they had the disadvantage of wasting fuel if not constantly at work, and they could not be set working at a moment's notice.

Since the development, within the last few years, of hydraulic pressure machinery, the introduction of travelling hydraulic cranes has been very rapid, so that now every thoroughly well-appointed dock is supplied with these machines. The cranes run on rails, and receive the water, under pressure, by means of telescopic copper pipes attached to the nearest hydrant on the main supply pipe, which is placed between the rails and is furnished with hydrants at frequent intervals. By this arrangement the hydraulic crane can be used at any point of the quay; and two or three can be concentrated at a single vessel, placed so as to suit the variable positions of the hatchways. They work very easily, and their motion can be regulated with great precision. Some cranes have been constructed with a passage for wagons underneath, as for instance those established on the Antwerp Quays; but generally the line of way is on the outside, next the shed or warehouse.

Hydraulic travelling cranes possess the advantages over portable steam cranes of occupying less space, of causing no waste when off work, and of being capable of being set to work at any time. Moreover, as the power is quite independent of the crane, it is possible to employ stronger cranes, with only a rather greater expenditure of water-pressure; whereas if more powerful steam cranes are required, it is necessary to increase the power of the engines as well. Also the man working a hydraulic crane has merely to attend to the crane; whereas in a steam crane, the engine also has to be looked after and maintained in repair.

Usually a powerful fixed crane, or shear legs, is placed in a convenient place on the quay for raising heavy weights, or for depositing boilers and machinery in steamers. Sometimes a floating derrick serves the same purpose; it has the advantage of not occupying the space of a berth, and of being able to be taken alongside a vessel in any part of the dock. Moreover by being placed outside a vessel lying close to the quay, the derrick in performing its work need not impede the operation of discharging or lading the vessel.

Coal Tips. In some ports, the shipping of coal constitutes one of the most important branches of the trade, as for instance on the Tyne and at Cardiff; and in many ports, some provision has to be made for coaling steamers. For this purpose special wharves are provided, or a portion of a dock is set apart for coal tips. There are two general forms of coal tips, depending upon the level at which the coal is delivered in the wagons on the quays, namely, balance, and hydraulic tips.

A Balance Tip is employed where the line by which the wagons arrive is situated at a high level. The line is prolonged up to the edge of the quay on the top of a staging. At the end of the staging there is a suspended platform, sliding in vertical guides, and held up by balance weights sufficiently heavy to sustain an empty wagon on the platform, but overbalanced by a loaded wagon. Accordingly, when a loaded wagon arrives on the platform it descends in a slanting position, the contents of the wagon are discharged into a shoot through the opened end of the wagon, and the shoot directs the coal into the hold of the vessel underneath. The descent of the platform is regulated by a brake; and directly the coal has been tipped the brake is released, and the platform rises up again with the empty wagon. At Cardiff, the increased height of the decks of the vessels has rendered it necessary to avoid the loss of height caused by the descent of the end of the wagon, so that now the other end is raised instead by a hydraulic ram. Sometimes the stage is made to overhang the quay; the wagon then descends over the hatchway of the vessel, and discharges its coal through a movable bottom; but this method, by discharging the coal over a wider area in the hold, breaks up the coal more than when it is deposited on the conical mound formed by a shoot.

A Hydraulic Tip receives the wagon on a platform at a low level, and raises it to the required height by the action of a hydraulic ram placed underneath; and another smaller ram, attached to the underside of the platform towards the back, tilts up the platform and discharges the wagon. The hydraulic tip is the best where the water-level is variable, as the height of the shoot can be exactly adjusted to suit the different levels by modifying the height of the lift; and when the main line is at a low level, it dispenses with the necessity of a high embankment and staging for raising the level of the line and sidings. When, however, the water-level is constant, and the approach railway is on a high level, the balance tip is the most convenient.

At the Tyne Docks, the shoot itself is made adjustable to suit the different levels of the vessels; and ten shipping places are provided at each long jetty, four at each side and two at the end. The coal is discharged through the bottom of the truck into a sort of hopper with inclined outlets, at different levels, to which the shoot is adjusted. At Cardiff and Penarth, the coals are shipped at the end of a series of short jetties ranged along one side of each of the docks. (Plate 11, Fig. 6.) A separate line, and a complete set of sidings are provided for the coal traffic at these docks.

Sheds. Long lines of sheds are erected along the quays for the purpose of protecting merchandise, when discharged from a vessel, till it can be sorted, or placed in a warehouse, or removed. Generally they form a merely temporary resting-place for the goods; but sometimes the sacks or bales are piled up and stored in them for a time. The sheds have wide openings, at frequent intervals, both facing the dock and on the opposite side, so that there may be no impediment to the conveyance of the goods from the quay to the sheds, or from the sheds to the sidings and roadway beyond. Each opening is closed by a pair of doors sliding on rollers, so that when open they are close against the wall and quite out of the way. The sheds are made at least high enough to pile up bales as high as a man can reach. Sometimes they are made lofty with high openings, if near the edge of the quay, and when, as at Liverpool, the height of the vessels above the quay varies considerably, so that there may be always headway for the gangway from the ship; and the trusses may be utilised for piling up the higher tiers of bales. Sheds for timber are also usually made high to afford space for high piles of planks and deals.

The sheds are generally roofed over in two bays running lengthwise, the rafters resting on the side walls and on a central row of columns. The side walls of the better class of sheds are built of brick or masonry; but sometimes wood or corrugated iron is substituted for cheapness, the trusses being supported on wooden or iron pillars. The cheapest form of roof is corrugated iron; but zinc provides a very light, cheap, and fairly durable covering. Slates are only used for the more important sheds, and for warehouses. The sheds are well lighted by glazed openings in the roof.

Warehouses. When goods have to be stored for a considerable time, they are placed in warehouses; and the rents obtained for storage often constitute an important return for the company. Warehouses are generally very solid brick structures, roofed with the largest slates; but some old wooden warehouses are still in existence at the Surrey Commercial Docks. They are built with three or more floors, according to the nature of the foundations and the requirements of trade. High warehouses are advantageous where land is scarce, and the same roof suffices for any number of floors; but the erection of several floors for the storage of heavy goods imposes a considerable weight on the supports of the lower floors; and the insurance companies generally place a limit on the cubical contents of the separate compartments enclosed within brick walls. Accordingly, where land is not very costly, the saving in land and roofing, realised by a number of floors, may be expended in the stronger supports and more frequent division walls which they entail; and the expense of lifting the goods the additional height must also be taken into account.

The floors are supported by rows of columns, rising one above another to the top floor. These columns are formed either of square upright oak beams, or more commonly of cast-iron pillars cylindrical or cross-shaped in section. The oak beams are considered to be less likely to fail than cast-iron in the event of a fire, as they are not quickly burnt through; whilst cast-iron, when raised to a high temperature, gives way under its load, or cracks when suddenly cooled by the water jets from the fire-engines Considerably more space, however, is occupied in the warehouse by the beams, and they are not so well suited for very heavy loads. Hollow columns are stronger than solid cross-shaped pillars with the same sectional area, owing to the greater moment of inertia of the annular section, and are therefore more economical for a given load; but the cross form is sometimes preferred, on account of a flaw in the solid casting being more readily detected than in a hollow cylindrical column, and as it also occupies a smaller space.

Wooden beams or iron girders are used for bearing the wooden joists; but wood is the preferable material, as when a fire breaks out the girders expand, bend up, and fall, and are, accordingly, a source of greater injury than the combustible wood. The joists support thick planking, necessitated by the heavy weights on the floors and the wear and tear over them. In some new warehouses at the Millwall Docks, Mr. Duckham has substituted 4-inch planking in place of joists with 1½-inch planking, so that the planking rests directly on the beams; this arrangement economises the height of the joists, less the increased thickness of planking, but the cost of renewal, in the event of considerable wear on the floor, would be increased.

The height between each floor varies some feet in different warehouses. The most economical clear height is that up to which a man can store goods, which is about 9 feet, and a foot or two should be allowed for clearance; so that, allowing 2 feet for the beams, joists, and planking, the requisite height would amount to 13 feet; and this is the distance between the floors at the Marseilles warehouses, which have six stories above the ground floor and vaults below. The bottom floor may with advantage be given some additional height, to provide adequate light and ventilation, especially as the quay side is generally blocked up with a shed, and the opposite side by trucks or carts.

Each block of warehouses is divided into compartments, separated by brick walls and fireproof iron doors, so that if a fire breaks out anywhere it may be confined, if possible, to one compartment. The ground floor is level with, or only slightly raised above the quay, and may be paved with flags or asphalted. The top floor is provided with lights in the roof, in addition to the side windows, which are placed along the front and back of the building. Special warehouses are frequently erected for jute, consisting of a ground floor only, and divided into small compartments owing to the great inflammability of that material. These warehouses can be given a greater height from the floor to the roof, as in this case the trusses of the roof can be used as supports for hoisting the upper bales.

A high warehouse, 250 feet long and 150 feet wide, was erected at the South West India Dock during the construction of that dock. It has a ground floor and three stories, and is divided into four compartments by three intermediate brick walls, built across the warehouse, and carried up some feet above the roof. Access is provided to the various floors by four staircases, at each end of the two side division walls, which can be shut off by iron doors; and double iron doors at each end of the centre division, with an interval of six feet between, afford communication between the compartments, and form a fireproof barrier when closed. Iron columns in longitudinal and transverse rows, 11¼ feet apart, centre to centre, support each floor, which is carried by wooden beams and joists; and wooden trusses support the roof. Iron girders and roofs were proposed in the first design, but were abandoned, as they were dearer than wood, and offered no additional security in the event of fire. The only fireproof construction consists of brick arches carrying a floor of concrete and asphalt between the girders, as adopted at Marseilles; but this forms a heavy floor, and though it would be serviceable in preventing the spread of a fire, in the first instance, as it proved at Marseilles, in 1872, yet if the fire got a thorough hold of the contents of one floor, the columns and girders would give way and bring down the floors above, unless the ironwork could be protected from the heat. The plan, accordingly, adopted at the South West India Dock warehouse consists in attempting to confine the fire to the compartment in which it originates, rather than erect a fireproof building, at greater cost, that could not be warranted to stand the heat of a fierce fire. The walls and columns were erected on brick foundations carried down several feet to the firm bed of gravel underlying the site. The cost of the warehouse was about 4½*d.* per cubic foot. The cost of two jute warehouses, each 250 feet long, 150 feet wide, and 20 feet high, amounted to 2½*d.* per cubic foot. The cost of the large warehouses at Marseilles, previously referred to, was 8¾*d.* per cubic foot.

By fixing cranes over the tiers of doors on the outside, and hydraulic lifts inside, the goods can be raised direct to any floor, or lowered into wagons, or shifted from one floor to another.

Granaries. Large warehouses have been erected for storing grain, and fitted with special machinery for its distribution through the warehouse. The grain in sacks is stored in the ordinary way by cranes and lifts. The grain in bulk is treated differently. It is removed from the hold of the vessel by means of large buckets, which are lifted by hydraulic cranes. At the Millwall Docks, some semicylindrical buckets in the form of grapple dredgers[1] are used, which open into two halves in the centre line of the bottom, and, sinking into the grain, close their jaws on being raised, and remove a bucketful of grain. The bucket is then swung round to the point of discharge, when its jaws are opened again, and the grain falls through. The grain is either conveyed at once by the cranes to the granary, which is the practice at the Surrey Commercial Docks, or it is deposited in special trucks, in which it is taken to the granary, a method adopted at the Millwall Docks. The latter system involves somewhat more trouble in removing the grain from the vessel to the warehouse, but it has the advantage of enabling several vessels to unload at the same time, and at any part of the docks.

The buckets empty their contents into a hopper, from which the grain passes through a shoot into one of the lower floors of the granary, from which it is raised to the top of the building by an elevator. The elevator consists of a series of buckets fastened to an endless india-rubber band, which runs from the bottom to the top of the granary, and, by revolving rapidly round rollers at the top and bottom, causes the buckets to scoop up the grain from a receptacle at the bottom and discharge it into a hopper at the top, just on the same principle as a bucket dredger excavates, raises, and discharges material.

The distribution of the grain, which has been conveyed to the top of the granary, on to the various floors, and to various parts of the same floor, is accomplished by two methods. At Liverpool and the Surrey Commercial Docks, the long granaries are provided with a series of endless

[1] An illustration of a grapple dredger is given in 'Rivers and Canals,' Plate I.

india-rubber bands, from 15 to 18 inches wide, placed horizontally several feet above the floor, running round rollers at each end, and supported on rollers at intervals. A very rapid motion of from about 6 to 8 feet per second can be imparted to these bands; and the grain which is led by a shoot on to one or other of these bands, on the upper floor, is carried along by the band in motion to any part of the floor. A movable machine, running on wheels underneath the band, is placed at the spot where the grain has to be discharged, and by raising and tilting the band, without arresting the motion, it throws off the grain into a hopper, from which it can be delivered, by a suitable arrangement of shoots, on to the part of any one of the floors lying vertically below it. The bands, accordingly, convey the grain to any horizontal position, which is regulated by the position of the throwing-off machine; and the arrangement of the shoots determines its vertical destination. In a high granary at the Millwall Docks, the horizontal bands are dispensed with; and each elevator serves a series of bins, on each floor, surrounding the downward tube through which the grain descends from the hopper above. A disc is inserted barring the downward passage in the pipe at whatever floor the grain is to be delivered on to; and the grain can be directed into any particular bin of the set by adjusting the corresponding shoot. The machinery for distributing the grain is worked by hydraulic power.

Hydraulic Machinery. Water-pressure stored up by steam power and the aid of an accumulator, and distributed through cast-iron pipes, has proved the most convenient agent for working the various machines employed at docks. The work carried on by machinery at docks consists of a number of intermittent operations, such as opening and closing the dock and sluice gates, turning the swing-bridges and capstans, and working the cranes and lifts.

The use of separate steam-engines for these numerous

machines would be costly and troublesome even for the cranes; and a great loss of power would be involved in the transmission of steam through pipes to the machines. Water under pressure, on the contrary, can be readily conveyed for considerable distances through pipes with comparatively little loss of power, and is peculiarly adapted for intermittent working, as no loss of power is undergone when the machines are at rest, whilst the power is at hand whenever they require to be set in motion.

A steam-engine is erected on a suitable site near the dock, which is continuously employed in pumping water into a vertical cylinder, from whence it flows, as required, to the various machines. Instead of effecting the pressure by a high column of water, necessitating a high tower, the water in the cylinder is pressed by a piston which can move up and down in the cylinder, and is weighted by a ballasted case suspended on it, at the top, by a cross-head. A constant pressure is thus imparted to the water pumped into the cylinder. The vertical cylinder with its weighted piston is called an accumulator, as the power is stored up in the cylinder, at a pressure of about 700 to 800 lbs. on the square inch. When the steam-engine pumps water more rapidly into the cylinder than it is drawn off for consumption, the piston rises in the accumulator, and the steam is shut off by the closing of the throttle valve when the weighted case reaches its full height. When, on the contrary, the consumption by the machines is greater than the supply, the piston falls in the accumulator, and opening the throttle valve causes the engine to work with its full effect. The accumulator, accordingly, serves both to store up and regulate the supply. Moreover the intermittent nature of the work of most of the machines prevents their ever being in action all together for any length of time continuously, so that any sudden unusual drain of water can be supplied by the reservoir in the cylinder of the accumulator; and the amount of power thus expended is soon stored up again directly some of the machines cease working. On this account, it is unnecessary to provide an engine of the same maximum efficiency as if the full amount of power, liable to be expended, had to be supplied simultaneously with its use. Frequently, when the machines are spread over a considerable area, or in groups some distance apart, the accumulators are distributed, so that the water for supplying the machines may not have to be drawn off through a long length of pipe at the moment of working, thereby diminishing the pressure and the efficiency of the machines.

Compressed-Air System. The use of compressed air as a motive power has proved very advantageous for underground operations, as it is more convenient in such situations than water-power, and also aids in ventilating the mines or headings. On the other hand, hydraulic power has been almost universally adopted at dock works. Whilst, however, hydraulic machinery has been provided for the new entensions of the Chatham Dockyard, compressed air has been resorted to at the Portsmouth Dockyard extensions, with the result, according to Mr. Colson [1], that for the special intermittent class of work at Portsmouth, the compressed air is the more efficient of the two systems. Compressed air, moreover, possesses the advantage of being exempt from the risk of stoppage during severe frosts, which might freeze the water in hydraulic pipes. In spite, however, of the favourable opinion expressed as to the working of compressed air at Portsmouth, hydraulic power is preferred at dock works. Compressed air requires larger pipes for its storage and distribution; and it is subject to considerable loss of power, owing to the heat which is developed in the process of compression. Accordingly, the loss of power, and the cost, are generally greater with compressed air than with water-pressure.

Maintenance of Depth in Docks. The water introduced

[1] 'Minutes of Proceedings, Inst. C. E.,' vol. lxiv, p. 231.

into most docks, at every tide, is frequently heavily charged with silt, which it deposits in the still water. Consequently, the removal of this deposit is necessary for the maintenance of the depth within the dock, which is effected by regular dredging operations. At some docks, the admission of silt is avoided by keeping up the water-level in the dock by the supply of fresh water. Thus, at Cardiff, the docks are supplied with land water from the river Taff; and the level inside is kept higher than the high-tide level outside, so that the muddy water of the Bristol Channel is excluded. A similar provision has been made at the Swansea, Grimsby, and Barrow docks; and pumps have been erected for supplying the new Alexandra Dock at Hull with water, instead of relying upon the very silty waters of the Humber estuary. The tidal water is also sometimes kept out, except when required for replenishing the docks, by means of reverse gates, a plan adopted at Barrow, Antwerp, and St. Nazaire. In spite of these precautions, the silt that has to be annually dredged from the Antwerp Docks amounts, on the average, to about 32,000 cubic yards.

At Liverpool, about 700,000 tons of deposit have to be removed from the docks every year by dredging, and conveyed out to sea; and large quantities of silt have to be dredged regularly from the docks of London, over 100,000 tons being removed annually from the East and West India Docks alone. The waters of the Loire at St. Nazaire contain so much matter in suspension that the depth of the St. Nazaire Dock would be reduced 3¼ feet every year if no work was undertaken for its removal. To provide against this enormous amount of accretion, the earliest sand-pump dredger[1] was designed at St. Nazaire, with the object of removing this accumulation, and proved perfectly successful; and the quantity of alluvium that has to be thus raised and carried away from the St. Nazaire Dock alone, of only 26 acres, in order to

[1] A drawing of this dredger is given in 'Rivers and Canals,' Plate 1, Figs. 5 and 6.

maintain the depth, is 170,000 cubic yards. In order to prevent a similar accumulation of silt in the Penhouët Dock, not only are reverse gates provided at the junction lock, but the necessary supply of water is only admitted, through a special conduit, when the water is particularly clear, at high water of spring tides and during calm weather.

The maintenance of the approach channels is effected partly by dredging, and partly by scour, the scouring current being provided either from the water in the docks escaping through the lock and gate sluices, or from special sluicing basins such as have been constructed at Calais and Honfleur. (Plate 1, Fig. 5, and Plate 12, Fig. 1.) The extent of the work of maintenance required necessarily varies considerably, depending upon the length, breadth, and depth of the entrance channel, the amount of sediment held in suspension, and the duration of slack tide. Owing to the very silty condition of the Loire, 300,000 cubic yards have to be dredged annually, for maintaining the short entrance channel to the St. Nazaire Docks.

CHAPTER XXIII.

DESCRIPTIONS OF DOCKS.

LONDON. LIVERPOOL. BIRKENHEAD.

Port of London: Growth; General Features; London and St. Katherine Docks; Surrey Commercial Docks; East and West India Docks; Millwall Docks; Victoria and Albert Docks; Tilbury Docks; Remarks. *Liverpool Docks:* Site; Growth; Description; Sandbanks in front of Docks; Sluicing Arrangements; Landing Stage; Appliances; Extensions; Remarks. *Birkenhead Docks:* Site; Low-Water Basin; Description of Docks; Landing Stages; Appliances; Remarks. Comparison between the Docks of London and Liverpool. Trade Statistics.

THE four most important estuaries of England, the Thames, the Bristol Channel, the Humber, and the Mersey, possess great natural advantages for navigation, which have been enormously extended during the present century by the construction of docks, either actually along their banks, or on those of their tributaries. No impediments exist in the first three estuaries to the approach of vessels; whilst the great rise of tide on the Mersey enables vessels to pass its bar, and to reach the very deep channel in front of Liverpool itself. The estuaries provide access and shelter to the ports; and the docks supplement these natural advantages by affording additional area, in still water at a uniform level, with extensive accommodation of quays furnished with the various appliances for facilitating traffic. Though the ports owe their existence to the waterway which nature has provided, they owe their development and their marvellous increase to the docks which have been created by engineering skill.

PORT OF LONDON.

Growth of the Port. Though the most important part of the city of London is situated on the north side of the Thames, and the chief development of the port has naturally taken place on that bank, the first dock was constructed on the southern shore. Excavations are recorded to have been made at Rotherhithe for diverting the course of the river, and were subsequently utilised for a dock, which existed as the Howland Great Wet Dock in 1660, and was the first dock in Great Britain. It was 1070 feet long and 500 feet wide, and had a depth of water of 17 feet; it is now known as the Greenland Dock of the Surrey Commercial Docks. (Plate 10, Fig. 1.) Although docks were commenced at Liverpool in 1709, no additional public docks were constructed on the Thames during the 18th century; and at its close, the only other accommodation for trade consisted in 1400 lineal feet of 'Legal Quays,' the same length as existed in 1660, and 3500 feet of 'Sufferance Quays,' which were wholly inadequate for the requirements of the port. The river also was at times blocked up with vessels; and very great difficulties, delays, and losses were experienced in unloading them. At last, the London merchants determined to put an end to this intolerable state of affairs by the construction of docks; and the West India Import and Export Docks were commenced in 1800, partially opened in 1802, and completed in 1805. The London Docks, commenced in 1802, were opened in 1805; and the East India Docks were begun in 1803, and completed in 1806. A large extension also of the docks on the south side of the Thames was carried out, between 1811 and 1815, by the formation of four docks to the north of the Greenland Dock, with an area of 46 acres, constituting together the Commercial Docks, and by the construction of adjoining docks in connection with the Grand Surrey Canal.

The ample accommodation for shipping afforded by these works proved adequate for a long period to meet the demands of trade; and, with the exception of the Regent's Canal, opened in 1820, and the St. Katherine Docks, opened in 1828, no important dock extension was undertaken till 1855.

In that year the Victoria Dock was opened, and this large addition of 90 acres was followed, in 1868, by the opening of the Millwall Docks with a water area of 35 acres; and in 1870, the reconstruction of the South West India Dock added 32 acres of deep water space. In 1880, the originally contemplated extension of the Victoria Dock was opened under the name of the Albert Dock, affording a further increase of 84 acres to the dock accommodation of London. (Plate 10, Fig. 1.)

The Commercial Docks have been gradually extended; and in 1864, the Surrey Docks were incorporated with them; and they have been since denominated the Surrey Commercial Docks.

Two small independent docks exist on the north side of the river, namely, the Limehouse Basin at the outlet of the Regent's Canal, having an area of 8 acres, and the Railway Dock at Blackwall with an area of 7 acres.

The present water area of the docks of the port of London is about 560 acres; and though an addition of 84 acres was made as recently as 1880, important dock works are now in progress at Tilbury, which will have an opening into the Gravesend reach of the river (Fig. 26, page 500). These docks will provide an addition of 76 acres of water to the port of London; and it is anticipated that they will be completed in 1885. Accordingly, in the short space of about six years, 160 acres of docks will have been added to the dock accommodation on the Thames, amounting approximately to one fourth of the whole.

General Features of the Docks of London. The basin of the Thames below London Bridge is peculiarly favourable for the formation of docks; for the low-lying lands bordering on the river render the amount of excavation comparatively small; and the thick gravel beds on the sites of the docks are readily excavated, afford excellent foundations, and supply capital material for concrete. Moreover the bends of the river enable entrances to be formed at each end of the docks, where the docks are situated on the land between the bends, which enable the ship and barge traffic to have independent access to the docks.

The plan of the Thames shows that all the bends of the river down to Gallion Reach below Woolwich have been utilised for docks, except Bugsby's Marsh which is inconveniently narrow and is away from a railway. (Plate 10, Fig. 1.) The London Docks stretch across the first bend below London Bridge; the Commercial Docks fully occupy the Rotherhithe bend; whilst two independent systems of docks, the West India and the Millwall docks, are established in the Isle of Dogs. The Victoria and Albert Docks extend right across the North Woolwich bend, nearly three miles in width. These docks have secured the advantage of an entrance at each end afforded by their sites, with the exception of the Millwall Docks, for which, however, a second entrance has been contemplated. The East India Docks, not being situated in a bend of the river, have been provided with two entrances a short distance apart.

The three parallel docks of the West India Docks alone present any symmetry of arrangement: the Commercial Docks, consisting partly of timber ponds, present every variety of size and shape; and the Victoria and Albert Docks, though forming a complete design, have been constructed according to two different types.

London and St. Katherine Docks. The docks highest up the river are the St. Katherine Docks, just below the Tower. (Plate 10, Fig. 1.) They consist of two small docks and a basin, having a total area of 10 acres, surrounded by high warehouses. These docks were constructed when the twenty-one years' monopoly, granted to the London, the East India, and the West India Docks, was expiring, with the hope of obtaining a share in the trade then thrown open. The warehouse accommodation is very extensive compared with the size of the docks; and the warehouses project over the quays, resting on pillars placed on the quays, so that the fronts of the warehouses are flush with the edge of the docks, and goods can be lifted straight out of the vessels and barges, and deposited in any floor of the warehouse.

The docks are connected with the river by one lock, 200 feet long, 45 feet wide, provided with three pairs of gates, and having its sill 28 feet below Trinity High Water. They are well supplied with hydraulic cranes and lifts; but owing to the early period at which they were constructed, and their enclosed position, they are unprovided with railway communication.

The London Docks were originally constructed, in opposition to the West India Docks, with the object of providing dock accommodation nearer to the City. The large Western Dock of 20 acres was opened in 1805; but the large jetty, which adds so much to its utility, was only built in 1838. It has direct access to the river through the Hermitage and Wapping basins and locks, the Wapping lock having its sill 23 feet below T. H. W. This dock is also in communication with the other docks to the east; and a more convenient outlet is provided by the larger Shadwell basin and lock, opened in 1858, with a sill 28 feet below T. H. W. and a width of entrance of 60 feet. (Plate 10, Fig. 1.) The total area of these docks is 40 acres. There is a railway connection with the Western Dock, on its western side, and with its jetty; but the railway accommodation is not much developed, and does not extend to the other basins. These docks are surrounded by extensive warehouses; and hydraulic power is employed for working the cranes, lifts, gates, and swing-bridges.

These two systems of docks have been amalgamated, and give their name to the London and St. Katherine Docks Company, who also own the Victoria and Albert Docks lower down the river. They have a combined area of 50 acres, and a length of quays of about 3 miles 1 furlong, which affords a proportion of 110 lineal yards of quay for each acre of water space.

Surrey Commercial Docks. It has been already mentioned that the Commercial Docks possess the oldest dock in Great Britain: this dock, however, called the Greenland Dock, had formerly merely sloping sides, but is now provided with quay walls, and, together with the adjacent South Dock, is surrounded by the principal warehouses of the Company. Each of these docks communicates directly with the river by a lock, and is in connection with a railway. A set of warehouses are devoted to the grain trade, being fitted with elevators and horizontal distributing bands for storing the grain which is lifted from the steamers by hydraulic travelling cranes. The remaining extensive and intricate network of docks and ponds is employed for the timber trade[1]. (Plate 10, Fig. 1.)

The largest entrance to the docks is the Surrey Lock, nearly opposite the Shadwell Lock, having a length of 250 feet, a width of 50 feet, and a depth on the sill of 28 feet below T. H. W.; it opens into a half-tide basin. The two lower entrances, opposite the Millwall Lock, are 42½ and 48 feet wide. The docks and ponds are all connected by passages, and a ship-channel enables vessels to pass through the Lavender and Acorn Ponds; whilst a passage with reverse gates, adjoining the Lavender Pond, enables the western series of docks to be independent of the rest as regards the water-level.

Owing to the large increase of steamers engaged in the timber trade, quays are being gradually substituted for the projecting jetties, on a sloping bank, which used to suffice for the unlading of sailing vessels. Large high sheds are placed along some of the quays, for sheltering certain kinds of timber; whilst a considerable portion of the timber is stacked on the open quays. A gantry, worked by hydraulic power,

[1] A plan of these docks was furnished me by Mr. James McConnochie, engineer to the Surrey Commercial Docks Company.

has been erected near the Albion Dock for lifting and stacking large balks of timber.

The area of the docks is 81½ acres, and of the ponds 78½ acres, giving a total water space of 160 acres. The length of quays is nearly 5 miles, which affords a proportion of only 53 lineal yards of quay for each acre of dock and pond

East and West India Docks. The three West India Docks are peculiarly well situated across the Isle of Dogs, with entrance-locks and basins at each end, and connected together by locks and a Junction Dock. (Plate 10, Fig. 1.) The earlier locks were only made 45 feet wide, and with sills 23¼ feet below T.H.W.; but the South Dock was opened in 1870, having an entrance lock 300 feet long, 55 feet wide, and a depth of 27 feet on the sill below T.H.W.

The Import Dock is the largest dock belonging to the Company, having an area of 30 acres; it is surrounded by very extensive warehouses. The South Dock is the deepest, and accommodates the larger class of vessels, and is provided with commodious warehouses along its southern quay. The actual cost of the South Dock, with its passage and entrance lock, was £447,000, which for 32 acres of water area amounts to £14,000 per acre[1]. A section of its quay wall is given in Plate 14, Fig. 3. The docks remained without proper railway accommodation till 1870, when the construction of the Blackwall and Millwall Railway enabled sidings to be brought alongside the quays. The quays are also well supplied with sheds for the temporary shelter of goods.

The East India Docks, consisting of two docks and a basin with two locks into the river, were originally started for accommodating the East Indian trade; whilst the West India Docks monopolised the trade of the West Indies. The two

[1] Full details about this dock will be found in my Paper on 'The New South Dock of the West India Docks.' 'Minutes of Proceedings, Inst. C. E.,' vol. xxxiv, p. 157. A recent plan of the East and West India Docks was given me by Mr. Manning, the Company's engineer.

systems have, however, been since 1838 in the hands of a single Company, who are extending their operations by the construction of the Tilbury Docks near Gravesend.

The old entrance lock to the East India Docks was adequate for the class of vessels in existence at the period of its construction, being 47½ feet wide, and having a depth of water over the sill of 24 feet 10 inches. In consequence, however, of the increase both in the width and draught of vessels since the commencement of the century, a second lock has been constructed, with a width of 65 feet, and a depth of 31 feet, which was opened in 1879; and the basin was at the same time deepened to 33 feet and enlarged.

The docks and basin are surrounded by sheds and warehouses; and the Export Dock and the basin are served by sidings from the Blackwall Railway. A large portion of the south wall of the Import Dock having subsided suddenly in 1879, another wall was erected in front, composed of concrete deposited in mass behind a row of timber piling secured by land ties. On this widened quay, sheds with two floors have been built, with iron rolling-up shutters in front instead of doors, so that any portion of the sheds can be opened for receiving goods from the hydraulic travelling cranes on the quay.

Hydraulic power is supplied to both systems of docks; and the more recent gates, swing-bridges, and capstans are worked by this means, as well as numerous cranes and lifts.

The West India Docks have an area of 95 acres, and the East India Docks 31 acres, giving a total of 126 acres for the docks and basins. The quays are about 5¾ miles in length, affording a proportion of 80 lineal yards of quay for each acre of water, without reckoning the additional accommodation gained by the timber jetties projecting out into the docks. The depth in the docks is regularly maintained by dredging, over 100,000 tons of deposit being annually removed and deposited on some land obtained for that purpose at Crossness, having an area of 107 acres[1].

Millwall Docks. A portion of the Isle of Dogs, south of the West India Docks, is occupied by the Millwall Docks[2]. (Plate 10, Fig. 1.) These docks consist of two basins at right angles, connected with the Limehouse reach of the river by a lock with three pairs of gates, having a length of 450 feet, a width of 80 feet, and sills 28 feet below T.H.W. A section of the quay wall is given in Plate 14, Fig. 8. The area of the docks is 35 acres. The length of quays is 1¾ miles, or 87 lineal yards per acre of water area. The docks cost about £7000 per acre[3].

These docks when opened in 1868 were unconnected with any railway, and were devoid of sheds and warehouses; and naturally therefore for some time they attracted very little trade. The construction, however, of the Blackwall and Millwall Railway, which was opened in 1870, joined their quays to the railway system. They had been projected with the object of accommodating the coal trade; but though failing in effecting this, they have acquired a large proportion of the grain trade of the port of London, as well as a certain amount of wool trade, besides other miscellaneous merchandise.

A large granary furnished with the most modern appliances for the distribution of grain, already referred to (p. 483), was completed in 1884; and a long depôt, with numerous sidings, has been erected for receiving the trucks containing grain till they are forwarded to their destinations. The quays are well supplied with sidings, cranes, capstans, and turntables. The

[1] The amount of mud which has to be dredged has greatly increased in the last three years; for whereas the removal of an annual average of 77,000 tons sufficed to clear the docks between 1878 and 1881, the amount dredged in 1882 rose to 107,000 tons, and to 125,000 tons in 1883; whilst during the first half of 1884 as much as 73,000 tons were dredged. This great increase is attributed to the increasing foulness of the river produced by the sewage discharged into it.

[2] A plan of the Millwall Docks was given me by the dock engineer, Mr. Duckham.

[3] 'The Thames and its Docks.' A. Forrow, p. 54.

machinery, the gates, the swing-bridge, and traversing bridge are all worked by hydraulic power. The ample equipment provided, and the good shed accommodation have developed a profitable trade.

An extension to the Blackwall Reach has been proposed, which besides affording increased dock and quay space, and a second entrance, would enable vessels to save the journey round by Greenwich, and place the entrance to the Millwall Docks further down the river than the entrances of its rival in the grain trade, the Commercial Docks. The difference between the past and present condition of the Company illustrates the advantages obtained by a well-equipped dock.

Victoria and Albert Docks. When, about the year 1850, a further extension of the dock accommodation of the port of London became necessary, and works on a large scale were proposed, it was natural that the unoccupied marsh land lying between Bugsby and Gallion reaches was selected. This site offered a long stretch of low-lying land of little value at that period, but in proximity to a branch of the Great Eastern Railway. Moreover, by placing the docks lower down the river, large vessels could avoid the tortuous bends and crowded reaches of the upper river. (Plate 10, Fig. 1.)

The construction of two series of docks right across the wide bend was contemplated in the original scheme, with half-tide basins and entrances at each extremity; but the Victoria Docks, forming the western portion, were alone carried out in the first instance. These docks consist of a main dock of 74 acres, joined to a basin of 16 acres, which communicates with the river, at Bugsby's Reach near Blackwall, by a lock 350 feet long, 80 feet wide, and with a sill 28 feet below T. H. W. The dock, which is about 3000 feet long, and 1050 feet wide, is the largest dock in the port of London; it is not bounded by a quay wall on its southern side, but has some small jetties, and is set apart for the guano trade. The length of quay on the northern side is considerably augmented by projecting walled jetties with two-storied warehouses upon them. Lines of rails have been laid along the north and south quays; and the jetties are provided with hydraulic cranes. The hydraulic lift graving dock, previously described (p. 461), opens out of this dock.

The cost of the dock works was about £706,500, or £7850 per acre. The length of quays is about 3 miles 7 furlongs, which is equivalent to 75 lineal yards per acre of water.

The Albert Dock with its basin and lock into Gallion's Reach, opened in 1880, completes the original design of the Victoria Docks, with which it is connected by a passage 80 feet wide. The Albert Dock is totally different in design to its predecessor the Victoria Dock, for it is a long narrow dock, about 6500 feet in length, and only 490 feet wide, without any jetties. The lock is 550 feet long, 80 feet wide, and has its sill 30 feet below T. H. W.; and an intermediate pair of gates enables the lock-chamber to be reduced to lengths of 400 or 150 feet. The quays are very fully supplied with hydraulic travelling cranes, and have lines of way and sheds all along the quays: they are also lighted at night by the electric light. Two graving docks on the southern side, 500 feet long and 84 feet wide, and 400 feet long and 76 feet wide respectively, serve to accommodate the longer class of vessels which have been constructed since the establishment of the hydraulic lift. (Fig. 24, p. 459.)

The quay walls of the Albert Dock have been constructed entirely of concrete (Plate 14, Fig. 7), forming an advance upon the cast-iron wharfing with concrete backing employed at the Victoria Dock; the walls of the entrance lock alone are faced with brickwork, 2½ feet in thickness.

The Albert Dock and its basin have a total area of 84 acres, whilst the length of their quays is about 3 miles; so that there are 62 lineal yards of quay to each acre of water, a notably smaller proportion than at the Victoria Docks, in spite of the much smaller width of the dock, resulting from the absence of jetties which so materially add to the length of quay. The convenience, however, of an unbroken quay, where every portion is available and more accessible, fully compensates for the smaller comparative length.

The total area of the Victoria and Albert Docks is 174 acres; and the combined quays, 6 miles 7 furlongs in length, afford 69 lineal yards of quay for each acre of water.

Since the opening of the Albert Docks, the Peninsular and Oriental steamers, which formerly started from Southampton, have been transferred to London.

Tilbury Docks. The construction of the Albert Dock, with its large convenient entrance in Gallion's Reach, has naturally tended to attract some of the larger class of vessels away from the docks higher up the river. The proprietors, however, of the East and West India Docks which have been to some extent deprived of the foremost position they originally held on the river, have determined to outstrip their rivals, and at the same time extend the port of London, by constructing docks at Tilbury, 26 miles below London Bridge. The Gravesend Reach, into which these docks will open, is so deep that vessels will be able to come up at any state of the tide. Moreover, this site is so far down the river, that large steamers will be enabled to escape a long distance of river navigation and some troublesome bends. The objection of the distance of the docks from London will be obviated by the same railway rates being charged as from the East and West India, or the Victoria and Albert Docks, and by an improved system of lighterage. Economy also will be effected in the transhipment of goods by providing the quays with a full equipment of hydraulic cranes and coal tips, together with ample sidings, for which it is difficult to make arrangements to an adequate extent in the older docks.

The Tilbury Docks, which were commenced in 1882, consist of a tidal basin, 19¼ acres in area, having direct access to the river by an opening 300 feet wide; a dock with projecting jetties, enclosing an area of 57¼ acres; and a lock, 700 feet long and 80 feet wide, connecting the tidal basin with the dock. (Fig. 26.)

TILBURY DOCKS.

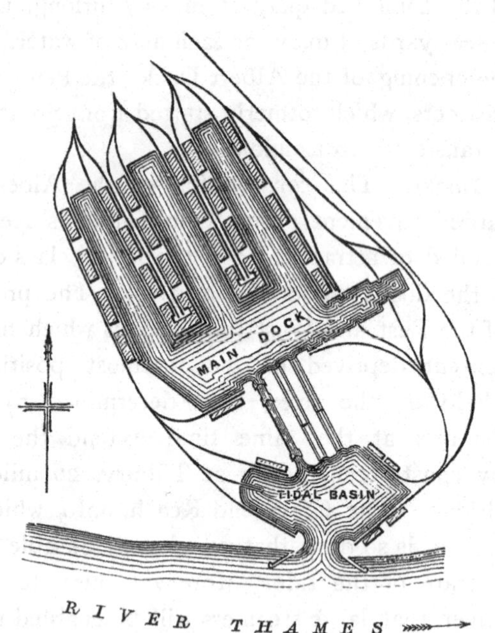

Fig. 26.

The tidal basin is to be excavated to a depth of 46 feet below T. H. W., giving a depth of 26 feet at low-water spring tides, so that, though open to the tidal oscillations, vessels will remain afloat in it at all states of the tide. The lock will have only one foot less depth than the basin, so that vessels will be able to enter it at any time. An intermediate pair of gates, 145 feet from the inner pair, will enable the length of the lock-chamber to be varied from 700 feet to 555 feet, and 145 feet. Two pairs of graving docks are placed parallel to the lock; and the larger and deeper pair will afford a supplementary lock for the passage of vessels in case of necessity. The main dock, having a width of 600 feet and an average length of 1600 feet, has three branch docks opening out of it, separated by broad jetties, and also a coaling dock at the side. The depth of the main and branch docks is to be 38 feet below T. H. W. Sidings and sheds will be placed on the adjacent quays and the wide jetties.

The plan of the dock, precisely similar to the arrangement at the Alexandra Dock, Liverpool, has the advantage of affording ample space for vessels to turn in the main dock, whilst the projecting jetties increase the length of quay. The dock itself affords 83 lineal yards of quay for each acre of water; but including the tidal basin, the proportion is reduced to 77 lineal yards per acre.

Remarks on the Docks of London. The docks on the Thames have had the benefit of the stimulus of competition ever since they were regularly established at the commencement of this century.

Most of the docks were originally constructed by independent companies, and though amalgamation has taken place to a certain extent, yet even now the principal docks are in the hands of four separate companies, namely, the London and St. Katherine Dock Company, who also own the Victoria and Albert Docks; the Surrey Commercial Dock Company; the East and West India Dock Company, who are constructing the Tilbury Docks; and the Millwall Dock Company. The docks are somewhat dispersed along the Thames, but this is a necessary result of the natural configuration of the river. The divided management must be naturally somewhat more costly than if all the docks formed part of a single concern; but this is compensated for by the Companies generally devoting special attention to particular classes of trade, and providing increased facilities with reference to them.

The dock extensions are carried down the river, partly on account of the comparative cheapness of land at a distance from London, partly owing to the better depth of the lower river, partly to avoid the more crowded reaches of the river and its bends, and partly to outstrip other companies in offering the earliest facilities for discharging to vessels coming up the river.

At the beginning of the century, the locks were made about 45 feet wide, 200 feet long, and with a depth of water of about 24 feet. The width reached its maximum of 80 feet at the lock to the Victoria Docks in 1855, and has not since been exceeded. The increase, however, in depth and length has been continued, from a depth of 28 feet at the Victoria and Millwall Docks, to 30 feet at the Albert Dock, 31 feet at the East India Docks, and 45 feet at Tilbury; and from a length of 350 feet at the Victoria Docks, to 450 feet at Millwall, 550 feet at the Albert Dock, and 700 feet at Tilbury.

Land for future extensions has been secured at Tilbury; and though prognostications have been hazarded at various times that the docks of London have reached their limit, they have been falsified by later experience. Till London itself ceases to increase, the growth of its trade is not likely to be arrested. The large additional accommodation afforded by the Albert Dock, and soon to be provided by the Tilbury Docks, may suffice for a time; but, eventually, further dock extensions will be needed on the Thames.

The large extensions which have been made to the port of London since the middle of the century have possessed the advantage of being fresh starting-points in the arrangement of docks. Unlike additions to old docks, they have been untrammelled by pre-existing arrangements; and they have been made on a large enough scale to afford every provision, by the most modern appliances, for expediting transhipment and traffic. Thus the Victoria Dock, by means of its jetties, cranes, sidings, and graving docks, offered advantages unprovided before: the Millwall Docks are provided with an unrivalled system of sidings for the grain traffic, together with ample mechanical appliances; the long unbroken lines of quays of the Albert Dock are furnished, at every point, with siding accommodation and hydraulic cranes; and the Tilbury Docks promise to afford siding and shipping facilities superior to any yet existing on the Thames

Important progress has, indeed, been made at the older docks in the introduction of hydraulic machinery and railway communication; but it is impossible to equip old docks to an adequate extent, at all points, within reasonable limits of cost, and consequently it is fortunate for the port of London that most of its recent extensions have been made where no docks previously existed.

LIVERPOOL DOCKS.

Site. The Liverpool Docks have been constructed along the foreshore of the right bank of the river Mersey, where the river flows in a deep narrow channel between Liverpool and Birkenhead. The depth of the channel in front of the docks is maintained by the tide, flowing into and out of the wide estuary above Liverpool; and the tidal scour from this wide expanse also preserves the shifting outlet channels between the sandbanks below Liverpool, and over the bar. As the depth over the bar is only about 9 feet at low-water spring tides, any notable shoaling of the bar would imperil the prosperity of the port. It is, therefore, of the utmost importance that nothing should be done to modify the condition of the upper estuary, and reduce its scouring capacity, as such a change would lead to a reduction in the depth of the channel over the bar. A description of the form and state of the estuary is given in 'Rivers and Canals[1].'

Growth of the Liverpool Docks. The first dock was commenced in 1709, in a pool communicating with the river, from which the town derived its name, and it had an area of about 3½ acres. It was filled up in 1826, and the Custom House now stands upon its site. The record, however, of its existence, as the Old Dock, is still preserved in the datum of Old Dock Sill, to which all the levels at Liverpool are referred. The sill itself has indeed disappeared, but its level has been transferred to a mark on the centre pier of the Canning half-tide Dock. This level is 10 feet above the level of low water of equinoctial spring tides, and the high water of these tides rises 21 feet above it.

The early progress of the port was slow, as the dock area had only reached 34 acres in 1816, which was increased to 47 acres in 1825. Subsequently the advance was more rapid, as the area had risen to 108 acres by 1846, and was doubled in the next fifteen years, rising to 220 acres in 1861; whilst now the dock space has attained to 354 acres, with an available length of quay of over 22 miles, affording the large proportion of 108 yards of quay for each acre of water. The docks have been extended both to the north and south of the original starting-point; but the northern extensions have been carried the furthest, partly because at the north end the entrances are beyond the influence of the Pluckington Bank, and partly to enable big steamers to avoid the danger and delay of passing up between the crowd of shipping in front of Liverpool.

Description of the Liverpool Docks. The docks present a narrow and almost continuous line along the river bank[2],

[1] 'Rivers and Canals.' L. F. Vernon-Harcourt, pp. 256–258, and Plate 14, Figs. 1 and 2.

[2] A plan of the Liverpool Docks, and sections of two dock walls and a dock-gate were furnished me by Mr. G. F. Lyster, engineer to the Mersey Docks and Harbour Board.

nearly six miles in length. (Plate 11, Fig. 1.) They used to communicate with one another from end to end through passages or locks; but they are now separated into two sections by the roadway leading to the great landing stage, which is placed over the site of George's tidal Basin. There are altogether 51 docks, communicating with the river by 30 entrances. The average area of the docks amounts only to 7 acres, owing to the small size of several of the older docks, and also of some docks set apart for special purposes, such as the Carriers and River Craft docks. The more recent docks, however, have naturally been made of larger dimensions. Thus the two new docks at the south end, between the Brunswick and Herculaneum docks, have areas of 11½ and 8½ acres respectively; and the Hornby Dock, recently opened at the extreme northern end, has an area of 17¾ acres. There are also still larger docks near the northern end, where the largest steamers are accommodated, and where a deep entrance is provided through the Canada Basin. For instance, the Huskisson Dock with its two branches has a total area of about 30 acres; the Canada Dock nearly 18 acres; the Langton Dock 18 acres; and the Alexandra Dock with its three branches, opened in 1880, has an area altogether of rather over 44 acres, forming thus the largest dock in Liverpool. Sections of the most recent dock walls at Liverpool are given in Plate 14, Figs. 1 and 2.

The internal communication between the docks is effected by means of 48 passages, provided in many cases with reverse gates, so as to enable the water-level of adjacent docks to be raised or lowered independently.

The docks communicate with the river, either through half-tide docks, such as the Herculaneum, Brunswick, Queen's, Prince's, Clarence, Salisbury, and Wellington, or through locks emerging into outer basins, such as the Sandon and Canada basins which are open to the river. These open basins facilitate the admission of vessels, at the exposed northern end, by affording a wider opening from the river and sheltering the entrances within them.

The depth of the sills of the entrances, passages, and locks, varies from $2\frac{1}{2}$ feet above the Old Dock Sill datum, down to 12 feet below it. The majority of the sills are about 6 feet below the datum; the Brocklebank Dock passages and the Canada Lock have sills $7\frac{3}{4}$ feet below datum; the passages of the Waterloo and Corn Docks, and the entrances to the Prince's half-tide Dock have sills 8 feet below; whilst the Langton Locks, and the passages leading to the Alexandra Dock and the Hornby Dock at the northern end, and the Herculaneum half-tide Dock at the southern end, have sills 12 feet below datum.

The widths of the entrances are very variable; thus the Brocklebank Dock alone has a lock opening out of it only 20 feet wide, and another of 32 feet, and passages 40, 60, and 80 feet in width. The Canada Lock has a width of 100 feet; one of the Canada Dock passages, the Huskisson Lock, and the northern entrance to the Herculaneum Dock are 80 feet wide; and the passages to the Sandon and Wellington Docks, and the larger entrances to the Wellington and Queen's half-tide docks, and the entrance to the Coburg Dock are 70 feet wide. The width, however, of the Langton Locks, affording access to the most recent docks at the northern end, is only 65 feet; whilst 60 feet has been adopted for the passages between the docks, as these widths are considered ample for the screw steamers now universally employed. A plan and sections of the gates of these locks are given in Plate 15, Fig. 2, and Plate 16, Fig. 2.

Sandbanks in front of the Liverpool Docks. The waters of the Mersey contain large quantities of sand and silt in suspension at certain states of the tide, which are deposited in the still water in the docks; so that 700,000 tons have to be annually removed by dredging in order to maintain the depth. In spite also of the rapid current in the narrow portion of the river in front of Liverpool, sandbanks have formed wherever the velocity is checked or eddies occur, so that, except at the narrowest part of the river, in front of Prince's Dock, shoals exist across the entrances to the docks. The most extensive of these shoals is Pluckington Bank, which extends from the southern extremity of the landing stage as far as the southern end of the Brunswick Dock, in front of the Canning, Queen's, Coburg, and Brunswick entrances. (Plate 11, Fig. 1.) It existed as a detached bank in 1765, in front of the site of the Canning Dock; but in 1820 it had assumed a more southerly position, and had become connected with the land, being somewhat similar in condition to its present aspect. The form of the bank has varied from time to time, owing to the changes of the channels in the upper estuary, and the modifications of the lines of the shores by the extension of the Liverpool Docks, and the construction of the docks at Birkenhead. In recent years, it has tended to travel northwards, protruding under the landing stage at its northern end, whilst its southern portion has been eroded. It is now about 8000 feet long, and extends about 900 feet into the river. The low-water area of Pluckington Bank has exhibited considerable fluctuations, for it increased from about 420,000 square yards in 1820, up to 750,000 square yards in 1871; and in 1882, it had decreased again to about 460,000 square yards. As the bank rises in places up to 8 feet above low water, it precludes the improvement of any of the entrances across which it stretches; and the expedient of lowering the sills of the Herculaneum half-tide Dock entrances has been resorted to for securing a deeper channel to the new docks beyond the southern limit of the bank.

The shoals to the north of the landing stage are less extensive, and less high; but they rendered useless any lowering of the sills beyond 6¾ feet below the datum, until the elaborate and powerful sluices constructed for scouring the Canada Basin enabled a depth of 12 feet below datum to be maintained.

Sluicing Arrangements at the Liverpool Docks. Besides the ordinary sluicing appliances for maintaining the entrances, two extensive systems of scouring sluices have been recently constructed, one at the northern end for maintaining the depth of the Canada Basin and its outlet, and the other at the southern end of the landing stage to prevent the stage being stranded on the tail of Pluckington Bank.

The level of about 6 feet below datum, to which most of the sills of the Liverpool Docks have been laid, and at which the entrance channels can be fairly maintained, affords an ample depth of about 25 feet at high water of ordinary spring tides. Owing, however, to the great tidal range, the high water of ordinary neap tides is considerably less, amounting to only 11½ feet above datum, which gives a depth of only about 17½ feet on these sills; and this is reduced at short neaps to 14½ feet. These depths are inadequate for many of the vessels frequenting the port; and several vessels have a draught of 23 feet when loaded. Accordingly, in order that these vessels may not be excluded from the docks for a great number of tides in the year, and thus forced frequently to unload in the river or to go to Birkenhead, the sills of the Langton Locks and of the new docks at the northern end have been placed 12 feet below datum, and the sills of the entrances to the Herculaneum Dock have been lowered to the same level, affording a minimum depth at high water of 23½ feet under ordinary conditions.

For the purpose of maintaining a depth of 12 feet below datum in front of the Langton Locks, and at the entrance to the Canada Basin into which these locks open, it has been necessary to establish an elaborate system of sluices, so that water drawn from the docks may be directed in a powerful sluicing current against the deposits from the river, which tend to accumulate in the basin, and would soon rise some feet above the sills if not frequently removed. Large culverts, 12 feet high and 12 feet wide, have been formed in the two side walls of the Langton Locks; and another culvert, 15 feet high and 13 feet wide, pierces the whole length of the central pier between the locks, the bottom of these culverts being at a depth of 11 feet below datum. These culverts convey the water from the docks, and pour it, beyond the outer gates, into a series of about seventy smaller branch culverts, from 4 to 6 feet in height and width, which discharge it into the approach channels to the locks, and into the Canada Basin and the northern side of its entrance channel. As the accumulations of silt in the centre of the basin would be less under the influence of the scouring currents issuing from the side walls, four culverts, consisting of iron pipes 8 feet in diameter, have been laid under the concrete floor of the basin, with fourteen openings altogether, 3 feet in diameter, on the top of the culverts, besides the end outlets, so that the central layer of silt may be broken up from below, and thus readily carried away by the outgoing current. The sluicing is conducted, during low water of spring tides, only at those parts where deposit has to be removed; but if the whole of the sluices were opened together, the discharge would amount to about 1,750,000 cubic feet per minute, and the velocity of the current flowing out of the entrance would attain about ten miles an hour[1]. These sluicing arrangements, which are the most extensive and effective that have hitherto been constructed, suffice to maintain the entrance channel to the deep sills of the Langton Locks.

The other important system of sluices at Liverpool has been constructed along the base of the river wall, opposite the southern end of the landing stage, with the object of removing the tail of Pluckington Bank, which had extended under the stage. A culvert, 11 feet in diameter, leads the water from George's Dock into an 8-foot culvert, carried along the wall at the level of low-water spring tides, from which the water is discharged against the bank through twenty-two sluices, $4\frac{1}{2}$ feet

[1] 'Transactions of the Liverpool Engineering Society,' p. 18.

in diameter, placed 20 feet apart. As the supply of water for sluicing can be drawn from the whole area of the southern docks, a very powerful scouring current is available, whose action extends more than 260 feet from the face of the wall. and it has proved most effectual in removing the bank. This system of sluices, together with the dredging required beyond the limits of the sluices, has cost over £20,000.

Landing Stage at Liverpool. A large floating stage has been constructed, in front of Prince's and George's docks, for the ferryboat service, and for vessels to come alongside to accommodate passengers. This landing stage is 2063 feet long, and from 80 to 110 feet wide, which affords a floating area of 4 acres; it is supported on 158 plate-iron pontoons, 80 feet long, 10 feet wide, and from 7 to 8 feet deep, and it cost £470,000. Seven hinged girder bridges and one floating roadway give access to the stage, which is used by about two million persons in a year. The approach of the ferry-boats to the southern end of the stage, at low water, was impeded by the shoal produced by the elongation of Pluckington Bank, and the stage was in danger of being strained by grounding at one end; but this source of inconvenience and danger has been removed by the sluices described above.

Various Appliances at the Liverpool Docks. Liverpool is well supplied with graving docks, having altogether twenty-five with a total floor length of 12,490 feet. They open into nine of the docks situated at various parts of the system, the most important being the Langton, Sandon, and Herculaneum graving docks. (Plate 11, Fig. 1.) They vary in width of entrance from 33 to 80 feet, and from 286 to 768 feet in length; but the majority are 60 feet and more in width of entrance, and from 450 to 565 feet in length. They are closed by reversed pairs of gates; and their total cost has amounted to nearly £940,000. Frequently two or more vessels are admitted into one of the larger graving docks at the same time.

Warehouses were first introduced at the Liverpool Docks in connection with the construction of the Albert Dock, in 1845; and most of the docks are now surrounded by warehouses or sheds. The floor area of the warehouses is 93 acres, and of the sheds 94 acres; and the total cost of their construction has been £3,336,000. Nevertheless, a large portion of the warehouse property is in the hands of private firms in various parts of the city.

A railway runs all along at the back of the docks, which is in communication with the various railway companies coming into Liverpool, and from which several branch lines diverge to the quays. The lines on the quays are, however, not extensively used, owing to the great proportion of cartage which takes place to and from the private warehouses in the city; and therefore the railway accommodation has not been largely developed.

There are only 100 cranes on the quays for the transhipment of goods, of which only fifteen are hydraulic; but there are also numerous cranes connected with the warehouses, and movable hydraulic jigger hoists. Most of the work of loading and unloading is performed by steam winches on board the vessels.

Hydraulic machinery is employed for working all the more recent dock-gates, and has been applied to several of the older gates; it is also much used in the warehouses belonging to the Dock Board, especially in the grain warehouses round the Waterloo Corn Dock.

Dock Extension at Liverpool. The docks have been gradually extended, north and south, to meet the increasing demands of trade; and with a view to further extensions, the Dock Board have acquired the rights over a length of about 3½ furlongs of the foreshore north of the new northern dock, and have purchased 50 acres of land at the southern extremity beyond the Herculaneum Docks. These areas will doubtless provide for the growing requirements of the dock for some time to come. Nevertheless the future prospects of

Liverpool are not quite devoid of difficulties. The extensions can only be made close alongside the river. On the north side, the foreshore is at a favourably low level for the construction of docks; but the new docks will need a nearer access to the river than the Canada Basin. A new entrance basin will require equally costly sluicing arrangements for the maintenance of its depth, and will be still more exposed to north-westerly storms. The southern site for extensions is well sheltered; but the dock excavations will have to be made in the red sandstone, which rises considerably above the level of the quays, and will have to be conveyed some miles out to sea to be deposited. The only compensating advantage in this situation will be that the sides of the docks will simply have to be faced with ashlar masonry, in place of regular dock walls, a plan which has been carried out on the east side of the Herculaneum Docks. (Fig. 23, p. 413.)

Remarks on the Liverpool Docks. The dock system at Liverpool, including both land and water, covers an area of 1075 acres; whilst the docks at Birkenhead, about to be described, cover an area of 506 acres; making a total of 1581 acres. The two systems of docks are under the control of a public trust called the Mersey Docks and Harbour Board, which devotes the revenues of the docks to their maintenance, to the payment of a fixed interest on the borrowed capital, and any surplus towards the reduction of the rates levied on vessels. This arrangement ought to conduce to low rates, as there are no shareholders whose interests have to be considered; but, nevertheless, the rates have been complained of as unusually heavy. This result can only be due to the large amount of dredging required for maintaining the depth in the docks; the number of entrances which have to be kept in repair, owing to the small size of many of the docks; the costly accessories of the sluices and landing stage; the high price of land for extensions in close proximity to Liverpool; and more especially the limited use of modern appliances for the rapid discharge of cargoes, and for their economical transit.

BIRKENHEAD DOCKS.

Site. Birkenhead is situated on the Cheshire shore of the Mersey, opposite the centre of the Liverpool Docks. Unlike those docks, the Birkenhead Docks run inland on the site of a large creek, 340 acres in area, called Wallasey Pool, which has been reclaimed, with the exception of that portion occupied by the docks. (Plate 11, Fig. 8.) The dock works were authorised on an extensive scale in 1844; but after the Egerton Dock and a portion of the present Morpeth Dock had been completed in 1847, the Company had come to an end of their funds; and little was done, beyond proposals for altering the designs, till, in 1855, the works were transferred to the rival undertaking on the opposite side of the Mersey.

Low-Water Basin at Birkenhead. An important feature of the original scheme had been the construction of a large Low-Water Basin, opening direct into the river, and whose bottom was to be maintained at a depth of 21 feet below Old Dock Sill datum, or about 12 feet below low water of ordinary spring tides, by means of the scour through large sluices leading from the Great Float. It was considered that such a basin would be very valuable to the shipping on the Mersey, by affording a place of shelter, at all states of the tide, to vessels of moderate draught; and it formed part of the modified scheme which was ultimately carried out by the Liverpool Docks Board.

The Low-Water Basin was commenced in 1858, and completed in 1863, having an area of 14 acres, a length of 1750 feet, and a width varying from 300 feet at the outer end, to 400 feet at the inner end where the outlets of the sluices were situated. The silting up in the sheltered area was so rapid, that in the seven months subsequent to the opening of the basin, and before the sluices could be worked, a layer of silt 32 inches in thickness had deposited over the bottom of the basin. More than half this deposit was removed by the sluicing operations, conducted during fifty-six low tides in 1864. The sluicing current, however, issuing from a series of sluices, with a total sectional area of 800 feet, and a mean head of over 20 feet, whilst scouring the basin, rendered it unapproachable for vessels, and also endangered the stability of the foundations of the sluiceways. The sluicing was, accordingly, discontinued; and the basin has been converted into the Wallasey Dock by closing its outlet to the river. During the years, however, that it remained open without being scoured, it silted up at the rate of 3¼ feet in a year.

Description of the Birkenhead Docks. When the Birkenhead Docks estate became the property of the Liverpool Corporation in 1855, there were only about seven acres of docks; but the works were soon prosecuted with vigour by the Liverpool Board, so that in 1869 the water area had reached 147 acres, and it has been subsequently raised by minor additions to 157 acres. The great East and West Floats, formed out of a portion of the old Wallasey Pool by dredging and surrounding them by quays, constitute the main part of the docks, comprising 59¾ and 52 acres respectively; and the other principal docks consist of the Wallasey Dock of 12¾ acres, the Alfred Dock of 8½ acres, and the enlarged Morpeth Dock of 11½ acres. (Plate 11, Fig. 8.) Owing to the large size of the Floats, the length of quay round the docks, amounting to 7⅜ miles, bears a much smaller proportion to the water space than at Liverpool, being 84 yards per acre.

The docks communicate with the river by four entrance locks, from 198 to 398 feet long, and from 30 to 100 feet wide. The locks have all their sills placed 12 feet below datum, which is the same level as the lowest dock sills at Liverpool; and they all have outer reverse gates, so that very high tides and waves can be excluded. The water is deeper on the

Birkenhead shore, so that deep entrances are more easily maintained than at Liverpool; and as that side is also more sheltered from the worst winds, no outer protecting basin is needed.

Landing Stages at Birkenhead. Two floating landing stages have been erected along the river side at Birkenhead for the accommodation of vessels and passengers, similar to the Liverpool stage, but of smaller dimensions. The northern, or Wallasey Stage, accommodates the larger class of steamers; it is 350 feet long and 70 feet wide, and cost £61,150. The Woodside Stage, to which the ferryboats run, is 800 feet long and 80 feet wide, and cost £154,700.

Appliances at the Birkenhead Docks. There are three large graving docks at the Birkenhead Docks, opening into the West Float, 750 and 930 feet in length along the floor, with entrances 50, 60, and 85 feet wide.

The Wallasey and Morpeth docks are surrounded by warehouses and railways; and there is good railway accommodation, and some warehouses alongside the Floats.

Hydraulic power is extensively used at Birkenhead, and the quays are well furnished with mechanical appliances.

Remarks on the Birkenhead Docks. The docks at Birkenhead were started in competition with the Liverpool Docks, but they have practically been converted into important adjuncts to that estate. They have generally been regarded as being a somewhat unprofitable portion of the Mersey Docks Board's property; but this has been largely due to the Liverpool merchants naturally taking the most profitable part of their trade to the Liverpool side, and also using Birkenhead as a place for vessels to lie up in. Accordingly, though not perhaps commercially profitable, they have greatly relieved the Liverpool Docks; and, till the deep Langton Locks were constructed, they offered 6 feet greater depth of entrance than Liverpool could provide.

The foreshore on each side has been appropriated by

private firms; so that the Birkenhead Docks are cut off from extensions along the river, as adopted at Liverpool.

Comparison between the Docks of London and Liverpool. The two most important ports of the United Kingdom exhibit some notable differences. At London, the docks belong to several independent companies, and are scattered along ten miles of river; whilst at Liverpool, they are combined under a single trust, and extend only six miles in a continuous line. The management might be expected to be more economical and uniform under the latter system; but the stimulus of competition has proved very useful on the Thames. Each dock company of the port of London strives to afford greater facilities for its special branches of trade than its neighbours; whilst, at Liverpool, there are no such incentives, and the tendency is rather to adhere too tenaciously to old established customs. As soon as the Millwall Dock Company established large granaries, and machinery and sidings for the storage and distribution of grain, the Surrey Commercial Dock Company hastened to afford equal facilities. Now that the London and St. Katherine Dock Company has extended its operations lower down the Thames, and secured a greater depth and wider entrances by means of the Victoria and Albert Docks, the East and West India Dock Company is endeavouring to anticipate the trade by an extension still further down the river, at Tilbury, with docks accessible at any state of the tide. The aim also of the London dock companies has been to afford the utmost facilities, by sidings and movable hydraulic cranes, for the rapid movement of merchandise.

Liverpool has been in advance of London in the matter of storage and distribution of grain, and still is in the provision of graving docks; but in siding accommodation, and facilities for the transhipment of goods, it is considerably behindhand.

The total water area of the docks of London is 482 acres, exclusive of 78½ acres of ponds at the Surrey Commercial Docks; but this will be raised to 558 acres when the Tilbury

Docks are completed. The combined area of the Liverpool and Birkenhead Docks amounts to 511 acres, or rather more than the existing water area of the docks on the Thames, but 47 acres less than they will attain with the addition of the Tilbury Docks. It must, however, be remembered, in comparing the ports of London and Liverpool, that there are numerous private wharfs and dry docks on the Thames; whereas the Liverpool Docks monopolise the whole frontage of the river in front of the city, and there are only a few private dry docks and a ship-building yard on the Birkenhead side.

The dimensions of the docks on the Thames are generally considerably larger than those on the Mersey, as may be readily observed by comparing Plate 10, Fig. 1, with Plate 11, Fig. 1; and though the East and West Floats at Birkenhead, and the Alexandra Dock at Liverpool, with its branches, exceed most of the London docks in area, whilst the Huskisson Dock is the same size as the West India Import Dock, yet these again are surpassed by the Victoria and Albert Docks. The maximum width of entrance on the Thames is 80 feet; and though one instance of a width of 100 feet exists at Liverpool, and one at Birkenhead, 65 feet is now the standard width adopted by the Mersey Docks Board.

The Thames possesses greater natural advantages than the Mersey, as it is free from a bar; it has less sedimentary matter in suspension, and brings down a larger volume of fresh water. It moreover maintains its channel, and has been improved by dredging in shoal places; whereas the outlet channels of the Mersey are liable to deteriorate, and would be imperilled by any works in the upper estuary tending to reduce its tidal capacity, for the upper estuary acts as a natural scouring reservoir to the port of Liverpool.

Trade Returns of London and Liverpool. There is not much difference between the export and import trade of Liverpool, including Birkenhead, either as regards the tonnage of vessels entering and clearing with cargoes, or the actual values of the goods, but the import trade is somewhat the largest. In London, the imports are also larger than the exports, but the difference between them is considerably greater; for whereas the import trade to London is decidedly larger than that of Liverpool, both in the tonnage of vessels with cargoes, and in the actual value, the export trade is less. In fact the values of the imports to London are about half as much again as the values of the exports; whilst the tonnage of the vessels entering with cargoes is double the tonnage of those clearing the port. (See Appendices V and VI.) Both London and Liverpool exhibit a steady rise, both in the total tonnage of vessels, and in the number of vessels frequenting these ports. (See Appendix VII.) The return for vessels clearing the port of London in ballast is not complete; but the tonnage of the vessels entering the port has risen from 7,888,000 tons in 1873 to 11,441,000 tons in 1883. A similar return for Liverpool gives 6,340,000 tons in 1873, and 8,194,000 tons in 1883. The increase in the number of vessels at these ports is equally noticeable; for whereas 28,561 vessels entered the port of London with cargoes in 1863, 38,597 entered in 1873, and 48,278 in 1883, the number having increased over two-thirds in twenty years. The increase at Liverpool is not so marked, the number of vessels entering with cargoes being 12,260 in 1863, 13,873 in 1873, and 14,769 in 1883; but this may be accounted for by the rise in the average tonnage of the vessels, from 340 tons in 1863, to 442 tons in 1873, and 510 tons in 1883; whereas the average tonnage for London has remained nearly stationary, at about 220 tons, during the same period[1].

The well-known change in the proportions of sailing and

[1] The figures given above relate to vessels with cargoes only, as these appear to provide the most correct standard of comparison; but the whole of the vessels, both entering and clearing the various ports, with cargoes and in ballast, are given in Appendix VII, so as to admit of comparison with other general returns.

steam vessels, which has occurred within recent years, has been more fully developed at Liverpool than in London; for whereas the proportion of steamers to sailing vessels entering the port of London with cargoes was about 1 to 4 in 1863, 1 to 3 in 1873, and 1 to $2\frac{5}{8}$ in 1883, it amounted at Liverpool to about 1 to $1\frac{2}{3}$ in 1863, 1 to 1 in 1873, and 1 to $\frac{5}{8}$ in 1883.

Turning to the actual values of the exports and imports at London and Liverpool, it appears that the imports to London have risen from £127,560,000 in 1873 to £145,140,000 in 1883, and the exports from £82,654,000 in 1873 to £95,219,000 in 1883; whilst the exports of Liverpool have remained nearly stationary, at about £103,000,000 during the same period, and its imports have slightly risen, having been £112,825,000 in 1873, and £114,626,000 in 1883.

From the above facts it will be observed that London is in advance of Liverpool, both in its total tonnage and in the value of its exports and imports, and also in the number of its vessels; but the exports and imports of Liverpool are much more evenly balanced, and the average tonnage of its vessels is more than double that of London.

CHAPTER XXIV.

DESCRIPTIONS OF DOCKS.

(*Continued.*)

HULL. GRIMSBY. BRISTOL. NEWPORT. CARDIFF. PENARTH. SWANSEA.

Hull Docks: Situation; Development; Description; Alexandra Dock; Comparison with the London and Liverpool Docks. *Grimsby Docks:* Position; Growth; Description; Appliances; Remarks. *Port of Bristol:* Description of the Bristol Docks; Appliances; Condition; Avonmouth Dock; Portishead Dock; Remarks. *Newport Docks:* Description. *Cardiff Docks:* Situation; Growth; Description; Appliances; Remarks. *Penarth Docks:* Description. *Swansea Docks:* Situation; Description; Appliances; Remarks. Comparison of the various Ports. Trade Returns of Hull and Grimsby; of the Bristol Channel Ports.

THE Humber, like the Mersey, possesses two important systems of docks on opposite sides of the estuary; but though the port of Grimsby has not obtained the same extent of trade as the port of Hull, it has not been absorbed, like Birkenhead, by its more powerful rival. The Humber, like the Thames, is free from a bar; and though its waters are densely charged with silt, which readily deposits in slack water, its channel is fairly maintained; and both Hull and Grimsby possess good communication with the sea.

HULL DOCKS.

Situation. The port of Kingston-upon-Hull is situated on the northern bank of the Humber, about 23 miles from its mouth, and at the mouth of the river Hull which constituted the old harbour and was the origin of the present port. (Plate 11, Fig. 4.) The Humber makes a sharp bend opposite the town; and as the port is situated on the concave bank, a deep channel is maintained near the shore. The approach is accordingly deep and well-sheltered; and the

North Eastern Railway furnishes the docks with railway communication.

Development of the Hull Docks. The earliest docks at Hull were constructed on the site of the ditches which formed part of the defences of the ancient town, situated along the right bank of the river Hull. The first, or Old Dock, now known as the Queen's Dock, was commenced in 1775, and completed in 1778; it opens into the river Hull, and its area is nearly 10 acres. (Plate 11, Fig. 4.)

The Humber Dock was next commenced in 1807, opening direct into the Humber through a tidal basin; its area is 7 acres. The Junction Dock, now called the Prince's Dock, covering 6 acres, was commenced in 1827, and opened two years later; it connects the Humber Dock with the Queen's Dock, and thus completes the circle between the Hull and the Humber enclosing the old town. This addition raised the total water area of the docks to a little over 23 acres. No further extension of dock accommodation was made till 1846, when the small Railway Dock of less than 3 acres was opened. As all the available space in the town had now been occupied, the next dock extension was formed on land on the east side of the river Hull, where the Victoria Dock of 12½ acres was opened in 1850, and was enlarged in 1863. This dock has now an area of 20 acres, and communicates with both the Hull and the Humber through half-tide basins of 1 and 3 acres respectively. The most recent additions to the docks have been formed along the foreshore of the Humber above the town; the first and largest of these docks is the Albert Dock of 24½ acres, opened in 1869; and two docks further to the west have since been constructed. The water area of the Hull Docks has thus been raised to 91 acres, exclusive of 25 acres of timber ponds communicating with the Victoria Docks, of which 68 acres have been added in the last forty years, and 41 acres of these within the last twenty years.

Description of the Hull Docks. The docks at Hull consist of three distinct sections; namely, the series of old docks in the town, the Victoria Dock with its timber ponds and yards on the east side of the river Hull, and, lastly, the line of docks along the reclaimed foreshore of the Humber to the west of the town. The old docks resemble somewhat in character the docks in some continental ports, such as Havre and Rotterdam, where, in the older docks, there is no separation between the quays and the streets of the town. These docks, having been established before the era of railways, were not well adapted for their introduction, though some sidings have been laid along the quays on one side. The narrow quays also, being hemmed in by the town, are not capable of being widened. Though the chain of old docks is provided with locks at both ends, their dimensions are small for the present requirements of trade; for these locks are only 121 and 158 feet long, 38 and 41½ feet wide, and have depths of 20½ and 26½ feet, respectively, on their sills at spring tides.

The Victoria Dock, having been more recently constructed, on an unoccupied piece of land, has ample quays well furnished with sidings; and its entrance from the Humber is 60 feet wide, with a depth of 27½ feet. It is chiefly devoted to the timber trade.

The Albert Dock is the largest dock, and it is accessible to the largest class of vessels; for its entrance lock is 80 feet wide and 320 feet long, and it has a depth of water on the sill at spring tides of 27¼ feet. The dock is only 200 feet wide at its western end, but widens out to 430 feet at the entrance, which enables vessels to turn. Sections of its quay walls are given in Plate 14, Figs. 5 and 6, and of its river embankment and wharfing in Fig. 22 (p. 392); whilst a plan and sections of its lock, sluices, and gates, will be found in Plate 15, Figs. 5 and 6, and Plate 16, Fig 6. It cost about £20,200 per acre of water area, without including the cost of the sidings, machinery, and other accessory works, which raised its actual cost to £22,600 per acre, exclusive of sheds or warehouses. The quays are fully supplied with sidings; and the extensive goods sidings of the North Eastern Railway stretch alongside on the north, just beyond the quays. Warehouses, sheds, cranes, and coal tips are also provided. An extension of 6 acres called the William Wright Dock, has been formed to the west of the Albert Dock, opening into it by an entrance, 50 feet wide, closed by a sliding caisson.

The St. Andrew's Dock (West Dock No. 2) of 10½ acres has also been constructed along the foreshore, still further westward, with a separate entrance lock, 250 feet long and 50 feet wide, having the same depth on its sill as the Albert Dock. The enlargement of this dock, by adding to its length, is contemplated at some future period. (Plate 11, Fig. 4.)

Hydraulic machinery is employed at all the docks, and the water-pressure is supplied in the usual manner at the Victoria and Albert docks; but, at the old docks, the water under pressure is delivered by an independent company, formed for supplying the town with hydraulic power.

The whole of the docks described above, and shown on Plate 11, Fig. 4, belong to the Hull Dock Company[1].

Alexandra Dock. A new dock is being constructed on the foreshore of the Humber, east of the Victoria Dock, in connection with the Hull and Barnsley Railway. This dock is designed to have an area of 46 acres, and an entrance lock 500 feet long, 75 feet wide, and with a depth of water on the sill of 34 feet at high-water spring tides. It is being formed, like the Albert Dock, by reclaiming a portion of the foreshore by an embankment, and thereby enclosing the site of the dock. The estimated cost of the Alexandra Dock is about £778,000, which is equivalent to £16,900 per acre. By the opening of this dock, the water area of the port will be increased by half its present amount, raising the total area to 137 acres. This

[1] A plan of the Hull Docks was supplied to me by Mr. Marillier, engineer to the Hull Dock Company.

dock is to be supplied with fresh water by pumps, so as to exclude the muddy deposits from the Humber.

Comparison of the Hull Docks with the London and Liverpool Docks. Hull is the third port in the United Kingdom, ranking next after London and Liverpool, in respect of the values of its imports and exports. It possesses at present eight docks and two timber ponds, two half-tide basins, seven entrances, three tidal basins, and three graving docks. The Hull Docks have a larger average area than those of Liverpool, amounting to nearly 11 acres; whilst, owing to their narrowness, they afford a greater length of quay per acre of dock, amounting to 113 yards at Hull, as compared with 108 yards at Liverpool. The Victoria and Albert docks on the Humber, are, moreover, larger than the docks at Liverpool, with the exception of the Huskisson and Alexandra docks; and the new Alexandra Dock at Hull will even exceed its namesake at Liverpool in area.

The entrance lock to the Hull Albert Dock has a width equal to the widest entrances on the Thames, and a depth nearly equal to the Victoria lock at Blackwall; but its length was reduced from 400 to 320 feet, owing to difficulties experienced in laying the foundations. The new Alexandra lock will be only 50 feet shorter and 5 feet narrower than the Albert lock on the Thames, and it will be 4 feet deeper.

The older docks at Hull, like those at London and Liverpool, are necessarily deficient in railway accommodation, and in adequate mechanical appliances on the quays; but the more recent docks are as well equipped as the later extensions on the Thames.

The port of Hull, accordingly, though possessing less than one-fifth of the dock accommodation of London and Liverpool, does not compare unfavourably with them in general arrangements; and it is more sheltered, and possesses a better channel to the sea than Liverpool. Its progress has been rapid of late years, and offers a good prospect of being maintained.

GRIMSBY DOCKS.

Position. The port of Grimsby is situated on the southern bank of the estuary of the Humber, which is seven miles broad at that part, and at a distance of about eight miles from the mouth of the river. The port is, accordingly, nearer to the sea, but more exposed than Hull; it is, however, protected from north-easterly gales by the projecting spit of Spurn Point, at the extremity of the northern bank of the estuary. The deep-water channel does not approach close to the shore at Grimsby, as it does at Hull; and it has been necessary to extend the docks across three-quarters of a mile of foreshore to get beyond low-water mark.

Growth of the Grimsby Docks. Though Grimsby was a port in early times, and possessed a dock of 19 acres previous to 1846, it owes its development, at that period, to the railway system. The Old Dock, formed out of a natural creek on the high-water margin of the river in 1801–2, was shallow, and had a very narrow entrance channel across the flat foreshore. The old port was purchased by the Manchester, Sheffield, and Lincolnshire Railway Company, who proceeded at once to extend and improve it.

The Royal Dock, with an area of 25 acres, was opened in 1853; and a Fish Dock of 12 acres was also formed, at the same time, on a portion of the reclaimed foreshore. (Plate 11, Fig. 7.) A second Fish Dock of 11 acres has since been added, adjoining the old one. The Old Dock was connected with the Royal Dock, in 1879, by the construction of the Union Dock of 1¼ acres; and it has been enlarged recently by a branch dock, at right angles, of 26 acres, the combined docks, now called the Alexandra Dock, having a total area of 48 acres. A timber pond of 19 acres is proposed to be formed to the north, and connected with the Alexandra Dock extension; and further dock extensions are contemplated, on the foreshore of 203 acres proposed to be reclaimed to the west of the Royal Dock.

Description of the Grimsby Docks. Previous to the construction of the Union Dock, the only outlet for the Old, or Alexandra Dock was by a lock, 145 feet long, 35 feet wide, and 18 feet deep at high-water spring tides. The old portion of the dock is somewhat irregular in form, having been doubtless regulated by the shape of the original creek; its entrance was adapted to the size of craft frequenting the port at the period of its construction, and to the depth of its shallow outlet channel; and it is only partially provided with sidings, having been made, and hemmed in by the town on the eastern side, before coming into the possession of the railway company. The enlarged Alexandra Dock has now an outlet into the Royal Dock by a lock, 230 feet long, 45 feet wide, and 21 feet deep. Its new branch is provided with three long coaling jetties; and the quays of this portion are fully supplied with sidings.

The Royal Dock, formed on the silty foreshore of the river, and surrounded by quays reclaimed from the river, has two parallel entrance locks, 300 and 200 feet long, and 70 and 45 feet wide, respectively, and a depth of 26 feet on the sill. These locks open into a tidal basin enclosed by timber jetties, through which a channel is maintained, down to about the level of the lock sills, by sluicing and dredging. The low-water line extends beyond the eastern pierhead; but the outlet channel is prolonged beyond the jetties to an inner roadstead, with a minimum depth of 6 or 7 feet at low water, across which access is attained to the deep river channel. Owing to the silty nature of the foundations, the dock walls were built on the arched system, like at the Hull Albert Dock and Penarth, but with spans of $27\frac{1}{3}$ feet between the piers[1]. The piers averaged 6 feet in thickness; but their length was made from 40 to 80 feet, to secure the wall on

[1] 'Minutes of Proceedings, Inst. C. E.,' vol. xxiv, p. 43.

a bad foundation against the pressure of the silty backing, and to provide a stable foundation for the warehouses and machinery on the quays. This great width, and the pile-work foundations raised the cost of the wall to £36 13*s.* per lineal foot.

The two Fish Docks, which are connected by a passage, have been made of sufficient depth to accommodate the vessels of greatest draught employed in the deep sea fishing trade, which has been greatly developed at Grimsby. Direct communication between the Old Fish Dock and the river is provided by two locks, 100 and 140 feet long, 20 and 30 feet wide, and 18 and 20½ feet deep respectively.

All the locks at Grimsby are provided with a pair of reverse gates.

There are two graving docks at Grimsby, one opening out of the Royal Dock for any class of vessel, and one connected with the Old Fish Dock, which is set apart for fishing craft. They are each 400 feet long, and have entrances 70 and 30 feet in width respectively.

Appliances at the Grimsby Docks. The Grimsby Docks, like all docks belonging to railway companies, are amply supplied with railway facilities, sidings, and turntables. There are also coal drops, on projecting jetties, both in the Royal and Alexandra docks, and hydraulic cranes on the quays.

The quays to the east of the Royal Dock are furnished with sheds and high warehouses, whilst the quays round the Alexandra Dock are mainly devoted to the timber trade.

The gates at the Royal Dock entrances were the first worked by hydraulic power; and the water-pressure was supplied by water pumped into a tank, placed on a water-tower, 200 feet above the ground. This tower was erected before accumulators had been thought of; but when the extension of hydraulic machinery rendered a further storage of power necessary, it was provided by accumulators, which afford a pressure seven or eight times as great as obtained by the column of water in the tower.

Remarks on the Grimsby Docks. The port of Grimsby possesses four docks, six locks, two graving docks, and a tidal basin. The total area of the docks is 97 acres; and there are 85 lineal yards of quay for each acre of water space. The acreage of the docks is, accordingly, actually larger than that of the Hull Docks; but the 23 acres of the Fish Docks are shallow, and the length of quays is about 1⅐ miles less than at Hull. Moreover the new dock at Hull will restore its predominance of water area over Grimsby. The Alexandra Dock, however, with its area of 48 acres, is larger than any dock at Liverpool, and even than its namesake now constructing at Hull.

Grimsby owes its advancement entirely to the fostering care of the railway company to which it belongs, as it is less favourably situated in every respect than Hull, except in nearness to the sea. It has a much shallower approach channel, a flatter foreshore, and a less sheltered position; and the conveyance of goods up the river by barges from the docks, which is so advantageously carried on at Hull and on the Thames, cannot be adopted at Grimsby on account of the exposure of the estuary. The possession of docks by a railway company ensures the utmost regularity, expedition, and development of their traffic; and though the railway company did not for many years obtain an adequate return from the docks for its large expenditure, yet doubtless the railway was benefitted by the additional traffic which was brought on to it by the docks.

The depth of water in the docks, both at Hull and Grimsby, has to be maintained by dredging out the large amount of deposit brought in by the muddy waters of the Humber.

PORT OF BRISTOL.

The Bristol Channel leads up to the river Avon upon which Bristol, formerly the second port in the kingdom, is situated, from whence it derives its name. It is also now the highway to other flourishing ports, such as Swansea, Cardiff, and Newport, and is noted for the great range of its tides, attaining their maximum of about 50 feet at Chepstow on the Wye, near the estuary of the Severn.

Description of the Bristol Docks. The old port of Bristol is situated about 7 miles up the river Avon. A floating harbour, between 2 and 3 miles long, was formed, in 1804–9, in the old circuitous bed of the river; and a straight cut was made, into which the river was diverted. (Plate 11, Fig. 3.) The floating harbour is lined with wharves and warehouses, and passes through the centre of the city. The water-level is maintained in this floating harbour by lock-gates, and it is therefore practically a dock, having a total area of 130 acres, of which 55 acres are available for large vessels. The harbour has two entrances; one through the Cumberland Basin of 4 acres, situated at the lower end of the harbour, 6 miles from the mouth of the river, and the other through the Bathurst Basin of 2 acres, about 2 miles up the cut. These basins serve as half-tide basins, having locks at each extremity; and the inner locks are provided with reverse gates. The locks at the Bathurst Basin are 36 feet wide, and about 150 feet long; and the northern locks at the Cumberland Basin are 62 feet wide, and 350 feet long. A plan of the gates of the Cumberland entrance lock, and sections of the gates, the lock, and the quay wall[1], are given in Plate 15, Figs. 3, 4, and 11, and Plate 16, Fig. 5. A small dock has been formed at the north side of the floating harbour, close to the Cumberland Basin, which is used for pilot boats and tugs.

[1] Particulars for these illustrations, and a map of the floating harbour at Bristol, were given me by Mr. Thomas Howard, late engineer of the docks.

Appliances at Bristol. There are six graving docks at Bristol, which belong to private firms. The quays on the right bank of the floating harbour are only from 50 to 100 feet wide; they have some small sheds, but are ill-adapted for convenient sidings. The quays on the left bank are much more commodious, being from 130 to 330 feet wide, and they are provided with large sheds, near the Bathurst Basin, by the railway companies, who have carried their sidings along a portion of the quay on the left bank, and purpose extending them to the Cumberland Basin. The warehouses belong to private firms. There are large timber yards along the harbour, and arrangements for cattle.

The gates and swing-bridges at the Cumberland Basin are worked by hydraulic machinery. Water-pressure, however, has not been applied to the cranes on the town quays, which are very inadequate both in number and power.

Condition of Bristol. On comparing Bristol with the other ports shown on Plates 10 and 11, it is difficult to realise that it was once the second port of Great Britain. It has been outstripped in accommodation, not merely by comparatively old ports like Liverpool and Hull, but even by ports of recent origin such as Cardiff and Barrow. Undoubtedly the growth of many ports has been due in great measure to the development of coal, mineral, and other trades. Bristol, however, might have retained a more important place amongst commercial ports if her merchants had been fully alive to the increasing requirements of modern maritime trade. The Avon, with its rise at spring tides of 39 feet at its mouth, and 33 feet at the Cumberland Basin entrance, afforded convenient access and good shelter for vessels of former times. The river, however, is too shallow at neap tides, and its bends are too sharp, for the larger vessels of the present day; and, moreover, the accommodation provided at Bristol for vessels and for the transhipment of goods was inadequate.

Two courses were open to the people of Bristol, either to

improve the Avon and their docks, like Newcastle and Glasgow, or to make docks near the mouth of the river and connect them by railway with Bristol.

The Avon would be difficult to improve sufficiently, except at very great cost, owing to the deep, narrow, and rocky gorge through which it flows. A plan has, accordingly, been proposed to dockise the river down to its mouth, by placing a dam and locks across its entrance, and thus convert the whole river up to Bristol into a floating harbour, with quays on the flat portions of its banks. The exclusion of the tidal ebb and flow from the river would, however, endanger the maintenance of its outlet, and consequently tend to shoal up the proposed approach to the floating harbour.

The second alternative, of making docks at the mouth of the Avon, is therefore the best, and has been carried out to a certain extent, though it met at the outset with opposition from the Corporation at Bristol who are the proprietors of the docks in the town. Two docks have been constructed at Avonmouth and at Portishead respectively, the former by an independent company, and the other with substantial aid from the Corporation.

Avonmouth Dock. The Avonmouth Dock is situated on the right bank of the river Avon, close to its mouth. It consists of a dock of 16 acres, with an opening into the outfall of the river through a lock, 454 feet long and 70 feet wide, and having a depth of 38 feet of water on the outer sills at spring tides, and 26 feet at neaps, so that vessels of the largest draught can be always admitted at high water. An intermediate pair of gates enables the lock-chamber to be reduced in length when desired. The dock was opened in 1877, and cost about £20,500 per acre. The addition of a timber pond and a graving dock is contemplated; and land to the north is available for extensions.

The dock is in communication with Bristol by railway; and it is fully supplied with sidings, and surrounded by warehouses and sheds. Hydraulic power is employed for working the gates, capstans, cranes, and a grain elevator. The dock in fact is fully equipped, and constitutes in every respect a notable advance on the Bristol Docks.

Portishead Dock. Another dock, of 20 acres, has been constructed close under Portishead Hill on the shore of the Bristol Channel, a little south-west of the mouth of the Avon, which opens through a lock into the anchorage of Kingroad; and the entrance is sheltered from the south-west by a pier, 1000 feet long, which serves as a landing place. A timber pond of 11 acres communicates with the dock. A railway has been made from Bristol to Portishead, with sidings on the quays, and a line on the pier. Some warehouses and sheds have been erected, and travelling steam cranes have been placed on the quay.

The Portishead Dock is both longer and wider (1600 feet by 590 feet) than its rival at Avonmouth (1400 feet by 500 feet); but it has a depth of water of only 33 feet on its sill at spring tides, as compared with 38 feet at Avonmouth. The length of quay is 72 lineal yards for each acre of dock at Portishead, and 77 yards at Avonmouth.

Remarks on the Port of Bristol. If the Corporation of Bristol had realised earlier the changed conditions of navigation, and had promoted the construction of deep docks at the mouth of the Avon, instead of first opposing and then competing with the projectors of the Avonmouth Dock, the financial state and the prospects of the port would have been more flourishing; and Bristol would not have been so much outstripped in the progress of commercial enterprise.

NEWPORT DOCKS.

Description. The river Usk affords access from the Bristol Channel to the town and harbour of Newport; and though the river is nearly dry in places at equinoctial spring tides, yet owing to the great rise of tide, amounting, at its mouth, to 40 feet at springs and 30 feet at neaps, it can admit vessels of the greatest draught. The port possesses two docks, namely, the Newport, or Old Dock close to the town, and the Alexandra Dock lower down the river, both on the right bank. The Newport Dock was opened in 1842, and enlarged in 1858; it has an area of 12 acres, and an entrance lock 220 feet long, 61 feet wide, and having 32 feet depth of water on the sills at springs, and 22 feet at neaps. Its quays are provided with sidings, and hydraulic machinery for the shipment of coal.

The Alexandra Dock was opened in 1875; it has an area of 28¾ acres, and an entrance lock 350 feet long, 65 feet wide, and having 37 feet depth of water on the sills at springs, and 27 feet at neaps. It has been furnished with ample sidings, and eight hydraulic coal tips for shipping coal. There are also several hydraulic and steam cranes for unloading vessels. A graving dock, 500 feet long and 56 feet wide, opens out of the dock; and there is a timber pond of 10 acres near the dock. The Dock Company has acquired 213 acres of land for future extensions to the south of the present dock; and a new dock connected with the Alexandra Dock, but with a second lock, has been designed. The docks afford 82 lineal yards of quay per acre of water area.

The docks alone do not represent the whole accommodation of the port, as there are several private wharves along the right bank, and two or three on the left bank, where large steamers lie on the soft mud of the foreshore, at low water, on levelled berths; and there are some private graving docks. There is ample space for the extension of both docks and wharves on both banks of the river.

CARDIFF DOCKS.

Situation. The port of Cardiff is situated on the left bank of the river Taff, and its entrance channel joins the river at a short distance from its outlet into the Bristol Channel. The outer portion of the river, traversing the Cardiff and Penarth Flats, and the approach channel to the docks are dry at low water of equinoctial spring tides; but the great rise of tide, amounting to 37½ feet at springs and 29 feet at neaps in the Cardiff and Penarth Roads, enables vessels to get up to the docks, near high water, at any state of the tides, across a flat foreshore which at low water would appear to bar the access of vessels[1]. (Plate 11, Fig. 6.)

The Cardiff Docks owe their existence to the late Lord Bute, and their prosperity to the construction of the Taff Vale Railway, which brings the mineral wealth of the surrounding district down to the port.

Growth of the Port of Cardiff. Till 1839, the Glamorganshire Canal, only available for vessels up to 200 tons, was the sole accommodation afforded by the port. In that year, however, the Bute West Dock having an area, with its basin, of 20 acres, was opened, having cost about £20,000 an acre; and after the opening of the Taff Vale Railway, in 1841, its trade rapidly increased, and an extension became necessary ten years later. Part of the Bute East Dock, and its basin were opened in 1855; and the dock was finally completed in 1859, adding 46 acres to the dock accommodation. This addition did not long suffice for the requirements of the increasing trade; and the Roath Basin, having an area of 12 acres, was opened in 1874 to form a new approach to fresh dock extensions. A dock in connection with this basin is in course of construction, and will raise the total area of the Cardiff Docks to about 113 acres; and the length of quay per acre of dock, which is at present 79 lineal yards, will be reduced to 71 yards, owing to the considerable width of the new dock.

Description of the Cardiff Docks. The two Bute Docks are both reached through half-tide basins with intermediate

[1] A plan of the Bute Docks, and other particulars, were furnished me by Mr. John McConnochie, the engineer of the port.

locks; and the Roath Basin will serve the same purpose for the large new dock. The West Dock is only 200 feet wide; but the East Dock is from 300 to 500 feet wide, and the Roath Basin is 550 feet wide; whilst the new Roath Dock will be 600 feet in width. (Plate 11, Fig. 6.)

A similar gradual increase has taken place in the dimensions of the locks. Thus the entrance to the West Bute Basin is 45 feet wide, and the inner lock is 152 feet long and 36 feet wide, the depth of water on their sills being 28⅔ feet at springs and 18⅔ feet at neaps: the outer lock of the East Dock is 220 feet long and 55 feet wide, with a depth of 31⅔ feet at springs; whilst the entrance lock to the Roath Basin is 350 feet long and 80 feet wide, with a depth of 35⅔ feet, which is larger than any lock at Hull, and equal in length and breadth to the Victoria Lock on the Thames, but exceeding it in depth. The lock leading from the Roath Basin to the new dock will have a length of 600 feet, and a width of 80 feet, and will therefore be larger than the Albert Dock lock on the Thames, both in length and depth.

A section of the quay wall, and a plan and sections of the gates of the Roath Basin lock, are given in Plate 14, Fig. 4, Plate 15, Fig. 1, and Plate 16, Fig. 4.

Appliances at the Cardiff Docks. The East and West Docks are fed with fresh water from the river Taff, which maintains the water-level in the docks some feet above the high-water level in the channel, so that the muddy waters of the Bristol Channel may be excluded, and the amount of deposit in the docks reduced.

The quays are very fully supplied with coal tips, or staiths, 13 balance tips at the West Dock, 12 balance, and 8 hydraulic tips at the East Dock, 8 hydraulic tips at the Roath Basin, and one at the entrance channel, to which reference has been previously made (p. 476). One movable hydraulic crane, capable of lifting 25 tons, is also used for shipping coal. About 12,000,000 tons of coal could be shipped in a year with the existing appliances at the Cardiff Docks. Water-pressure is used for working 30 cranes on the quays, and also the gates, bridges, and other machinery at the Roath Basin entrance. Ample siding accommodation has been provided for the coal trade; and warehouses have been erected for grain, with hydraulic appliances to expedite its storage and delivery. Eight graving docks afford provision for the repair of vessels, having lengths of from 220 to 600 feet.

A low-water pier, extending down the entrance channel, was erected in 1868, to afford means of communication with vessels in the Roads whilst the tide is low. It is 1400 feet long, 34 feet wide, and carries a roadway and railway; and a floating landing stage is placed at the end of the pier, where there is a minimum depth of 6 feet at low water.

Remarks on the Cardiff Docks. The docks at Cardiff comprise at present one entrance, five locks, two half-tide basins, one large basin, two docks, and a timber float; but another lock and dock will soon be added. They present a remarkable contrast to the docks at Bristol in their development and facilities. The docks, however, can hardly keep pace with the increase of trade, which has doubled in tonnage in the ten years between 1873 and 1883. Rivals have consequently sprang up, in the Penarth Docks which have been established several years, and in the Barry Docks which were authorised in 1884. Cardiff, however, shows how rapidly a port can be developed, under favourable conditions, by energy and good management.

PENARTH DOCKS.

Description. Penarth is only about three miles from Cardiff, and its docks emerge into the river Ely which joins the river Taff close to its mouth. (Plate 11, Fig. 6.) The docks were constructed by the Taff Vale Railway Company, in opposition to the Cardiff Docks, and were opened in 1864. They consist of an entrance from the river, 60 feet wide, with reverse gates; a half-tide basin of 3 acres; a lock, 275 feet long and 60 feet wide; and a dock which originally had an area of 17 acres, but was extended in 1883 to 22 acres[1]. Sections of the dock walls, and a plan and sections of the lock, sluices, and gates, are given in Plate 14, Figs. 11 and 12, Plate 15, Figs. 7 and 8, and Plate 16, Fig. 3. The depth of water on the sills is $32\frac{1}{6}$ feet at springs and $25\frac{2}{3}$ feet at neaps. The dock has the rather large proportion of 98 lineal yards of quay for each acre of water space, owing to the small width of the dock. The whole of the south quay is occupied with coal tips and their accompanying sidings, and the north quay is provided with hydraulic and steam cranes.

It was not feasible to furnish the Penarth Docks with a supply of fresh water, as was done for the Bute Docks, so the muddy waters of the Bristol Channel have to be admitted.

SWANSEA DOCKS.

Situation. Swansea is the nearest port in the Bristol Channel to the sea, being situated near its entrance. It lies at the mouth of the river Tawe, within the shelter of Swansea Bay, and protected from the west by the Mumbles Head. The tide rises $27\frac{1}{4}$ feet at springs and $20\frac{1}{4}$ feet at neaps. The entrance to the river was formerly impeded by a bar; and the few wharves along its banks were exposed to a swell during south-westerly gales.

Development of the Port of Swansea. The first improvement of the port consisted in constructing two piers, at the close of last century, which protected the entrance and lowered the bar[2]. A new cut was next made for the river, in 1840, which protected the shipping lying at the quays of the town from floods, and led the current more directly across the bar. In 1851, the old channel of the river was converted

[1] A plan and detailed drawings of the Penarth Docks were lent me by Mr. J. C. Hawkshaw, one of the engineers of the port.

[2] 'Minutes of Proceedings, Inst. C. E.,' vol. xxi, p. 310.

into a dock, with a lock and tidal basin at the lower end, resembling in fact the works carried out at Bristol. A second tidal basin and lock were added at the upper end of the old channel in 1861. In the meantime a dock was constructed on the foreshore south of the town, and to the west of the river mouth, with a total area of 17 acres, at a cost of nearly £9000 per acre. These works afforded a water area of about 34 acres; and this has been since raised to 63½ acres by the construction of another dock, on the east foreshore of the river, which was opened in 1881.

Description of the Swansea Docks. There are three separate docks at Swansea, each with its half-tide basin and independent entrance; namely, the North Dock in the old river channel, with an area of 14 acres, and a basin of 2¼ acres, together with a small private dock of about 1 acre; the South Dock along the western foreshore, having an area of 13 acres, and a basin of 4 acres; and the Prince of Wales Dock, on the east foreshore, whose area is 23 acres, with a basin of 6¼ acres. The tidal basins communicate with the docks through locks, whilst they have entrances at their outer ends. The entrances of the North and South basins, 60 and 70 feet wide respectively, open into the river Tawe; but the entrance of the Prince of Wales Basin, 65 feet in width and 32 feet deep, opens direct into Swansea Bay under shelter of the West Pier. The lock of the Prince of Wales Dock is the largest in the port, being 500 feet long and 60 feet wide, and having a depth of 32 feet of water on the sill at spring tides, and 25 feet at neaps.

Appliances at the Swansea Docks. The port is well provided with railway accommodation, as the Great Western, the London and North Western, and the Midland Railway Companies have each their sidings and coal tips on the quays.

There is a graving dock, 500 feet long and 50 feet wide, at the Prince of Wales Dock, as well as several belonging to private merchants.

Hydraulic machinery has been employed for working the gates. bridges, coal tips, cranes, and capstans, ever since the establishment of the docks; and the new Prince of Wales Dock has been fully equipped with hydraulic appliances like the other docks.

Remarks on the Swansea Docks. The docks are in the hands of a public Harbour Trust, like the Liverpool Docks. The aim of the trustees is the benefit of the town; their finances are in a flourishing condition; and both the amount of trade, and the size of the vessels are continually increasing. Swansea has the advantage over Cardiff of being nearer the sea and easier of access; whilst it is nearer to the coalfields of South Wales, and draws its traffic from the same districts. Its docks, however, were commenced some years later than the Cardiff Docks; and hitherto it has not been able to rival the dimensions, or equal the growth of that port.

Comparison of the various Ports. The docks of Hull and Grimsby have very nearly the same acreage at present; and Cardiff is already only 20 acres below Grimsby, and will be 15 acres ahead when the new dock is completed. Hull, however, will be well in advance when the new Alexandra Dock is opened. The dock area of Swansea is about two-thirds the existing area of Hull or Grimsby; and Newport bears the same relation to Swansea. The combined dock area of the above-mentioned ports of the Bristol Channel amounts to three-fifths of the area of the Liverpool and Birkenhead Docks, and is half as much again as the combined area of the docks at Hull and Grimsby. The most recently constructed locks at Cardiff, Swansea, Newport, and Avonmouth, compare favourably in general dimensions with the locks at other ports, and are unequalled in depth.

Trade Returns of Hull and Grimsby. Hull is the third port in the kingdom in respect of the values of its exports and imports, being next to London and Liverpool, though a long way behind them. Grimsby occupies a lower place, as the values of its exports and imports are exceeded by Glasgow, the Tyne ports, and Leith. The values of the exports and imports of Hull are very nearly balanced, the former being slightly in excess; but at Grimsby, the exports are about double the value of the imports. The exports at Hull have fallen off from £28,018,000 in 1873 to £22,032,000 in 1883, and at Grimsby from £10,725,000 in 1873 to £8,223,000 in 1883; whilst the imports have risen from £15,909,000 to £21,625,000 at Hull, and from £3,154,000 to £4,100,000 at Grimsby in the same period. (Appendix VI.)

Judged merely by the total tonnage of the vessels frequenting the various ports, Hull and Grimsby would be relegated to inferior positions amongst the ports of the United Kingdom; for the tonnage of the vessels trading with Hull is much less than that of the Tyne ports or Cardiff, and is exceeded by Glasgow, Sunderland, and Dublin. Grimsby, in this respect, is inferior also to most of the principal ports, with the exception of Barrow. (Appendix V)

Though the combined values of the exports and imports at Hull and Grimsby have fallen off between 1873 and 1883, the total tonnage of the vessels entering these ports has increased, having risen at Hull from 1,600,000 tons in 1873 to 2,055,000 tons in 1883, and at Grimsby from 574,000 tons to 715,000 tons in the same period.

The number of vessels with cargoes entering the port of Hull has undergone little alteration in the twenty years from 1863 to 1883, having been 4034 in 1863 and 4380 in 1883[1]. The class of vessel, however, has been completely altered; for the proportion of steamers to sailing vessels was about 1 to 1½ in 1863, nearly equal in 1873, and 2 to 1 in 1883; whilst the average tonnage has risen from 212 tons in 1863, to 310 tons in 1873, and 448 tons in 1883. The vessels with cargoes entering Grimsby increased from 2439 in 1873 to 2752 in 1883;

[1] The total numbers of vessels, both entering and clearing, with cargoes and in ballast, at the various ports are given in Appendix VII.

but the steamers, which were only 525 in number in 1873, were doubled in the ten years, whilst the sailing vessels were somewhat diminished. The average tonnage only slightly increased between 1873 and 1883, rising from 176 to 194 tons.

Trade Statistics of the Bristol Channel Ports. Bristol, the oldest of the ports on the Bristol Channel, comes next to Grimsby amongst British ports in the values of its exports and imports. It is essentially an importing place, as its imports amount to seven times the value of its exports. The imports of Swansea also are larger in value than its exports. Cardiff and Newport, on the contrary, are mainly exporting harbours, and owe their existence to the mines and coalfields of the adjacent districts. The exports of Cardiff are two and a quarter times its imports, and at Newport the proportion is about two to one.

The exports of Bristol have doubled in value since 1873, having been £577,000 in that year and £1,182,000 in 1883; and there has been a moderate rise in the exports from Newport, which were £1,840,000 in 1873 and £2,063,000 in 1883. On the other hand, the exports of Cardiff and Swansea have suffered a reduction, from £4,966,000 in 1873 to £4,859,000 in 1883 at Cardiff, and the larger diminution from £1,856,000 to £1,612,000 at Swansea in the same period. The imports of Bristol and Cardiff have risen from £7,607,000 and £1,885,000 in 1873, to £8,482,000 and £2,193,000, respectively, in 1883; whilst the imports of Swansea have fallen from £2,557,000 in 1873, to £1,931,000 in 1883. (Appendix VI.) The imports at Newport have doubled in the same period, having risen from £489,000 in 1873 to £957,000 in 1883. The imports, however, of Cardiff, Swansea, and Newport, were larger in 1880 than in 1883, and are liable to considerable fluctuations in value according to the state of trade.

Considering now the tonnage of the vessels frequenting these ports, Cardiff is far ahead, with a total tonnage of vessels leaving the port of 5,117,000 tons in 1883, having about doubled since 1873. Newport comes next, with a tonnage of 1,949,000 tons in 1883, having more than doubled since 1873, when the tonnage of the vessels leaving the port only amounted to 871,000 tons. Swansea stands third, with 1,505,000 tons in 1883, having gained half a million tons since 1873: whilst Bristol is last, with 1,228,000 tons in 1883, though in 1873 it was almost equal to Swansea with 1,059,000 tons for vessels entering, and ahead of Newport. (Appendix V.)

The number of vessels entering Bristol with cargoes was 7,544 in 1863; it rose to 9,072 in 1873, but fell to 8,260 in 1883; whilst at Swansea, it rose from 7,289 in 1873 to 7,542 in 1883. At Cardiff, the number of vessels going out with cargoes was 10,786 in 1873, and 12,678 in 1883, and 8,369 and 9,625 respectively at Newport.

The reduction of sailing vessels, and the increase of steamers, have taken place in the Bristol Channel as elsewhere. Thus at Bristol, the proportion of steamers to sailing vessels with cargoes was 1 to 3 in 1863, 1 to 2 in 1873, and 1 to $1\frac{1}{4}$ in 1883; at Cardiff, it was 1 to 2 in 1873, and 1 to $1\frac{1}{10}$ in 1883; at Newport it was 1 to 8 in 1873, and 1 to 2 in 1883; and at Swansea, 1 to $4\frac{1}{3}$ in 1873, and 1 to $1\frac{1}{2}$ in 1883.

The average tonnage of the vessels with cargoes has risen at all the ports. At Bristol, it was 97 tons in 1863, 114 tons in 1873, and 143 tons in 1883; at Cardiff, it has risen from 228 tons in 1873 to 392 tons in 1883; and at Newport and Swansea, it has risen from 97 to 183 tons, and from 137 to 184 tons, respectively, in the same period.

From the above statistics it will be seen that Bristol, with its appendages of Avonmouth and Portishead, is considerably ahead of Cardiff with Penarth in the value of its trade, though far below it in the total tonnage, as well as in the number and average tonnage of its vessels. Cardiff comes second to Bristol, amongst the Bristol Channel ports, in the value of its merchandise, and is by far the first port in other respects. Swansea is the third port as regards the values of its exports and imports, though its total tonnage and the number of its vessels are less than at Newport; whilst the value of the exports alone from Newport is higher than at Bristol and Swansea.

CHAPTER XXV.

DESCRIPTIONS OF DOCKS.

(*Continued.*)

LEITH. GREENOCK. DUBLIN. BELFAST. BARROW. HARTLEPOOL. SUNDERLAND.

Leith Docks: Site; Description; Appliances. *Port of Greenock:* Description. Trade Statistics of Glasgow, Leith, and Greenock. *Port of Dublin:* Description. *Port of Belfast:* Improvement of Entrance Channel; Description. Trade Statistics of Dublin and Belfast. *Barrow Docks:* Site; Progress; Description. *Hartlepool Docks:* Situation; Description. *Sunderland Docks:* Description. Trade Statistics of Barrow, Hartlepool, Sunderland, the Tyne, and the Tees.

THE three principal ports of Scotland are Glasgow, Greenock, and Leith, of which Glasgow is a river port, improved, like Newcastle and Middlesborough, by the training and dredging of the Clyde[1]; whilst Leith and Greenock are situated upon the large Firths of the Forth and the Clyde, similarly to the English ports on the Thames, the Mersey, the Humber, and the Bristol Channel, previously described.

The two most important Irish ports, at Dublin and Belfast, are situated at the mouths of rivers, opening into extensive estuaries, or bays, which provide them with naturally sheltered approaches.

Though docks are generally established on the banks of tidal estuaries, where shelter and access to the sea are combined, yet in some cases, owing to special requirements of trade, ports are formed on the open sea-coast, with only such natural advantages as the configuration of the shore may offer. Hartlepool on the east coast of England, and Barrow on the west, are examples of such ports; and Sunderland may be included in this category, for though the port ori-

[1] 'Rivers and Canals.' L. F. Vernon-Harcourt, p. 280, and plate 17.

ginated on the river Wear, its principal docks and main entrance are situated on the coast.

LEITH DOCKS.

Site. The port of Leith is situated at the mouth of the Water of Leith, which flows into the Firth of Forth. This river flows across the flat foreshore in a narrow shallow channel which, except in flood time, is mainly maintained by the ebb and flow of the tide, the rise of spring tides being about 17½ feet, and of neaps 13 feet[1]. (Plate 10, Fig. 2.)

Rise and Progress of the Port of Leith. The port is of ancient origin, and appears to have been first improved, like the jetty harbours of Calais, Dunkirk, and Ostend, by wooden and stone piers on each side of the channel, which were erected previous to 1600, one of them having been extended in 1607, when a stone wharf was also constructed. Towards the end of the 17th century, and early in the 18th century, the two piers on the west and east sides of the present inner harbour were partially reconstructed and prolonged to afford better shelter. In 1828, the eastern pier was extended 1500 feet; and a breakwater was formed, along the present high-water limit north-west of the docks, converging towards the pier on the opposite side of the channel, and leaving an opening between their ends of about 230 feet. These works provided improved protection, and the depth was increased by dredging. The east pier was again extended, in 1842, for 1000 feet, running nearly due north, to prevent the sands travelling along from the east from falling into the channel, and to secure a better depth. In 1852, the east pier was finally carried to its present termination, turning more towards the west; and a new west pier was built out, nearly parallel to the other, with an opening between their heads of about 250 feet. (Plate 10, Fig. 2.)

[1] A description of the port was given me by Mr. George Broadrick, the superintendent of the port.

A graving dock was built as early as 1720; but the first wet docks were only commenced in 1800, and completed in 1808. These two docks, known as the Old Docks, have each an area of 5 acres. Another dock, of 5 acres, called the Victoria Dock, was constructed in 1852. The next addition was the Albert Dock of 11 acres, on the east side of the channel, which was completed in 1868. The Edinburgh Dock to the east of the Albert Dock, having an area of 16⅔ acres, was opened in 1881, raising the dock area of Leith to 42⅔ acres. These last two docks have been formed on the eastern foreshore, of which 170 acres have been reclaimed.

Description of the Leith Docks. The accommodation at Leith comprises five docks, two locks, one entrance, and two passages; and also quays along the entrance channel of the river, and four public graving docks, together with three private ones. The quays round the docks have a length of 116 yards per acre of dock, an unusually large proportion, being greater than that of Liverpool or of Hull, which is due to the small area and width of the older docks, and also to the excellent arrangement of the quays at the Edinburgh Dock with its projecting jetty.

The earliest lock, forming the entrance to the Old East Dock, is only 170 feet long and 36 feet wide, and has a depth of 17 feet on the sills at spring tides, which is reduced to 12½ feet at neaps. This depth had become inadequate for shipping in 1852, when the Victoria Dock was made; so that the entrance to this dock was given a width of 60 feet and a depth of 23 feet on the sill at springs. When the Albert Dock lock was constructed in 1868, the width of 60 feet was maintained, but the depth was increased to 25 feet. This lock is 350 feet long, and opens into a tidal basin of 2½ acres. As the approach to the Edinburgh Dock is through the Albert Dock, the same lock serves for both docks. Many vessels frequenting the port draw 22 feet, so that they are unable to enter at neaps.

The entrance channel has been deepened to about 8 feet below low water of extreme spring tides; but this depth can only be maintained by the removal of about 110,000 tons of silt every year, for which purpose dredging plant, costing £13,000, was provided in 1879.

The Old Docks cost £28,500 per acre, the Victoria Dock £27,000, the Albert Dock £20,400, and the Edinburgh Dock £24,000 per acre.

The largest graving dock, constructed in 1864, is 372 feet long, 70 feet wide at the entrance, and 22 feet deep at high-water spring tides; it cost £100,000.

Appliances at the Leith Docks. Hydraulic power is employed for working the gates, swing-bridges, capstans, sluice-gates, coal hoists, and cranes. The hoists can raise 15 tons of coal a height of 30 feet.

The quays are well provided with railway accommodation, being served by the Caledonian and North British Railway Companies, the total length of lines laid down being 14 miles.

PORT OF GREENOCK.

Description of the Port. Greenock is situated on the Firth of Clyde, and enjoys the advantage of excellent shelter and deep water, as well as a small range of tide amounting to only 10 feet at springs and 6½ feet at neaps. The port has, accordingly, been formed with open tidal basins along the coast, surrounded by quays, which are called harbours, and possess independent entrances opening direct into the Firth. There are four of these basins; the East and West in the centre, and the Victoria and Albert on each side. The East Harbour of 6¾ acres, and the West Harbour of 7½ acres have a depth at their entrances of 20 feet at high-water spring tides, 18 feet at neaps, and 10 and 11½ feet, respectively, at low water of springs and neaps. The Victoria Harbour, having an area of 5½ acres, opened in 1851, and the Albert Harbour of 10½ acres, built about 1863, have 4 feet greater depth of water than the earlier basins. An ordinary wet dock, called the James Watt Dock, has been constructed to the east of the basins, with an area of about 13½ acres, raising the total water area of the port of Greenock to about 44 acres.

Greenock possesses three public graving docks, the largest being 650 feet long and 60 feet wide, with a depth of 18 feet on the sill at high-water spring tides.

Trade Statistics of Glasgow, Greenock, and Leith. Glasgow is the first port in Scotland, both in respect of the value of its exports and imports, and the tonnage of the vessels frequenting it; whereas Greenock is in advance of Leith as regards the total tonnage of its vessels, but inferior in the value of its merchandise. The export and import trade at Glasgow are nearly balanced, the export trade being somewhat the largest, both in the tonnage of the vessels with cargoes clearing the port, and also in the actual value of the goods. The export trade has risen in tonnage from 1,721,700 tons in 1873 to 2,906,400 tons in 1883, or an average increase of nearly 120,000 tons annually; and in value, it has increased from £10,583,200 in 1873 to £14,988,000 in 1883, being a rise of about £440,000 annually. (Appendices V and VI.)

The import trade is in excess of the export, both at Greenock and Leith; but whilst at Leith the tonnages of the vessels entering and clearing with cargoes present no great difference, at Greenock the vessels entering with cargoes have more than two and a half times the tonnage of the vessels clearing. The value of the imports was more than three times the value of the exports at Leith in 1883, and about twelve times the value of the exports at Greenock.

The numbers of vessels frequenting Glasgow and Greenock have steadily increased; the total number of vessels leaving Glasgow with cargoes amounting to 5,266 in 1863, 6,008 in 1873, and 8,681 in 1883; whilst those entering Greenock during the same years were 2,065, 3,856, and 5,975 respectively. There was a slight increase at Leith in 1873 in the number of vessels entering with cargoes as compared with 1863, but in 1883 they had fallen again slightly below the number in 1863. In all these ports there has been a regular decrease in the number of sailing vessels; whilst between 1863 and 1883, the increase of steamers with cargoes has been over threefold at Glasgow, sevenfold at Greenock, and double at Leith. There has been a corresponding rise in the average tonnage of the vessels, being half as much again at Glasgow and Leith; whilst at Greenock the change has been trifling[1].

PORT OF DUBLIN.

Description. The entrance to the port of Dublin, on the river Liffey, is across Dublin bar, which has been considerably lowered by the construction of the north and south walls already referred to (p. 168). The chief docks are the Spencer Dock, connected with the Royal Canal, on the north bank, and the Grand Canal Docks on the south bank; and there are three smaller docks, on the north bank, higher up the river. (Plate 11, Fig. 5.) As, however, the range of tide is only about 13 feet at springs and 7 feet at neaps, a considerable portion of the trade of the port is transacted along the river quays, which have been extended and deepened so as to admit vessels drawing 22 feet alongside at all states of the tide; and the North Wall Basin is being added, with an area of 51 acres and a minimum depth of 26 feet[2]. (Plate 14, Fig. 15.) The dock area amounts to nearly 38 acres, which with the North Wall Basin gives 89 acres of water space, exclusive of the accommodation of the river quays. Owing to the very narrow width of the Spencer Dock, the quays bear the unusually large proportion of 149 lineal yards of quay per acre of water space for the docks alone, but including

[1] The total numbers of vessels, both entering and clearing, with cargoes and in ballast, at the various ports, are given in Appendix VII.

[2] A plan of this basin and a section of its quay wall were sent to me by Mr. B. B. Stoney, engineer of the port.

the North Wall Basin this is reduced to 86 lineal yards per acre of water.

The Spencer Dock, being the property of the Midland Great Western Railway Company, is well supplied with siding accommodation on both quays; but no railways communicate with the other docks. The Great Southern and Western Railway, however, has sidings on the north quay, and can easily supply the wants of the North Wall Basin which will add so largely to the capabilities of Dublin as a port.

PORT OF BELFAST.

Improvement of Entrance Channel. Belfast is situated on the river Lagan, about two miles above the point where the river flows into Belfast Lough. The outlet channel was winding, and had silted up considerably towards the close of last century, so that vessels drawing more than 10 feet of water could not come up at neap tides. The river was first deepened by dredging, and shoals were removed; and the channel has been straightened by two cuts, one, 3000 feet long, completed in 1841, and the second, 3300 feet long, opened in 1849[1]. These cuts, forming the Victoria Channel, were given a depth of about 23 feet at high water[2]. (Plate 10, Fig. 4.) Since 1858, regular dredging operations have been carried on for deepening the channel, and the depth is now sufficient to admit steamers trading with the port at any state of the tide. The average depth is about 12 feet at low water, and the rise of tide is 8 feet 10 inches at springs, and 8 feet 1 inch at neaps. A further improvement of the approach to the port has been contemplated, by deepening the existing channel, so that vessels drawing 24 feet could get up at high-water neap tides, and by prolonging the channel to deep water in the Lough.

[1] Minutes of Proceedings, Inst. C. E.,' vol. lv, p. 25.

[2] A plan of Belfast Harbour, in 1883, was given me by Mr. T. R. Salmond, the engineer of the port.

Description of the Belfast Docks. At the end of last century the port possessed some small tidal basins, and 590 lineal yards of quays, only suitable, for the most part, for vessels of small draught. The state of the port in 1883 is shown on Plate 10, Fig. 4. The Dufferin Dock, of 3¼ acres, is the only dock closed by gates, and it has an entrance 60 feet wide. The rest of the docks and basins are tidal; the largest is the Abercorn Basin of 12½ acres, which is quite open to the river, and the next in size is the Spencer Dock of 7½ acres, with an entrance 80 feet wide. A section of the Spencer Dock wall is given in Plate 14, Fig. 17. The Prince's Dock of nearly 4 acres was the earliest dock, opened in 1832; next in order of time comes the Clarendon Dock, having nearly the same area, opened in 1851; then the Abercorn Basin, opened in 1867; and, lastly, the Dufferin and Spencer docks, opened in 1872. These five docks, with a total area of 31 acres, afford 103 lineal yards of quay per acre of water; and there is an additional length of about a mile and a half of river quays. There are also six timber ponds, two being of large dimensions, and three graving docks, of which the Hamilton Dock, opening out of the Abercorn Basin, is 450 feet long on the floor, and has an entrance 60 feet in width. Good railway facilities are provided for the quays on the north-western side of the river.

Belfast, like Dublin, having a small tidal range, can utilise quays along the river banks. Both ports possess a well-sheltered approach from the sea; and the access to both has been improved by training the channel and by dredging.

Trade Statistics of Dublin and Belfast. The value of the export and import trade of Dublin is larger than that of Belfast; but whilst the imports of Dublin are half as much again as those of Belfast, the exports of Belfast were ninefold those of Dublin in 1883, but the trade of Dublin has been falling off considerably during the last few years. The imports both of Dublin and Belfast have fallen in value since 1873, having been

£3,800,000 and £2,236,000, respectively, in that year; whilst in 1883 they were £3,258,000 and £2,171,000, though they had reached a maximum in 1877 and 1876. The exports have fallen at Dublin, namely, from £117,000 in 1873, to £82,000 in 1883, a great fall having occurred in that year; but at Belfast they have quadrupled, having been only £189,000 in 1873, and £782,000 in 1883.

There has been a moderate rise in the total tonnage of vessels entering these ports between 1873 and 1883, from 2,052,000 tons in 1873 to 2,175,000 tons in 1883 for Dublin, and from 1,430,000 tons to 1,805,000 tons for Belfast, though a maximum was attained at Dublin in 1878, and at Belfast in 1879.

A great change is noticeable in the class of vessels frequenting these ports between 1863 and 1883, though the actual number of vessels has been little altered; for the steamers entering Dublin with cargoes increased from 1,536 in 1863, to 4,280 in 1883, and the sailing vessels with cargoes diminished from 6,313 to 2,854; whilst at Belfast, the corresponding numbers were 1,983 and 4,849 for steamers, and 5,223 and 3,684 for sailing vessels. The average tonnage of the vessels with cargoes has risen at Dublin in this period from 140 tons to 246 tons; whilst the rise at Belfast is from 146 tons to 203 tons.

BARROW DOCKS.

Site. The port of Barrow is situated on the west coast of Lancashire, under the shelter of Walney Island which stretches in front of it for a length of about 8 miles; and the channel to the sea is round the southern end of the island. Barrow, accordingly, though on the sea-coast, possesses excellent natural shelter. Its entrance channel, between Walney Island and the shore, was shallow and winding; but a straight channel has been dredged, from the docks to Piel Island at the southern end of the harbour, to a depth of 13 feet at low-water spring tides, so that with a rise of tide of 28 feet at springs, and 20½ feet at neaps, the largest class of vessels can enter the port.

Progress of Barrow. Barrow is one of the most remarkable instances of rapid growth, for in 1847 it was a mere hamlet containing only 325 inhabitants; in 1861 the population had increased tenfold; in 1871 it was 18,000; whilst in 1881 it had attained about 49,000. The first docks at Barrow were opened in 1867; and in 1883 there were 234 acres of docks, inferior only in this respect to London and Liverpool. This marvellous progress is due to the rich veins of hæmatite iron ore which exist in the vicinity, the working of which has developed a flourishing trade. The dock accommodation, however, at Barrow is very large in proportion to its existing trade.

Description of the Barrow Docks. The port of Barrow-in-Furness possesses four docks, a basin, two locks, an entrance, and a graving dock[1]. (Plate 11, Fig. 2.) All the docks are of large dimensions, ranging from the Devonshire Dock of 30 acres, up to the Cavendish Dock of 102 acres. The docks are all in communication, and can be entered through a lock at one extremity, or through an entrance at the other. The Devonshire Dock, and the Buccleuch Dock, of 31 acres, were constructed in a channel between Barrow Island and the mainland, and were opened in 1867; the Ramsden Dock of 63 acres, with its basin of 8 acres (Plate 14, Fig. 13), was opened in 1879; and the large Cavendish Dock has since been constructed. The Devonshire Dock is entered through a basin lock, closed by gates at one end, and by a caisson at the other which also serves as a bridge; and the available width of entrance is 60 feet. The entrance into the Ramsden Basin, and the lock leading into the Ramsden Dock have been made 100 feet wide, with the object of facilitating the entrance of

[1] A plan of the Barrow Docks and other particulars were given me by Mr. F. C Stileman, engineer of the port.

vessels from the tideway, and to enable them to pass through in opposite directions at the same time. The passages into the Buccleuch and Cavendish Docks are 80 feet in width. The Devonshire Lock has a length of 500 feet, and a depth of 18 feet on the sill at high-water neap tides; the Ramsden Lock is 700 feet long, and its outer sill, as well as the sill of the entrance, is 24 feet below high-water neap tides.

There are only 53 lineal yards of quay per acre of water in the docks, in spite of the well-arranged branches of the Ramsden Dock. This low proportion is due to the large size of the docks, and more particularly to the unusual width of the Cavendish Dock. This dock, indeed, is larger in area than the Victoria Dock on the Thames; and the Liverpool Docks appear insignificant in dimensions beside it. (Compare Plate 11, Fig. 1, with Fig. 2.) The main object, however, of this dock is the accommodation of timber, for which a large area of water is advantageous; and the conditions of the site were favourable for a large enclosure.

The water in the docks is maintained at the level of spring tides (or a depth of 24 feet in the docks, and 30 feet over the deeper sills) by means of the flow from the Abbey Beck.

The docks belong to the Furness Railway Company, and are well supplied with sidings, cranes, warehouses, and other facilities for trade. Hydraulic power is employed for working the gates, bridges, capstans, sluices, cranes, grain elevators, and other machinery.

A depositing dry dock has been recently provided, thus adding to the accommodation of the port for the repair of vessels. (See p. 463.)

HARTLEPOOL DOCKS.

Situation. The port of Hartlepool is situated in Hartlepool Bay, which is sheltered from the north by the coast, and from the north-east by a breakwater projecting from the shore at the northern extremity of the bay. (Plate 10, Fig. 3.) The docks are approached through two tidal harbours, one situated under the shelter of the coast, at Old Hartlepool, on the northern shore of the bay, and the other in front of the newer town of West Hartlepool. The rise of tide is 15 feet at springs and 11½ feet at neaps; and there is a depth of water in the old harbour of 25 feet at high-water spring tides.

Description of the Hartlepool Docks. The first dock at West Hartlepool was opened in 1847, the Jackson Dock in 1852, and the Swainson Dock in 1856. The docks have since been extended northwards, and now join the Old Harbour. (Plate 10, Fig. 3.) The only dock of Old, or East Hartlepool is the Victoria Dock of 19 acres, which communicates with the Old Harbour by a lock. The port now possesses seven docks, and the North and South half-tide basins, opening into the old and west harbours respectively, having a total area of about 79 acres; and there are 91 lineal yards of quay per acre of water. The Old Harbour has an area of 20 acres, and the West Harbour 44 acres; and there are 57 acres of timber ponds in connection with the docks The old entrance to the Coal Dock, and the lock of the Victoria Dock, have a width of 42 feet, and a depth of water on the sills at springs of 21½ feet; but the newer entrances to the North and South basins have a width of 60 feet, and depths of 26 and 21½ feet respectively.

The docks, having been constructed by the North Eastern Railway Company, are amply supplied with sidings, as well as with cranes, coal staiths, and other appliances.

SUNDERLAND DOCKS.

Description. Sunderland, situated on the sea-coast, at the mouth of the Wear, between Hartlepool and the Tyne, mainly owes its prosperity as a port to coal and ship-building. The river has hitherto afforded only a moderate depth over its bar; so that when the docks were constructed south of the river, a direct entrance was formed into the sea by two converging piers across the beach[1]. (Plate 12, Fig. 6.) There is a single dock on the north side of the river, six acres in area. The rest of the docks, three in number, extend in a continuous line to the south, parallel to the coast. They communicate with the river by a half-tide basin with two entrances, and with the southern outlet by a lock and an entrance. The sea lock is 480 feet long, with a width of 90 feet, which is reduced to 65 feet at the gates; it has a depth of 27 feet of water on the sills at high-water springs, and 23½ feet at neaps; the entrance has a width of 60 feet, and a depth of 26½ feet at springs. The North Dock belongs to the North Eastern Railway Company, and the Southern Docks belong to the River Wear Commissioners. The docks, with the half-tide basin, have an area of 52½ acres, and afford 81 lineal yards of quay per acre of water.

The quays are amply equipped with sidings, coal tips, hydraulic and steam cranes, and other appliances for the rapid landing and shipment of cargoes.

Trade Statistics of Barrow, Hartlepool, Sunderland, the Tyne, and the Tees. The Tyne ports are far in advance of the other north-eastern ports in the value of their trade, as they rank next after London, Liverpool, and Glasgow, their exports and imports in 1883 having reached £15,547,000. Hartlepool stands second amongst these ports with a value of £3,624,000 in 1883; the Tees comes next with £3,194,000; Sunderland ranks fourth amongst the ports described with only £1,519,000, having been nearly reached by Barrow, in 1883, with a value of exports and imports of £1,411,000.

The imports of the Tyne and Tees ports were larger in 1883 than in 1873, having risen from £5,765,000 to £9,826,000 at the former, and from £664,000 to £694,000 at the latter. The imports of Barrow, Hartlepool, and Sunderland have, on the contrary, declined.

[1] A plan of the docks and other particulars were furnished me by Mr. H. H. Wake, engineer to the River Wear Commissioners.

The exports of the Tyne and Tees have been considerably reduced, together with those of Hartlepool and Sunderland; whilst the exports of Barrow have risen from £676,000 in 1873, to £879,000 in 1883. (Appendix VI.)

The total tonnage of the vessels frequenting these ports has risen in every case since 1873; on the Tyne, it has risen from 4,677,000 tons in 1873 to 6,669,000 tons in 1883; at Sunderland, from 2,125,000 tons to 2,800,000 tons; on the Tees from 618,000 tons to 1,276,000 tons; at Hartlepool from 861,000 tons to 913,000; and at Barrow from 348,000 tons in 1873, to 575,000 tons in 1883. (Appendix V.)

The number of vessels with cargoes was larger on the Tees and at Barrow in 1883 than in 1873: but was reduced on the Tyne, and at Hartlepool, and Sunderland. The number of sailing vessels with cargoes has decreased in every instance, having been supplanted by steamers[1]. The most notable example of this change occurs on the Tyne, where the sailing vessels clearing with cargoes, in 1863, were 17,694 in number, and only 6,349 in 1883, having been reduced to nearly a third in a period of twenty years; whilst the steamers with cargoes increased in the same interval from 1,410 to 8,680, being multiplied sixfold.

The average tonnage of the vessels has risen at all these ports to the extent of about one half as much again between 1873 and 1883, except on the Tees, where it has nearly doubled.

[1] The total numbers of vessels, both entering and clearing, with cargoes and in ballast, at the various ports, are given in Appendix VII.

CHAPTER XXVI.

GOVERNMENT DOCKYARDS.

PORTSMOUTH. CHATHAM. QUEENSTOWN.

Sites for Dockyards. Differences between Dockyards and Commercial Docks. *Portsmouth Dockyard:* Rise and progress; Extension Works, site and description; Graving Docks. *Chatham Dockyard:* Extensions proposed; Extension Works adopted, site and description; Graving Docks. *Queenstown Dockyard:* Site of Dockyard and Extension; Description of Extension Works. General Remarks.

NATIONAL dockyards are generally established where deep water and good natural shelter are found combined with easy access to the sea. Dockyards, unlike commercial docks, are not dependent on the caprices of trade and the channels of traffic, so that the best natural sites can be selected without reference to commercial considerations. Dockyards are, accordingly, formed in sheltered bays or creeks, as at Portsmouth, Pembroke, and Brest, or near the mouths of estuaries, as at Chatham, Sheerness, and Toulon. Occasionally additional shelter has to be provided by a breakwater, as at Plymouth and Cherbourg.

Differences between Dockyards and Commercial Docks. The objects of dockyards are the construction and repair of vessels of war. Accordingly graving docks and slips form very essential portions of a dockyard; and workshops and yards are established on the quays. The trade of a dockyard is confined to the materials required for construction or repair, and to the necessary stores for the equipment of sea-going vessels.

Wet docks are comparatively modern adjuncts to dockyards, for the first wet dock at Portsmouth was opened in

1848, and the first at Chatham in 1871; whilst none exist at Plymouth and Sheerness. The sheltered water area in front of the dockyards serves as anchorage-ground for vessels; and the older graving docks and slips at Portsmouth and Chatham open direct into Portsmouth Harbour (Plate 12, Fig. 5), and the Medway, respectively. Basins have been constructed at Cherbourg Dockyard (Plate 3, Fig. 5); but the Penfeld river furnishes the sole accommodation at Brest Dockyard for vessels afloat.

The locks at dockyards are sometimes constructed so that in case of emergency they may be temporarily used as graving docks. They are, moreover, given a greater width than is usual at commercial docks, to admit vessels of novel types: thus one entrance at Portsmouth has a width of 94 feet, and a lock at Chatham a width of 94½ feet; whereas, since the abandonment of paddle steamers, the standard width at Liverpool has been reduced to 60 feet, and no lock has been built on the Thames exceeding 80 feet in width. The depth also of the sills at dockyards are made exceptionally great, in order to be capable of admitting disabled vessels at any state of the tide. Thus the sills of the Chatham Locks are 32 feet below the level of high-water spring tides; and the outer sills of the Portsmouth Locks are 41⅔ and 40⅔ feet below the same level. Caissons are employed for closing the entrances and locks at dockyards, in place of the gates almost invariably adopted at commercial docks, because the entrance and exit of vessels is less frequent at dockyards; and caissons are more suitable than ordinary movable bridges for bearing the heavy loads which have sometimes to be carried over them.

Dockyard basins are made wide so as to enclose a considerable area of water, since ample water space for the accommodation and movement of vessels is of more importance than a large proportion of length of quays, which is so valuable at trading ports. At Portsmouth there are only 68 lineal yards of quay per acre of water, and at Chatham 57 yards, as compared with 108 yards at Liverpool, and 113 yards at Hull.

PORTSMOUTH DOCKYARD.

Progress of the Dockyard. Portsmouth became an important naval station in early times, in consequence of the excellent shelter and depth of its harbour, and because of its position in regard to the Channel and the French coast. The port and fortifications were improved by Edward IV. In 1540 the dockyard covered only 8 acres of land; 2 acres were added in 1658; and at the close of the last century it had been increased to 95 acres. No further additions were made till 1843, when the first wet dock and four graving docks were commenced, to meet the increased requirements occasioned by the introduction of steamships. The dock, called the Steam Basin, having an area of 7 acres, was opened in 1848. These additions, however, only sufficed for a few years; and in 1864 a large extension of the docks was approved[1]. (Plate 12, Fig. 5.)

Site of Portsmouth Dockyard Extension. The extension has been carried out by reclaiming an area of 95 acres from some tide-covered mud-lands in Portsmouth Harbour, on the north side of the island of Portsea and bordering Fountain Lake[2]. The reclamation was effected by forming a shallow inner timber dam, and an outer timber dam, joining the land at each end and encircling the site. These dams were constructed of sheet-piling strutted behind, and they were placed from 230 to 280 feet apart. The outer harbour wall (Plate 14, Fig. 9), the northern wall of the Rigging Basin, and the northern and north-western walls of the Fitting-out Basin were built in the interval between the dams, thus providing a

[1] A plan of Portsmouth Dockyard was supplied me by Mr. Colson, late resident engineer of the extension works.

[2] 'Minutes of Proceedings, Inst. C. E., vol. lxiv, p. 118.

permanent walled embankment round the reclaimed site. The entrance to the new Tidal Basin was temporarily closed by a segmental cofferdam.

This low-lying land formed a very suitable site for the excavation of basins; and the surplus earthwork was conveyed by railway, over a viaduct, to form land at Whale Island.

Description of the Portsmouth Dockyard Extension Works. The works comprise a tidal basin open to the harbour and forming the entrance to the other basins, three floating docks or basins, two entrance locks, three entrances, five graving docks, and a line of outer quay walls bordering the harbour and Fountain Lake. (Plate 12, Fig. 5.)

The Tidal Basin has an area of 10 acres, and has been carried to a depth of 30 feet at low-water spring tides, so that ships are always able to enter it. Direct access has been provided from the Tidal Basin to the Steam Basin, the Repairing Basin, and the Fitting-out Basin, and also to a deep graving dock, besides the two locks which can be used as graving docks. A section of the quay wall of the Tidal Basin is given in Plate 14, Fig. 9.

The Repairing Basin, having an area of 22 acres, is entered from the Tidal Basin through two locks, each 458 feet long, and furnishes access to the Rigging Basin through an entrance 80 feet wide, and having a depth of 32½ feet on the sill below high-water spring tides. The depth of the Repairing Basin is 32½ feet below high-water spring tides, and the depth of the Rigging Basin is the same.

The Rigging Basin has an area of 14 acres; and communication has been provided from it to the Fitting-out Basin, by an entrance of the same dimensions as the one previously mentioned connecting it with the Repairing Basin. Direct access to the harbour has also been furnished by an entrance, on the northern side of the basin, with the same depth of sill, but having a width of 94 feet. This basin forms the second instalment of the extension, the Tidal Basin and Repairing Basin having been opened a few years earlier.

The Fitting-out Basin, having an area of 14 acres, will form the final instalment of the extension, the works of which are still in progress. Communication is afforded for this basin with the harbour, through the Tidal Basin, by an entrance 80 feet wide, and having a depth on the sill of $32\frac{1}{2}$ feet at high-water spring tides.

All the entrances to the basins, and the outer ends of the two locks, are closed with sliding caissons (Plate 16, Fig. 8); but the inner ends of the locks are provided with ship-caissons. The north lock has an entrance width of 82 feet, and a depth of $41\frac{2}{3}$ feet on the outer sill, and 33 feet on the inner sill, below high-water spring tides; the south lock is 80 feet wide at each end, with depths of $40\frac{2}{3}$ and $32\frac{1}{2}$ feet on the sills. The lock-chambers are widened out to 100 feet.

Graving Docks at Portsmouth Dockyard. There were eleven graving docks and three slips in existence at Portsmouth previous to the extension works; three graving docks communicating with the Steam Basin, and the remainder opening direct into the harbour. Out of the five additional graving docks contemplated, three only have hitherto been constructed; but either of the two locks, with their deep outer sills, can be used, on an emergency, as a graving dock. Two of the graving docks have their entrances on the south side of the Repairing Basin; and the entrances of two others have been constructed along the same quay.

The largest graving dock, called the Deep Dock, opens into the Tidal Basin, so as to be immediately available for docking a disabled ship. This dock is 428 feet long; and its entrance is 82 feet wide, and has a depth on the sill of $40\frac{2}{3}$ feet at high-water spring tides, only 2 feet less than the Tidal Basin. The Deep Dock is closed by a sliding caisson, whereas the other graving docks have ship-caissons.

The sliding caissons, cranes, capstans, and penstocks are worked by means of compressed air.

CHATHAM DOCKYARD.

Proposed Extensions. Till recently the Medway furnished the only floating accommodation for Chatham Dockyard. The inconvenience, however, of the river with its tortuous course, rapid currents, and shoals, for vessels to lie in for repairs, led to the suggestion of forming a floating dock as far back as the reign of Charles II, as recorded in Pepys' Diary; but nothing resulted from the proposal, though additions were made at various periods to the dry docks. In 1814, Mr. Rennie proposed the diversion of a portion of the river channel by a cut across a sharp bend opposite Chatham, and the conversion of the existing river bed, at that part, into a wet dock. Though, however, the increase in the draught and length of vessels aggravated the defects of the river as anchorage-ground, no decided steps were taken towards the construction of floating basins till 1861, when an extension of the dockyard, comprising the formation of basins across the low land between Upnor and Gillingham, was approved by the Admiralty[1].

Site of the Chatham Dockyard Extension Works. The site selected for the floating docks, or basins, was along the line of St. Mary's Creek, forming a short cut across a sharp bend and connecting the Upnor and Gillingham reaches of the river. This site admitted of the construction of a line of basins with entrances into the river at both ends; whilst the low-lying land reduced the amount of excavation, and provided ground on which the earthwork could be deposited for the formation of quays. The river was shut off from the works by an embankment dam, at the lower, or Gillingham end, formed with the excavated material (alluded to in page 399),

[1] 'Chatham Dockyard Extension Works.' Lectures by E. A. Bernays, 1879.

and by a cofferdam at the Upnor extremity. The exact line of St. Mary's Creek was adopted for the basins, to diminish the excavation, which occasioned the irregularity in their form[1]. (Plate 12, Fig. 3.)

Description of the Chatham Dockyard Extension Works. The works consist of three basins; namely, the Repairing Basin of 20½ acres at the Upnor end; the Factory Basin of 20 acres in the centre, and communicating with the other two basins by a passage at each end; and the Fitting-out Basin of 27¾ acres at the Gillingham end. Four graving docks also, opening out of the Repairing Basin, formed part of the scheme.

Communication is provided between the Repairing Basin and the upper reach of the river through an entrance, 79½ feet wide, and 30 feet deep at high-water spring tides, which is closed by a ship-caisson. The Fitting-out Basin is reached from the river through two locks, with a depth of 32 feet at high-water spring tides over the four sills. The north lock has a length of 436 feet and a width of 94½ feet, and the south lock is 438 feet long and 84½ feet wide. Both locks have been constructed so as to be available for graving docks, and they are closed at each end by sliding caissons. The passages between the basins are 84 feet wide and 32 feet deep, and are closed by ship-caissons. (Plate 12, Fig. 3.) Hydraulic machinery has been erected close to the locks for working the sliding caissons and other appliances.

The section of the earlier quay wall at Chatham is precisely similar to the Portsmouth wall (Plate 14, Fig. 9); but as the work proceeded concrete was more largely used, a portion of the walls of the Fitting-out Basin having merely a brick facing above the water-level, and the later portions being constructed entirely of Portland cement concrete. (Plate 14, Fig. 10.)

The Repairing Basin was opened in 1871; and the Fitting-out Basin, with its locks, has been recently completed. The

[1] A plan of the Chatham Extension Works was given me by Mr. Bernays, the superintending engineer.

total cost of the works amounted to £1,950,000. The area of the dockyard has been increased from 95 acres to 500 acres by the extension works.

Graving Docks at Chatham Dockyard. There were four graving docks and seven slips at Chatham Dockyard, previous to the extension works, situated along the right bank of the Medway, higher up the river than the new basins. In addition to the basins, the extension works included four graving docks, opening out of the Repairing Basin and completed between 1871 and 1873. All these new graving docks are 416 feet long; the entrances of two are 80 feet wide, and 31½ feet deep below high-water spring tides, and the other two are 82 feet wide and 32 feet deep. The details of one of the larger ones are given in Plate 16, Figs. 11 and 14.

QUEENSTOWN DOCKYARD.

Site. The harbour of Queenstown is a fine natural harbour. in the bay of Cork, sheltered by the Spit Bank and Haulbowline and Spike Islands. A dockyard, covering 22¼ acres, exists on Haulbowline Island; and a small cove on the north side of the island is protected by a breakwater. In order to provide better accommodation for shipping, and to increase the dockyard, a portion of the Spit Bank, adjoining Haulbowline Island on the east, has been reclaimed by surrounding it with an embankment, and has been connected with Spike Island by another embankment forming a sort of long bridge between them[1]. (Plate 12, Fig. 4.) The area thus reclaimed and added to the dockyard amounts to 33¼ acres.

Description of the Extension of Queenstown Dockyard. A floating basin of 9¾ acres is being formed on the reclaimed land just mentioned. The basin will have access to Queens-

[1] A plan of Queenstown Dockyard and a section of the basin wall, were given me by Col. Percy Smith, Director of works to the Admiralty.

town Harbour on the north-western side, through an entrance 94 feet wide, having a depth of water on the sill of 32⅔ feet. A graving dock is being constructed on the south side of the basin, with the same width and depth of entrance; and space has been left for the construction of a second graving dock, parallel to the first, if required at a future time. The rest of the reclaimed area will be used for quays, factories, and workshops. The entrances both to the basin and graving dock will be closed with caissons.

A section of the basin wall is given in Plate 15, Fig. 12. It is founded on rock underlying a thick layer of muddy sand and gravel, and is composed of rubble masonry with a facing of ashlar.

Remarks on the foregoing Dockyard Works. The extension works at Portsmouth, Chatham, and Queenstown present several points of resemblance. Floating basins as well as graving docks have been constructed at all the three dockyards. The area of the three basins at Chatham exceeds that at present provided at Portsmouth; but when the extensions at Portsmouth are completed, the floating accommodation at both dockyards will be nearly equal, amounting to about 68 acres. The entrances at all the dockyards have been given a maximum width of about 94 feet; and the standard depth of the main entrances is about 32 feet, though the two outer lock sills at Portsmouth have been given a depth slightly exceeding 40 feet. All the locks and entrances are furnished with sliding or ship caissons; and the pairs of locks both at Portsmouth and Chatham are convertible into graving docks. The locks at Chatham not only provide a deep entrance into the basins, but also enable vessels to avoid a sharp and circuitous bend of the river which had to be traversed to reach the old dockyard. Convict labour has been employed on the works; but this assistance must be regarded rather as affording occupation for the convicts than as effecting much economy in the execution of the works.

The extensions at Portsmouth and Queenstown have been effected by reclaiming shallow portions of the sheltered areas of the harbours; whilst waste low-lying land has been utilised at Chatham. The works at the three places have very largely increased the accommodation and efficiency of the dockyards, and afford facilities for construction and repairs which were previously wanting.

CHAPTER XXVII.

FOREIGN DOCKS.

HAVRE. HONFLEUR. ST. NAZAIRE. BREST. MARSEILLES. TRIESTE.

Port of Havre: Early History; Progress; Description; Extensions in progress; Appliances; Maintenance; Statistics; Remarks. *Port of Honfleur:* Early History; Progress; Extensions; Description; Maintenance; Statistics; Remarks. *St. Nazaire Docks:* Recent origin of Port; Site of Docks; Development; Description; Quay Walls on Well Foundations; Accessory Works; Maintenance; Cost; Statistics; Remarks. *Port of Brest:* Description; Statistics. *Port of Marseilles:* Difference from Tidal Ports; Description; Quays; Graving Docks; Statistics; Remarks. *Port of Trieste:* Description; Extensions; Remarks.

THE ports of Havre, Honfleur, and St. Nazaire closely resemble the English docks previously described, for they possess floating basins, surrounded by quays, which have been constructed on low-lying or reclaimed lands, and are approached through tidal basins, and by locks and entrances closed by gates. The tideless ports of Marseilles and Trieste are different in construction, being sheltered by breakwaters and moles, as indicated in Chapter XII, enclosing a space of water without reclamation, for their entrances do not need to be closed by gates owing to the absence of fluctuations in the water-level. The commercial port of Brest, though exposed to the rise and fall of the tide, has been similarly formed; but it is a comparatively small port, and the proposed extension is to be constructed as an enclosed floating dock, on a site reclaimed along the foreshore, like ordinary tidal ports.

PORT OF HAVRE.

Early History and Progress. Havre was founded by Francis I in 1516; but it only possessed a tidal basin till 1667, when the King's Basin, which had been excavated and surrounded by quay walls in 1628, was converted into a wet dock[1]. The entrance to the port is through a jetty channel, formed originally in a natural creek, and subsequently maintained by sluicing basins. The jetties have been periodically prolonged to prevent the inroad of shingle from the adjacent beach to the north of the entrance channel; so that the early history of the port is similar to that of the jetty harbours of the Channel and North Sea.

A large extension of the port was determined upon in 1787. The works were commenced forthwith; but their final completion was delayed till 1834, by the troubles of the revolution, the wars of the empire, and the subsequent want of funds. They consisted in the construction of two docks, namely the Bar and Commercial docks; an enlargement of the tidal harbour; and the formation of a large sluicing basin, reclaimed from the foreshore of the river, to the south of the tidal harbour. The port at this latter date comprised an outer harbour of 26 acres, three docks with a combined area of 29⅔ acres, and three sluice-ways for scouring the channel.

A further extension was authorised in 1839, consisting of the Vauban Dock on the site of the old fortifications, and the Florida Dock, for steamboats, formed out of a portion of the large Florida sluicing basin. (Plate 12, Fig. 2.)

The increase of trade was so rapid that an additional enlargement of the port was approved in 1844. The works comprised the completion of the Vauban and Florida docks; the construction of the large Eure Dock communicating with the outer harbour, the Vauban Dock, and the Florida Dock; the construction of the Warehouse Dock, opening into the Eure Dock and parallel to the Vauban Dock; the improvement of the entrance channel; the formation of a graving dock, and other minor works. The Eure Dock was opened in 1855, and the adjoining Warehouse Dock in 1859; whilst the large Transatlantic entrance connecting the Eure Dock

[1] 'Le Port du Havre.' Baron Quinette de Rochemont, 1882, p. 3.

with the outer harbour, and the large graving dock opening into the Eure Dock were opened at the beginning of 1864.

Recent Additions to the Port of Havre. The construction of a dock for coasting steamers was authorised in 1865, and opened at the close of 1871. This dock is situated on the site of the former citadel, and hence is called the Citadel Dock; it has an entrance into the Eure Dock, and communicates with the outer harbour by a lock, and is provided with three graving docks.

Before these works were completed, further improvements were determined upon in 1870, namely, the enlargement of the outer harbour, and the construction of a river wall towards the east for reclaiming some of the river foreshore. The construction of a ninth dock, on the reclaimed site, was approved in 1879, together with two graving docks on either side of the large Eure graving dock.

Lastly, the Tancarville Canal, for connecting the port of Havre with the regulated channel of the Seine at Tancarville, was authorised in 1880; and a tenth dock, to the east of the Ninth Dock, has been designed [1]. (Plate 12, Fig. 2.)

Description of the Port of Havre. There are eight docks open for traffic at Havre, having a total area of 131½ acres; the Ninth Dock, with an area of 23 acres, is approaching completion, which will raise the water space to 154½ acres; and works for the construction of the Tenth Dock, designed to have an area of nearly 27 acres, have been commenced. The proportion of quay length to the water area of the eight docks is 75 lineal yards per acre; but on the addition of the Ninth and Tenth Docks, the proportion will be reduced to 70 lineal yards per acre, owing to the width of the new docks. A section of the wall of the Eure Dock is given in Plate 14, Fig. 22.

There are eleven entrances, of which seven serve as passages between the docks, and one lock forming the approach to the

[1] A plan of the extension works at Havre was given me by M. E. Widmer, one of the engineers of the port.

Citadel Dock. Seven of the entrances are closed with a single pair of gates; but the two entrances to the Vauban Dock, and the entrance to the Warehouse Dock have reverse gates, so that the water-levels may be independent if required; whilst the main entrance to the Eure Dock is provided with two pairs of gates pointing inwards. This latter entrance is the largest in the port, having a width of 100 feet and a depth on the sill of 35 feet below high-water spring tides; the remainder vary from the Vauban entrance, with a width of 40 feet and a depth of 20½ feet, up to the Florida and St. Jean entrances of the Florida Dock, having a width of 69 feet and a depth of 25¼ feet. The lower end of the lock has a width of 53 feet, and a depth of 31 feet, 7½ feet lower than on the upper sill in order to admit and let out vessels at a lower state of the tide. All the gates are constructed of wood, and are studded with nails to protect them from the teredo.

The movable bridges crossing the entrances, which were originally formed of various types, are being gradually replaced by swing-bridges, in one span, with a double line of way, which can be opened or closed in less than two minutes.

Four graving docks, a floating wooden graving dock, and a gridiron have been provided for repairing vessels; and two more graving docks are to be constructed. The largest graving dock has a length of 426 feet at the bottom, a width at the entrance of 99 feet, and a depth of 28½ feet at high-water spring tides.

Extension Works at Havre. The new works in course of construction along the foreshore to the east of Havre consist of two docks, having a combined area of about fifty acres, connected with the Eure Dock by an entrance, 98½ feet wide, with its sill 33½ feet below high-water spring tides. The two docks will communicate by a passage of the same width as the entrance. There will be an intake at the far end of the Tenth Dock, so that water may be admitted at high tide, and the currents at the entrances and in the outer harbour may be thereby diminished. The intake is designed to have a width of 98½ feet, and a depth of 12⅜ feet below high-water spring tides; and the opening will be closed by a double set of sluice-gates.

The embankment of St. Jean, for enclosing the site, is being constructed by tide work in the river; it is formed by a quay wall with a curved batter, and founded on a layer of concrete deposited within a double line of sheet-piling.

The construction of the quay walls of the Ninth Dock has not been delayed till the reclamation was complete and the tide excluded, according to the usual custom, but they have been prosecuted simultaneously with the embankment. This has necessitated the execution of the greater part of the walls by tide work, on a foundation of rectangular brick wells sunk to a firm bottom and filled in with concrete. (Plate 14, Fig. 23.) The remainder of the walls at the western end of the dock, built on the solid ground out of reach of the tide, have been carried out in the ordinary manner within a timbered trench. (Plate 14, Fig. 24.)

Appliances at the Port of Havre. The more recent docks are provided with railway sidings, with the exception of the small Florida Dock; but the King's and Commercial docks, and one side of the Bar Dock are too closely surrounded by the streets and houses of the town to admit of railway communication. Hydraulic power is to be supplied for working the gates, bridges, and sluice-gates at the new docks; but hitherto the port has been devoid of hydraulic machinery. The quays are provided with warehouses, sheds, and cranes; and a large shear legs has been erected at the west end of the Commercial Dock.

Maintenance of Depth at the Port of Havre. The system of sluicing for maintaining the entrance channel has been abandoned for many years at Havre; for the sluicing basins have been replaced by other works, and the outer harbour has been deepened, so that the upper layer of water in the docks, available for sluicing, would be powerless to improve the depth of the channel. Dredging has, accordingly, been resorted to for maintaining and deepening the approach to the port.

The yearly accumulation of deposit in the outer harbour amounts to nearly 100,000 cubic yards, whilst outside the jetties it attains 26,000 cubic yards. The thickness of the average annual layer of deposit in the docks, having direct communication with the outer harbour, exceeds 1½ inches; it is about 1⅕ inches in the other docks, and reaches 6½ inches in the lock. The deposits are small in calm weather, and attain their maximum when a storm is coincident with a flood of the Seine.

The approach channel to the port, for a distance of 1640 feet from the northern pierhead, and the navigation channels in the outer harbour, are maintained by dredging at a depth of 32 feet below high-water spring tides. The annual cost of maintenance was fixed at £8000, previous to the construction of the Citadel Dock and the enlargement of the outer harbour; but this sum is insufficient now, especially as the amount of deposit appears to have increased of late years.

Havre not only possesses a rise of tide amounting to 23⅔ feet at springs, but it also enjoys the great advantage of the continuance of high water for a considerable period. The actual period of slack water, at high tide, is indeed only 11 minutes, on the average, in the open sea off Havre; but during the space of an hour the change of level is only 1 inch, and for over two hours, during both spring and neap tides, the variation does not exceed 6 inches, so that the docks can remain open for about three hours.

Considerable anxiety is felt as to the maintenance of the approaches to the port, owing to the changes that are occurring in the estuary of the Seine in consequence of the embankments which, though stopped 12 miles above, and not prolonged since 1869, are still producing accretions extending as far down the bay as the meridian of Havre. The prospect of a deterioration in the approach channels to the port is considered so imminent, that the locks and bridges on the Tancarville Canal, connecting Havre and Tancarville, are being made of dimensions suitable for a ship-canal; and proposals have been made for constructing a new entrance to the harbour going straight into the sea, and thus avoiding altogether the estuary of the Seine.

Commercial Statistics of the Port of Havre. The trade of Havre has steadily increased; the tonnage of vessels entering the port having risen from 1,042,000 tons in 1860 up to 2,267,000 tons in 1880, or more than doubled in twenty years. The actual tonnage of the imports was nearly twice that of the exports between 1860 and 1875; but in 1878 and 1879, the imports increased so much as to approximate to three times the exports.

The number of vessels frequenting the port has remained stationary since 1860, at a little over 6,400; whilst their average tonnage has more than doubled, having risen from 161 tons in 1860 to 353 tons in 1880. The above figures do not include steam tugs, or the small passenger steamboats to Honfleur and Trouville. The number of sailing vessels was nearly double the steamers in 1860, but since then the sailing vessels have considerably diminished; whereas the steamers have exhibited a corresponding increase, together with an increase in their tonnage.

Remarks on the Port of Havre. The trade of Havre is second only to Marseilles amongst French ports. Havre is a little ahead of Hull in the total tonnage of the vessels entering the port, and it possesses a much larger area of docks. In fact its enclosed water space, including the Ninth Dock, is very nearly equal to that of the Surrey Commercial Docks, and of the Birkenhead Docks; and when the extension works in progress are fully completed, its acreage of docks will exceed the Albert and Victoria Docks, and will only be surpassed by Liverpool and Barrow besides London.

The Transatlantic entrance to the Eure Dock, having a width of 100 feet, and a depth of 35 feet below high water, is equal in width to the largest dock entrance in England, namely the Canada Lock at Liverpool, and superior to any entrances at London or Liverpool in depth; though Portsmouth, Tilbury, and some of the Bristol Channel ports have deeper sills. The entrance also to the Ninth Dock has been given very nearly the same dimensions, the width being practically the same, and the sill only 1½ feet higher.

The docks at Havre present some analogy in position to the Hull Docks, for the earlier docks at both places stretch into the town, and are hemmed in by houses, whilst the later docks have been placed along a site reclaimed from the foreshore of the river.

The first floating dock at Havre was constructed only seven years after the first dock on the Thames, and more than forty years before Liverpool possessed a dock. Nevertheless the port made little progress till the present century, and even in 1834 the dock area only amounted to 30 acres; but since that period its advance has been very rapid, so that in the last fifty years the dock accommodation has increased fivefold. This result is doubtless due to its position, for it is close to the sea, whilst its entrance is within the shelter of the Seine estuary; and it is within easy reach of England, whilst it is conveniently situated for transatlantic steamers, for it lies in the direct road to Paris, and possesses ample depth of water. A deficiency, however, from which it suffers, is the absence of hydraulic machinery for facilitating the working of the machinery of the docks, and the unloading and loading of vessels, in which respect it contrasts unfavourably with English ports of similar extent, where hydraulic power works the gates, bridges, and capstans, and is applied to movable cranes along the quays. Hydraulic machinery, indeed, is to be supplied to the new extension; but it should also be furnished to the older docks, so as to increase their efficiency, for a rapid transfer of merchandise on the quays is equivalent to a large increase of dock accommodation.

PORT OF HONFLEUR.

Early History and Progress. The port of Honfleur is of early origin, for mention was made of it as far back as the 11th century, and by the close of the 12th century it had become a port of some importance. This prominence was due to the existence of a creek at Honfleur, forming the outlet of the little river Claire on the left bank of the Seine estuary, under the secure shelter of a projecting cliff to the west. This creek formed the earliest port, in which the ships sought shelter and grounded at low tide. When the town was fortified, the river was led into the trenches; these trenches served as tidal sluicing basins for scouring the creek, and towards the close of the 15th century the trenches were enlarged, and sluice-gates were placed at their outlet to render the sluicing current more efficient. The port at that period consisted of a tidal harbour, 400 feet long and 160 feet wide, communicating with the estuary by an opening, from 50 to 65 feet in width, on the present site of the West Dock entrance[1].

Towards the close of the 17th century, when the fortifications, having become antiquated, were demolished, the port was enlarged; and the tidal harbour was converted into a floating dock, which was completed in 1690.

As the trade of the port increased, an enlargement of the accommodation became necessary, and preparatory works for a new tidal harbour were commenced in 1725; and after considerable delays the basin was finally completed, and converted into a floating dock in 1772. About the same time a large wooden jetty was constructed beyond the new dock, forming the north-eastern boundary of a sort of outer harbour which, together with the site of the new dock, was abstracted from the estuary of the Seine.

[1] 'Ports Maritimes de la France,' vol. ii, p. 281.

The port in 1772 consisted of the old and new docks, subsequently called the East and Central Docks; a sluicing basin between the two docks, formed out of the old trenches, filled up in 1861; a small tidal harbour to the north of the old dock, which now forms an annex to the outer harbour; and the outer harbour of the new dock, which still forms the entrance channel from the enlarged outer harbour to the reconstructed Central Dock. (Plate 12, Fig. 1.) Cachin, the engineer of the Cherbourg breakwater, proposed about this period the construction of a ship-canal along the left bank of the Seine, for avoiding the shifting and dangerous channels in the estuary between Honfleur and Villequier. He also hoped, by means of the large reservoir of water formed by the canal, to provide a powerful sluicing current adequate to maintain the approach channel to Honfleur, which periodically silted up when the main channel of the river wandered away from the port. No fresh works, however, were carried out till 1837; though the jetties were repaired, the quays reconstructed, and the entrance to the old dock rebuilt.

At length, after the lapse of over fifty years, extension works were started; a west jetty was completed in 1841, and an east jetty in 1842, whilst the East Dock was opened in 1848. (Plate 12, Fig. 1.)

Recent Extensions at Honfleur. The next important extension of the port of Honfleur was commenced in 1874, comprising the prolongation of the west jetty, the formation of a large sluicing basin on the foreshore of the estuary to the north-east of the port, and the construction of a Fourth Dock alongside the sluicing basin and opening into the East Dock. The extension of the west jetty was completed in 1875; the sluicing basin was finished in 1881, and the Fourth Dock is approaching completion. A prolongation of the Fourth Dock to the east has been proposed[1].

[1] A plan of the docks, including the extensions, was given me by M. Picard, engineer of the port.

Description of the Port of Honfleur. The entrance channel is guided between the embankment of the sluicing basin and the west jetty, the inner portion of the jetty being formed of masonry and the outer portion of timberwork. The channel is 130 feet wide at its outer end, and it leads into the outer tidal harbour, having an area of 10¼ acres, from whence direct access is afforded to three of the docks.

There are four docks, namely, the West Dock of 2⅜ acres, the Central Dock of 3⅛ acres, the East Dock of 5⅓ acres, and the Fourth Dock of 8 acres, giving a total area of nearly 19 acres. The first three have entrances from the outer harbour, closed by single pairs of gates, with widths of 33¾, 40, and 54 feet respectively. The passage between the East and Fourth docks has a pair of reverse gates in addition. The lowest sill, namely the East Dock entrance sill, has a depth of 22 feet at low-water spring tides. Portions of the quays of the outer harbour, Central Dock, and East Dock, are provided with sidings. Owing to the narrowness of the East and Fourth docks, and the small area of the other two, the proportion of quays to water space amounts to 144 lineal yards per acre.

Maintenance of Depth at Honfleur. The sluicing basin, having an area of 143 acres, constitutes the most extensive work of the port. It was constructed in order to keep open the approach channel. The inlet, situated on the north side of the basin, has been made wide enough to enable the basin to be filled in the 2½ hours of nearly slack water just after high tide, whilst only admitting the upper layer to a depth of two feet, so as to introduce the least turbid water, and thus avoid the rapid silting up of the basin. There are ten bays in the inlet, each 33 feet wide, and closed by three gates turning on horizontal axes on the sill of the opening. The outlet has four openings, each 16½ feet wide, and each closed by a sluice-gate revolving upon a nearly central vertical axis, whose motion is regulated by a valve in the larger panel of the gate. A volume of 654,000 cubic yards of water can be discharged through the sluice-ways in forty-five minutes, attaining a velocity of 26 feet per second. The total cost of the sluicing basin amounted to about £200,000.

The outer harbour is maintained by sluicing, during low water, at all but the feeblest tides; and so rapid is the accretion in the three docks communicating with the outer harbour, that though their total area is only about 11 acres, their maintenance by dredging involves a yearly expense of from £1,600 to £2,800.

Trade Statistics of the Port of Honfleur. The number of vessels frequenting Honfleur, like at other ports, has not increased, but rather decreased in the twenty-five years between 1844 and 1869; whilst the average tonnage of the vessels rose from 44½ tons to 117½ tons, and the total tonnage from 171,000 tons to 434,000 tons in the same period. Since that time the trade of the port appears to have somewhat declined; for whereas the total weight of exports and imports reached a maximum of 290,000 tons in 1865, and was 262,400 tons in 1869, it had fallen to 226,200 tons in 1878; and in this respect Honfleur ranked at that time as the thirteenth port in France. The weights of its exports and imports are nearly balanced, though the latter maintain a varying preponderance.

Remarks on the Port of Honfleur. There is some resemblance at the present time between the ports of Calais and Honfleur, with their jetty channels, large sluicing basins reclaimed from the foreshore, and their limited areas of docks; but Calais, owing to its position, has the larger trade. Honfleur is an older port than Havre, and possessed a wet dock before Liverpool; but it was more suited by nature to prosper when the draught of vessels was small, and artificial shelter was difficult to obtain, and has been far outstripped by comparatively recent ports. The accretions in the estuary of the Seine, and the shifting of the channels, must render the maintenance of the approach channel to Honfleur precarious. It is possible that the extensions of the docks, and the deepening of the channel by scour from the sluicing basin, may improve the trade of the port; but a material increase in the depth of the outer channel could only be secured by costly training works and systematic dredging.

It is curious that the proposition made for Honfleur a hundred years ago, of connecting the port with the Seine higher up by a lateral canal, and thus avoiding the difficulties of navigation in the shifting shallow channels of the estuary, is being actually carried out with the same object, on the opposite bank, for the port of Havre by means of the Tancarville Canal.

ST. NAZAIRE DOCKS.

Though the port of St. Nazaire ranked sixth amongst the French ports in 1878, in respect of the total weights of its exports and imports, being in advance of the ancient ports of Boulogne, Calais, and Honfleur, it is of quite recent origin. In fact, though many of the French ports have made very rapid progress during the last quarter of a century, St. Nazaire is the only port of France which has sprung into existence in the middle of the present century, and rapidly attained a foremost place.

Site. The river Loire, on whose estuary St. Nazaire is situated, has long possessed an inland port at Nantes; but the increase in the draught of vessels, the difficulties experienced in improving the shallow irregular channel of the river up to Nantes, and the blocking up of the river by ice in the winter, rendered the establishment of a more accessible port an urgent necessity if the trade of the Loire was to be retained.

St. Nazaire is situated in a small bay, on the right bank of the Loire, close to its mouth, but protected by a projecting headland from westerly storms which are so severe in the Bay of Biscay. The site enjoys the important natural advantages of good and safe anchorage-ground, and proximity to the open ocean, whilst sheltered from its waves. The mouth of the Loire is, indeed, encumbered by a bar; but vessels drawing 22 feet can pass over it during 2 hours at the lowest neap tides, 3½ hours at ordinary neap tides, and 4¾ hours at ordinary spring tides; so that at St. Nazaire, like at Liverpool, the bar is merely a cause of delay under certain conditions of tide, and not an impediment to the entrance of the largest class of vessels.

Development of the Port of St. Nazaire. The first idea of forming a port at St. Nazaire appears to have been started at the beginning of the present century. A mole for sheltering the roadstead of St. Nazaire was commenced in 1828, extending straight out from the projecting promontory upon which a small portion of the town of St. Nazaire is built. This mole, 592 feet long, with a landing quay along its inner side, was completed in 1835, at a cost of £12,800[1]. Though various proposals and designs were made for constructing docks at St. Nazaire during the early part of the century, it was not till 1845 that the creation of a port was definitely sanctioned; and the dock works were commenced in 1847.

The St. Nazaire Dock, having an area of 26 acres, was opened in 1856[2]. (Plate 13, Fig. 4.) It was constructed on the foreshore of the estuary, within the shelter of the promontory and the projecting mole. As soon as this dock was opened, the construction of a second dock in the cove of Penhouët was advocated in the interests of trade; and this opinion was justified by the number of vessels frequenting the port in 1857 having trebled in 1858. After several projects of dock extension had been submitted, the design of the existing Penhouët Dock and of future extensions was approved in 1861; and the works were commenced in the following year. The Penhouët Dock, having an area of 55½ acres, was opened

[1] 'Ports Maritimes de la France,' vol. v, p. 18.

[2] I am indebted to M. Pocard Kerviler, engineer-in-chief of the port, and to the 'Ports Maritimes de la France,' for a plan of the docks and a section of the Penhouët Dock wall.

in 1881; the great difficulties experienced in the foundations of the walls, and the war of 1870, having delayed its progress. This addition has raised the dock area of St. Nazaire to 81½ acres in the short space of twenty-five years. Owing to the very large development of the trade of the port, fresh extensions are contemplated as shown by dotted lines on the plan. (Plate 13, Fig. 4.) This scheme, however, will not be taken in hand till the Penhouët Dock indicates symptoms of becoming overcrowded; but already the construction of a second entrance, from the mouth of the small river Brivet, by a canal to the Penhouët Dock in the line of the future extensions, has been proposed.

Description of the St. Nazaire Docks. The mole protects a small tidal harbour, between its quay and the southern jetty of the St. Nazaire entrance, which is used by pilot boats; but the St. Nazaire Dock opens directly into the estuary, its short entrance channel being protected and guided by two jetties curving out towards their extremities. The entrance to the two docks forming the port is at present through the lock and entrance connecting the St. Nazaire Dock with the estuary; and a lock, with reverse gates at each end, provides access to the Penhouët Dock. When, however, the extensions are carried out, a second entrance will be provided at the upper end of the chain of docks. The site of the works was enclosed by embankments previous to the formation of the docks.

The entrance channel to the St. Nazaire Dock was excavated through the silt which overlies the stratum of gneiss rock met with at variable depths in that locality. It has been given a depth of 23 feet below the lowest high tides, and is maintained at that depth by dredgers and sand-pumps.

There are two entrances to the St. Nazaire Dock, placed side by side. The largest is a simple entrance provided with two pairs of gates pointing inwards, and only 92 feet apart, with the object of having a pair in reserve in case of repairs or accident. This entrance is 82 feet wide; and the lowest part of its inverted sill is 29 feet below high water of spring tides, and 24 feet below high water of neaps, so that all vessels which can pass over the bar can enter the docks. A lock forms the second entrance; it is 197 feet long, 42$\frac{2}{3}$ feet wide, and its sills are 4 feet higher than the sill of the entrance. The entrance was designed to accommodate the wide paddle-wheel Atlantic steamers; whilst the lock admits smaller vessels at half tide. All the gates of both entrance and lock have been renewed within the last few years; they were originally constructed of wood, with some iron in the entrance gates, but now iron gates have been placed in the entrance and at the lower end of the lock, whilst wood has been again adopted for the upper lock-gate. The object of having one pair of iron gates for the lock, and one pair of wood, is to compare their durability.

The lock connecting the Penhouët Dock with the St. Nazaire Dock has the same width of 82 feet as the St. Nazaire entrance, and the same depth on the sill. The distance between the two pairs of gates pointing the same way is 492 feet. For the five years preceding the opening of the Penhouët Dock, this lock was used as a graving dock for the transatlantic steamers. Recesses have been formed at each end of the lock for receiving a ship-caisson, so that the lock could be closed at both extremities and pumped dry if any repairs should be required. The four pairs of gates have been made of iron; and each gate has been provided with two rollers for supporting it, though the gates have been given buoyancy by watertight compartments, and the flotation has been made self-adjusting to the varying water-level. Reverse gates have been provided in order to render the water-level of the two docks independent, and also to prevent the introduction of silt into the Penhouët Dock, which enters the St. Nazaire Dock at every tide from the densely charged water of the estuary.

The St. Nazaire Dock has been given the ample width

of 525 feet, which has been increased in the central portion to 820 feet. The foundations of the masonry quay walls have been everywhere carried down to the schistous gneiss, underlying the layer of silt, of variable depth, which had accumulated in the sheltered cove of Penhouët.

The Penhouët Dock has only been given the width of 525 feet in the central portion, the greater width of 755 feet having been necessitated towards each end of the dock, owing to the dip of the rock, and the great depth of silt in the line which would have been preferred for the walls. The length of the dock is 3610 feet, as the great variations in the depth of silt necessitated its being extended further than would have been desired, in order to secure a good foundation for the passage contemplated in the future, at its northern end, to connect it with additional docks.

Quay Walls at the Penhouët Dock. The preliminary borings had indicated that the depth at which the rock was met with on the site of the Penhouët Dock was very variable. It appears that the site is traversed by the bed of an ancient watercourse which had eroded its bed through the stratum of rock, and had subsequently been obliterated by the general deposit of silt over the whole of the sheltered area. Where the rock was found at a depth not exceeding 13 feet below the dock bottom, the foundations of the walls were excavated in the ordinary manner. When, however, the rock dipped lower than this limit, the system of masonry well foundations was adopted, as described on page 405. Where the surface of the rock was fairly level, a single well was sunk through the silt down to the rock; and the quay wall was built on the top of the well after it had been filled in with masonry. (Plate 15, Fig. 13.) In some places, however, the dip of the rock was so great, at the sides of the old river channel, that two narrower wells were sunk in a line, at right angles to the line of the quay, so as to reduce the amount of rock excavation needed to form a level base for the well, and to diminish the depth to which one of the wells had to be sunk. In a few instances it was found advisable to sink three wells in a line. The length of the wells, in the direction of the line of the quay, was invariably made 16½ feet. The sinking of the large wells through the silt was accomplished without any difficulty, with the exception of the occasional influx of silt; but as soon as a point of the sloping rock was touched, great care was necessary to prevent the unsupported portion of the well from sinking down through the silt and causing the well to heel over from the perpendicular. Accordingly, the well was supported by a series of piles driven into the silt; and as the rock was excavated, blocks were inserted to support that side of the well, and as soon as the rock had been formed to a level bench, the blocks were exploded by dynamite, the well descended on the one side to the levelled base of rock, and its motion in the silt, on the opposite side, was controlled by the supporting piles. The excavation of the rock was then resumed, and the sinkage accomplished, in the same manner, in a series of stages, till the whole well rested on the solid rock. The smaller wells were more troublesome to sink, both owing to their smaller weight in proportion to their external surface, which rendered it often necessary to weight them with 80 to 100 tons of pig iron before the adhesion of the silt could be overcome, and also owing to the smaller proportion of base to height, which rendered them more liable to swerve from the vertical, and the unequal lateral pressure on the back row of wells tending to draw them towards the wells in front. The well foundations contained about 46,000 cubic yards of masonry, and cost £1 14*s*. 10*d*. per cubic yard, including excavation.

These wells served as piers, on which semicircular arches, 20 feet in diameter, were turned; and the quay wall was built on these piers and arches, so that the extent of these exceptionally difficult foundations was thereby reduced. About a third of the west wall has been founded on wells, and also the eastern half of the northern return wall, and about 900 feet of the east wall. Altogether, the masonry walls round the dock, founded either directly on the rock or on wells, have a total length of 7040 feet.

The depth of the silt along the line of the remaining 1900 feet of quays exceeded 60 feet below dock bottom, and in some places reached 100 feet before the rock could be attained in the centre of the bed of the ancient watercourse. For these excessive depths, the construction of masonry quay walls was abandoned; and pitched slopes, with timber wharfing and projecting jetties, were placed on the silty foundations; and the quays behind were reserved for light goods.

Owing to the width of both the docks, the proportion of the length of quays to water space is small, amounting to only 57 lineal yards per acre of dock.

Accessory Works at St. Nazaire. In order to secure the Penhouët Dock against the rapid deposit of silt which occurs in the St. Nazaire Dock, it was necessary, in addition to shutting out the tidal water by means of reverse gates in the lock, to provide a supply of clear water for maintaining the water-level in the dock. An intake was, accordingly, constructed in the eastern quay, 82 feet wide, and communicating by a short cut with the estuary. The sill of the intake is 3¼ feet below high-water spring tides; and the opening of the intake is divided into three bays by two piers, 6½ feet wide, and is closed by three gates turning upon horizontal axes placed on the sill, thus resembling the method adopted for regulating the inlet to the large sluicing basin at Honfleur. The water is only drawn from the estuary three times a month through the intake, during calm weather, at high-water spring tides, when the upper layers of water are fairly free from silt.

Three graving docks have been constructed in connection with the Penhouët Dock, on a site where the foundations were suitable. The largest has an entrance 82 feet wide; it is 460 feet long, and the depth on its sill is 24 feet at high-water neap tides. They are closed by ship-caissons very similar to those in use at Chatham Dockyard.

The first example in France of a rolling, or traversing, movable bridge has been erected for crossing the Penhouët Lock. It carries two lines of railway, and a footpath on each side.

The gates, rolling bridge, capstans, and sluice-gates of the Penhouët Lock are all worked by hydraulic power, and also the gates of the intake for the dock, so that in this respect St. Nazaire is in advance of Havre.

The railway station is close to the west quays of the docks; and sidings branching off from the main line, and provided with connecting turntables, furnish ample railway accommodation along the quays.

Maintenance of Depth at St. Nazaire. Considerable deposits of silt occur, not merely in the entrance channel between the jetties, but also in the St. Nazaire Dock, to such an extent, that a layer of silt, 3¼ feet in depth, would accumulate annually in the dock if not removed. The yearly deposit in the entrance channel has been estimated at about 250,000 cubic yards, and 150,000 cubic yards in the dock, giving a total of about 400,000 cubic yards of silt which have to be removed every year to maintain the depth. To cope with this formidable amount of deposit, the first sand-pump dredger was constructed at St. Nazaire, and proved perfectly successful[1]. One ordinary dredger, and three sand-pump hopper dredgers have been employed for the work of maintenance.

Cost of the Works at St. Nazaire. The total cost of the mole, together with the removal of some neighbouring rocks, amounted to £16,000. The St. Nazaire Dock cost, originally, in construction, £191,400; the entrance and lock £84,000; the entrance channel and jetties £41,800; and the dredging

[1] 'Annales des Ponts et Chaussées,' 4th Series, vol. xviii, p. 15; and 'Rivers and Canals,' p. 54, and plate I, Figs. 5 and 6.

plant £29,800, giving a total of £347,000. The total cost, however, of the works connected with the St. Nazaire Dock and its approach had been raised by improvements, reconstructions, and large repairs, to £444,800 by the end of 1882. The Penhouët Lock cost £239,700, and the dock £699,800, or together £939,500, which, with some small works, raises the total amount expended on the port of St. Nazaire to about £1,400,000. The actual cost per acre of dock, after making some deductions, has been estimated roughly at £16,000.

Commercial Statistics of the Port of St. Nazaire. The increase in the trade of St. Nazaire has been very rapid, and has, moreover, been very steadily maintained. The number of vessels frequenting the port in 1860, four years after the opening of the St. Nazaire Dock, was 4,833, with an average tonnage of 66 tons; and in 1880, the number of vessels had risen to 9,463, with an average tonnage of 173 tons; so that in twenty years, the number of vessels was nearly doubled, whilst their tonnage was almost trebled. The total tonnage of the vessels actually increased more than fivefold in the above period, having been 319,600 tons in 1860, and 1,642,200 in 1880; and with only three exceptions, namely, in 1866, 1873, and 1874, the progress has been maintained from year to year. The actual weights of the total exports and imports increased from 277,400 tons in 1860, to 1,242,700 tons in 1880, or a nearly fivefold increase in twenty years.

St. Nazaire has essentially an importing trade as regards its maritime commerce; for the sea-borne imports in 1880 amounted to 750,900 tons, of which 426,400 tons were English coal, and there were only 145,400 tons of exports. On the contrary, the exports are about double the imports in the river traffic; but this trade is little more than a fourth of the whole.

Remarks on the St. Nazaire Docks. It is remarkable that the great increase in trade, with the large tonnage reached in 1880, as noted above, was all accomplished by a single dock of 26 acres; for the Penhouet Dock was not opened till 1881.

In spite of the much larger area of the Havre Docks at that period, the total tonnage of the vessels frequenting that port was not quite double the tonnage at St. Nazaire, which is intermediate between the tonnage at Grimsby and Hartlepool. The dock accommodation at St. Nazaire has been trebled by the opening of the Penhouet Dock; so that now it stands on a par with Hartlepool in that respect, and with Cardiff also till the extension there is completed. St. Nazaire owes its prosperity to its facility of access and its convenient situation for the distribution of imported goods, for it possesses no rich mineral resources in its neighbourhood, like Barrow and Cardiff, to develop its trade.

PORT OF BREST.

The new harbour of Brest has already been described in Chapter X. The site possesses excellent natural shelter; but as a commercial port, Brest has been overshadowed by the naval dockyard, and has only recently attained an independent position.

Description. The port of Brest, known also as the commercial port of Porstrein, consists at present merely of a harbour open to the tide, enclosed by an outlying detached breakwater, and a mole to the east and a jetty to the west projecting from the shore, resembling in principle the tideless ports of Marseilles and Trieste, with the inconvenient accompaniment of a rise of tide of 21 feet at spring tides. (Plate 3, Fig. 6, and Plate 6, Fig. 5.) The sheltered area contains three small basins for the local craft, formed by short broad jetties projecting from the shore, a gridiron, and a large basin in the north-eastern corner for sea-going ships; the outer portion is left open for the movement of vessels, which can enter or leave at either end. The harbour has been dredged to a depth of 24½ feet below low water, so that vessels remain afloat at all states of the tide. A graving dock is to be constructed, opening into the north-eastern corner of the harbour;

it is designed to be 525 feet long and 59 feet wide. The beach between the hill behind the harbour and the quays has been filled up with dredged material, so as to afford space for the various buildings and roads in connection with the port. A railway has also been laid along the quay.

A wet dock, proposed to be constructed along the foreshore to the east of the harbour, is shown in dotted lines on the plan. (Plate 3, Fig. 6.) The dock is designed to be entered from the harbour through a lock, 426 feet long and 82 feet wide, with a depth of 20⅓ feet below low-water spring tides on its lower sill. The dock is to have a length of 1640 feet, and a breadth of 656 feet, giving an area of 24¾ acres. The execution, however, of this work has been postponed; so that at present the commerce of Brest is subjected to all the inconveniences of considerable tidal fluctuations. Moreover, though the harbour has been designed on the same principle as the ports of Marseilles and Trieste, its sheltered space has not been nearly so well utilised. (Compare Plate 3, Fig. 6 with Figs. 4 and 7.)

Trade Statistics of the Port of Brest. In spite of the excellent approach nature has provided for the port of Brest, it does not appear that Brest is destined to enjoy a great commercial prosperity. The trade returns exhibit little increase in the total weights of exports and imports between 1869 and 1876, amounting to 184,000 and 189,000 tons respectively in these years. The imports amount to nearly treble the exports, as Brest distributes to the other little ports of the extensive roadstead; but it does not possess an inland trade, and consequently few ocean steamers trade with the port.

PORT OF MARSEILLES.

Marseilles enjoys the important position of being the first seaport of France, being quite as much in advance of Havre as Havre is in advance of Bordeaux. The opening of the Suez Canal route has given an additional impulse to its trade, by placing it in a direct line of communication with the east. Though Marseilles is an old port, its development is of recent date, having only commenced in 1844. Till that period the natural shelter of the Old Harbour had alone been utilised; but now artificial methods of protection have entirely transformed the port, as described in Chapter XII. The mode of formation of the protecting breakwater has been already given in detail (page 230); and the general form of the works is fully indicated by the plan and sections. (Plate 3, Fig. 4, and Plate 7, Figs. 15 and 16.) Accordingly, it is only necessary now to add a few particulars relating to Marseilles as a port.

Difference between Marseilles and Tidal Ports. Most of the ports previously described in Part II are situated at the mouths of tidal rivers, or in their estuaries, where shelter is afforded for docks, whilst a deep channel and easy access to the sea is secured. Rivers, however, flowing into tideless seas, such as the Mediterranean, present great impediments to navigation, owing to the shallow channels formed by the deltas at their mouths; and therefore, except in the rare instances where extensive training works have been carried out for preserving the river navigation, as at the mouths of the Danube and the Mississippi, tideless rivers are avoided as sites for ports. Accordingly, the Mediterranean ports are situated at suitable points on the sea-coast, where some natural shelter exists; and they combine harbour and port together, the area sheltered by the breakwaters forming the basins or docks, which are more or less surrounded by quays, of which Alexandria, Algiers, Genoa, Barcelona, and Civita Vecchia are examples. Marseilles differs from these ports, merely in having its basins more systematically arranged, so as to afford greater accommodation within a given space, in being amply supplied with quays all round the basins, and in being adapted for successive extensions to meet the increasing requirements of trade.

The absence of tide at Marseilles secures it an almost unvarying water-level, without the necessity of dock-gates and their costly accessories.

Description of Basins at Marseilles. The Old Harbour at Marseilles is a naturally sheltered creek, which has been surrounded by quays to render it suitable for trade. It has an area of 67 acres; and the small Customs Canal of 1½ acres, constructed last century, opening out of its south-east corner, and a careening basin of 3½ acres near its entrance, added in 1829, raised its total area to 72 acres. This was the whole of the accommodation provided for shipping till 1852.

The Joliette Basin, or dock, which formed the first instalment of the new port, was opened in 1852; it has an area of 54 acres, and is surrounded by 2,300 lineal yards of quays. It communicates with the Old Harbour by a canal, 440 yards long, and is approached from the south through an outer harbour of 56½ acres, partially sheltered by the end of the detached breakwater. This portion of the works cost £640,000, and raised the sheltered area of the basins to 121 acres.

The Lazaret and Arenc Basins were completed in 1863. They have an area of 51 acres, and 2700 lineal yards of quays. The cost of these two basins amounted to £200,000, considerably less in proportion to the area than the Joliette Basin, owing partly to no outer harbour or connecting canal having to be provided, and partly to the formation of the quays by the Dock Company to whom they were leased.

The Maritime Station Basin, or Railway Dock, was completed in 1863; it has an area of 41 acres, and 2700 lineal yards of quays. This addition cost £600,000, and with the Lazaret and Arenc Basins opened the same year, brought up the area of the basins to 213 acres.

The latest extension is the National Basin, completed in 1881; it is the largest of the basins, having a length of 2985 feet, a width of 1663 feet, and an area of 105 acres. It has three wide projecting jetties, or moles, in addition to the two which enclose it at each end; it possesses 4400 lineal yards of quays, and it cost £1,000,000. This addition makes the total area of the basins 318 acres.

The extension of the detached breakwater, for a distance of 2460 feet beyond the National Basin, provides a partly sheltered outer harbour of 44 acres; this serves as an approach to the entrance into the National Basin, which has a width of 330 feet. The two passages between the National Basin and the Railway Basin have each a width of 100 feet, spanned by a single swing-bridge resting on a central pier. The passage between the Railway and Arenc Basins is 100 feet wide, and the passage between the Lazaret and Joliette Basins is 70 feet wide; whilst the main southern entrance into the Joliette Basin is 220 feet in width.

Quays at Marseilles. All the quay walls of the different basins are constructed according to one special type, with modifications merely in detail. This type is shown on the harbour side of the sections of the Marseilles breakwaters (Plate 7, Figs. 15 and 16), and in a section of the quay wall of the National Basin. (Plate 15, Fig. 14.) It consists of a rubble-mound foundation, upon which a concrete quay wall is built, faced with stone above the water-line. The Joliette quay wall (Plate 7, Fig. 16) consists of concrete-in-mass, $8\frac{1}{4}$ feet wide, founded on the rubble base at a depth of 13 feet below the water-level, the width of the rubble mound at the top being $21\frac{1}{3}$ feet. The rubble mound rises too high for vessels to lie alongside the wall, so the vessels lie end on and are unloaded and laden by the help of steam winches. This method of procedure adds about $5\frac{3}{4}$*d.* per ton to the cost, but it enables more vessels to be discharged at the same time. The quay walls of the Lazaret and Arenc Basins are founded $19\frac{2}{3}$ feet below the water-level, so that vessels can lie alongside. These walls, 11 feet wide, are raised to the surface of the water by four tiers of blocks. The quay walls of the Railway and National Basins are built in a similar manner; but owing to the increased draught of vessels, the foundations have been placed 23 feet below the water-level. The object of this type of construction was to enable the walls to be built below the water, without cofferdams, on a site where there is no fall of tide that can be taken advantage of for laying foundations at a low level.

The Dock and Warehouse Company of Marseilles, to whom the Lazaret and Arenc Basins have been conceded, have formed quays round these docks by land reclaimed from the sea, having an area of about 50 acres, on which they have erected sheds and warehouses, and formed roads and sidings; and they have amply provided the quays with cranes and other machinery, worked by hydraulic power, at a cost of about £880,000. The quays of the Joliette Basin are not suitable for the introduction of hydraulic machinery; but the Chamber of Commerce are equipping the Railway and National Basins with sheds, sidings, and hydraulic machinery, at a cost of £280,000.

In spite of the absence of cranes and hydraulic machinery, the Joliette Basin has a larger traffic in proportion to its length of quays than any of the other basins, owing to its close proximity to the town, amounting to 700 tons per lineal yard of quay in 1882.

The Lazaret and Arenc Basins, though devoted almost exclusively to the import trade, had a traffic of 570 tons per lineal yard of quay in the same period, and, like the Joliette Basin, have reached the limit of their possible traffic. The capabilities of the Railway and National Basins are being largely augmented by the equipment of their quays.

Graving Docks at Marseilles. The graving docks and basins of the port of Marseilles open out of the National Basin, with which they are connected by a passage 92 feet wide. (Plate 3, Fig. 4.) They were constructed, between 1864 and 1874, on a site originally reclaimed from the sea at the back of the National Basin; and the site of the works was enclosed by a rectangular concrete dam over 3300 feet long. The passage from the National Basin opens into a basin, having an area of 11½ acres and a depth of 26¼ feet, in which vessels can await their turn for entering a graving dock, or undergo such repairs as can be executed when they are afloat. Four graving docks have been constructed in connection with this basin; the entrances for two more have been completed, so that inner works can be readily carried out as soon as the need for additional graving docks arises; and space has been set apart for three more, which will eventually raise the number of graving docks to nine. The largest graving dock, having a length of 464 feet and a depth of 23 feet, serves also as an approach channel to the basin for vessels to be repaired on pontoons. These docks have been leased to the Dock and Warehouse Company for 90 years, with the exception of the larger basin, which is open to the public for vessels repaired afloat. The cost of the works amounted to £320,000.

Trade Statistics of the Port of Marseilles. The total weight of the imports at Marseilles in 1882 amounted to 3,019,500 tons, and the exports to 1,450,500 tons, giving a total of 4,470,000 tons. As the length of quays utilised for trade amounts to about 14,200 lineal yards, the average traffic was only 315 tons per lineal yard of quay in 1882, much less than at the Joliette, Lazaret, and Arenc Basins, owing to the comparatively small average at the other basins, reaching only 123 tons at the large National Basin. The number of vessels frequenting the port in 1882 was 8,568.

Remarks on the Port of Marseilles. The growth of the port has been very rapid within the last thirty-five years, the area of its basins having risen from 69 acres in 1851 up to 318 acres in 1881, exclusive of two outer harbours having a combined area of 100 acres. Its accommodation for vessels has, indeed, more than quadrupled in thirty-six years, being now twice that of Havre, and inferior only to London and Liverpool amongst British ports.

The provision of warehouses, machinery, and sidings, at the Lazaret and Arenc Basins, is considerably in advance of other French ports, contributing largely to the efficiency of these basins. The extension of these facilities to other basins will add to the capabilities of the port; but if the progress of the last thirty years is to be maintained, it will be necessary to start some fresh extension of the basins, as already indicated (page 232), for the previous works have each occupied from five to seven years in construction.

PORT OF TRIESTE.

The situation of Trieste, and the works that have been carried out for providing it with sheltered basins have been already described (p. 234); and the resemblance of the works to those at Marseilles has been pointed out. (Plate 3, Fig. 7, and Plate 7, Fig. 7.)

Description of the Port of Trieste. Austria is mainly an inland country, and possesses only a small extent of sea-coast, on the north-eastern shore of the Adriatic; and, consequently, its seaports are few in number. Its principal ports, since the cession of Venetia to Italy, are Trieste and Fiume, of which Trieste is much the most important. Trieste possesses an excellent natural roadstead, being situated on a bay at the head of an inland sea; and its capabilities were not overlooked by the Romans, who formed a mole, on the south-west side of the small bay in which Trieste stands, to afford protection from the exposed quarter. Traces of the Roman mole still exist; and the mole St. Theresa, at the present day, affords a similar protection. Some jetties and moles were built out from the shore in front of the town, to the north-east of the St. Theresa mole, being partially sheltered by this mole and the projecting St. Andrea Point. Two basins also were formed to the north of the town, one of which has been filled up in the construction of the new works.

Extension of the Port of Trieste. The detached breakwater, parallel to the shore, described on page 235, shelters an area which is divided into three basins by four moles projecting from the land, and is provided with quays exactly on the same principle as the basins and moles at Marseilles. (Plate 3, Fig. 7.) Quay walls, moreover, have been constructed on the inner side of the breakwater, and round the moles, precisely like the quay walls at Marseilles, with four tiers of concrete blocks from the top of a rubble mound, 20 feet below the water-level, up to the water-line, and with concrete faced with stone above. (Plate 15, Fig. 14.)

The basins enclosed by the moles are 985 feet long and 705 feet wide; and the width between the ends of the moles and the breakwater is 560 feet, except at the northern end, where a jetty from the breakwater reduces the width of the entrance to 310 feet.

Remarks on the Port of Trieste. The extension works have largely augmented the capabilities of the port; but much remains to be done before these capabilities can be fully developed. The railway runs near the new basins, and therefore it would be easy to provide suitable siding accommodation on the quays and moles. By equipping also the quays and moles with warehouses, sheds, and hydraulic cranes, as accomplished at Marseilles, the facilities for trade would be much increased. Moreover the quay along the breakwater cannot be utilised till it is connected with the land by one or more swing-bridges, involving the reduction of the openings between the moles and the breakwater.

The depth in the basins will probably have to be maintained by systematic dredging operations, owing to the tendency of silt to be deposited in sheltered places within the Bay of Trieste.

CHAPTER XXVIII.

FOREIGN DOCKS AND RIVER QUAYS.

ANTWERP. HAMBURG. ROTTERDAM. ROUEN. NEW YORK.

Docks and Quays. Sites of Ports with River Quays. *Port of Antwerp:* Early History; Progress; Description of Docks; Quays along the Scheldt, their reconstruction and extension; Appliances for Trade; Extension of Docks; Development of the Port; Statistics; Remarks. *Port of Hamburg:* Site; Description; Appliances; Statistics. *Port of Rotterdam:* Improved access; Description of Old Port; Extensions on Feyenoord Island, description, appliances on quays; Remarks. *Port of Rouen:* Improvement of Approach Channel; Situation; Extension of Quays, and Regulation of River Seine; Quay Walls; Compared with Bordeaux. *Port of New York:* Situation; Old Wharves; System of the Department of Docks, Bulkheads and Piers, cost; Appliances; Remarks. Concluding Remarks on Docks, comparative trade of ports, requirements.

IN most of the ports hitherto described, the trade is mainly carried on at quays surrounding sheltered basins or docks, as at Liverpool, Hull, Cardiff, Marseilles, Havre, and St. Nazaire, though at London, Dublin, and Belfast, considerable use is also made of quays and wharves bordering the rivers on which these ports are situated. There are some foreign ports which, like Dublin and Belfast, though provided with docks or basins, afford equally important accommodation for trade along their river quays, of which Antwerp, Hamburg, and Rotterdam are instances; whilst others carry on the whole of their trade at quays or wharves along the river banks, such as Rouen and New York.

Docks possess the advantage of concentrating the trade in a special locality, where space is available for sidings, warehouses, and other facilities for tradé; whereas river quays necessarily extend along a considerable distance, where the space for quays is restricted, and where it is difficult to introduce sidings and other accessories for trade, unless land is reclaimed from the river in the construction of the quay walls. Moreover, river banks are not convenient for the transhipment of merchandise, unless the river is small and well sheltered, or the port is situated at a considerable distance from the sea; and a good depth of water, and a small rise of tide are also important requisites for river quays. In suitable sites, however, quays on rivers possess the great advantage of saving the delay occasioned in entering and leaving docks.

The ports about to be described combine the various requirements necessary for the successful establishment of river quays. (Plate 13, Figs. 1, 2, 3, and 5.) Antwerp is situated on the right bank of the river Scheldt, at a distance of 50 miles from its mouth, where the depth is ample and the rise of spring tides is 15⅛ feet. The city of Hamburg is situated on the right bank of the river Elbe, at a distance of 93 miles from the sea; its port extends along both sides of the river; the depth in the main channel, in front of the basins, is from 13 to 18 feet at low water; and the rise of tide at springs is only 6¼ feet. Rotterdam is situated on the right bank of the river Maas, at a distance of 20 miles from the North Sea; its port extends also to an island in front of the town, and the rise of tide averages 4 feet. Rouen, on the right bank of the river Seine, has quays along both banks; it is 78 miles from the mouth of the Seine, and the tide there has a rise of only 6½ feet at springs. New York stands at the upper end of New York Bay; and as it is mainly situated on Manhattan Island, which separates the Hudson and East rivers above their confluence, it possesses facilities for wharves for some miles along both these rivers in front of the city. The tide, moreover, has a mean rise of only 4¾ feet at New York; and the nearest portions of the wharves are about 12 miles distant from the Atlantic Ocean, and in a well-sheltered position.

Of all the above examples, Antwerp alone has a rise of tide which could be inconvenient for the transhipment of goods; and this port is amply provided with docks closed by gates, as well as with river quays, so that vessels can select either position according to circumstances, and have the choice between an easy access, at any time, to a berth with a fluctuating level, or the longer operation of entering a dock at high water with the advantage of an unvarying water-level.

PORT OF ANTWERP.

The port of Antwerp consists of two distinct parts, namely, the docks belonging to the municipality, and the new quays along the right bank of the Scheldt, which have been constructed by the Government.

Early History and Progress. Antwerp was one of the principal ports of Europe as early as the 13th century, and increased in importance up to the 16th century. The Scheldt, however, was closed to vessels by the treaty of Westphalia in 1648, for the benefit of the Dutch ports of Amsterdam and Rotterdam; and it was not opened again to trade till 1795, by the treaty of the Hague, when it had been annexed to France. Napoleon I, recognising the advantageous position of Antwerp, ordered the establishment of a naval arsenal there, both for purposes of defence and also for an attack upon England. He, moreover, aided the commercial progress of the port by the construction of quays along the river, and the creation of the earliest docks, now known as the Little and Great Docks. (Plate 13, Fig. 3.) These docks were opened for traffic in 1811 and 1813 respectively; and graving docks were in course of construction, when the fall of Napoleon, in 1814, put a stop to the works, and Antwerp became part of the kingdom of the Netherlands. During the French occupation, about £520,000 were expended upon the formation of quays and docks. In 1815, the docks were handed over to the city authorities; and in 1819, the river quays were also entrusted to them. The unfinished quays were completed by the municipality; and a large block of warehouses was erected at the east end of the Great Dock in 1828. When Belgium separated from Holland in 1830, Antwerp at once ranked as the principal port of the new kingdom. The commercial progress of Antwerp dates from that period, though its development was at first comparatively slow. The Kattendyk Dock, to the north of the Great Dock, was commenced in 1856, and was opened in 1860. By this time the trade of Antwerp was beginning to increase so rapidly that additional accommodation was soon required; and the advancement further out of the lines of fortification afforded clear space for the extension of the docks, as well as the enlargement of the town. Extension works were commenced in 1865; the Junction Dock, connecting the Great Dock with the Kattendyk Dock, was completed in 1869; the Timber Dock, opening out of the Kattendyk Dock, which had been formed in 1864, was extended; and the Canpine and Canal Docks, to the east of the Kattendyk Dock, were opened in 1873. Seven years later the Kattendyk Dock was prolonged northwards, and three additional graving docks were constructed, raising the dock accommodation to its present position.

Description of the Antwerp Docks. The docks at present consist of seven docks and a half-tide basin, two entrances into the Scheldt, six passages, and six graving docks [1]. (Plate 13, Fig. 3.) The docks are all placed in communication by means of passages, and these passages are open, with the exception of the passages between the old docks, and between the Junction and Kattendyk Docks, which are provided with reverse gates. The passages into the Timber, Canpine, and Canal Docks are only 50 feet wide; but the passages connecting the old and Kattendyk Docks have a width of 59 feet. The docks range in size from the Little Dock of $6\frac{1}{4}$ acres up to the Kattendyk Dock of 32 acres, omitting the

[1] Plans and some particulars of the Antwerp Docks were sent me by Mr. Royers, engineer to the Municipality of Antwerp.

small Junction Dock which mainly serves as a wide passage. The total area of the docks is about 105 acres, exclusive of the barge docks; and there are 81 lineal yards of quays per acre of dock.

The entrance to the Little Dock from the Scheldt is 57½ feet wide, and is provided with reverse gates. The half-tide basin, serving as a large lock to the Kattendyk Dock, has an area of nearly two acres, and has reverse gates at each end, and at the outer entrance there is a second pair of inner gates; the width at its entrances is 81⅓ feet. The depth at high-water spring tides on the sill of the Little Dock entrance is 24 feet, and on the sills of the other entrance is 25⅝ feet; but these sills are only 21 feet and 22⅝ feet below the ordinary water-level of the docks.

The Canpine Canal joins the Canal Dock; and a barge dock of 4½ acres communicates with the canal, a short distance from its termination, to accommodate the barges navigating the canal. A barge dock, comprising three small basins, having a total area of 10 acres, and entered through a lock, has been constructed by the Government, near the southern end of the new quays, for the accommodation of small river craft.

All the six graving docks of the port communicate with the Kattendyk Dock, being situated near together along the west quay of the dock, and they are each closed by a pair of reverse gates. The largest of the graving docks has a width of entrance of 81⅓ feet, corresponding to the Kattendyk Dock entrance, and a length of 410 feet. There are two graving docks on the left bank of the river.

River Quays at Antwerp. Quays existed along the Scheldt previous to the French occupation of Antwerp; but the quays were extended and improved during the French rule, and subsequently they were prolonged gradually, partly by the Government and partly by the municipality. The quay walls were constructed of masonry, and had a total length, in

1876, of 2370 yards. The total available quay space was 13 acres. The quays had a width of only from 60 to 100 feet, including the roadway, and were intersected by five small tidal creeks, which were dry at low water. The depth, moreover, of the river alongside the quays was very slight at low water, and low-water mark extended beyond them in many parts, so that several jetties were built out from the quays for the accommodation of the larger vessels. The line of the quays was irregular; and the width of the river was contracted opposite the town by the projecting point of Werf, which interfered with the flow of the currents.

The reconstruction of the quays, combined with a rectification of the right bank of the Scheldt along the front of the town, was decided upon in 1874, as a greater depth was required to enable large vessels to come alongside, and increased quay accommodation was urgently needed for the rapidly growing trade of the port.

The extension works consist of an entirely modified line of quays, having a total length of about 3850 yards, and forming a continuous curve concave to the stream, with a sharper curvature towards the extremities, so as to conform to the bends of the river whilst regulating its width. (Plate 13, Fig. 3.) In addition to the quays, a south Barge Dock of 10 acres, already mentioned, has been constructed, also three landing places along the quay between the docks and the South Barge Dock, and an embankment, about 700 yards long and pitched with stone on the river slope, connecting the projecting line of quays, at the entrance to the South Barge Dock, with the natural bank further up the river. These works were let to MM. Couvreux and Hersent, in 1877, for £1,531,000.

The rectification of the river necessitated the construction of the portion of the quay walls south of the central landing place within the bed of the river; the cutting off of the projecting point of Werf, together with the enlargement of the river, north of the central landing place, for a distance of about 550 yards; and the extension of the quay, a little in advance of the old quays, from a point a little south of the northern landing place, and about 380 yards from the entrance to the Little Dock, to the termination of the quays at the entrance to the Kattendyk Dock. Thus a large area was reclaimed from the river at the southern part of the quays, which is very valuable in providing ample width for the quays; whilst in the central portion, it has been necessary to encroach upon land previously occupied by the town, in order to widen the river, and to provide adequate width for the new quays. Dredging had also to be carried out in the bed of the river, to attain the proposed depth of 26¼ feet in front of the quays at low tide; and the dredged material could be utilised in filling up the space at the back of the projecting southern quays, which extend as much as 500 feet into the river. The plan of Antwerp shows the line of quays as executed[1]. (Plate 13, Fig. 3.) The rectification of the opposite bank of the river by quays has been proposed, so as to give the river a uniform width of 1150 feet.

The quay walls, which have been built on the variable and silty bed of the river, were founded on caissons sunk to a solid foundation by aid of compressed air. (Plate 8, Figs. 5 and 6.) The method of their construction has been previously described (p. 408), and a section of the walls is given on Plate 14, Fig. 21.

The quays have been given a width of 300 feet from the southern end to the southern landing place opposite the Waës Station, and 330 feet for the remainder of their length. The available area of the quays amounts to 74 acres, so that whilst the quays have gained an increase of two-thirds of their former length, their area has been augmented nearly sixfold.

The total cost of the quay works, including dredging, pur-

[1] Space did not admit of the insertion of the northern Barge Dock connected with the Canpine Canal, nor the southern extremity of the quays.

chase of land and buildings, and the equipment of the quays, amounted to £3,200,000.

Appliances at the Port of Antwerp. The docks are furnished with sidings communicating with the railway; and swing-bridges are placed over the entrance to the Little Dock and the passages between the docks, for maintaining the communication between the different quays. (Plate 13, Fig. 3.) The bridge across the Kattendyk Basin entrance channel, having a span of 88½ feet, is a traversing movable bridge, rolling backwards and forwards: it is somewhat larger than the similar bridge at Millwall (p. 467), and weighs 360 tons.

The quays of the Great Dock were so narrow that, about 1874, the south and east quays were enlarged by building new quay walls, 60 and 80 feet respectively in front of their original walls, making their widths 130 and 140 feet. The walls were founded on the dredged bottom by means of bottomless caissons, which were filled at the base with concrete when sunk in place; and the walls were then built up within the shelter of the upper portions of the caissons. These quays have been provided with sheds, but do not possess sidings.

Hydraulic power has been introduced for working the gates, capstans, and bridges of the docks: movable hydraulic cranes have been established on the quays of the Large Dock; and two 40-ton hydraulic cranes, and a shear legs of 120 tons, worked by hydraulic power, have been fixed on the east quay of the Kattendyk Dock. Electric lights have been placed at the entrances, and proposals have been made to adopt this system of lighting throughout the docks for facilitating the working of the docks by night. The electricity for producing the lights at the entrances was, till quite recently, generated by hydraulic power applied to dynamo machines, which was probably the first instance of this application of hydraulic power.

The river quays have been fully equipped with sidings,

sheds, and hydraulic travelling cranes. The sheds are placed 20 feet back from the edge of the quay. A line of rails, having a guage of 13 feet, is laid along this space, in front of the sheds, upon which the hydraulic cranes travel. These cranes are so constructed that trucks can pass underneath their staging on an inner line of rails. The sheds are arranged in groups, as shown on the plan, being separated by roadways 40 feet wide, and they have a width of about 150 feet. Three lines of rails have been placed at the back of the sheds; one of these lines is under the shelter of the sheds, and is devoted to trucks which are being loaded or unloaded; the centre line is for outgoing trucks; and the outer line is for incoming trains of trucks. The four lines are connected by turntables and crossings, so that the trucks can be readily shunted on to the various lines. The sidings run along the whole length of the quays; and communication is maintained across the entrance to the South Barge Dock by two swing-bridges, carrying two lines of railway over each end of the lock.

The hydraulic cranes can be adjusted for loads of 14 cwts. or 1½ tons; they weigh about 17 tons, of which 5½ tons consist of ballast, and they cost £450 each.

Extension of Docks at Antwerp. The site of the North Citadel, forming part of the abandoned fortifications of Antwerp, has provided a very convenient open space for dock extension. This land was purchased by the municipality in 1881; and in 1883, Mr. Royers' design for two new docks, to the north of the existing docks, was approved, and is being carried out. (Plate 13, Fig. 3.)

The two docks will have a total area of about 50 acres, thus affording an addition equal to half the existing area of the Antwerp Docks; and they will furnish about 3700 lineal yards of quays, almost equivalent in length to the river quays. The Africa Dock will communicate with the Kattendyk Dock, through a basin lock with entrances 59 feet wide; and a passage of the same width will lead from the Africa Dock into the America

Dock. The depth of the sills at the entrance to the Africa Dock will be 22¾ feet below the ordinary water-level of the docks, corresponding with the sill at the entrance to the Kattendyk Dock. The docks, however, are being excavated to a depth of 30 feet below the ordinary water-level, in order to provide an ample depth at all times for the largest class of vessels, and as it is intended ultimately to construct a new entrance direct into the Scheldt from the Africa Dock, below the other entrances, and having sills placed at a lower level, as shown on the plan of the port of Antwerp. (Plate 13, Fig. 3.)

The Africa Dock is intended to accommodate transatlantic steamers; and the America Dock is destined for the petroleum trade. The junction basin lock will have quay walls similar in construction to the walls of the Kattendyk Dock, consisting of a brick wall with counterforts, faced along the upper half with ashlar, and resting upon a foundation of rubble stone, supported in front by a row of sheet-piling. The Africa Dock and three sides of the America Dock will be surrounded by quay walls, similar in construction, but thicker and without counterforts. The west side of the America Dock will be terminated by a pitched slope, with the object of facilitating future extensions in that direction.

These works were let by contract, in 1883, for a sum of £231,500, and are to be completed in 1886. Assuming that the docks are constructed for the above sum, the works, exclusive of land, would cost about £4800 per acre of water area.

Development of the Port of Antwerp. The port of Antwerp is not of recent origin, like Barrow and St. Nazaire, but its recent development is quite as remarkable, and, moreover, it holds a much more important position in the trade of the world. Previous to the present century it possessed only a small length of narrow quays, alongside the river, in shallow water. Though the quays were slowly extended, the only dock accommodation up to the opening of the Kattendyk Dock, in 1860, consisted of the two old basins having an area of 23½ acres. Thirteen years later, the area of the docks had been raised to 100 acres, so that the accommodation was quadrupled in this short period; and the length of river quays amounted to about 2370 yards. At the present time (1884), the extension of the Kattendyk Dock has increased the dock area to about 105 acres, and the North and South Barge Docks have added 15 acres more of shallow water area, suitable for river craft; whilst the river quays are 3850 yards in length, capable of berthing the largest steamers at any state of the tide. In spite of these considerable extensions, the accommodation afforded cannot keep pace with the growth of trade; and the north docks, in course of construction, will raise the dock accommodation to about 170 acres, including the barge docks.

The formation of quays along the left bank of the river has been proposed; and sufficient land has been purchased by the municipality to provide for extensions of the north docks at a future period.

The figures given above do not even afford a full measure of the increased accommodation; for the widening of the quays, and the provision of ample sidings, and hydraulic cranes add largely to the capabilities of the quays for traffic.

Trade Statistics of Antwerp. The rapid extension of the accommodation for vessels at Antwerp, just described, serves as an indication of the development of its trade. Up to 1834, the trade of the port had been almost stationary during several years; for the tonnage of the sea-going vessels entering the port was 128,800 tons in 1816, and 127,200 tons in 1833. From this period, however, when Belgium had recently gained her independence, the advance has not only been steady, with the exception of occasional fluctuations in unfavourable years, but it progressed for some years at a rapidly increasing rate. Thus in the ten years from 1833 to 1843, the yearly increment of growth for sea-going vessels averaged 11,500 tons; in the following ten years the advance was slightly less; but in the ten years from 1853 to 1863, the yearly increment reached 27,400 tons, nearly treble that of the preceding ten years; whilst the most rapid advance was experienced from 1863 to 1873, when the yearly increment of growth averaged 145,200 tons, more than five times the preceding increment. Since that period the average yearly increase has been maintained with a moderate advance, for in the ten years up to 1883 it amounted to 179,600 tons.

Taking now the actual increase, and comparing it with the total trade in these same periods, the tonnage of vessels entering the port rose from 127,200 tons in 1833 to 242,500 tons in 1843, or nearly double; to 335,300 tons in 1853, an increase of two-fifths; and to 609,700 tons in 1863, an increase of four-fifths over the tonnage of 1853. Then in 1873 it had bounded up to 2,062,200 tons, more than treble the tonnage of 1863; and in 1883 it reached 3,858,000 tons, an increase greater actually than between 1863 and 1873, but less relatively to the tonnage of 1873, being an augmentation of about five-sixths. Adding the tonnage of the river craft, the total tonnage of vessels entering the port reached about 6,087,500 tons in 1883.

As noticed with regard to other ports, the increase in the tonnage has been effected by the increased size of the vessels frequenting the port, not by an increase in the number of vessels. In fact the number of sea-going vessels entering the port has actually decreased, from 4,797 in 1873, to 4,689 in 1883; but their average tonnage has risen from 430 tons to 822 tons in the same period. This result is due to the great decrease in sailing vessels, and the substitution of steamers of larger tonnage; for the sailing vessels have fallen from 2,182 in 1873, to 989 in 1883, whilst the steamers have increased from 2,615 to 3,700.

Remarks on the Port of Antwerp. The value to a port·

of being favourably situated for communication with a large tract of country in the interior is fully exemplified by the case of Antwerp. In spite of the distance of Antwerp from the sea, as compared with Flushing, Rotterdam, and Amsterdam, and the good depth which has been provided for these two latter ports by the works at the mouth of the Maas, and the Amsterdam Ship-Canal, the trade of Antwerp continues to make rapid progress, the port being conveniently situated for the trade of Germany and the east of France, and being connected with the interior by canals as well as by the river and railways. The two things necessary to secure such a development are a good and safe navigable channel up to the port, and facilities for trade at the port itself; the former has been supplied to Antwerp by nature in the river Scheldt, and the latter has been gradually provided by the enterprise of its inhabitants and the assistance of the Belgian Government.

The prosperity of Antwerp presents a striking contrast to the comparatively deserted state of Flushing, more favourably situated as regards access to the sea, and fostered by the Dutch Government. In this respect the two principal ports on the Scheldt have experienced a precisely opposite fate to the ports of St. Nazaire and Nantes on the Loire; for whereas the inland port of Antwerp has got far in advance, on the Scheldt, of its rival on the coast, it is the seaport of St. Nazaire which has so rapidly developed on the Loire. This difference is due to the bad condition of the Loire for navigation as compared with the Scheldt.

Comparing the position of Antwerp with other ports, the dock accommodation at Antwerp is only surpassed by London, Liverpool, and Barrow amongst British ports, being slightly larger than Hull in this respect. The French ports of Marseilles and Havre have a larger area of docks; and when the works in progress, both at Havre and Antwerp, are completed, Havre will still possess a slight advantage in this respect but the river quays at Antwerp give it a decided superiority over Havre in quay accommodation, and it is in advance of Havre in the adoption of hydraulic machinery. In respect of the total tonnage of the vessels frequenting the port, Antwerp is only exceeded by London, Liverpool, and the Tyne, and is somewhat in advance of Cardiff (Fig. 27, p. 628); and it possesses a larger trade than Marseilles, the principal port of France.

PORT OF HAMBURG.

Site. The city of Hamburg was originally situated at some distance from the main branch of the Elbe; but partly from natural causes, and partly by aid of artificial cuts, made in the 16th century, the main channel now passes close to Hamburg. The port of Hamburg is even further from the sea than Antwerp, its competitor for the trade of Germany; but it possesses the great advantage, for trading operations, of a small tidal range.

Description of the Port of Hamburg. The river Elbe itself and the adjoining creeks were, till quite recently, the only places where vessels could lie, at a distance from the banks, being moored to piles placed in rows in the water for the purpose (called locally Dukes of Alba); and the vessels were discharged by means of lighters, which conveyed the goods to warehouses on shore. The lighters can penetrate to various parts of the town along canals which are very shallow and become dry at low tide. These canals, however, cannot be deepened, owing to the high level of the foundations of the adjacent warehouses which rest upon piles.

When the extension of trade and the introduction of railways necessitated a more rapid method of transhipment, the formation of quays round the old basins, or havens, leading into the river was commenced. The first basins that have been dealt with in this manner are the Sandgate and Grasbrok Havens, which are the basins lowest down the river on the right bank, and run parallel to one another, leaving an interval between very suitable for a quay. (Plate 13, Fig. 5.) The north quay of the Sandgate Basin was opened in 1872; and the southern Sandgate, or Kaiser quay, together with the northern, or Dallmann quay of the Grasbrok Basin, were opened in 1876.

The Sandgate and Grasbrok Basins are quite open to the river, owing to the small tidal range, only reaching 6¼ feet at ordinary spring tides; they have a depth of 18 feet below low water, which corresponds with the depth of the main river channel. The Sandgate Basin has a width at the entrance of 256 feet, and a maximum width of 414 feet: the Grasbrok Basin is 275 feet wide at the entrance, and has a maximum width of 472 feet.

A quay has been formed along the Elbe itself, south of Grasbrok Basin; and a siding has been laid along the right bank, south of the Beacon Harbour, higher up the river. A petroleum dock, and a large and a small timber pond have been constructed on the left bank, opposite the other basins.

Besides the basins and canals referred to, there are other basins, or havens, on the right side of the river, which accommodate the river craft and lighters, but which have not a sufficient depth to receive larger vessels.

Appliances on the Quays at Hamburg. The quays surrounding the Sandgate and Grasbrok Basins have been fully supplied with sidings, sheds, and steam cranes for the rapid transhipment of cargoes. The northern Sandgate quay, having a width of 160 feet, is provided with a line of rails near the edge of the quay for the 1½-ton travelling steam cranes; and there is another line of rails between the crane road and the sheds, and four lines of way behind the sheds with a roadway beyond. The sheds on this quay are only 46 feet wide; but the sheds on the central quay, between the Sandgate and Grasbrok Basins, have been given the more convenient width of 75 feet. This central quay, with a width of 335 feet, has in other respects been arranged similarly to the northern quay, on each side of a central roadway 33 feet wide. The southern quay of the Grasbrok Basin, and the adjacent quay bordering the river have been similarly equipped. A large warehouse has been erected at the end of the central quay, and has been provided with hydraulic cranes.

There are two graving docks, three floating docks, and four slipways for the repair of vessels at the port of Hamburg.

Statistics of the Port of Hamburg. The number of ships entering the port has remained nearly stationary; for whereas, in 1865, the number was 5,186, it had only risen to 5,260 in 1875. The total tonnage of the vessels entering the port has increased steadily, with only occasional slight relapses: it amounted to 1,216,500 tons in 1865; it reached 2,084,700 tons in 1875, and rose to 2,243,800 in 1877, having thus nearly doubled in twelve years. In this respect the trade of Hamburg approximates to the trade of Glasgow and Sunderland; it is in advance of Hull and Dublin, and about equivalent to that of Havre.

There is ample scope for the extension of the quays at Hamburg; but the depth hitherto attained is not adequate for the largest class of vessels, so that in this respect it is at a disadvantage as compared with its rival Antwerp; and further improvements will be needed to maintain its growth.

PORT OF ROTTERDAM.

The trade of Rotterdam has received of late years a fresh impulse, by the opening of the new channel at the mouth of the Maas[1], and by the construction of the Dort and Rotterdam Railway, placing it in more direct communication with the continent. The new channel across Hoek-van-Holland was opened in 1872; and the total tonnage of the vessels

[1] A description of this work is given in 'Rivers and Canals,' pp. 272–279, and plate 15.

frequenting the port, which was 2,333,000 tons in 1869, had risen to 3,463,000 tons in 1875. The railway communication was finally completed in 1877.

Description of the Old Port of Rotterdam. Several narrow basins, or canals, intersect Rotterdam on the right bank of the Maas; and there is a considerable length of river quays in front of the town. (Plate 13, Fig. 1.) These basins are only adapted for small vessels; and the river quays, having their foundations situated at mean low-water level, have to be protected by slopes and rubble mounds in front, at their base, to prevent undermining, and consequently cannot receive large vessels alongside.

The quay walls, moreover, are so unsuited for supporting heavy weights that in only two places, where the foundations happen to be lower, have two cranes been erected capable of lifting 10 and 25 tons respectively.

The quays along the river and round the basins and canals are narrow, and are used as ordinary roadways; and the swing-bridges, spanning the various entrances and passages intersecting the quays, are unsuitable for carrying lines of rails; so that the quays are quite unprovided with sidings or sheds, except at the east end of the town where these facilities have been established by the Rhenish Railway Company in connection with their station since 1856.

A considerable portion of the trade of Rotterdam consists of goods which are despatched by river; and the tranship-ment of this merchandise is effected by the river craft, going alongside the large sea-going steamers moored to buoys in the middle of the river. This traffic is not affected by the insufficiency of the quays; and for this purpose the river itself affords a deep and sheltered roadstead.

Extension of Basins at Rotterdam. When the Dort and Rotterdam Railway was constructed by the Dutch Govern-ment, the Minister for Public Works decided to follow the wise example of the Rhenish Railway Company, and provide adequate quay accommodation in connection with the railway. As the railway passes at a high level across Rotterdam, it was necessary to construct the new quays on the left bank of the Maas, in the island of Feyenoord, where access to the railway could be gained at a lower level, and where there was a large area of unoccupied land only separated from the town by the river. This island has, accordingly, been connected with Rotterdam by a low-level roadway bridge, which can be opened for river traffic by means of a swing-bridge across the King's Haven branch of the stream. The goods station has been established on this island, along the south-west quay of a basin constructed in the island, and called the Railway Harbour. (Plate 13, Fig. 1.)

The Railway Harbour, or basin, has been constructed without gates, owing to the small tidal range of the Maas at Rotterdam; it cuts right across the island of Feyenoord, opening direct into the lower end of the King's Haven branch of the river opposite the town, and communicating also with the river higher up by a channel, 100 feet wide, at its other extremity. The width of the basin is 377 feet, for a length of 3600 feet from its lower end; its depth is 15½ feet at low tide.

A large portion of the remainder of the island of Feyenoord has been leased by the town, for 99 years, to the Commercial Company of Rotterdam. This Company, besides forming a quay wall, 4600 feet in length, along the north-east side of the Railway Basin, opposite the quay occupied by the railway company, has constructed a second basin, parallel to the first and separated from it by a width of quay of 430 feet, opening into the King's Haven. This Inner Basin has a length of about 3300 feet, a width of 260 feet, decreasing to 130 feet at its inner end, and a depth of 19 feet at low tide. Another basin, called the Warehouse Basin, opens out of the Inner Basin, on its north-eastern side, near its entrance. It has been excavated to the same depth as the Inner Basin, and has a length of 720 feet, and a width of 200 feet, and, like the Inner Basin, is surrounded by quay walls.

River quays have been constructed by the Commercial Company along the left bank of the King's Haven, on each side of the entrance to the Inner Basin, having a total length of about 1120 feet, which are accessible for large vessels.

The quays of the Feyenoord Basins have been raised 8½ feet above ordinary high-water level, so as to be above the highest rise of the Maas, which occasionally, on the concurrence of a storm with a flood of the river, attains a height of 7½ feet above the same level; whereas, at these periods, the quays of Rotterdam, which are not more than 6½ feet above high tide, are all submerged.

Appliances on the Feyenoord Quays at Rotterdam. The State and the Commercial Company have supplied at Feyenoord, on the left bank of the Maas, the accommodation which is wanting on the right bank at Rotterdam, with the single exception of the Rhenish Company's quays, and have equipped the new quays with sidings, sheds, warehouses, and cranes.

The width of the railway quay, along the south-west side of the Railway Basin, is about 400 feet. A line of rails close to the basin serves for the steam travelling cranes; and two sidings for trucks run alongside. A shed has been erected behind these lines, having a length of 500 feet and a width of 40 feet; seven lines of rails have been laid at the back of the shed, and the goods station has been established beyond.

The equipment of the north-eastern quay of the Railway Basin, belonging to the Commercial Company, is somewhat similar. The outer line of rails is reserved for the 1½-ton travelling cranes, and there is a siding between this line and the sheds which have been built alongside. The sheds have a width of 60 feet between their walls; but they cover a total width of 80 feet by the overhang of their roofs. There are three lines of way at the back of the sheds, one for arrivals, another for departures, and the third for trucks to stand along. The various lines are connected by crossings and turntables.

The arrangement of the quays of the Inner Basin differs only from that just described in dispensing with the siding in front of the sheds, and in the construction of cellars below the line on which the travelling cranes run.

A large warehouse has been erected on the northern quay of the Warehouse Basin; it is 650 feet long, 120 feet wide, and 53 feet high. It has been divided into five compartments by four fireproof cross walls; it has a cellar and four floors, and is furnished with lifts and travelling cranes.

The Commercial Company have adopted hydraulic power for working all their machinery, with the exception of twelve steam travelling cranes; and they have provided adequate appliances for unloading and loading four thousand steamers in the course of a year.

There are no graving docks at Rotterdam, but facilities for repairing vessels exist higher up the river.

Remarks on the Port of Rotterdam. The transhipment of cargoes into river craft in the middle of the stream, and the absence of good railway communication, prevented, for a long time, the want of quays, accessible for large vessels, to be seriously felt at Rotterdam. Moreover, the shallow outlets of the Maas formerly barred the entrance of vessels of large draught. The basins at Feyenoord are the direct results of the opening of the Dort Railway, and the improvement of the Maas; and they should open out a fresh era of commercial prosperity for Rotterdam. The proximity of suitable unoccupied land was most fortunate for the town; and the basins have been amply provided with facilities for traffic, and have been carried to a depth sufficient for any vessel that can navigate the Maas. The depth attained, however, is not equal to that procured at the more important ports, and is not adequate for the largest class of vessels, a matter of considerable moment considering that the draught of vessels has been steadily increasing. If, therefore, Rotterdam is to compete with Antwerp for trade, it will be necessary for the channel of the Maas below Rotterdam to be improved, and a still better depth at its outlet secured. There is ample space on Feyenoord Island for the extension of basins, as soon as the development of the trade of Rotterdam demands fresh accommodation; and quays could be easily extended along the left bank of the river opposite the town.

PORT OF ROUEN.

Improvement of Approach Channel. The growth of trade at Rouen is entirely due to the improvement of the river Seine, below La Mailleraye, by means of training works[1]. Before the commencement of these works in 1848, the navigation of the lower portion of the Seine below La Mailleraye was difficult, and even dangerous, owing to the shifting of the channel, and the existence of shoals, particularly at Aizier and Villequier, where the depth was only about 10 feet at spring tides. The bore also aggravated, at times, the perils of navigation.

The approach to Rouen was, accordingly, barred formerly to large vessels, by the shoals and other dangers in the lowest reaches of the Seine; though there was a good deep channel above, right up to Rouen. The situation of Rouen, however, has been entirely transformed by the regulation of the river from La Mailleraye to Berville; for whereas, previous to the training works, vessels of from 100 to 200 tons experienced difficulties in traversing the lower reaches, at the present time vessels of 2000 tons can navigate the river, and ships drawing 21 feet of water have managed to accomplish the passage.

Situation of Rouen. The town of Rouen, standing upon the right bank of the Seine, is about 78 miles from the sea, and about 16 miles below the termination of the tidal Seine

[1] A description of these works is given in 'Rivers and Canals,' pp. 266-272, and plate 16.

at Martot Weir, near St. Aubin. Since the improvement of its access to the sea, it has developed into a flourishing commercial port; though its only facilities for trade consist of the quays along both banks of the river, in front of the town and its suburb St. Sever on the opposite bank. The small rise of tide at Rouen, amounting to only 6½ feet at spring tides and 3½ feet at neaps, renders river quays quite convenient for trade; but the old quays are so largely devoted to the general traffic of the town, that adequate accommodation in the form of sheds and sidings cannot be allotted to the shipping traffic. In spite of this disadvantage, and though vessels drawing more than 17 feet cannot get up the river at ordinary neap tides, Rouen had attained the rank of the fifth port of France in 1878, with a tonnage of imported and exported goods of 1,026,000 tons, being only inferior to Marseilles, Havre, Bordeaux, and Dunkirk in this respect.

Extension of Quays at Rouen. The passage of large vessels further up the river is stopped at Rouen, by the stone bridge at the upper end of the town. Rouen, accordingly, is the limit of sea-going traffic on the Seine; and the extension of quays can only be effected down the river.

At the lower end of the town, the river became wider, and was intersected by islands. This natural conformation has proved very convenient in facilitating the extension of quays; for the river has been regulated by connecting the islands, on either side, with the banks, and dredging away any projecting portions; so that the wasted area of the islands has been converted into very valuable quay space, whilst the course of the river has been improved[1]. (Plate 13, Fig. 2.) The islands of Little Gay and Méru have been already connected with the shore, on the right and left banks respectively; and the new river banks have been lined with quay walls. The cost of the dredging, filling in, and construction of quay walls

[1] Plans of the works and a section of the quay wall were given me by M. Juncker, the engineer in charge of the works.

required for this work was estimated at £201,000; and the work was let by contract in 1880.

Further extensions are proposed by connecting Rolet Island with Méru Island, and by prolonging the line of quay also along the right bank, as indicated by dotted lines on the plan. (Plate 13, Fig. 2.) The water space between Little Gay Island and the original bank is to be filled up; but the creek left on the opposite side, between Rolet and Méru islands and the old left bank, will be retained as a timber pond. Plans have been drawn up for equipping the new quay space with sheds, sidings, and hydraulic cranes; and it is proposed to extend the cranes along the front of the old quays. Additional works are in contemplation further down the river, such as graving docks and slips on the left bank opposite the end of Rolet Island, grain elevators and sheds on Rolet Island, a petroleum basin between Elie Island and the left bank, and petroleum warehouses alongside. If all these proposals are carried out, the port of Rouen will be quite transformed, and will possess excellent capabilities for traffic.

Construction of Quay Walls at Rouen. The new quay walls which are being constructed at Rouen have been already described (p. 428); and their section is given on Plate 15, Fig. 15. It was originally intended that the portions of the walls situated on the islands should be built in a trench, kept dry by pumping; but eventually one method was followed for all the walls, whether on the islands or in the river. This method consisted in placing watertight wooden caissons, or boxes, open at the top, on the foundation piles which were cut off at a level of 14 feet below high-water spring tides; and the top of the caissons rose to within 1½ feet below high-water spring tides. Each caisson had a total length of 68 feet, and an internal width of 11½ feet. The walls were built in separate lengths inside these caissons, and were connected above the top of the caisson by small arches spanning the interval between two adjoining caissons, or by oak flooring when the water-level permitted; and a continuous wall was built on the top up to quay level.

The bed of the river has been dredged to a depth of 16½ feet below the lowest observed water-level, or 19 feet below low-water neap tides which, at Rouen, falls 2⅓ feet lower than spring tides.

Comparison between Rouen and Bordeaux. There are several points of resemblance between Rouen and Bordeaux, the only other French interior river port which has attained a leading position. Bordeaux, like Rouen, is many miles from the sea, being about 70 miles from the mouth of the Gironde. Like Rouen, its trade is carried on along quays on both sides of the river Garonne; though owing to the greater tidal oscillation at Bordeaux, amounting to 14¾ feet at spring tides, it is not so favourably situated in this respect, and therefore, like Antwerp, is being provided with floating dock accommodation, the first dock of about 25 acres having been constructed within the last few years. Moreover, like the Seine at Rouen, the river Garonne possesses a good natural depth at Bordeaux; whereas in the lower river there are shoals which it is proposed to lower by regulating the river.

Bordeaux, however, possesses some important natural advantages over Rouen, which account for its having obtained a larger trade, and ranking, in advance of Rouen, as the third of the French ports in respect of the tonnage of its exports and imports. For instance, vessels drawing 24½ feet can reach Bordeaux at spring tides, though this depth is reduced to 19½ feet at neaps. The Gironde, moreover, possessed a better natural depth than the Seine; so that there has not existed the same hindrance to the early development of trade as at Rouen. Bordeaux also commands a considerable internal trade, being the principal port for the wine producing districts of France.

The only other interior river port of France which is of any consequence is Nantes; but the difficulties in the way of improving the Loire have led to the diversion of its natural traffic to St. Nazaire.

PORT OF NEW YORK.

Situation. The naturally well-sheltered and deep-water approach to New York was doubtless the cause of the rise of that city; and the great increase of the city necessitates a corresponding growth of its port. The insular position of New York, with its long lengths of river frontage bordering the two large rivers which flow into New York Bay, affords ample opportunities for quays; whilst the small rise of tide, averaging only about 4¾ feet, renders river quays very suitable for trade. Manhattan Island, on which the business portion of the City of New York is situated, is 13½ miles long, with a maximum width of 2¼ miles, and has an area of 22 square miles. It is surrounded by three rivers, or tidal estuaries, namely, the North, or Hudson River on the west; the East River on the east, which joining the North River at the south of the island opens into New York Bay; and the Harlem River, on the north side of the island, connecting the North and East rivers. Manhattan Island has a shore frontage of 13 miles along the North River, 8¼ miles along the East River, and 8 miles along Harlem River, giving a total water frontage of about 29¼ miles. A portion, however, of the frontage along the Harlem River, amounting to about four miles, does not possess a channel suitable for navigation; so that the actual available frontage is about 25 miles, but it is proposed to improve the shallow portion of the Harlem channel so as to make the whole river frontage available for commerce. At present the shipping is mainly concentrated along the lower 6 miles of the North River, and the lower 5 miles of the East River, as the main business quarter of the city and the storehouses are situated at the lower end of the island; and the piers are only extended northwards from want of accommodation below, or where necessitated by local interests. The shores of Manhattan Island, accordingly, afford ample space, not merely for the existing trade of New York, but also for extensions equivalent to more than double its present requirements; and as the banks shelve rapidly, deep water is reached close inshore.

The port is approached through the well-sheltered New York Bay, which provides a safe anchorage area of 88 square miles in its lower part, and 14 square miles in its upper portion; and these areas, together with 13½ square miles of anchorage-ground in the North and East Rivers adjacent to the city, form a harbour 115½ square miles in extent, where vessels can lie in perfect security.

The currents in the estuaries surrounding Manhattan Island are peculiar, and vary in velocity at different points. Their rate of flow is generally moderate, ranging between 1½ and 2¾ miles an hour in the harbour and North River; but on the East River, the ebb current has a velocity of 4⅗ miles an hour, and the flood current attains to 8½ miles an hour. The currents in the East River are intensified by the obstructions at Hell Gate, some miles up that estuary; but the removal of the rocks there, which is still in progress (as described on p. 383), has already effected a notable improvement, which will be still more marked when the works are finished.

Docks at New York. The small range of tide at New York, and its sheltered site render floating docks with gates quite unnecessary; and river quays fully supply the requirements of trade. The southern shores of Manhattan Island are lined with bulkhead walls, from which numerous piers project at right angles; and the intervals left for vessels between the piers are termed docks, resembling in fact the basins formed by the projecting moles at Marseilles and Trieste, with the suppression of the outer protecting breakwater provided at those ports, which is rendered unnecessary by the sheltered position of New York.

Original Construction of Wharves at New York. The river front was formerly, for the most part, in the hands of numerous private individuals; and only a small portion belonged to the city. The owners erected cheap cribwork walls and wooden piers in any directions, and to almost any distance out that suited them, without any concert between themselves. The result was that the river front consisted of a series of very irregular lines of walls, from 100 to 200 feet long, with numerous projections and recesses, and with projecting piers, at various angles and of different shapes, at irregular intervals apart. The whole system of wharves, in fact, presented no systematic arrangement, and was almost free from the control of the city authorities, having grown up to satisfy conflicting private interests without any regard to the general public advantage.

System of the Department of Docks at New York. Demands for improved public facilities for trade, and the substitution of an organised system, in place of the uncontrolled method which prevailed, were commenced early in the present century; but it was not till 1870 that the Department of Docks was established, to reconstitute the system of wharves, and to reconstruct them according to a general definite plan [1].

The system adopted consists of a regular continuous bulkhead, or quay wall, of masonry and concrete, resting upon piles with a rubble foundation (Plate 14, Fig. 18), already described (p. 426), with timber piers projecting into the water and placed at such intervals apart as prove most convenient for the shipping.

The bulkhead line is being carried out into the water to a sufficient distance to form a street by the river side, 250 feet wide along the North River, and from 200 to 150 feet wide along the East River, for the convenience of trade. A lighter

[1] I am indebted to Mr. T. J. Long, engineer of the Department of Docks at New York, for numerous particulars relating to these docks.

section of bulkhead was erected as a temporary structure along a short length of the East River frontage, on a sloping bed of rock, at a cost of £6 5*s*. per lineal foot. (Plate 15, Fig. 16.)

The piers are from 400 to 500 feet long, and from 60 to 80 feet wide along the North River, but are somewhat smaller along the East River; the intervals between the piers are from 150 to 200 feet in width. The piers, with only one exception, are constructed very substantially of wood, calculated to last 10 years before needing repairs, and to have a life of about 20 years. This course was estimated to be more economical than the construction of a permanent structure of stone.

Hitherto the timberwork of the piers has not been defended in any way against the teredo, with the exception of the oak fenders placed on the outside of the piles, which have been creosoted. The teredo, however, has been in a great measure driven away by the sewage discharged at the bulkhead, and remaining between the solid cribwork of which many of the old piers were formed. Accordingly, the ravages of the teredo are likely to be much increased when the water between the piers becomes cleansed by the free flow of the current through the open pilework piers, and by the extension of the sewers to the ends of the piers where the rapid current will carry off the discharge. When this great sanitary improvement has been effected, it will become necessary to protect the timberwork by creosote or other means.

There are about eighty piers on each side of the island, in continuation of the lines of the streets. Some of these are to be extended, several have been rebuilt, and additional ones are designed. The progress of the work of reconstruction and regulation is somewhat slow, as the city can only gradually acquire possession of the river frontage; but work done by private owners has to be executed now in conformity with the general design.

The increase obtained by the new system, both in length of

wharfing and quay space, is considerable; for whereas, in 1870, the length of wharfing was 150,400 feet, and the area of the piers amounted to 2,322,670 square feet, there will be afforded by the new system 195,000 lineal feet of wharfing, and 5,105,600 square feet of pier area, which is equivalent to a gain of nearly one-third in length, and affords more than double the area. The piers are mostly leased; but only wharfage dues are charged for the bulkheads.

Cost of Bulkhead and Piers at New York. The actual cost of the bulkhead, built in accordance with the most recent approved design (Plate 14, Fig. 18), averaged £49 16*s*. per lineal foot. The cost of the piers varies between 5*s*. and 5*s*. 5*d*. per square foot of area. The following estimate, furnished me by Mr. T. J. Long, of the Department of Docks, gives the approximate cost of a pier with the corresponding length of bulkhead enclosing a complete dock, or slip, assuming the pier to be 500 feet long and 80 feet wide, and the slip to have a width of 200 feet between the piers:

		£ *s*.	£
Bulkhead, including removal of old work, superintendence, &c.:			
280 lineal feet of wall, at		55 4	15,456
Pier:			
40,000 square feet, at	5*s*. 2½*d*.	10,417 0	
Dredging (about)		1,562 0	
Total for Pier			11,979
Total cost of Pier and adjacent Bulkhead			£27,435

Appliances at the New York Quays. Several small cranes are ranged along the piers, having a power of only one or two tons, which are worked by shafting from an engine on the pier. A few large cranes have been erected at private wharves, worked by hand or steam. Floating derricks are usually employed for lifting the larger weights; these derricks, of which there are several, belong generally to the large iron foundries, and have capacities ranging up to 50 or 60 tons. The derricks are preferred at New York to large stationary cranes, as they can come to the outer side of a vessel, lying alongside a pier, and lift the heavier weights without impeding the unlading of the lighter cargo. The port of New York, however, does not possess the facilities of sidings, travelling cranes, and hydraulic power, introduced at many European ports.

Remarks on the Port of New York. The harbour of New York affords unrivalled natural advantages for commerce, which have made it by far the foremost port of the United States. It is close to the Atlantic, whilst possessing excellent and spacious shelter, and an ample depth of water from the ocean to the port; and it communicates with the interior of the country by river navigation, as well as by railways. The organised system of piers and bulkheads recently introduced appears well suited to the special requirements of the port; though the extension of mechanical appliances, and the introduction of hydraulic power, would enlarge the capabilities of the wharves to the great benefit of trade. The port possesses, moreover, the immense advantage of ample space for extensions of quays; though naturally the most convenient positions have been already fully occupied.

More than half the foreign trade of the country passes through New York Harbour; for the value of the imports to New York, in 1882, amounted to £102,720,000, out of a total to the whole country of £150,970,000, and its exports, in the same year, amounted to £71,770,000, out of a total of exports of £156,360,000. The only ports whose exports and imports exceed those of New York in value are London and Liverpool; and although the imports to London are nearly half as much again in value as those to New York, the exports from London exceed those from New York by only about a fourth; whilst the proportion in value of the Liverpool imports to those of New York is as 9 to 8, and its exports as 10 to 7.

Concluding Remarks. The tendency of shipping traffic is to concentrate itself at a few ports, in the most direct lines of communication with extensive districts, and possessing special advantages for navigation, and where important centres of

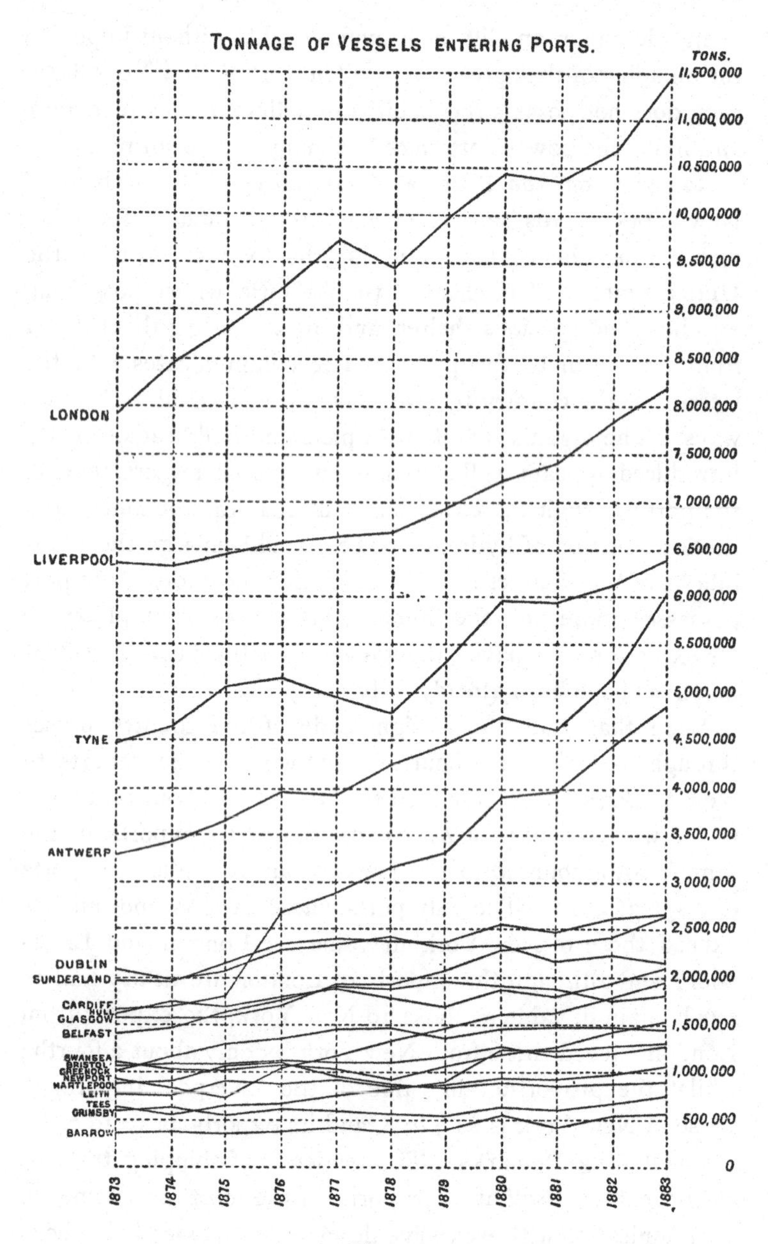
TONNAGE OF VESSELS ENTERING PORTS.
TONS.
11,500,000
11,000,000
10,500,000
10,000,000
9,500,000
9,000,000
8,500,000
8,000,000
7,500,000
7,000,000
6,500,000
6,000,000
5,500,000
5,000,000
4,500,000
4,000,000
3,500,000
3,000,000
2,500,000
2,000,000
1,500,000
1,000,000
500,000
0
LONDON
LIVERPOOL
TYNE
ANTWERP
DUBLIN
SUNDERLAND
CARDIFF
HULL
GLASGOW
BELFAST
SWANSEA
BRISTOL
GREENOCK
NEWPORT
HARTLEPOOL
LEITH
TEES
GRIMSBY
BARROW
1873
1874
1875
1876
1877
1878
1879
1880
1881
1882
1883

Fig. 27.

commerce have grown up. Thus London and Liverpool monopolise a large proportion of the maritime trade of the United Kingdom[1]; New York secures more than half the total shipping trade of the United States; Marseilles and Havre possess about one-sixth of the maritime trade of France; whilst Antwerp, belonging to the small kingdom of Belgium, and the town of Hamburg are the principal emporiums of the trade of central Europe.

The accompanying diagram (Fig. 27) exhibits graphically the relative importance, and the growth within the last ten years, of most of the principal ports of Great Britain, and of Antwerp as well, in respect of the total tonnage of the vessels entering these ports.

The table on p. 630 indicates the order in which the ports of Great Britain, described in Chapters XXIII to XXV, stood in 1883, as regards the tonnage of the vessels both entering and leaving with cargoes only, and the total tonnage of the vessels entering or clearing with cargoes and in ballast; and in respect of the values of the imports and exports, both separately and together. (See also Appendices V and VI.) The Tyne and the Tees ports have been also added to the table. The figures in brackets in the third and sixth columns represent the proportionate amount of trade, at the different ports, as regards tonnage and values, respectively, where the ports lowest on the list are represented by unity.

New York, as already stated, would occupy the third place in the last column of the following table; for its imports, though less in value than those of London or Liverpool, are nearly five times the value of those of Hull; whilst its exports are

[1] It will be seen, from Appendices V and VI, that the tonnage of vessels entering the ports of London and Liverpool is equal to the combined tonnage of the Tyne, Cardiff, Hull, Bristol, Glasgow, and Belfast; whilst the values of the exports and imports at either London or Liverpool are very much larger than at all the other British ports of which statistics are given, the combined values of the London and Liverpool imports and exports being nearly three times the values of the trade of all the rest of the ports named in the tables.

BRITISH PORTS, 1883.

Table giving most of the principal British Ports, placed in the order of the Tonnage of their Vessels, and the Value of their Merchandise.

Number in Rank.	TONNAGE OF VESSELS.				VALUES OF MERCHANDISE.			
	Entering with Cargoes.	*Clearing with Cargoes.*	*Entering or Clearing with Cargoes and in Ballast.*		*Imports.*	*Exports.*	*Exports and Imports.*	
1	London.	Liverpool.	London.	(20)	London.	Liverpool.	London.	(170)
2	Liverpool.	London.	Liverpool.	(14)	Liverpool.	London.	Liverpool.	(154)
3	Glasgow.	Tyne Ports.	Tyne Ports.	(11½)	Hull.	Hull.	Hull.	(31)
4	Hull.	Cardiff.	Cardiff.	(9)	Glasgow.	Glasgow.	Glasgow.	(20)
5	Tyne Ports.	Glasgow.	Glasgow.	(5)	Leith	Grimsby.	Tyne Ports.	(11)
6	Dublin.	Sunderland.	Sunderland.		Tyne Ports.	Tyne Ports.	Leith.	(9)
7	Belfast.	Newport.	Dublin.	(3¾)	Bristol.	Cardiff.	Grimsby.	(8½)
8	Greenock.	Hull.	Hull.	(3½)	Greenock.	Leith.	Bristol.	(7)
9	Bristol.	Swansea.	Newport.		Grimsby.	Tees Ports.	Cardiff.	(5)
10	Cardiff.	Belfast.	Belfast.	(3¼)	Dublin.	Newport.	Greenock.	(4)
11	Leith.	Dublin.	Swansea.	(2¾)	Hartlepool.	Swansea.	Hartlepool.	(2½)
12	Newport.	Tees Ports.	Greenock.	(2½)	Cardiff.	Hartlepool.	Swansea.	
13	Swansea.	Hartlepool.	Tees Ports.	(2¼)	Belfast.	Bristol.	Dublin.	(2¼)
14	Tees Ports.	Leith.	Bristol.		Swansea.	Barrow.	Tees Ports.	
15	Grimsby.	Bristol.	Leith.	(1¾)	Newport.	Belfast.	Newport.	(2)
16	Hartlepool.	Grimsby.	Hartlepool.	(1½)	Sunderland.	Sunderland.	Belfast.	
17	Sunderland.	Greenock.	Grimsby.	(1¼)	Tees Ports.	Greenock.	Sunderland.	(1)
18	Barrow.	Barrow.	Barrow.	(1)	Barrow.	Dublin.	Barrow.	

more than three times the value of the Hull exports. Antwerp would occupy the fourth place in the third column; for the total tonnage of its vessels is slightly below that of the vessels frequenting the Tyne ports, though it amounts to three-fourths of the Liverpool tonnage, owing to its large inland navigation trade. Marseilles would rank between Cardiff and Glasgow in the third column; whilst Havre and Hamburg would come between Sunderland and Dublin.

The principal ports, such as London, Liverpool, and New York, maintain a fairly steady progress, owing to the growth of their cities, and the natural increase in population and development of the extensive districts which they serve. Other ports are more or less dependent upon certain trades, and undergo fluctuations according to the varying demands for the goods which form the staple of their traffic. For instance the coal and iron ports, such as Cardiff, Newport, Swansea, and Middlesborough on the Tees, are affected by the state of the iron trade, which reacts upon the coal trade; and they are also exposed to competition amongst themselves. Several of these mineral ports are essentially places of export, as for example Cardiff, Newport, Swansea, the Tyne, and the Tees. These ports are placed at some disadvantage in being mainly ports of supply, and not of demand, where vessels have to enter in ballast. Some ports possess the great advantage of having a fairly balanced trade, such as Liverpool, Hull, Glasgow, and Barrow. Grimsby receives as great a tonnage of vessels with cargoes as it dispatches; but the value of its exports are double those of its imports. The exact opposite occurs in the case of the Tyne ports; for though, owing to their large coal trade, the tonnage of the vessels leaving with cargoes is more than three times the tonnage of the entering vessels, nevertheless the value of the imports is nearly double the value of the exports. A similar anomaly also exists with regard to the trade of Sunderland and Hartlepool. Lastly, in some commercial centres distributing to a large district, but possessing comparatively few manufactures, and no minerals to supply in return, the imports exceed the exports, of which London, Bristol, Leith, Greenock, Dublin, and Belfast are instances.

The primary requirements for a port are a sufficient depth of water and good shelter, which may exist naturally as at London, Hull, New York, and Antwerp, or may be obtained by artificial means as at Glasgow, the Tyne, the Tees, Marseilles, and Dunkirk. The cost, however, of the improvement works forms a burden which must be defrayed by the shipping dues, from which ports adequately endowed by nature are exempt; and to that extent artificial ports are weighted in the competition. If, however, improvements in depth, to meet the increasing draught of vessels, are disregarded, a port may be left behind in the general progress, like Bristol, or may have its trade diverted to another port with greater facilities, as has happened to Nantes.

Commodious quays are the next essentials, and their cost of construction depends largely upon the natural conditions of the site; quay walls being cheaply constructed of concrete on the banks of the Thames, and of concrete, stone, and timber at Rouen; whereas a great expenditure is entailed on silty sites, like St. Nazaire. The amount also of excavation necessary for basins, and the price of land, materially affect the cost of quays, except where the quays are extended into the water, as at New York, Marseilles, and Antwerp. The extent of the quays must be regulated by the amount and nature of the trade; for it is as unwise to embark on a large expenditure of capital in forming quays before there is a prospect of trade to afford a return, as it would be to check a growing trade by neglecting timely extensions. Flushing serves as an example of a port where dock accommodation has been extended far in advance of its trade; and though Barrow is the third port of Great Britain in respect of dock area, it holds a much lower position as regards trade. On the other hand, a given length of quay has limits to its capabilities; and the limit at which trade has been found to become seriously hampered is a proportion of tonnage of about 350 tons per lineal yard of quay, which was reached at Antwerp in 1876, since which time large extensions have been accomplished, and hydraulic and other appliances introduced. The proportion of the tonnage of the vessels to the length of quays is decidedly less than this at London, and only about half the amount at Liverpool. The tonnage amounts to about 200 tons at Hull per lineal yard of quay, 140 tons at Grimsby, 120 tons at Hartlepool, and only 46 tons at Barrow; it was 640 tons at St. Nazaire previous to the opening of the Penhouët Dock, and has already reached 420 tons at Antwerp. The proportion is exceptionally large at Cardiff, amounting to nearly 600 tons per lineal yard of quay, if only the quays of the docks at Cardiff and Penarth are considered; but there are also coal staiths outside, and as the trade is mainly export, and the vessels are very rapidly laden by the tips, a smaller length of quay suffices than would be adequate under ordinary conditions.

The capabilities of quays may be greatly augmented by providing ample mechanical appliances, such as sidings, turntables, hydraulic travelling cranes, and capstans, together with sheds, warehouses, and lifts. In old docks with quays of limited dimensions, the introduction of these facilities for trade is not always feasible; but by their adoption at new docks, the value of the extensions is greatly enhanced, and the necessity for further extensions is deferred. Many ports, at present hampered by their trade, would be fully equal to the task if they had been provided with suitable accessories. Hydraulic power is one of the most important adjuncts to the efficient working of a dock; and it has been more or less adopted at most of the important English ports, though it has not been adequately developed at Liverpool. Hydraulic machines have been used in the general working of docks for about a quarter of a century; but the extension of the system abroad has been very slow. Marseilles, Antwerp, and Hamburg have, indeed, been furnished with hydraulic appliances; but they have not yet been introduced at Havre, and have only recently been established at St. Nazaire; whilst most of the foreign ports are devoid of these important aids to traffic, and even the quays of New York are destitute of hydraulic power.

Adequate depth, good shelter, sufficient quays, ample equipment, and scope for extension are necessary for the smaller ports, to enable them to compete successfully for traffic and to maintain their position; and even the larger ports cannot attain their full development, unless these requirements are maintained so as to keep pace with the increasing exigencies of trade.

APPENDIX I.

TIDES AT HARBOURS AND DOCKS.

The first two columns give the Rise of Spring and Neap Tides above Low-Water Spring Tides, which is the Datum from which the fathom lines on the Plans are reckoned. The third column gives the Range of Neap Tides, which is the difference in level between High and Low Water of Neap Tides.

TIDES AT HARBOURS.	*Spring Tides Rise.*		*Neap Tides Rise.*		*Neap Tides Range.*	
ENGLAND.						
ENGLISH CHANNEL.	ft.	in.	ft.	in.	ft.	in.
Dover	18	9	15	0	11	0
Newhaven	20	0	15	0	10	0
Portland	6	9	4	6		
Plymouth	15	6	12	0	7	6
WEST COAST.						
Holyhead	16	0	12	6	9	0
Whitehaven	26	0	19	0	12	0
EAST COAST.						
Tynemouth	14	9	11	4	8	4
Tees, Mouth	15	0	12	3	10	0
Lowestoft	6	6	5	3		
Ramsgate	15	0	12	0	9	0
SCOTLAND.						
EAST COAST.						
Wick	10	0	7	6	5	0
Fraserburgh	11	0	8	6	6	0
Aberdeen	12	9	9	3	5	6
IRELAND.						
EAST COAST.						
Dublin	13	0	10	0	7	0
Kingstown	11	3	8	10	6	4
Rosslare, near Wexford	5	0	3	6		
CHANNEL ISLANDS.						
ENGLISH CHANNEL.						
Alderney	17	0	12	9	8	0
St. Catherine's, Jersey	32	0	23	0		
HOLLAND.						
NORTH SEA.						
Ymuiden (Amsterdam Ship Canal) ...	5	5				

TIDES AT HARBOURS (*continued*).	*Spring Tides Rise.*		*Neap Tides Rise.*		*Neap Tides Range.*	
	ft.	in.	ft.	in.	ft.	in.
BELGIUM.						
NORTH SEA.						
Ostend	15	7	11	6	8	6
Nieuport	16	0	13	0		
FRANCE.						
NORTH SEA.						
Dunkirk	17	10	14	6	10	6
Gravelines	19	1½	14	11	10	7
Calais	20	6	16	3	12	4
ENGLISH CHANNEL.						
Boulogne	28	0	22	0	15	0
Dieppe	28	0	21	3	15	6
Cherbourg	19	2	13	6	8	4
WEST COAST.						
Brest	21	0	14	9	8	6
Biarritz	12	0				
St. Jean-de-Luz	12	3	8	0		
SPAIN.						
SOUTH COAST.						
Malaga	3	0				
AUSTRIA.						
ADRIATIC SEA.						
Trieste	2	0				
ITALY.						
ADRIATIC SEA.						
Malamocco	2	3				
INDIA.						
EAST COAST.						
Madras	3	4				
NORTH WEST COAST.						
Kurrachee	8	9	7	0		
Mormugao	6	0				
CEYLON.						
WEST COAST.						
Colombo	2	0				
SOUTH AFRICA.						
WEST COAST.						
Table Bay	5	6				

TIDES AT HARBOURS (*continued*).	*Spring Tides Rise.*	*Neap Tides Rise.*	*Neap Tides Range.*
AMERICA.			
NORTH ATLANTIC COAST.	ft. in.	ft. in.	ft. in.
Newburyport, Mass.	7 8½	(*Mean Rise*)	
Delaware	6 0	"	"
Charleston	5 11	5 1	4 3
GULF OF MEXICO.			
Galveston, Texas	1 4	0 10	
TIDES AT DOCKS.			
ENGLAND.			
SOUTH COAST.			
Portsmouth	13 6	10 9	8 6
BRISTOL CHANNEL.			
Avonmouth	39 0	29 0	19 0
Bristol	33 0	21 0	12 8
Newport	35 9	25 9	22 0
Cardiff	38 0	28 0	18 8
Swansea	27 3	20 3	
WEST COAST.			
Liverpool	27 6	20 3	13 0
Barrow	28 0	20 6	
EAST COAST.			
Sunderland	14 5	11 0	7 9
Hartlepool	15 0	11 6	8 0
RIVER HUMBER.			
Hull	20 10	16 4	11 10
Grimsby	19 2	15 2	11 1
RIVER THAMES.			
London	20 8	17 3	15 1
Tilbury	17 6	14 0	10 6
RIVER MEDWAY.			
Chatham	17 6	14 0	10 6
SCOTLAND.			
FIRTH OF FORTH.			
Leith	17 6	13 0	
FIRTH OF CLYDE.			
Greenock	10 0	8 3	6 6
IRELAND.			
BELFAST LOUGH.			
Belfast	8 10	8 1	7 4
SOUTH COAST.			
Queenstown (Haulbowline Island) ...	11 9	9 0	

TIDES AT DOCKS (*continued*).	*Spring Tides Rise.*	*Neap Tides Rise.*	*Neap Tides Range.*
FRANCE.			
RIVER SEINE.	ft. in.	ft. in.	ft. in.
Havre	23 7	18 0	11 6
Honfleur	23 9	18 10	13 1
Rouen	6 7	1 3	3 7
RIVER LOIRE.			
St. Nazaire	16 5	11 6	7 6
BELGIUM.			
RIVER SCHELDT.			
Antwerp	15 2	13 8	12 10
GERMANY.			
RIVER ELBE.			
Hamburg	6 3	5 11	
AMERICA.			
HUDSON AND EAST RIVERS.			
New York	4 9	(*Mean Rise*)	

APPENDIX II.

ABERDEEN HARBOUR.

[Memorandum by Mr. W. Dyce Cay, Engineer to the Harbour Commissioners.]

Extension of North Pier. The submarine part of this work was carried out by depositing concrete in bags, containing 50 tons each, from a hopper barge.

Hopper Barge. An iron barge was specially designed for the work; its length was 55 feet, breadth 20 feet, depth 8 feet, rise of deck 4 inches, and sheer forward and aft 1 foot. The hopper box, fixed in the well, had an average length of 24 feet, breadth 6 feet, and depth 5 feet 7 inches, thus holding 50 tons of concrete, at 16 cubic feet to a ton. The box had also an elm fender, 4 ins. × 3 ins., rounded at the top edge, which increased its depth. Its sides converged slightly towards the top, and its bottom was in two leaves which were hollow tubular caissons; they were hinged to the sides, and, when open, did not project below the bottom of the barge. The sides of the well were formed to the shape of the outside of the doors, so that when the latter were open, their inside surfaces formed continuations of the sides of the box and offered no obstruction to the descent of the concrete bag. The doors were supported at each end, outside the box; and the supporting catches were so arranged, that by the withdrawal of one pin, the supports at both ends were relieved simultaneously, to avoid straining the doors. The barge was fitted with crab winches for raising the doors, and for working the mooring chains, also with a towing post, rudder, bulkheads, cabin, and bilge pump.

Deposit of Concrete Bags. The barge was towed alongside a wharf in the harbour, where a previously prepared bag of jute cloth was fitted into the box; the bag was larger than the box, to allow for the shrinkage of the cloth when wet, and had an open top, the cover, or lid, of which was turned back while the bag was being filled with freshly mixed concrete. When full, the barge was towed to the place of deposit, and on the way the cloth lid was sewn over the mouth of the bag. The barge was then attached to previously laid down moorings, and warped by means of the deck crabs into the required position. This position, and the exact alignment of the bag were fixed by cross bearings, marked by two lines laid out with ranging poles on the pier and on the opposite south shore. The bag, which while in the hopper was partially immersed, was then dropped from the box, through the water, into its place in the work; this was signalled to those in charge of the concrete mixers, who commenced to mix a supply for the next bag.

Rate of Deposit. In this way a maximum of 9 fifty-ton bags were laid by this barge in a day in summer, 6 or 7 bags being a common number. The whole operation of filling the barge, towing it out, depositing the bag, and bringing the barge back to the wharf, occupied two hours.

Concrete. The concrete was made of the best quality of Portland cement, tested on the work; and the sand and shingle were coarse, granitic, and sea-washed, being used, approximately, in the proportion of 1 part of Portland cement, 3 of sand, and 4 of shingle. It was mixed in concrete mixers revolved by steam power; there were 8 half-yard mixers fixed on a staging with storage room for 3600 tons of sand and gravel, and a small cement shed close around their supply hoppers.

Bag. The jute cloth was 36 inches wide, and described as 36-inch, 24-oz. double warp, fine jute pocketing. It cost about 5*d.* per yard.

Quantities. The quantities of concrete used in the works were:—

	Cubic yards.
Concrete in bags	16,694
" frames	12,881
" parapets	390
" apron	747
Total cubic yards	30,712

Cost of Concrete in Bags. The cost of concrete in bags was as follows:—

COST PER CUBIC YARD.

	£	*s.*	*d.*
Sand and gravel delivered at the mixers	0	6	0
Cement	0	7	6
Jute	0	2	5
Mixing and depositing, coals, stores, repairs of plant, etc.	0	4	11
Total cost of materials and labour	1	0	10
Plant, establishment of the works, superintendence, etc., say	0	9	0
Total cost per cubic yard	£1	9	10

The expense of the plant, of course, would have been less per cubic yard if more cubic yards had been required; also the establishment of the works was expensive, as land had to be reclaimed from the sea to form a yard for storing materials, for workshops, and for making concrete.

The sand and gravel was also expensive, as it had to be brought from some distance, on the other side of the harbour, by tramways and by barges, and had partly to be washed before it was fit for use.

No building work could be done at the Aberdeen Pier when the wind was easterly, involving expense in making depôts and in moving the stuff twice.

Cost of Concrete-in-Mass. The expense of the concrete-in-mass deposited in frames above the bag work, amounting to 12,881 cubic yards, was as follows:—

COST PER CUBIC YARD.

	£	*s.*	*d.*
Labour and materials	1	0	6
Plant, establishment of work, superintendence, etc., as before . . .	0	9	0
Total	£1	9	6

This is almost exactly the same as the cost of the bag work.

APPENDIX III.

PROPORTIONS OF MATERIALS TO ONE OF PORTLAND CEMENT, OR THEIL LIME, FOR CONCRETE IN BLOCKS, BAGS, AND MASS; AND THE COST PER CUBIC YARD.

Breakwater.	*Sand.*	*Gravel.*	*Shingle.*	*Stone.*	*Sand and Stone.*	*Total Proportion.*	*Cost per Cubic Yard.* £ s. d.	*Remarks.*
Aberdeen, N. Pier	3	...	4	...	...	7 to 1	1 9 10	In bags.
							1 9 6	In mass.
Aberdeen, S. Pier	4	5	...	...	...	9 to 1	0 16 0¾	In frames. (Exclusive of plant.)
	3	4	...	...	...	7 ,, 1		
	4	5	...	...	...	9 ,, 1	0 15 10¾	In blocks. (Exclusive of plant.)
	2½	3½	...	...	...	6 ,, 1	1 5 3	In bags. (Exclusive of plant.)
	2½	3½	...	...	...	6 ,, 1	2 4 6¾	In apron.
Alderney	2	...	4	4	...	10 ,, 1	1 9 6	Blocks under low water.
							1 8 6	Blocks above low water.
Alexandria	...	...	...	...	5	5 ,, 1	1 8 10	Blocks with Theil lime.
Buckie	...	...	...	...	4	4 ,, 1	1 0 0	In mass under low water.
	...	...	...	...	7½ & 12	$10\frac{9}{10}$,, 1	0 15 0	In mass above low water.
Dover	...	...	...	...	9	9 ,, 1	... ...	Blocks.
Kustendjie	2½	...	...	7½	...	10 ,, 1	... ...	Blocks.
Manora	4	...	6	3	...	13 ,, 1	0 15 0	Blocks, exclusive of Plant.
Marseilles	2	...	5	...	...	7 ,, 1	... ...	Blocks with Theil lime.
Newhaven	2	...	5	...	...	7 ,, 1	... ...	In bags and mass.
Port Said	3½	...	...	...	...	3½ ,, 1	1 5 1	Blocks with Theil lime.
Ymuiden	3	...	5	...	...	8 ,, 1	... ...	Blocks.

APPENDIX IV.

DIMENSIONS OF LOCKS AND ENTRANCES AT VARIOUS PORTS.

Docks. With Name of Lock or Entrance.	*Depth on Sill.* H.W.O.S.T.		*Depth on Sill.* H.W.N.T.		*Length.*	*Width.*
	Ft.	In.	Ft.	In.	Feet.	Feet.
ST. KATHERINE.						
St. Katherine Entrance Lock ...	28	0	24	7	200	45
LONDON.						
East Dock Entrance Lock	28	0	24	7	300	60
Shadwell Basin Lock	28	0	24	7	350	60
LIMEHOUSE BASIN.						
East Entrance Lock	28	0	24	7	320	60
SURREY COMMERCIAL.						
Surrey Lock	28	0	24	7	250	50
Lavender Lock	20	8	17	3	320	34
Greenland Lock	27	0	23	7	209	42½
South Lock	28	0	24	7	220	48
MILLWALL.						
Entrance Lock	28	0	24	7	450	80
WEST INDIA.						
Limehouse Entrance Lock	22	0	18	7	155	36
South Dock, West Lock	23	5	20	0	192	45
South Dock, East Lock	27	0	23	7	300	55
Blackwall Entrance Lock	23	3	19	10	191	45
EAST INDIA.						
Import Dock Lock	24	10	21	5	209	47¾
West Entrance Lock	24	10	21	5	210	47½
East Entrance Lock	31	0	27	7	100	65
VICTORIA AND ALBERT.						
Victoria Dock Entrance Lock ...	28	0	24	7	350	80
Albert Dock Entrance Lock ...	30	0	26	7	550	80
TILBURY.						
Entrance Lock (in progress) ...	45	0	41	6	700	80
CHATHAM.						
North Entrance Lock	32	0	28	6	436	94½
South Entrance Lock	32	0	28	6	438	84½
Repairing Basin Entrance	30	0	26	6		97½

Docks. With Name of Lock or Entrance.	*Depth on Sill.* H.W.O.S.T.		*Depth on Sill.* H.W.N.T.		*Length.*	*Width.*
	Ft.	In.	Ft.	In.	Feet.	Feet.
LIVERPOOL.						
Langton West Lock	30	10	23	7	238	65
Canada Lock	26	7	19	4	498	100
Huskisson East Lock	25	4	18	1	338	80
Herculaneum Dock Entrance, North	30	10	23	7		80
" " " South	30	10	23	7		60
BIRKENHEAD.						
Outer North Lock	30	10	23	7	348	100
Outer South Lock	30	10	23	7	398	50
Morpeth Lock	30	10	23	7	398	85
HULL.						
St. Andrew's Dock Entrance Lock (West Dock No. 2)	27	3	22	9	250	50
Albert Dock Entrance Lock... ...	27	3	22	9	320	80
Half Tide Basin Entrance	27	6	23	0		60
Alexandra Dock Entrance Lock (in progress)	34	0	29	6	500	75
GRIMSBY.						
Royal Dock, West Entrance Lock	26	0	22	0	300	70
Union Dock Lock	21	0	17	0	230	45
BRISTOL.						
Cumberland Northern Entrance Lock	33	0	21	0	350	62
AVONMOUTH.						
Entrance Lock	38	0	26	0	454	70
NEWPORT.						
Newport Lock	32	0	22	0	220	61
Alexandra Lock	37	0	27	0	350	65
CARDIFF.						
Roath Basin Entrance Lock	35	8	25	8	350	80
East Basin Entrance Lock	31	8	21	8	220	55
PENARTH [1].						
Entrance	32	2	25	8		60
Lock	32	2	25	8	275	60
SWANSEA.						
Prince of Wales Dock	32	0	25	0	500	60
South Dock	23	0	16	0	300	60

[1] The depths at Penarth have been taken from the working drawings; but there is probably some error in the level of high water, as the difference between springs and neaps only amounts to $6\frac{1}{2}$ feet, according to the drawings; whereas, according to the Admiralty Chart, it is 10 feet at Cardiff, 9 feet at Penarth, and $8\frac{1}{2}$ feet in the Cardiff and Penarth Roads.

Docks. With Name of Lock or Entrance.	*Depth on Sill.* H.W.O.S.T.		*Depth on Sill.* H.W.N.T.		*Length.*	*Width.*
	Ft.	In.	Ft.	In.	Feet.	Feet.
BARROW.						
Ramsden Basin Entrance	31	6	24	0		100
Ramsden Lock	31	6	24	0	700	100
Devonshire Dock Entrance	25	6	18	0		60
HARTLEPOOL.						
Victoria Dock Entrance Lock ...	21	6	18	0		42
North Basin Entrance	26	0	22	6		60
Coal Dock Entrance	21	6	18	0		40
SUNDERLAND.						
South Dock, New Lock	27	0	23	6	480	65
PORTSMOUTH.						
North Lock	41	8	38	11	458	82
South Lock	40	8	37	11	458	80
Rigging Basin Entrance	32	6	29	9		94
LEITH.						
Albert Dock Entrance Lock... ...	25	0	22	9	350	60
Victoria Dock Entrance	23	0	20	9		60
BELFAST.						
Spencer Dock Entrance...	23	10	23	1		80
Dufferin Dock Entrance	23	10	23	1		60
HAVRE.						
Eure Dock Entrance	35	0	29	5		100
Florida Dock Entrance...	25	3	19	8		69
HONFLEUR.						
East Dock Entrance	22	0	17	1		50
Central Dock Entrance...	20	4	15	5		35
ST. NAZAIRE.						
St. Nazaire Dock Entrance	28	10	23	11		82
Penhouët Lock	28	10	23	11	492	82
ANTWERP.						
Kattendyk Entrance	25	8	24	2		$81\frac{1}{3}$
Little Dock Entrance	23	10	22	4		$57\frac{1}{2}$

APPENDIX V.

TOTAL TONNAGE OF VESSELS WITH CARGOES, BOTH ENTERING AND CLEARING MOST OF THE PRINCIPAL BRITISH PORTS, IN 1873 AND 1883; AND ALSO THE TOTAL TONNAGE OF VESSELS, EITHER ENTERING OR CLEARING THE SAME PORTS, WITH CARGOES AND IN BALLAST, IN 1873 AND 1883.

Note.—The largest Tonnage is given for the Vessels with Cargoes and in Ballast, which sometimes appears under the head of entered, and sometimes as cleared, according to circumstances. The Tonnage of both entered and cleared is given for Vessels with Cargoes only, in order to indicate the relative importance of the Imports and Exports at the various Ports.

PORT.	1873. With Cargoes.		With Cargoes and Ballast.		1883. With Cargoes.		With Cargoes and Ballast.	
	Entered.	*Cleared.*			*Entered.*	*Cleared.*		
ENGLAND.	Tons.	Tons.	Tons.		Tons.	Tons.	Tons.	
Barrow	267,152	262,843	347,613	Entered.	331,587	444,678	575,350	Cleared.
Bristol	1,034,938	520,438	1,059,071	,,	1,185,919	616,113	1,228,083	Entered.
Cardiff	716,256	2,462,259	2,544,285	Cleared.	1,081,054	4,978,250	5,116,571	Cleared.
Grimsby	429,489	419,393	575,193	,,	535,787	526,895	715,529	Entered.
Hartlepool ...	383,729	631,372	860,624	Entered.	423,384	769,263	912,903	Cleared.
Hull...	1,455,293	1,270,738	1,603,544	Cleared.	1,963,324	1,516,767	2,055,091	Entered.
Liverpool ...	6,143,659	5,508,231	6,339,376	Entered.	7,540,656	6,755,447	8,194,129	,,
London	7,776 040	3,967,504	7,887,764	,,	11,002,768	5,483,380	11,440,707	,,
Newport	380,430	817,875	870,940	Cleared.	789,517	1,761,994	1,948,600	Cleared.
Sunderland ...	319,248	1,991,046	2,124,696	,,	391,736	2,649,823	2,800,466	,,
Swansea	596,194	1,003,645	1,076,146	,,	639,306	1,393,270	1,505,064	Entered.
Tees...	255,329	486,889	617,639	,,	570,598	902,263	1,275,867	,,
Tyne	1,467,653	4,393,648	4,677,526	,,	1,827,632	6,183,864	6,668,887	Cleared.
SCOTLAND.								
Glasgow	1,421,077	1,636,320	1,721,690	Cleared.	2,273,677	2,653,899	2,906,395	Cleared.
Greenock... ...	975,453	319,981	1,047,812	Entered.	1,301,232	500,994	1,444,433	Entered.
Leith	683,348	587,675	769,889	Cleared.	961,186	759,036	1,036,371	Cleared.
IRELAND.								
Belfast	1,397,407	844,666	1,430,137	Entered.	1,733,617	1,208,569	1,850,537	Cleared.
Dublin	1,692,990	1,050,330	2,052,202	,,	1,757,752	1,050,836	2,174,951	Entered.

APPENDIX VI.

VALUES OF THE IMPORTS AND EXPORTS AT MOST OF THE PRINCIPAL BRITISH PORTS IN 1873 AND 1883; AND THE MAXIMA VALUES OF IMPORTS IN ANY ONE YEAR BETWEEN 1873 AND 1883.

Note.—The values of the exports for the earlier years can be only approximately calculated, so their maxima values could not be precisely determined.

PORT.	1873.			Maximum Value between 1873 & 1883.	Date of Maximum.	1883.		
	Imports.	*Exports.*	*Total.*	*Imports.*		*Imports.*	*Exports.*	*Total.*
ENGLAND.	£	£	£	£		£	£	£
Barrow	732,605	676,522	1,409,127	790,916	1880	532,207	878,596	1,410,803
Bristol	7.607,258	576,709	8,183,967		1883	8,482,091	1,182,264	9,664,355
Cardiff	1,885,146	4,965,871	6,851,017	2,338,133	1880	2,193,320	4,858,925	7,052,245
Grimsby	3,154,005	10,725,049	13,879,054		1883	4,099,663	8,223,396	12,323,059
Hartlepool	2,468,803	2,449,958	4,918,761	2,620,700	1874	2,277,942	1,345,559	3,623,501
Hull	15,908,603	28,018,148	43,926,751		1883	21,624,724	22,031,682	43,656,406
Liverpool	112,824,613	102,736,175	215,560,788		1883	114,625,869	102,713,186	217,339,055
London	127,560,447	82,654,024	210,214,471		1883	145,139,505	95,219,073	240,358,578
Newport	489,420	1,839,933	2,329,353	994,143	1880	956,934	2,062,638	3,019,572
Sunderland	904,036	1,615,190	2,519,226	939,524	1874	761,296	757,831	1,519,127
Swansea	2,556,850	1,855,712	4,412,562	2,895,390	1880	1,930,706	1,612,405	3,543,111
Tees	663,591	3,444,260	4,107,851	870,206	1882	693,847	2,500,266	3,194,113
Tyne	5,764,612	7,475,559	13,240,171		1883	9,825,734	5,721,146	15,546,880
SCOTLAND.								
Glasgow	9,596,023	10,583,169	20,179,192		1883	13,972,464	14,987,786	28,960,250
Greenock	5,587,728	663,245	6,250,973	7,947,491	1877	5,102,802	435,692	5,538,494
Leith	8,149,184	4,980,405	13,129,589		1883	9,854,263	2,950,426	12,804,689
IRELAND.								
Belfast	2,236,560	189,226	2,425,786	2,705,295	1876	2,171,377	782,246	2,953,623
Dublin	3,800,644	117,402	3,918,046	4,231,790	1877	3,258,424	82,369	3,340,793

APPENDIX VII.

Sailing and Steam Vessels, Entered and Cleared, with Cargoes and in Ballast, at most of the Principal British Ports, and the Average Tonnage of the Vessels, in 1863, 1873, and 1883.

Port.	1863.				1873.				1883.			
	Sailing Vessels.	*Steam Vessels.*	*Total.*	*Average Tonnage of Vessels.*	*Sailing Vessels.*	*Steam Vessels.*	*Total.*	*Average Tonnage of Vessels.*	*Sailing Vessels.*	*Steam Vessels.*	*Total.*	*Average Tonnage of Vessels.*
England.	No.	No.	No.	Tons.	No.	No.	No.	Tons.	No.	No.	No.	Tons.
Barrow ...	...	...	...	...	2,466	1,379	3,845	179	1,766	2,819	4,585	241
Bristol ...	8,362	3,602	11,964	102	11,870	6,343	18,213	115	9,502	7,846	17,348	141
Cardiff ...	...	...	...	...	14,404	6,880	21,284	196	14,007	12,609	26,616	374
Grimsby ...	...	...	...	...	5,058	1,306	6,364	180	4,895	2,467	7,362	192
Hartlepool	...	...	...	...	8,544	1,970	10,514	160	4,934	2,114	7,048	255
Hull... ...	4,386	3,452	7,838	212	5,094	4,993	10,087	317	3,031	6,032	9,063	445
Liverpool...	16,296	9,281	25,577	322	15,835	14,275	30,110	418	12,719	21,251	33,970	475
London[1] ...	34,860	10,245	45,105	228	43,379	15,279	58,658	...	52,377	21,895	74,272	...
Newport ...	...	...	...	...	15,509	2,202	17,711	98	13,059	7,047	20,106	193
Sunderland	...	...	...	...	13,985	4,802	18,787	218	9,634	8,409	18,043	303
Swansea ...	...	...	...	...	12,989	2,881	15,870	135	9,513	6,464	15,977	186
Tees... ...	...	...	...	...	3,651	3,518	7,169	169	2,961	5,370	8,331	303
Tyne... ...	24,474	2,296	26,770	163	23,297	11,168	34,465	264	14,307	18,604	32,911	396
Scotland.												
Glasgow ...	5,015	3,800	8,815	199	5,089	7,120	12,209	269	3,972	13,624	17,596	315
Greenock...	1,985	1,118	3,103	178	2,553	4,740	7,293	210	2,509	9,703	12,212	185
Leith ...	3,061	1,697	4,758	204	4,252	2,971	7,223	213	2,501	4,213	6,714	308
Ireland.												
Belfast ...	6,024	3,992	10,016	175	10,818	5,641	16,459	167	7,910	10,693	18,603	196
Dublin ...	8,591	3,393	11,984	162	9,842	7,121	16,963	236	5,319	10,387	15,706	271

[1] The returns of the vessels clearing the Port of London are not complete, as it has not been found possible to register all the coastwise craft.

INDEX.

D.

F.

K.

L.

S.

END OF VOL. I.

Clarendon Press Series

HARBOURS AND DOCKS

VERNON-HARCOURT

London

HENRY FROWDE

Oxford University Press Warehouse

Amen Corner, E.C.

Clarendon Press Series

HARBOURS AND DOCKS

THEIR

PHYSICAL FEATURES, HISTORY, CONSTRUCTION EQUIPMENT, AND MAINTENANCE

WITH

STATISTICS AS TO THEIR COMMERCIAL DEVELOPMENT

BY

LEVESON FRANCIS VERNON-HARCOURT, M.A.

MEMBER OF THE INSTITUTION OF CIVIL ENGINEERS
AUTHOR OF 'RIVERS AND CANALS'

VOL. II.—PLATES

Oxford
AT THE CLARENDON PRESS
1885

CONTENTS OF PLATES.

PLATE I.

Jetty Harbours. Channel and North Sea.

Scales: Plans $\frac{1}{20000}$, Sections $\frac{1}{200}$.

Dieppe Harbour. (Plan of Harbour and Docks, and Sections of Jetties.)
Calais Harbour. (Plan of Harbour and Docks, and Section of East Jetty.)
Dunkirk Harbour. (Plan of Harbour and Docks, and Section of East Jetty.)
Ostend Harbour. (Plan of Harbour and Docks, and Section of East Jetty.)
Aberdeen Harbour. (Plans of Harbour and Docks in 1867 and 1879.)

PLATE II.

Jetty Harbours.

Scales: Plans $\frac{1}{80000}$, Sections $\frac{1}{400}$.

Charleston Harbour. (Plan of Harbour, and Section of South Jetty.)
Newburyport Harbour. Mass. U. S. (Plan of Harbour, and Section of North Jetty.)
Kurrachee Harbour. (Plans of Harbour in 1858 and 1879.)
Malamocco Harbour. (Plan of Harbour, and Section of Jetty.)
Dublin Harbour. (Plan of Harbour, and Sections of North and South Walls.)
Galveston Harbour. Texas, U. S. (Plan of Harbour, and Sections of Jetty.)

Plate III.

Harbours with Detached Breakwaters.

Scale $\frac{1}{30000}$.

Plate IV.

Harbours. British and Colonial.

Scale $\frac{1}{20000}$.

PLATE V.

Harbours. American and Foreign.

Scale $\frac{1}{30000}$.

Algiers Harbour. (Plan.)

Ymuiden Harbour. Amsterdam Ship-Canal. (Plan.)

Genoa Harbour. (Plan.)

Odessa Harbour. (Plan.)

Chicago Harbour. Lake Michigan. (Plan.)

Bastia Harbour. Corsica. (Plan.)

Port Said Harbour. (Plan.)

Boulogne Harbour. (Plan of Harbour and Docks.)

Malaga Harbour. (Plan.)

Barcelona Harbour. (Plan.)

Oswego Harbour. Lake Ontario. (Plan.)

Buffalo Harbour. Lake Erie. (Plan.)

Mormugao Harbour. India. (Plan.)

Tarragona Harbour. (Plan.)

Oran Harbour. Algeria. (Plan.)

Cette Harbour. (Plan of Harbour and Docks.)

PLATE VI.

Sections of Breakwaters. Rubble and Concrete-Block Mounds, and Upright Walls.

Scale $\frac{1}{800}$.

Mound Breakwaters.—Plymouth.—Portland (Outer Breakwater).—Kingstown.—Table Bay.—Brest.—Delaware.—Cette.—Tees (South Gare Breakwater).—Algiers.—Port Said (Inner, Centre, and Outer Sections of West Breakwater).—Biarritz.—Alexandria.

Upright-Wall Breakwaters.—Aberdeen (South Breakwater, and North Pier).—Newhaven.—Buckie.—Ymuiden (Amsterdam Ship-Canal).—Whitehaven (West Pier).—Dover.

PLATE VII.

Sections of Breakwaters. Rubble or Concrete-Block Mounds with Superstructure.

Scale $\frac{1}{600}$.

Cherbourg.—Holyhead.—Genoa.—Bastia (Corsica).—Ile Rousse (Corsica).—Leghorn. — Trieste. — Boulogne (New Harbour). — Tynemouth (North Pier).—St. Catherine's (Jersey).—Portland (Inner Breakwater).—Oran.—Alderney (at 1000 feet, and 4200 feet from the shore).—Marseilles (Main and Joliette Breakwaters).—Madras (Old, and Proposed New Breakwaters).—Manora.—Mormugao.—St. Jean-de-Luz (Socoa and Artha Breakwaters).—Colombo.—Odessa.

PLATE VIII.

Staging; Foundations; and Block-Setting Machines.

Scale $\frac{1}{300}$.

Titan. Manora Breakwater. (Elevation and Cross Section.)

Cranes and Staging. Aberdeen South Breakwater. (Elevation and Cross Section.)

Compressed-Air Foundations. Antwerp Quay Walls. (Longitudinal and Cross Sections.)

Fifteen-Ton Crane. (Elevation.)

Goliath. Tynemouth Breakwater. (Elevation.)

Staging. Holyhead Harbour. (Elevation and Cross Section.)

Floating Shears. Dublin Quay Walls. (Side and End Elevations.)

PLATE IX.

Lighthouses; Beacons; and Buoys.

Scales: Lighthouses and Beacons $\frac{1}{400}$, Buoys $\frac{1}{150}$.

Lighthouses.—Spectacle Reef (Lake Huron, U. S.).—Minot's Ledge (off Boston, U. S.).—Wolf Rock.—Bell Rock.—Old Eddystone.—New Eddystone.—Bishop Rock.—Des Barges.—Ar-men.—Walde.

Beacons.—Iron Beacon.—Lavezzi.—Stone Beacon.

Buoys.—Beacon Buoy.—Light Buoy.—Bell Buoy.—Whistling Buoy.—Boat Bell Buoy.—Mooring Buoy.—Can Buoy.—Cask Buoy.—Mast Buoy.

PLATE X.

Docks.

Scale $\frac{1}{20000}$.

London Docks. (Plan.)
Leith Docks. (Plan of Harbour and Docks.)
Hartlepool Docks. (Plan of Harbour and Docks.)
Belfast Docks and Quays. (Plan of Harbour and Docks.)

PLATE XI.

Docks.

Scale $\frac{1}{20000}$.

Liverpool Docks. (Plan.)
Barrow Docks. (Plan.)
Bristol Docks and Quays. (Plan.)
Hull Docks. (Plan.)
Dublin Docks and Quays. (Plan.)
Cardiff and Penarth Docks. (Plan.)
Grimsby Docks. (Plan.)
Birkenhead Docks. (Plan.)

PLATE XII.

Docks.

Scale $\frac{1}{15000}$.

Honfleur Port and Docks. (Plan.)
Havre Port and Docks. (Plan.)
Chatham Dockyard. (Plan.)
Queenstown, Haulbowline Island Dockyard. (Plan.)
Portsmouth Dockyard. (Plan.)
Sunderland Docks. (Plan.)

PLATE XIII.

Docks and Quays.

Scale $\frac{1}{20000}$.

Rotterdam Docks and Quays. (Plan.)
Rouen Quays. (Plan.)
Antwerp Docks and Quays. (Plan, and small Plan of New Docks.)
St. Nazaire Docks. (Plan.)
Hamburg Docks and Quays. (Plan.)

PLATE XIV.

Dock Walls.

Scale $\frac{1}{200}$.

Liverpool (Herculaneum Dock, and New Hornby Dock, North End).—West India (Plan and Section).—Bute Docks, Cardiff (Roath Basin).—Hull (Albert Dock).—London (Albert Dock).—Millwall.—Portsmouth (Tidal Basin Extension).—Chatham (Fitting-out Basin).—Penarth (Old Dock, and Extension).—Barrow (Tidal Basin, and Wall outside Entrance).—Dublin (North Quay).—Whitehaven (East Quay).—Belfast (Spencer Dock).—New York (North River Quay).—Dieppe (Duquesne and Half-Tide Basins).—Antwerp (Quays).—Havre (Eure Basin, Ninth Dock, and Ninth Dock tidal work).—Algiers.—Dunkirk.

PLATE XV.

Locks. Dock and Quay Walls.

Scales: Dock and Quay Walls $\frac{1}{200}$, Locks $\frac{1}{400}$.

Locks.—Cardiff, Entrance Gates, Roath Basin. (Plan.)—Liverpool, Entrance Gates, Langton Dock. (Plan.) — Bristol, Entrance Lock, Inner Gates. (Plan and Section.) — Hull, Albert Dock, Dock end of Lock. (Plan and Section.)—Penarth, Dock end of Lock. (Plan and Section.)—Dunkirk outer end of Lock, No. 1 Dock. (Plan and Section.)

Dock and Quay Walls.—Bristol (Wall outside Entrance). — Queenstown (Haulbowline Island Dock).—St. Nazaire (Penhouët Dock).—Marseilles (National Basin).—Rouen Quays.—New York Quays.

PLATE XVI.

Lock-Gates. Caissons. Graving Docks.

Scales: Gates and Caissons $\frac{1}{100}$, Graving Docks $\frac{1}{800}$.

Sections of Lock Gates.—Sunderland.—Liverpool.—Penarth.—Cardiff.—Bristol.—Hull.—Dunkirk.

Caissons.—Sliding Caisson, Portsmouth. (Section.)—Ship Caisson, Devonport. (Section.)

Graving Docks.—Graving Dock, Devonport. (Plan and Sections.)—Graving Dock, Chatham. (Plan and Section.)

ERRATA.

Plate 11, Fig. 1 (Liverpool), *for* New Dock (North end) *read* Hornby Dock.

" " Fig. 4 (Hull), *for* West Dock No. 2 *read* St. Andrew's Dock.

" 12, Fig. 2 (Havre), *for* Dock *read* Warehouse Dock.

" 13, Fig. 3 (Antwerp), *for* Timber Pond *read* Timber Dock.

" " Fig. 5 (Hamburg). In basins on northern bank of River Elbe, west end, *read* Sandgate Basin, and Grasbrok Basin.

" 14, Fig. 2 (Liverpool), *for* New Dock, North End, *read* Hornby Dock.

" " Figs. 5 and 6 (Hull). H. W. O. S. T. should be 1 foot lower, and H. W. O. N. T. $1\frac{1}{2}$ feet higher than shown.

" 15, Fig. 6 (Hull). H. W. O. S. T. should be 1 foot lower than shown.

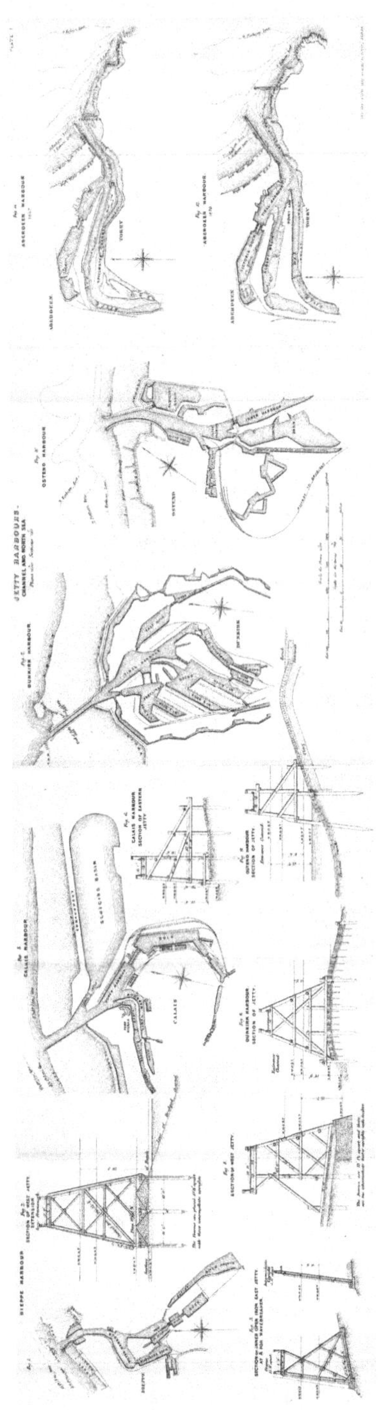

The material originally positioned here is too large for reproduction in this reissue. A PDF can be downloaded from the web address given on page iv of this book, by clicking on 'Resources Available'.

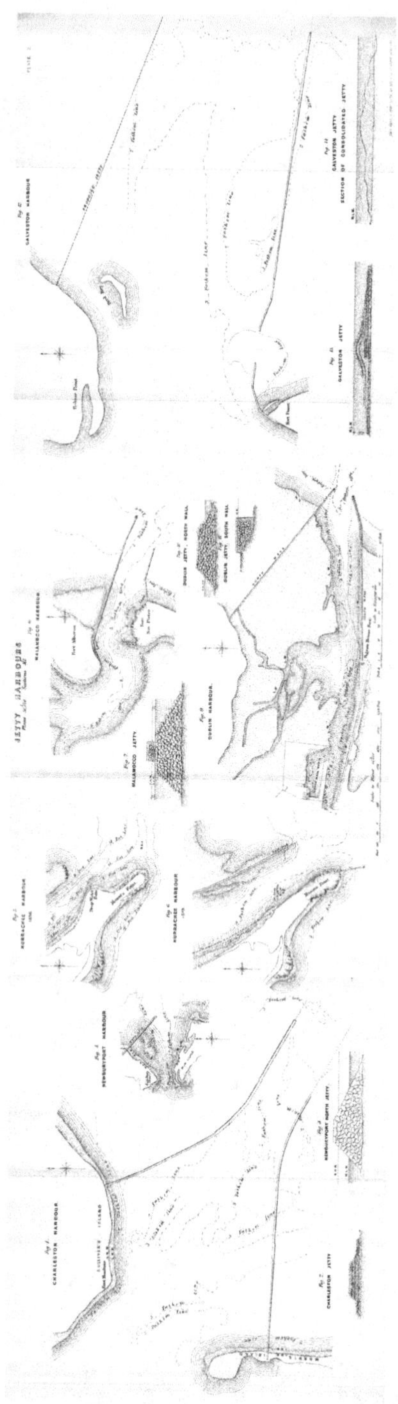

The material originally positioned here is too large for reproduction in this reissue. A PDF can be downloaded from the web address given on page iv of this book, by clicking on 'Resources Available'.

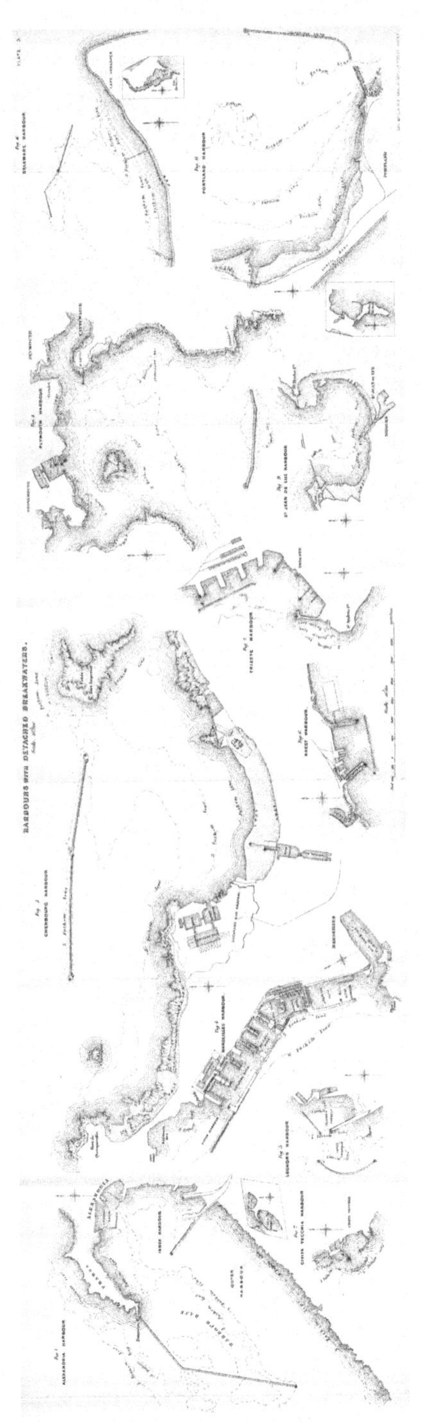

The material originally positioned here is too large for reproduction in this reissue. A PDF can be downloaded from the web address given on page iv of this book, by clicking on 'Resources Available'.

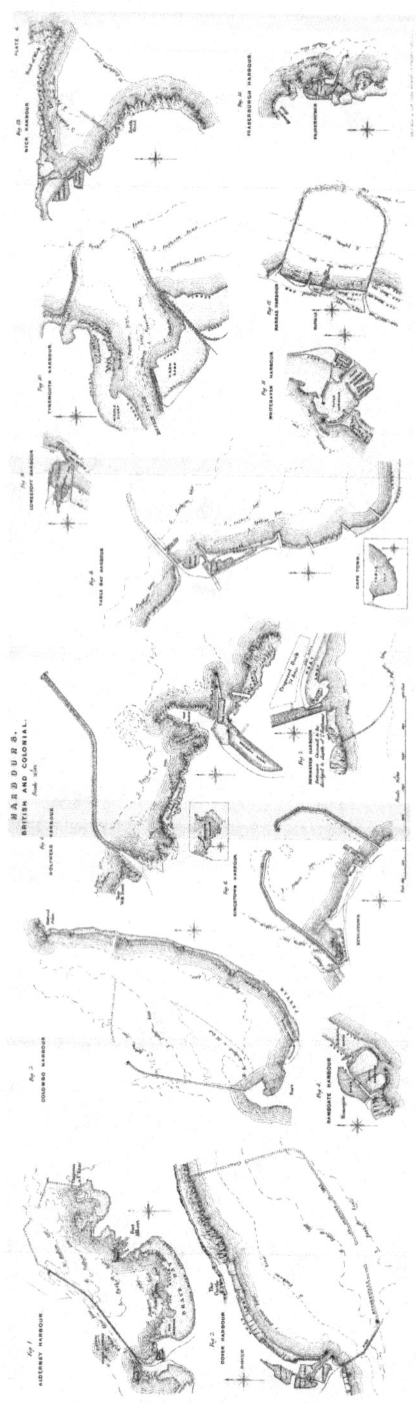

The material originally positioned here is too large for reproduction in this reissue. A PDF can be downloaded from the web address given on page iv of this book, by clicking on 'Resources Available'.

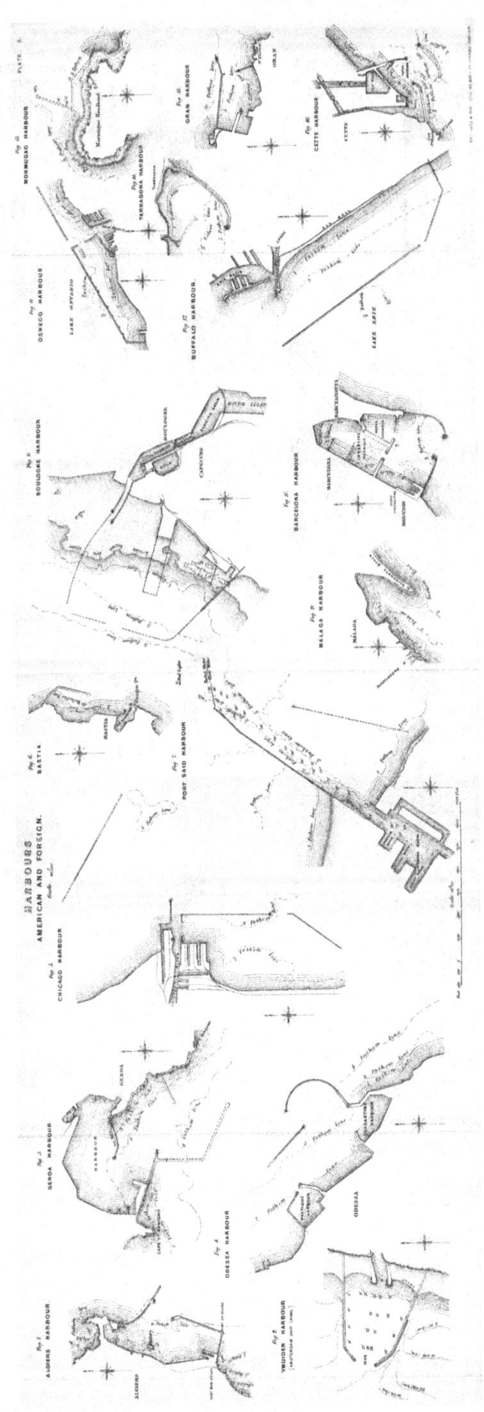

The material originally positioned here is too large for reproduction in this reissue. A PDF can be downloaded from the web address given on page iv of this book, by clicking on 'Resources Available'.

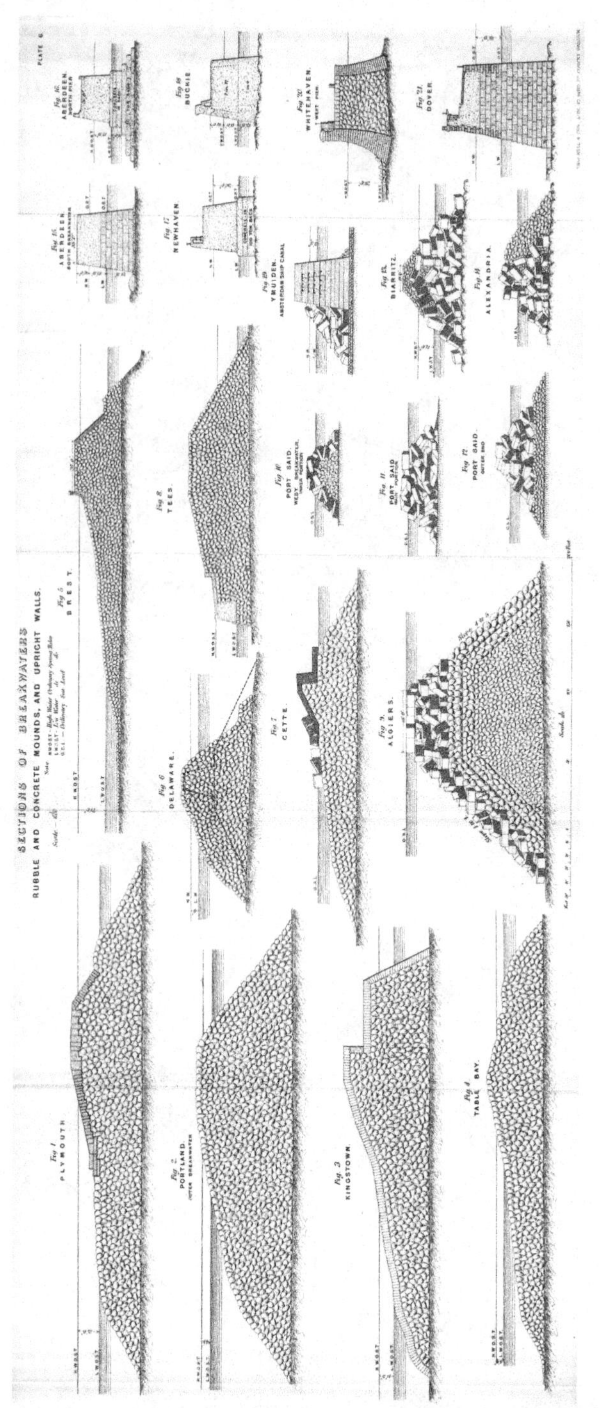

The material originally positioned here is too large for reproduction in this reissue. A PDF can be downloaded from the web address given on page iv of this book, by clicking on 'Resources Available'.

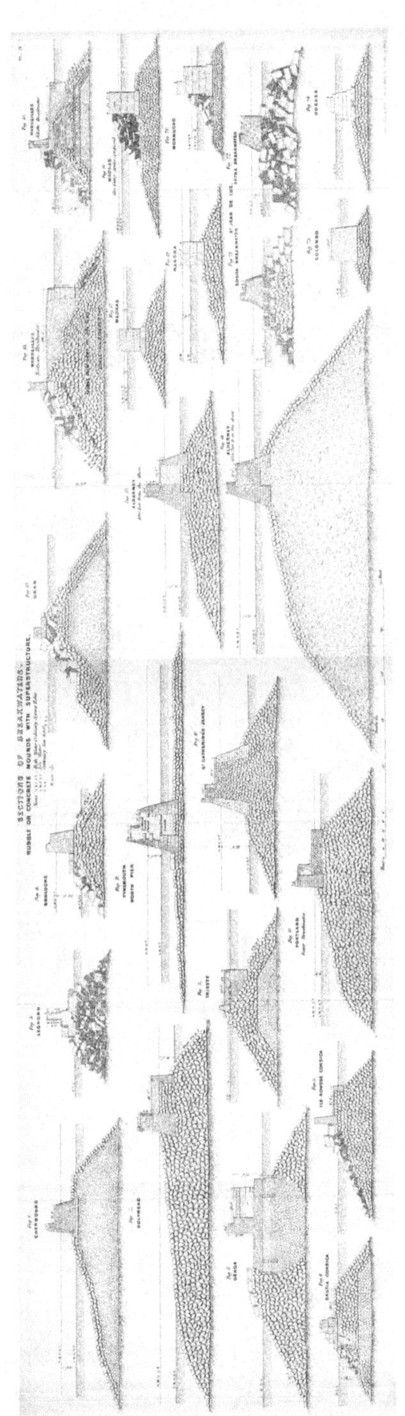
SECTIONS OF BREAKWATERS.
RUBBLE OR CONCRETE MOUNDS WITH SUPERSTRUCTURE.

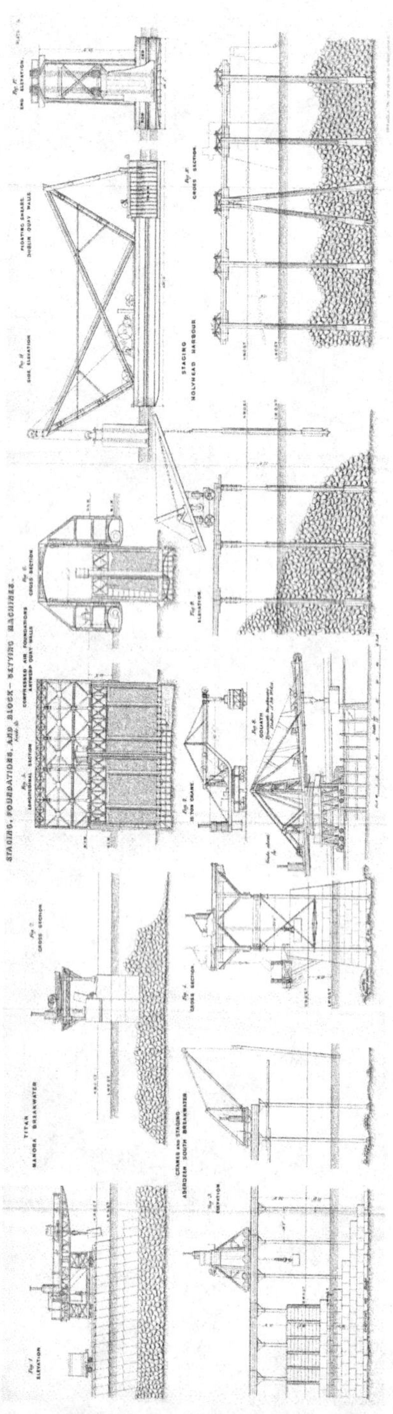

The material originally positioned here is too large for reproduction in this reissue. A PDF can be downloaded from the web address given on page iv of this book, by clicking on 'Resources Available'.

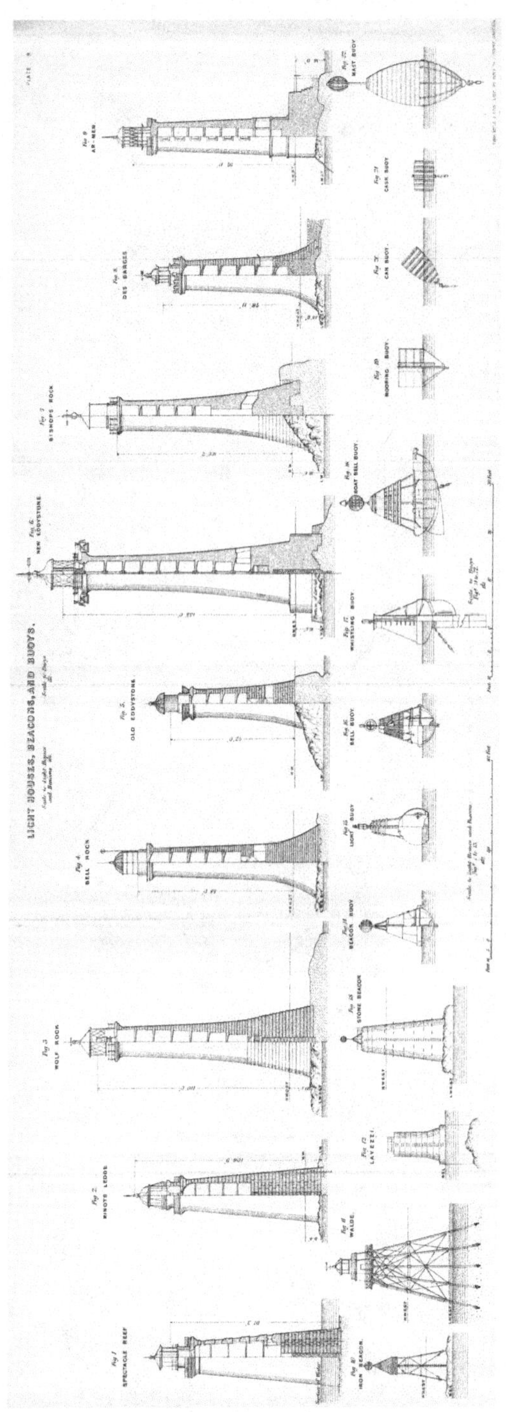

The material originally positioned here is too large for reproduction in this reissue. A PDF can be downloaded from the web address given on page iv of this book, by clicking on 'Resources Available'.

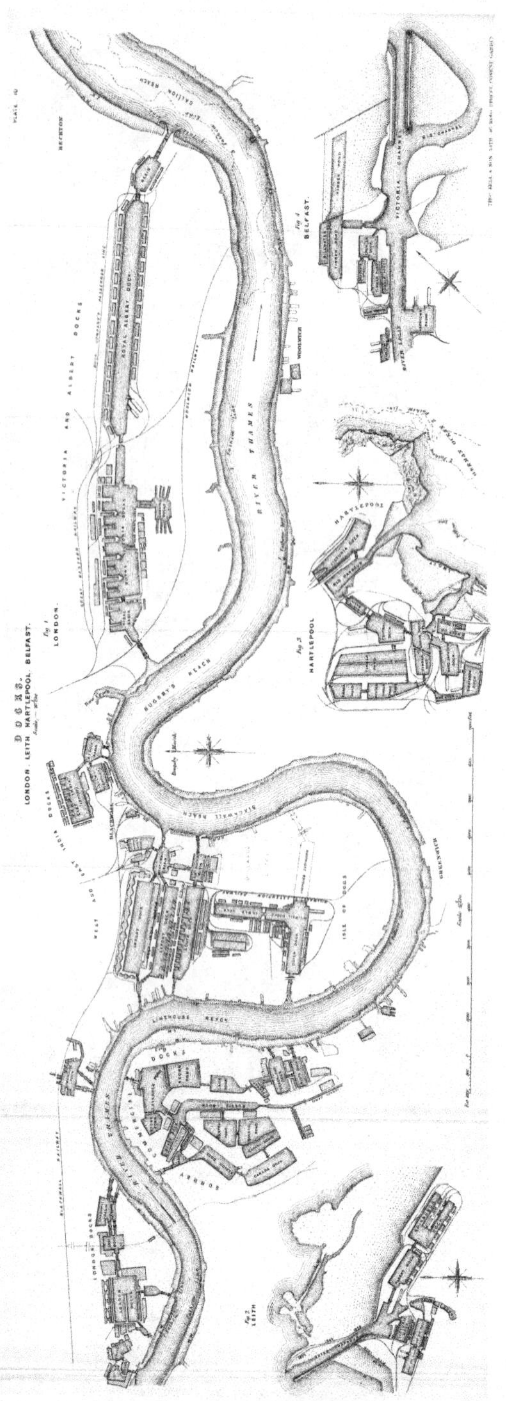

The material originally positioned here is too large for reproduction in this reissue. A PDF can be downloaded from the web address given on page iv of this book, by clicking on 'Resources Available'.

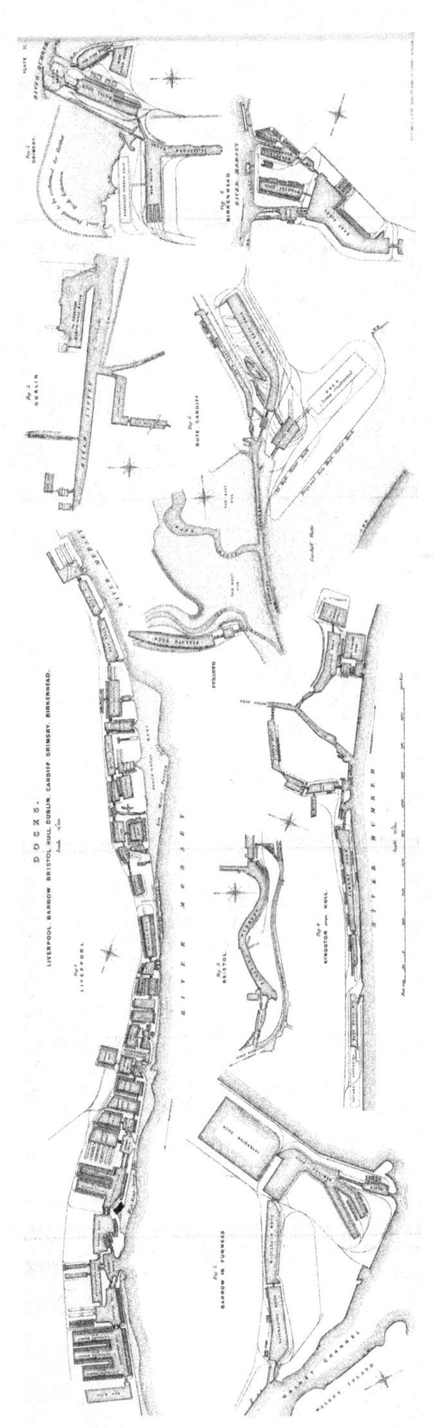

The material originally positioned here is too large for reproduction in this reissue. A PDF can be downloaded from the web address given on page iv of this book, by clicking on 'Resources Available'.

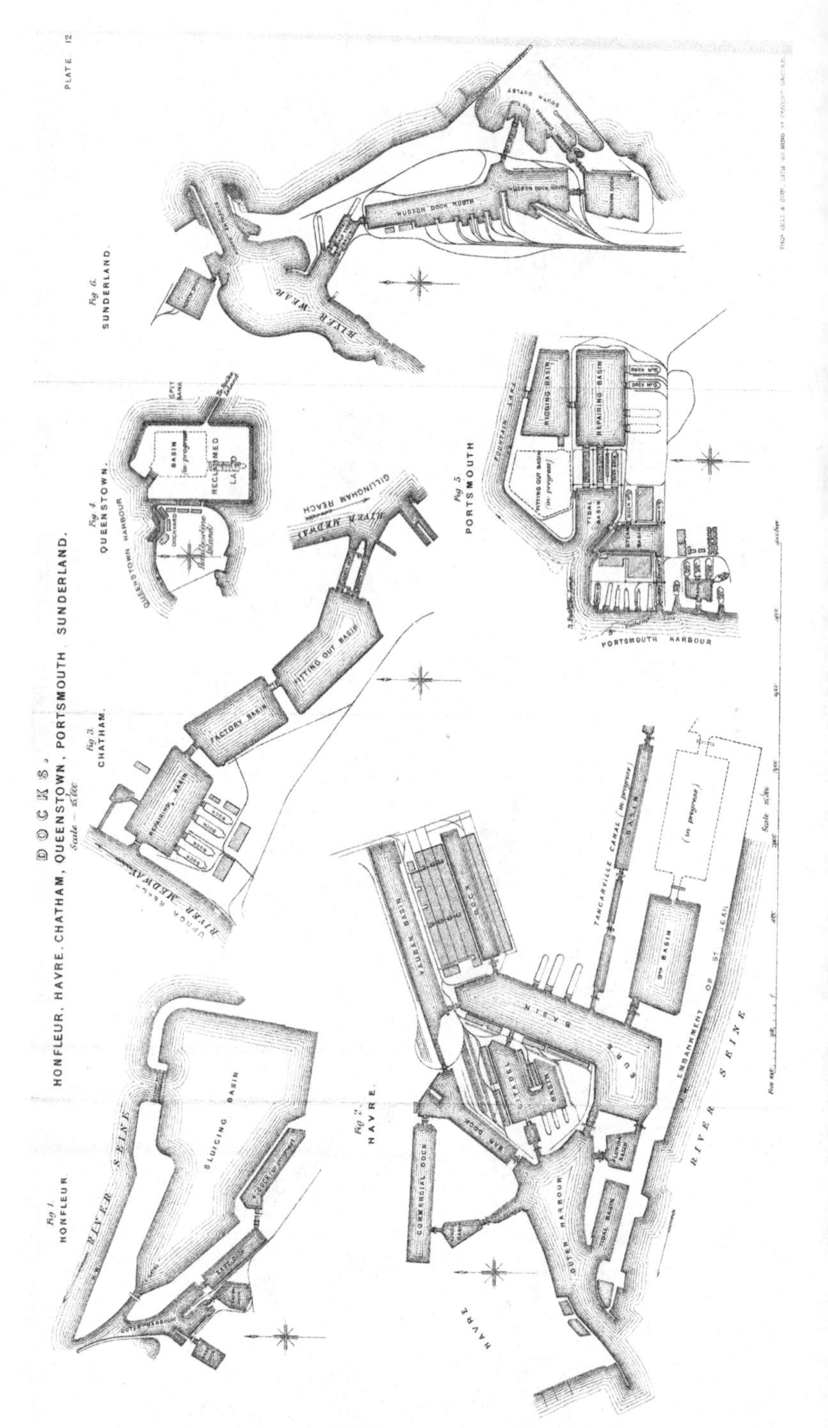

The material originally positioned here is too large for reproduction in this reissue. A PDF can be downloaded from the web address given on page iv of this book, by clicking on 'Resources Available'.

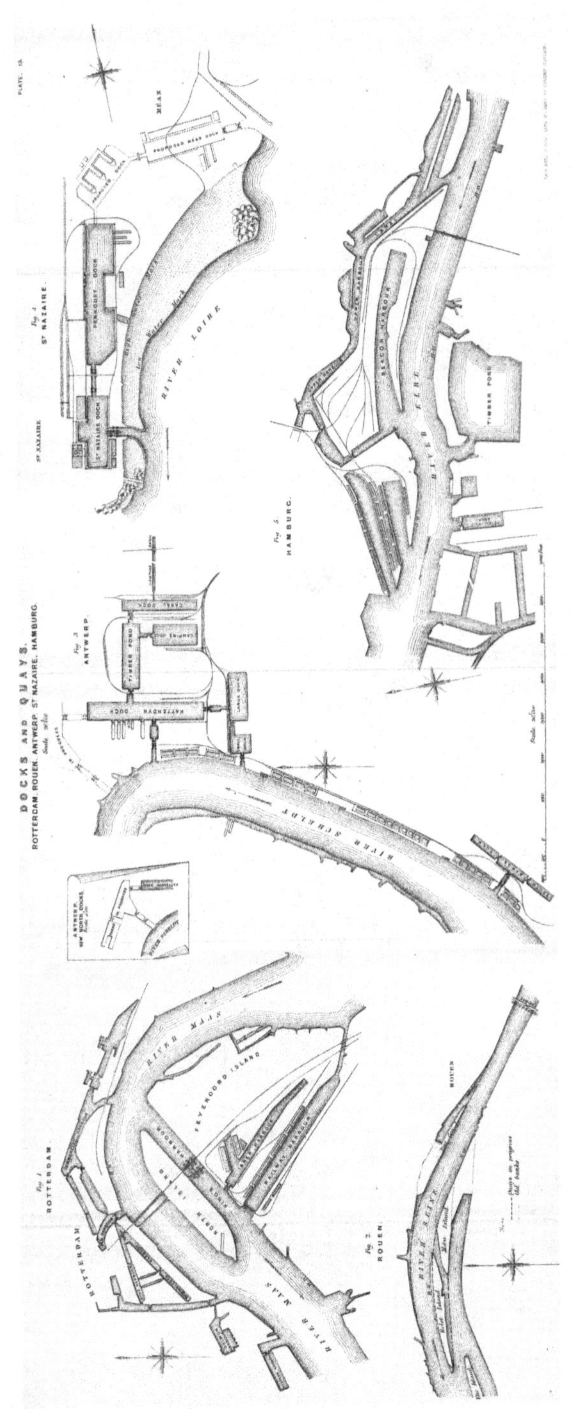

The material originally positioned here is too large for reproduction in this reissue. A PDF can be downloaded from the web address given on page iv of this book, by clicking on 'Resources Available'.

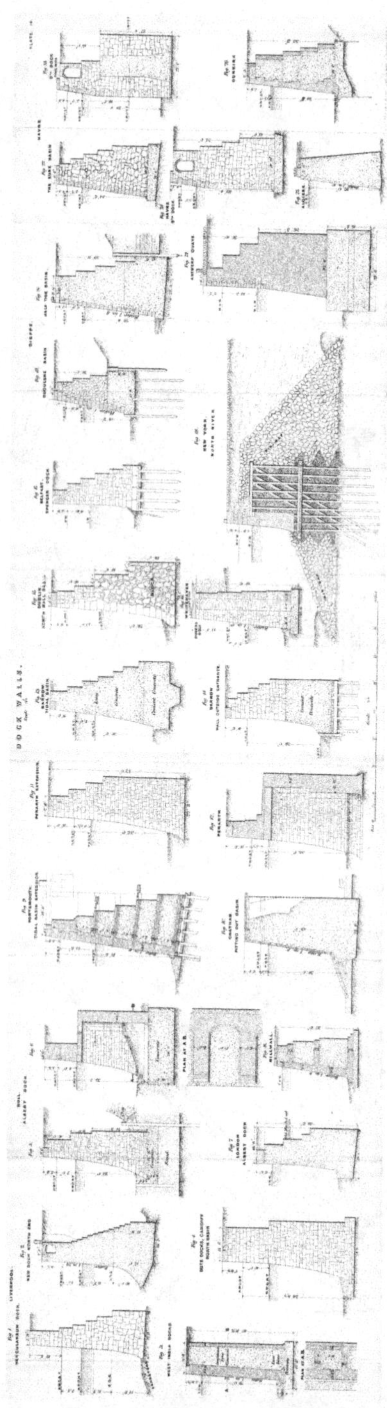

The material originally positioned here is too large for reproduction in this reissue. A PDF can be downloaded from the web address given on page iv of this book, by clicking on 'Resources Available'.

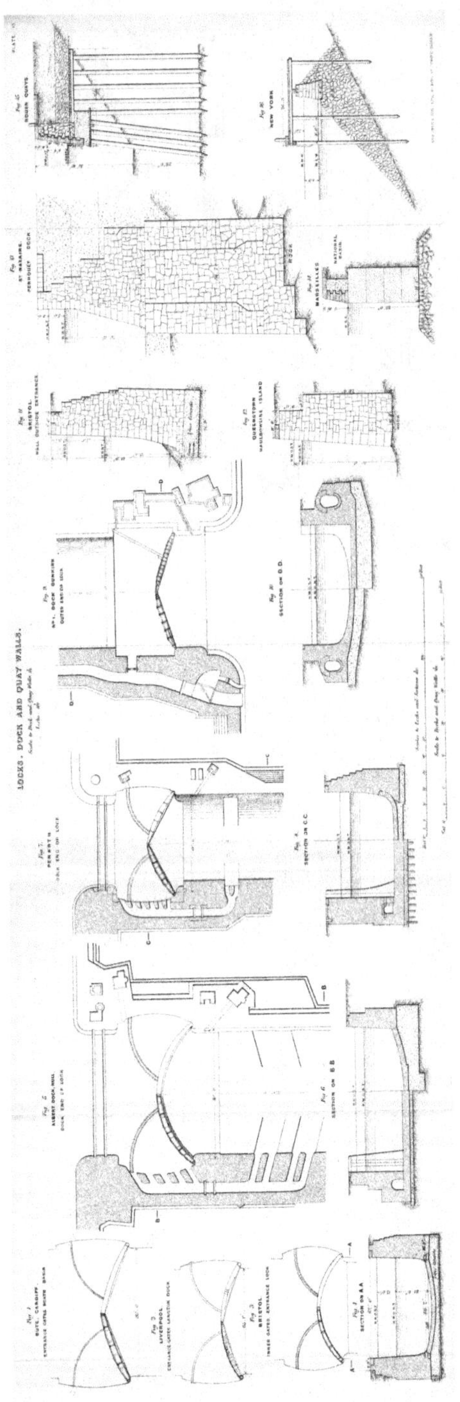

The material originally positioned here is too large for reproduction in this reissue. A PDF can be downloaded from the web address given on page iv of this book, by clicking on 'Resources Available'.

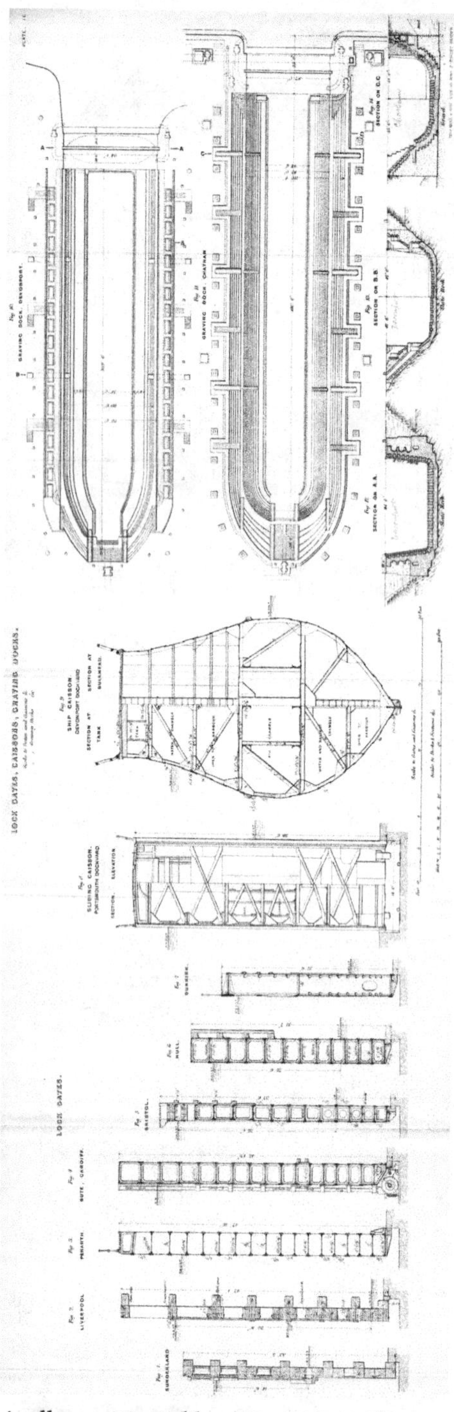

The material originally positioned here is too large for reproduction in this reissue. A PDF can be downloaded from the web address given on page iv of this book, by clicking on 'Resources Available'.

Printed in the United States
By Bookmasters